AF389106

Additive Manufacturing of Bio and Synthetic Polymers

Additive Manufacturing of Bio and Synthetic Polymers

Editors

Emin Bayraktar

S. M. Sapuan

R. A. Ilyas

MDPI • Basel • Beijing • Wuhan • Barcelona • Belgrade • Manchester • Tokyo • Cluj • Tianjin

Editors

Emin Bayraktar
Mechanical and Manufacturing
Engineering School
ISAE-Supmeca-Paris
Saint-Ouen
France

S. M. Sapuan
Department of Mechanical and
Manufacturing Engineering,
Faculty of Engineering
Universiti Putra Malaysia
Serdang, Selangor
Malaysia

R. A. Ilyas
School of Chemical and Energy
Engineering, Faculty of
Engineering
Universiti Teknologi Malaysia
Skudai
Malaysia

Editorial Office
MDPI
St. Alban-Anlage 66
4052 Basel, Switzerland

This is a reprint of articles from the Special Issue published online in the open access journal *Polymers* (ISSN 2073-4360) (available at: www.mdpi.com/journal/polymers/special_issues/additive_manufacturing_of_bio_and_synthetic_polymers).

For citation purposes, cite each article independently as indicated on the article page online and as indicated below:

LastName, A.A.; LastName, B.B.; LastName, C.C. Article Title. *Journal Name* **Year**, *Volume Number*, Page Range.

ISBN 978-3-0365-3320-9 (Hbk)
ISBN 978-3-0365-3319-3 (PDF)

Contents

About the Editors

Emin Bayraktar

Emin Bayraktar (Habil., Dr (PhD), DSc- Dr es Science) is an academic and research staff-member in Mechanical and Manufacturing Engineering at SUPMECA, Paris, France. His research areas include the manufacturing techniques of new materials (basic composites and hybrids), the metal forming of thin sheets (design + test + FEM), static and dynamic behaviour and the optimisation of materials (experimental and FEM, the utilisation and design of metallic and non-metallic composites, powder metallurgy, and energetic materials for aeronautical applications), powder metallurgy and the metallurgy of steels, welding, heat treatment and the processing of new composites, sintering techniques, sinter-forging, thixoforming, etc. He has authored more than 200 publications in international journals and international conference proceedings, and has also authored more than 90 research reports (European Steel Committee projects, Test + Simulation). He has advised 32 PhD, and 120 MSc theses, with 7 currently ongoing. He is a Fellow of WAMME (World Academy of Science in Materials and Manufacturing Engineering), an Editorial Board Member of *JAMME* (*International Journal of Achievement in Materials and Manufacturing Engineering*), and an Advisory Board Member of AMPT-2009 (Advanced Materials Processing Technologies), APCMP-2008 and APCMP-2010. He was a Visiting Professor at Nanyang Technology University, Singapore, in 2012; Xi'an Northwestern Technical University, Aeronautical Engineering, in 2016; and at the University of Campinas, UNICAMP-Brazil from 2013 until 2023. He is a recipient of the Silesian University Prix pour STAUB Golden Medal-2009"by the Academy of WAMME (World Academy of Science) in Poland, materials science section, and a recipient of the William Johnson International Gold Medal in 2014, AMPT academic association.

S. M. Sapuan

S.M. Sapuan is a Professor of Composite Materials at Universiti Putra Malaysia. He earned his B.Eng degree in Mechanical Engineering at the University of Newcastle, Australia, in 1990, his MSc from Loughborough University, UK, in 1994, and his Ph.D. from De Montfort University, UK, in 1998. His research interests include natural fiber composites, material selection, and concurrent engineering. To date, he has authored or co-authored more than 1521 publications (730 papers published/accepted in national and international journals, authored 16 books, edited 25 books, 153 chapters in books and 597 conference proceedings/seminar papers/presentation (26 of which are plenary and keynote lectures, and 66 of which are invited lectures). S.M. Sapuan was the recipient of the Rotary Research Gold Medal Award in 2012, The Alumni Medal for Professional Excellence Finalist, 2012 Alumni Awards, University of Newcastle, NSW, Australia, and the Khwarizmi International Award (KIA). In 2013, he was awarded with 5-Star Role Model Supervisor award by UPM. S.M. Sapuan was recognized as the first Malaysian to be conferred a Fellowship by the U.S.-based Society of Automotive Engineers International (FSAE) in 2015. He was the 2015/2016 recipient of SEARCA Regional Professorial Chair. In 2017, he was conferred the IOP Outstanding Reviewer Award by the Institute of Physics, UK; the National Book Award; The Best Journal Paper Award, UPM; the Outstanding Technical Paper Award, Society of Automotive Engineers International, Malaysia; and the Outstanding Researcher Award, UPM. He also received a Citation of Excellence Award from Emerald, UK; SAE Malaysia's Best Journal Paper Award; the IEEE/TMU Endeavour Research Promotion Award; the Best Paper Award by Chinese Defence Ordnance; and

Malaysia's Research Star Award (MRSA), from Elsevier in 2017. In addition, in 2019, he was named the Top Research Scientist Malaysia 2019 (TRSM 2019) and presented with a Professor of Eminence Award from AMU, India.

R. A. Ilyas

R.A. Ilyas is a Senior Lecturer at the School of Chemical and Energy Engineering, Faculty of Engineering, Universiti Teknologi Malaysia, Malaysia. He received his Diploma in Forestry, a Bachelor's Degree (BSc) in Chemical Engineering, and a Ph.D. degree in the field of Biocomposite Technology and Design at Universiti Putra Malaysia. R.A. Ilyas was the recipient of the MVP Doctor of Philosophy Gold Medal Award UPM 2019, for the Best Ph.D. Thesis and the Top Student Award, INTROP, UPM. In 2018, he was deemed an Outstanding reviewer by *Carbohydrate Polymers*, Elsevier, United Kingdom, and presented with the Best Paper Award (11th AUN/SEED-Net Regional Conference on Energy Engineering), Best Paper Award (Seminar Enau Kebangsaan 2019, Persatuan Pembangunan dan Industri Enau Malaysia), and the National Book Award 2018. R.A. Ilyas was also listed as Among the World's Top 2% of Scientists (Subject-Wise) due to his citation impact during the calendar years of 2019 and 2020 by Stanford University, US. He achieved a PERINTIS Publication Award 2021 from Persatuan Saintis Muslim Malaysia, an Emerging Scholar Award by Automotive and Autonomous Systems, 2021, Belgium, and Young Scientists Network–Academy of Sciences Malaysia (YSN-ASM). His main research interests are: (1) polymer engineering (biodegradable polymers, biopolymers, polymer composites, and polymer gels); and (2) material engineering (natural-fiber-reinforced polymer composites, biocomposites, cellulose materials, and nanocomposites). To date, he has authored or co-authored more than 325 publications (published/accepted): 123 journal articles indexed in JCR/Scopus, 14 books, 89 book chapters, 67 conference proceedings/seminars, 2 research bulletins, and 10 conference papers (abstract published in a book of abstract); he has been the Guest Editor of 10 Journal Special Issues, and the Editor/Co-Editor of 10 conference/seminar proceedings on green-material-related subjects.

Preface to "Additive Manufacturing of Bio and Synthetic Polymers"

Additive manufacturing technology offers the ability to produce personalized products with lower development costs, shorter lead times, less energy consumed during manufacturing and less material waste. It can be used to manufacture complex parts and enables manufacturers to reduce their inventory, make products on-demand, create smaller and localized manufacturing environments, and even reduce supply chains. Additive manufacturing (AM), also known as fabricating three-dimensional (3D) and four-dimensional (4D) components, refers to processes that allow for the direct fabrication of physical products from computer-aided design (CAD) models through the repetitious deposition of material layers. Compared with traditional manufacturing processes, AM allows the production of customized parts from bio- and synthetic polymers without the need for molds or machining typical for conventional formative and subtractive fabrication.

Today, AM offers numerous advantages, i.e., little assembly required, supply chain efficiencies, sustainability, geometric flexibility, low Buy-to-Fly ratios, shortened times to market, environmentally friendly, and the production of single and multiple components, offering incomparable design independence with the facility to manufacture components from various bio- and synthetic polymers. These advantages make AM a major player in the next industrial revolution of polymers. Polymers and their composites are one of the most widely used materials in today's industry and are of great interest in AM due to the vast potential for various applications, such as in the apparel, art and jewelry, electric and electronic, healthcare, biomedical, robotics, military defense, sensor and actuators, construction, aerospace and automotive industries. Polymers that are utilized in AM include hydrogels, elastomers, thermosets, thermoplastics, functional polymers, polymer composites, polymer hybrid composites, polymer nanocomposites and polymer blends. Over the past 30 years, much research has been carried out on developing new polymeric materials for AM processes, such as material jetting (MJ), drop-on-demand (DOD), sand binder jetting, vat photopolymerization, fused deposition modelling (FDM), stereolithography (SLA), digital light processing (DLP) and selective laser sintering (SLS).

In this Special Issue, we aimed to capture the cutting-edge state-of-the-art research pertaining to advancing the additive manufacturing of polymeric materials. The topic themes include advanced polymeric material development, processing parameter optimization, characterization techniques, structure–property relationships, process modelling, etc., specifically for AM.

Emin Bayraktar, S. M. Sapuan, R. A. Ilyas
Editors

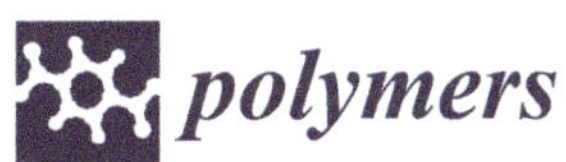

Review

Polylactic Acid (PLA) Biocomposite: Processing, Additive Manufacturing and Advanced Applications

R.A. Ilyas [1,2,*], S.M. Sapuan [3,4,*], M.M. Harussani [4], M.Y.A.Y. Hakimi [4], M.Z.M. Haziq [4], M.S.N. Atikah [5], M.R.M. Asyraf [6], M.R. Ishak [6], M.R. Razman [7,*], N.M. Nurazzi [8], M.N.F. Norrrahim [9], Hairul Abral [10] and Mochamad Asrofi [11]

[1] School of Chemical and Energy Engineering, Faculty of Engineering, Universiti Teknologi Malaysia, UTM Johor Bahru 81310, Johor, Malaysia

[2] Centre for Advanced Composite Materials, Universiti Teknologi Malaysia, UTM Johor Bahru 81310, Johor, Malaysia

[3] Laboratory of Biocomposite Technology, Institute of Tropical Forestry and Forest Products (INTROP), Universiti Putra Malaysia, UPM Serdang 43400, Selangor, Malaysia

[4] Advanced Engineering Materials and Composites (AEMC), Department of Mechanical and Manufacturing Engineering, Faculty of Engineering, Universiti Putra Malaysia, UPM Serdang 43400, Selangor, Malaysia; mmharussani17@gmail.com (M.M.H.); yuzihakimi23@gmail.com (M.Y.A.Y.H.); haziqz.dev@gmail.com (M.Z.M.H.)

[5] Department of Chemical and Environmental Engineering, Universiti Putra Malaysia, UPM Serdang 43400, Selangor, Malaysia; sitinuratikah_asper7@yahoo.com

[6] Department of Aerospace Engineering, Faculty of Engineering, Universiti Putra Malaysia, UPM Serdang 43400, Selangor, Malaysia; asyrafriz96@gmail.com (M.R.M.A.); mohdridzwan@upm.edu.my (M.R.I.)

[7] Research Centre for Sustainability Science and Governance (SGK), Institute for Environment and Development (LESTARI), Universiti Kebangsaan Malaysia, UKM Bangi 43600, Selangor, Malaysia

[8] Centre for Defence Foundation Studies, Universiti Pertahanan Nasional Malaysia (UPNM), Kem Perdana Sungai Besi 57000, Kuala Lumpur, Malaysia; mohd.nurazzi@gmail.com

[9] Research Center for Chemical Defence, Universiti Pertahanan Nasional Malaysia (UPNM), Kem Perdana Sungai Besi 57000, Kuala Lumpur, Malaysia; faiznorrrahim@gmail.com

[10] Department of Mechanical Engineering, Andalas University, Padang 25163, Sumatera Barat, Indonesia; habral@yahoo.com

[11] Department of Mechanical Engineering, University of Jember, Kampus Tegalboto, Jember 68121, East Java, Indonesia; asrofi.teknik@unej.ac.id

* Correspondence: ahmadilyas@utm.my (R.A.I.); sapuan@upm.edu.my (S.M.S.); mrizal@ukm.edu.my (M.R.R.)

check for **updates**

Citation: Ilyas, R.A.; Sapuan, S.M.; Harussani, M.M.; Hakimi, M.Y.A.Y.; Haziq, M.Z.M.; Atikah, M.S.N.; Asyraf, M.R.M.; Ishak, M.R.; Razman, M.R.; Nurazzi, N.M.; et al. Polylactic Acid (PLA) Biocomposite: Processing, Additive Manufacturing and Advanced Applications. *Polymers* **2021**, *13*, 1326. https://doi.org/10.3390/polym13081326

Academic Editor: Pietro Russo

Received: 17 March 2021
Accepted: 15 April 2021
Published: 18 April 2021

Publisher's Note: MDPI stays neutral with regard to jurisdictional claims in published maps and institutional affiliations.

Abstract: Over recent years, enthusiasm towards the manufacturing of biopolymers has attracted considerable attention due to the rising concern about depleting resources and worsening pollution. Among the biopolymers available in the world, polylactic acid (PLA) is one of the highest biopolymers produced globally and thus, making it suitable for product commercialisation. Therefore, the effectiveness of natural fibre reinforced PLA composite as an alternative material to substitute the non-renewable petroleum-based materials has been examined by researchers. The type of fibre used in fibre/matrix adhesion is very important because it influences the biocomposites' mechanical properties. Besides that, an outline of the present circumstance of natural fibre-reinforced PLA 3D printing, as well as its functions in 4D printing for applications of stimuli-responsive polymers were also discussed. This research paper aims to present the development and conducted studies on PLA-based natural fibre bio-composites over the last decade. This work reviews recent PLA-derived bio-composite research related to PLA synthesis and biodegradation, its properties, processes, challenges and prospects.

Keywords: polylactic acid (PLA); natural fibres; biocomposite; mechanical properties

1. Introduction

Since the early 1970s, biodegradable polymers have been attracting the focus of researchers on their development as a result of increasing concern about material resources as well as plastic disposal issues [1–4]. Composite materials are formed within the complete structure of more than one material. The bulk of composites have solid and rigid fibres with a low density in a matrix. Polylactic acid (PLA) is a flexible polymer that is fermented into a carboxylic acid and made from sustainable agricultural waste [5,6]. The lactic acid is then polymerized through a cyclic dilactone, lactide, and ring for product modification. The growing awareness of environmental sustainability and new laws and policies has pushed companies to produce environmentally friendly products [7].

A lot of studies are performed that aim to produce entirely biodegradable composite structures with PLA and natural fibres combination [8]. Since both PLA and natural fibres are sourced from renewable sources, biodegradable, as well as biocompostable, natural/PLA composites are recyclable green materials. Thus, the biocomposites have significant advantages due to their reduced manufacturing and waste disposal treatment costs, as natural fibre reinforced composites can be disposed of easily via landfill, incineration, or by green treatment of pyrolysis, as stated by Harussani et al. [2]. In addition, biopolymers are suitably utilised in various methods of composite fabrication, for example, injection moulding, extrusion, compression moulding, and so forth, whereas less research had been conducted on composites derived from recycled raw materials with matrixes. Biopolymers also satisfy the long-term characteristics of sustainable materials because they are not single-use products. The investigation by Oksman et al. [9] showed that the strength indicated by the stiffness of glass fibre composites was lower than the natural fibre composites. Moreover, the adhesion of the fibre/matrix is a dynamic process with several overlapping variables. Graupner [10] believed that lignin has the ability to strengthen the fibre and matrix bond that is proven by the production of new advanced types of natural fibre reinforced PLA composite.

The reinforcement of PLA composite with natural fibre is a key factor in increasing biocomposites applications in the mechanical field. In addition, among biodegradable polymers, bio polyester PLA is recorded with the highest volume applications for industrial demands. This is attributed to the effective life cycle assessment of PLA, where the supply chain of PLA requires less transportation as well as emits fewer greenhouse gases. The significant properties, such as biodegradability, renewability and lower CO_2 emissions, of PLA-based biocomposites led to their performance in the market. Thus, this review aims to investigate the inherent properties of PLA itself and natural fibre, as well as their synergistic effects when hybridized together. This paper comprises a compilation of previous works and the development of PLA/natural fibre composites in industrial fields.

2. Natural Fibre

Natural fibre can be obtained from plants, animals or the environment. Figures 1 and 2 show basic information about the branches of biocomposites and natural fibres classification. The advantages of natural fibres over synthetic fibres are their recyclability and biodegradability [11–13]. The natural fibre market and production are drastically progressing, shifting the focus of science and engineering to PLA composites. Nowadays, natural fibres are well known in reinforced polymeric materials for industrial developments, for example, glass fibre as a matrix material [14–17]. The polymer is obtained from the fermentation of corn, potato, sugar, beet, and other agricultural sources. Despite their biodegradability properties, natural fibres exhibit several main drawbacks that hinder their developments, including differences in consistency, sensitivity to moisture intake due to their hydrophilic nature, and low thermal stability [18–21].

Studies on the processing effects and natural fibres properties show an improvement in mechanical properties parallel to the National Policy on Industry 4.0 (Industry 4WRD). Natural fibres have sparked great interest among researchers and industry players for their applications in the military [22], automotive [23–27], industrial [28–33], furniture [34], civil [35,36], and biomedical fields [37]. It is reported that the natural fibres global production exceeded 105 million metric tons in 2018 [38]. Mignaeault et al. [39] studied the applications of wood-plastic composites that are typically manufactured as hybrid products by combining wood flour with recycled plastic. The studies on the fibre length and processing effects on the composite's properties revealed enhanced mechanical properties with increasing fibre content. The non-wood fibres include plant straw, leaf, bast, fruit, seed, or grass. Straw fibres originating from resources such as maize, wheat, and rice hull are comparatively solid, rigid, low density, and sustainable. These natural fibre reinforced composites are used as deck boards in housing construction materials [40].

Biopolymer PLA has long been a matter of considerable concern in multiple sectors. This is because, in fibre or matrix adhesion, the type of fibre plays a significant role and that influences the biocomposite's mechanical performance. Recent evidence by Mukherjee and Kao [40] showed that the cellulose content and the microfibrils angle are the two variables that govern the mechanical properties of natural fibre. The better the mechanical properties of the fibre, the better the mechanical properties of the biocomposites. The mechanical properties of natural fibres depend on the cellulose content in them. Table 1 shows the chemical composition of the natural fibres, where the natural fibres' chemical composition and cell structures are quite complex and differ in plant parts and origins. Depending on the cellulose crystallinity, the physical, chemical, and mechanical behaviours of the lignocellulosic fibres vary from one another [41]. Type of fibre is an important parameter in determining their chemical composition. The primary natural fibres components comprise cellulose, hemicellulose and lignin [42–44]. Each fibre exists as a composite by nature, where the rigid cellulose microfibrils are protected in the amorphous matrix containing hemicellulose and lignin [45–49]. Thus, natural fibres can also be identified as cellulosic or lignocellulosic fibres [43]. Generally, the main natural fibres' constituent is cellulose with 30%–80% followed by hemicellulose of 7%–40%, and 3%–33% lignin as shown in Table 1.

Table 1. Chemical composition of selected common natural fibres.

Fibres	Holocellulose (wt%)		Lignin (wt%)	Ash (wt%)	Extractives (wt%)	Crystallinity (%)	Ref.
	Cellulose (wt%)	Hemicellulose (wt%)					
Sugar palm fibre	43.88	7.24	33.24	1.01	2.73	55.8	[46]
Wheat straw fibre	43.2 ± 0.15	34.1 ± 1.2	22.0 ± 3.1	-	-	57.5	[50]
Soy hull fibre	56.4 ± 0.92	12.5 ± 0.72	18.0 ± 2.5	-	-	59.8	[50]
Arecanut husk fibre	34.18	20.83	31.60	2.34	-	37	[51]
Helicteres isora plant	71 ± 2.6	3.1 ± 0.5	21 ± 0.9	-	-	38	[52]
Pineapple leaf fibre	81.27 ± 2.45	12.31 ± 1.35	3.46 ± 0.58	-	-	35.97	[53]
Ramie fibre	69.83	9.63	3.98	-	-	55.48	[54]
Oil palm mesocarp fibre (OPMF)	28.2 ± 0.8	32.7 ± 4.8	32.4 ± 4.0	-	6.5 ± 0.1	34.3	[55]
Oil palm empty fruit bunch (OPEFB)	37.1 ± 4.4	39.9 ± 0.75	18.6 ± 1.3	-	3.1 ± 3.4	45.0	[55]
Oil palm frond (OPF)	45.0 ± 0.6	32.0 ± 1.4	16.9 ± 0.4	-	2.3 ± 1.0	54.5	[55]
Oil palm empty fruit bunch (OPEFB) fibre	40 ± 2	23 ± 2	21 ± 1	-	2.0 ± 0.2	40	[56]
Rubber wood	45 ± 3	20 ± 2	29 ± 2	-	2.5 ± 0.5	46	[56]
Curauna fibre	70.2 ± 0.7	18.3 ± 0.8	9.3 ± 0.9	-	-	64	[57]
Banana fibre	7.5	74.9	7.9	0.01	9.6	15.0	[58]
Sugarcane bagasse	43.6	27.7	27.7	-	-	76	[59]
Kenaf bast	63.5 ± 0.5	17.6 ± 1.4	12.7 ± 1.5	2.2 ± 0.8	4.0 ± 1.0	48.2	[60]
Phoenix dactylifera palm leaflet	33.5	26.0	27.0	6.5	-	50	[61]

Table 1. *Cont.*

Fibres	Holocellulose (wt%)		Lignin (wt%)	Ash (wt%)	Extractives (wt%)	Crystallinity (%)	Ref.
	Cellulose (wt%)	Hemicellulose (wt%)					
Phoenix dactylifera palm rachis	44.0	28.0	14.0	2.5	-	55	[61]
Kenaf core powder	80.26		23.58	-	-	48.1	[62]
Water hyacinth fibre	42.8	20.6	4.1	-	-	59.56	[63]
Wheat straw	43.2 ± 0.15	34.1 ± 1.2	22.0 ± 3.1	-	-	57.5	[64]
Sugar beet fibre	44.95 ± 0.09	25.40 ± 2.06	11.23 ± 1.66	17.67 ± 1.54	-	35.67	[65]
Mengkuang leaves	37.3 ± 0.6	34.4 ± 0.2	24 ± 0.8		2.5 ± 0.02	55.1	[66]

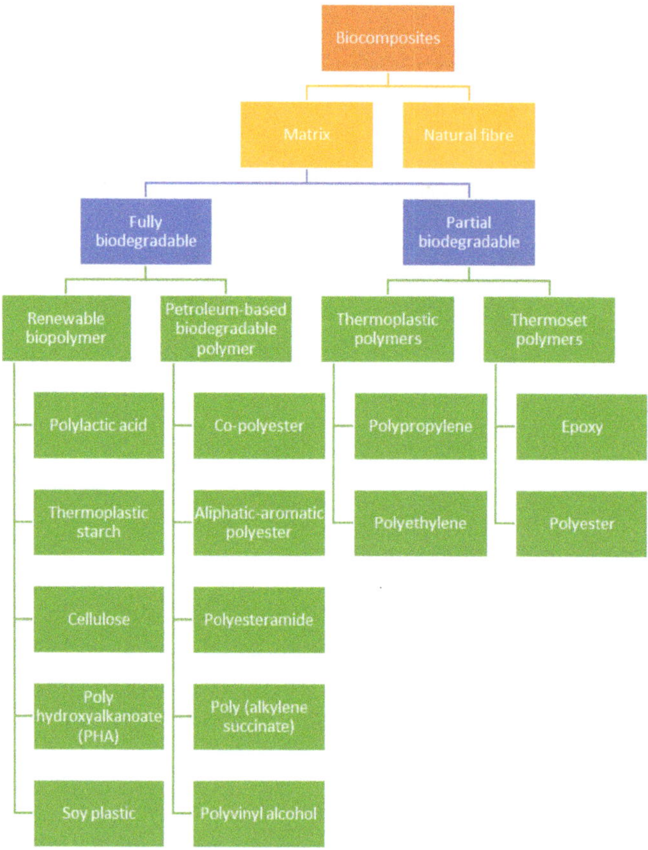

Figure 1. Classification of biocomposites. Adapted with copyright permission from Mohanty et al. [67].

Besides this, the surface structure of the natural fibre plays a vital role in the adhesion behaviour of the matrix polymer and the fibres. The higher and stronger the fibres and matrix adhesion, the better the natural fibres' mechanical properties. The composite intensity depends on the various level of the fibres in use and is approximately equal to its matrix binding mechanism.

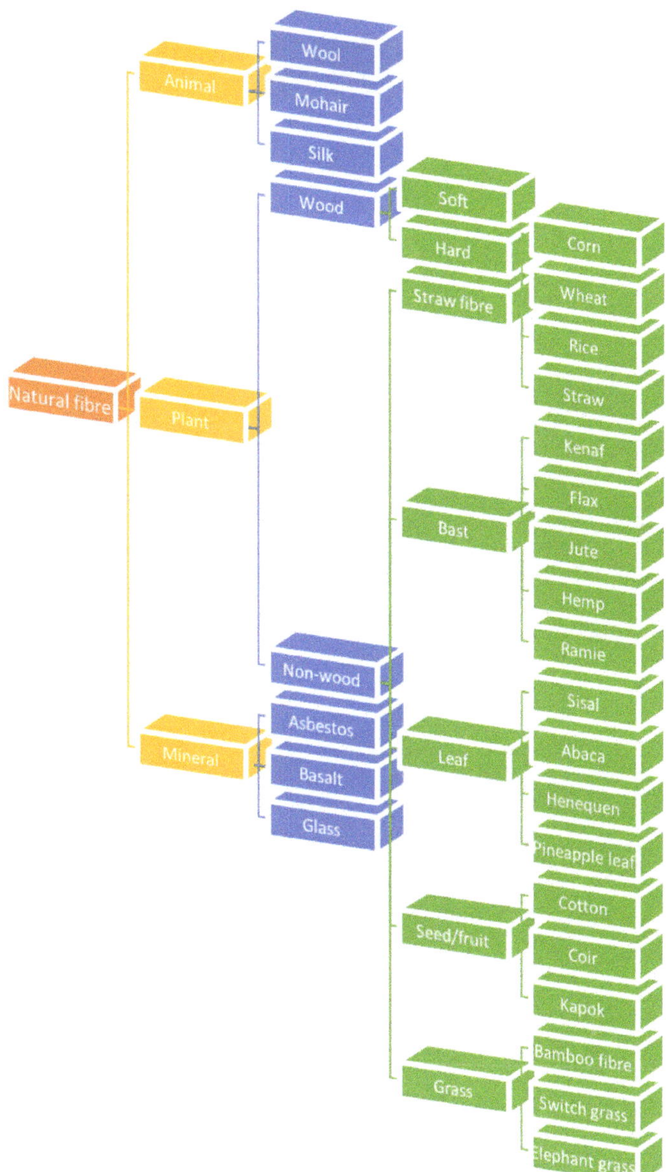

Figure 2. Classification of natural fibres.

Chemical Treatment of Natural Fibres

Various studies have assessed the efficacy of reinforced natural fibre polymer composite in a variety of fields. In recent years, a number of treatments were developed to enhance interfacial bonding and to improve mechanical properties and water-resistance characteristics of natural fibres. The most critical downside of these natural fibres in polymer composites is their hydrophilicity that creates a weak interfacial bonding between fibres and matrix. Numerous physical impurities and hydroxyl groups are found on the surface of the fibres that affect their performance as reinforcing agents. Thus, several surface modifications and treatments have been explored in order to obtain the desired properties of polymer composites.

To identify a knowledge gap in the field of study in natural fibres reinforced PLA composites, a review by Sawpan et al. [68] on the improvement of mechanical performance of industrial hemp fibre reinforced polylactide biocomposites was conducted. Recently, simpler and more rapid tests of PLA have been developed for a variety of fibre material properties of random fibres. Interestingly, aligned PLA composites reinforced with long hemp fibre (0–40 wt.%) were studied. It was revealed that 30% alkali-treated fibre reinforced PLA composite (PLA/ALK) had the highest mechanical strength with a tensile strength of more than 70 MPa, a Young's modulus of more than 8 GPa module, and flexural toughness of 2.64 kJ/m^2. The use of qualitative case studies is a well-established approach in the alignment of long fibres along with the improvement of PLA/ALK composites.

A previous literature review by Li et al. [69] focused on the effect of chemical treatments on natural fibres for the application of natural fibre-reinforced composites. The authors identified poor fibre-matrix compatibility and relatively high absorption of moisture as the key drawbacks of natural fibres in composites. Thus, modifying the fibre surface properties using chemical treatments were being contemplated. Chemical treatments, for example, alkali, silane, acetylation, benzoylation, and so forth were found to improve fibre-matrix surface adhesion and reduce the water absorption properties of the fibres [19]. In order to facilitate the adhesion, chemical coupling of the adhesive with certain compounds, for example, sodium hydroxide, acetic acid, silane, acrylic acid, isocyanates, maleate coupling agents, potassium permanganate and peroxide was applied. Data from this study showed that chemical treatments should be applied when modifying the properties of natural fibres.

Reinforcing natural fibre with polymer composite has been a major focus of previous research into the natural fibre. In a study on natural fibre surface modifications and performance of the subsequent bio-composites, Mohanty et al. [67] stated that an adequate degree of adhesion between the surface of hydrophilic natural fibre cellulose and the polymer matrix resin is usually needed to ensure the desirable properties of bio-composites. Useful methods to enhance fibre-matrix adhesion in natural fibre composites are alkali, peroxide, coupling agents and isocyanate treatments, dewaxing, bleaching, vinyl grafting and acetylation. By replacing the hydroxyl group with some chemical functionality, the crystalline structure of cellulose can be broken. This process of decrystallization enhances the thermoplasticity of the cellulose because the role of the plasticiser is performed by the substituted groups [70]. The findings of this study suggested that a few levels of success in creating a superior interface were achieved by surface modifications of hydrophilic natural fibres, but cheaper cost surface modification needs to be emphasised in many applications in the future for biocomposites to replace glass fibre composites.

3. Polylactic Acid (PLA)

Poly (lactic acid) (PLA) exits in the form of polymers that are biodegradable polyesters obtained from lactic acid (LA) or 2-hydroxy propionic acid, typically obtained from agricultural crops such as maize, potatoes, and cassava through bacterial fermentation of carbohydrates. As shown in Figure 3, PLA is categorised as a biopolymer originated from chemically treated bio-based monomers, LA [71]. Owing to its biocompatibility with the human body, PLA has been the choice of material for medical applications [72]. The discovery of the latest polymerisation pathways that enable high molecular weight PLA to be produced economically, together with the rise of awareness on environmental issues have led to the broadening of the use of PLA as alternative raw materials [73].

PLA is the most widely researched and promising biopolymer that is capable of substituting conventional petroleum-based polymers due to its renewability, recyclability, biodegradability and compostability. In addition, PLA has an excellent manufacturing ability as it is suitable to be processed with various methods. Injection moulding, film extrusion, blow moulding, thermoforming, fibre spinning and film-forming are among the

PLA manufacturing processes [74]. PLA-derived products had utilized in several industrial applications, including in packaging [8,75], textile [76], biomedical [77], structural [78,79] and automotive [80].

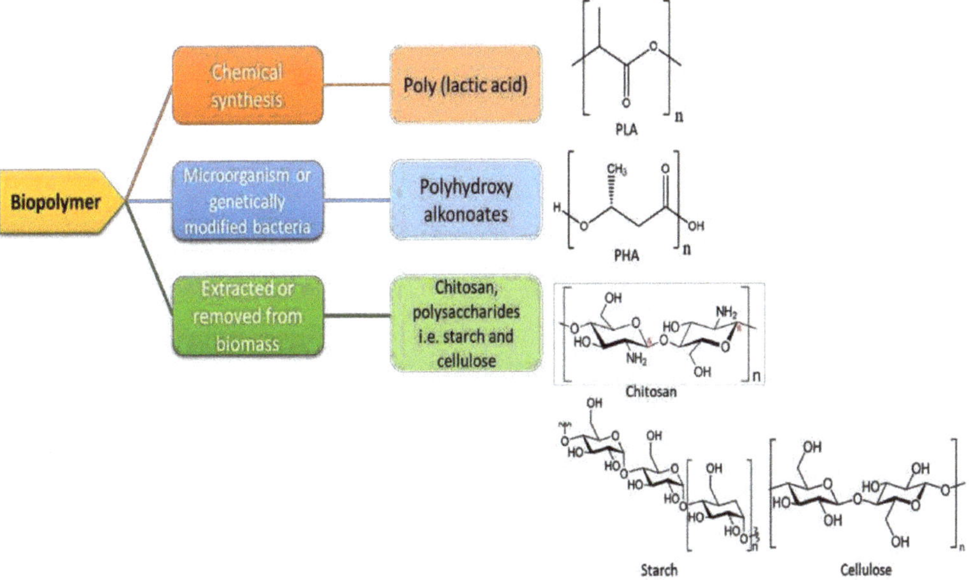

Figure 3. Chemical structure and classification of significant biopolymers based on their origin. Reproduced with copyright permission from Mohanty et al. [81].

To determine the properties of PLA, Sodergard et al. [72] stated that PLA polymers are obtained from agricultural crops such as maize and potatoes through bacterial fermentation of carbohydrates. Three synthesis processes of PLA were used by Sodergard et al. [72] to collect data. Then, there are a lot of advantages of using PLA as composites, for example, minimized weight, decreased thermal conductivity and suitable for medical implant applications as identified in this study. PLA with greater mechanical and flexural characteristics has improved the temperature of heat distortion and improved membrane characteristics.

4. Composite

Composite materials could be described as materials that consist of two or more phases separated by different interfaces that are chemically and physically dissimilar [82–84].

The various systems are carefully integrated in order to attain a system with more valuable structural or functional properties which cannot be obtained by either of the components alone [85–87]. In general, composite materials exhibit special features in which the strength-to-weight ratio is high. One other benefit of composite material is that it offers structural versatility since it is possible to form composites into complex shapes [88]. Composite materials obtain the majority of their advantageous properties through a tight bond between a strong rigid reinforcement—commonly fibres (filaments) or reinforcements with other geometric forms, such as particles, platelets—and a weaker, less rigid matrix [89]. The basic composite components and features giving the mechanical properties of the composite are shown in Figure 4.

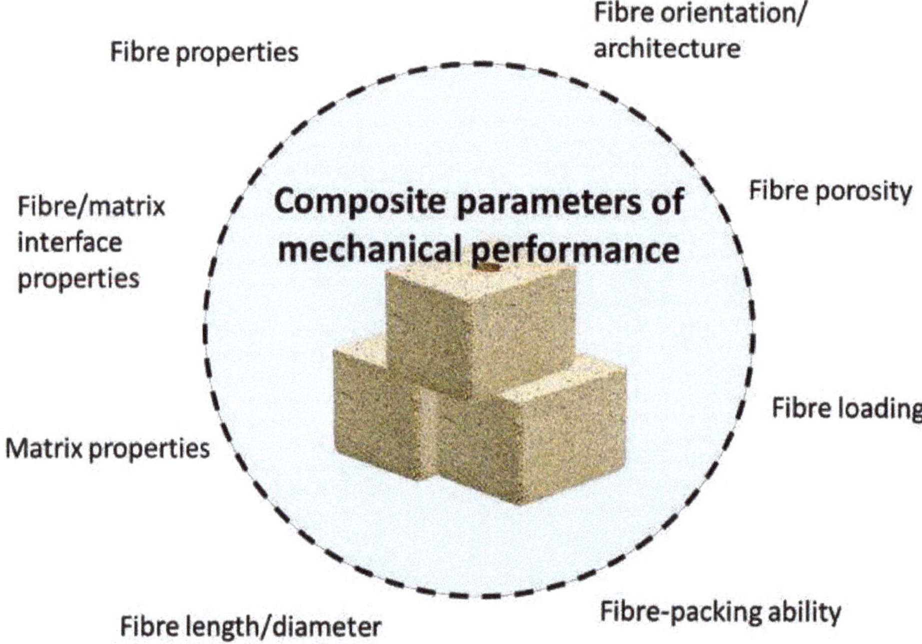

Figure 4. Mechanical properties of the composite.

A number of studies have begun to examine natural fibres reinforced PLA composites. The analysis was based on the factors influencing the performance of tensile strength. Based on the article written by Ku et al. [90], the composites' tensile strength with fibres was found to be 20–40% lower in the perpendicular direction than that of composites with fibres in the parallel direction. The mean score for natural fibre was it might be used to substitute traditional fibres in plastic content reinforcement. For natural fibre reinforcements, matrix interfacial bonding, new manufacturing methods and chemical and physical alteration techniques are developed, and the tensile properties of the composites are improved.

5. Natural Fibre Reinforced PLA Composite

5.1. Brief on PLA Composite

PLA is a naturally resourced thermoplastic polymer produced globally with a capacity of approximately 211,000 tons in 2020 [91]. In addition, chitosan and chitin derivatives productions were about 107,000 tons [92], while cellulose and PHA were produced globally with a capacity of over 580,000 tons and 30,000 tons, respectively, in 2020 [93]. These data had been represented as graph bar in Figure 5. The global market for PLA is estimated to double every four years according to Jem's law [91]. This is supported by the fact that environmental pollution due to excessive petroleum-derived plastic production and global warming pressure had attracted the public's attention to extensively explore the utilisation of biodegradable products.

According to Nazrin et al. [8], biodegradable polymers PLA are brittle, sensitive to moisture and have low-impact strength. Thus, a possible way to further reinforce the polymers is by hybridising them with natural fibres to yield the enhanced mechanical properties of biocomposite. Moreover, biocomposites are biodegradable, reusable, and organic-based, assisting to reduce both the reliance on depleting petroleum sources and the environmental burdens from petroleum-based products. The PLA's excellent mechanical properties and barrier capability can be used to manufacture biomaterials that suit a range of applications.

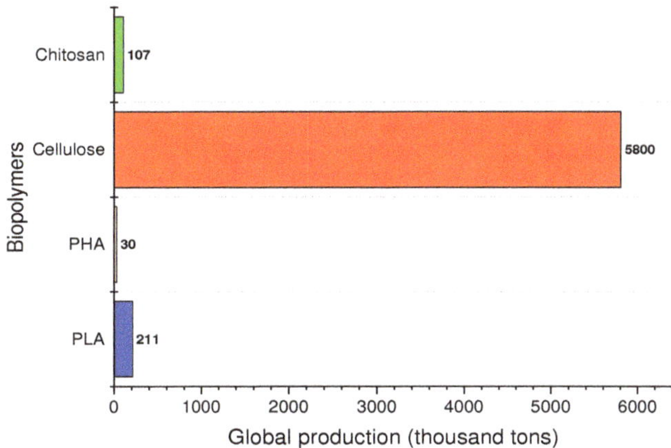

Figure 5. Global production of biopolymers in the year 2020 [91–93].

A variety of studies have begun to investigate PLA-based composite with natural fibres reinforcement. A recent systematic literature review on natural fibre-reinforced PLA composites by Siakeng et al. [94] successfully demonstrated the developments in the study of natural fibre composites based on PLA and neat PLA over the past couple of years. A comprehensive investigation was conducted to understand the characteristics of natural fibre and PLA and their activities after a variety of processing techniques. From the study, it was shown that natural fibre reinforced PLA composite can provide a possible alternative to replace the utilisation of synthetic polymers that are unsustainable and hard to dispose. However, there has been limited progress in attempting to build entirely biodegradable composites from biopolymers. This study points out the efficient replacement of synthetic polymer composites with fully biodegradable composites would need an improvement in terms of PLA manufacturing cost.

Hinchcliffe et al. [95] integrated the theoretical and experimental researches on pre-stressed natural fibre reinforced PLA composite components. In the study, the mechanical properties of PLA were enhanced with additive loading. Using both tensile and flexural mechanical tests, the effect of fibre form, matrix cross-section geometry PLA, and post-tension level of the PLA were mechanically investigated. In contrast to solid non-reinforced PLA composites, the PLA reinforced composite showed increment in tensile strength by 116% and flexural specific strength by 12%. This study revealed the possibility of basic tensile and flexural strengths' improvements of PLA composites by creating increasingly efficient structural shapes using additive manufacturing (3D printing) and by processes such as initial post-tensioning fibre.

Many researchers have utilised natural fibres reinforced polylactide (PLA) composites. Several developments focused on PLA due to properties greater than conventional polymers. The mean score for natural fibres was most of the layer and mass modification techniques were optimized to change a specific property and had also ignored the effect of the technique on other primary properties. The effectiveness of bulk and surface modification methods of widely used thermoplastic polyesters on 3-dimensional (3D) scaffolds (PHA) is reported in [74]. The findings will be of interest to investigate the work of PLA shape memory properties.

5.2. Development of Natural Fibres/PLA Biocomposite

5.2.1. Brief Study on PLA Biocomposite

A lot of studies have investigated natural fibres reinforced polylactide (PLA) composites in any systematic way. There is a growing body of literature that recognises the importance of combination biodegradable polymers with plant fibres to produce composites. Table 2 summarises works on natural fibre reinforced PLA composites. A seminal study from Holbery and Houston [96] carried out the application of natural fibre reinforced polymer composites in the automotive industry. Natural fibres, for example, kenaf, hemp, flax, jute and sisal were found to offer benefits such as weight, cost, CO_2 reduction, less dependency on external oil sources and recyclability. The production of the composite involved the combination of natural fibre preform or mats along with a thermoplastic binder system. The most popular system currently used in vehicle industries is thermoplastic polypropylene, especially for non-structural components. In a variety of automotive components, using natural fibre reinforcement is proven feasible. The applications of seatback linings, door cladding, and floor panels are refined from flax, sisal, and hemp. Coconut fibre is used in the manufacturing of seat bottoms, head restraints, and back cushions, while cotton can be used for sound barrier and wood fibre is often used in the backseat. Only if problematic areas such as moisture issues, compatibility, durability, and continuous fibre source could be tackled by vehicle manufacturers, can natural fibre composites compete as reliable materials in the automotive industry.

In a review conducted by Dunne et al. [97], the sustainability and automotive applications of natural fibres were compiled. Based on previous research, there is a lot of new advancement for humanity in the production of natural fibres for industrial use. This trend is attributed not only to an increased awareness of the environment but also to the fact that natural fibres have excellent properties. At the moment, there is still a lot of ongoing research on natural fibres and their ability to replace petroleum-based products, and only time will tell which of the two will be the better option. Faruk et al. [98] developed a study on biocomposites reinforced with natural fibres. The physical, thermal and mechanical properties of the biocomposites could be affected by various factors, including the type of fibre, conditions for the climate, manufacturing processes and surface modification of the fibre. A broad range of biocomposite processing techniques and variables (moisture quality, form and content of fibre, binding agents and their effect on the properties of composites) influence these processes.

Results from several studies conducted by Herrera-Franco and Valadez-Gonzalez [99] about mechanical properties of continuous natural fibre-reinforced polymer composites showed positive results. Strong evidence of natural fibres reinforced PLA composites was found when ASTM D 638-99 was tested using an Instron 6025 100 kN test machine. Fibres samples analysed (10% of waste fibres) using the silane concentration struggled in a brittle manner to facilitate a chemical interrelationship and fibre-matrix binding. The result took three weeks for the equilibrium conditioning at 65% relative humidity and final moisture. Whereas, for transverse tensile and flexural properties, the rise was greater than 50% corresponding to the untreated fibre composite direction. The findings of this analysis showed that it was found from the failure surfaces that the fibre matrix enhanced the association of the failure mode.

The topic of natural fibre reinforced PLA composites was investigated a long time ago. Research by Alkbir et al. [100] was carried out about natural fibre-reinforced composite structure fibre characteristics and chassis rigidity parameters. From the findings, materials, for example, jute, sisal, flax, kenaf and hemp can be used in various industrial applications due to their outstanding mechanical features, low cost, high strength, eco-friendliness and biodegradability, ease of processing and strong rigidity of the structure. The total tensile strength and bending strength characteristics of natural fibre-reinforced polymer hybrids were strongly focused on the ratio of water uptake. The findings will be of interest to geometric structures between two things that were found to have a major effect on the parameters of crashworthiness and basic load transfer of polymer composite tubes reinforced with natural fibre.

5.2.2. Non-Wood Natural Fibre/PLA Composite
Leaf

The efficacy of natural fibre reinforcements with PLA composite was tested in various studies. To date, several studies were carried out on strengthening the PLA-based composites using leaf originated natural fibres including sisal and leaf fibre. From previous research by Asaithambi et al. [101] on reinforced PLA hybrid composites with banana/sisal fibre (BSF), the impact of benzoyl peroxide (BP) surface treatment on the mechanical properties of BSF-reinforced PLA composites was investigated. BSF reinforced PLA hybrid composites were manufactured using the twin-screw extrusion method followed by injection moulding. Using the Universal Testing Machine (UTM), tensile strength and flexural strength were evaluated. The results of the study suggested that a cross-linking approach is a good alternative to raise the compatibility of the PLA matrix with BSF. In addition, the production method using extrusion and injection moulding of BSF/PLA composite was found to have excellent mechanical properties. It was concluded that a prolonged analysis would help pave the way for an extended scope and prospect for the possible end-use of BSF reinforced PLA composite applications.

Bajpai et al. [102] studied the tribology of natural fibre-enhanced PLA composites and found that the incorporation of natural fibre sheets into the PLA matrix had significantly boosted the properties of neat polymer wear. In order to create a laminated composite using the hot compression process, natural fibres, for example, nettle, *grewia optiva*, and sisal were integrated with the PLA polymer. The operating parameters were set such that the loads applied were between 10 N and 30 N, the sliding speeds were between 1 m/s and 3 m/s, and the sliding distances were between 1000 m and 3000 m. The specimens were examined for friction and wear characteristics of the produced composites under dry surface conditions. The results of the experiment showed that there were decrements in the friction coefficient of 10%–44% relative to neat PLA and the real wear rate of composites produced by more than 70%. This research has established and demonstrated that natural fibres can be an efficient alternative to tribological material reinforcement implementations.

Bast

To date, several studies have investigated the reinforcement of natural fibre with PLA composite. In this section, reinforced PLA composite using natural fibres originated from non-wood basts, for example, kenaf, jute, hemp and flax are reviewed. Graupner et al. [103] investigated the mechanical characteristics of natural and artificial cellulose fibres reinforced with PLA composite. Various types of natural fibre composites, for example, hemp, cotton, kenaf and man-made cellulose fibres with different characteristics were manufactured by compression moulding with a fibre mass of 40% and the use of PLA. Kenaf and hemp/PLA composites showed promising results in their tensile strength and young modulus values, while cotton/PLA exhibited strong collision features. Meanwhile, lyocell/PLA composites displayed promising result in all three features. In a nutshell, the study revealed the distinct characteristics of the composites that are able to be used for various technical applications, each meeting unique standards.

A few studies have started to investigate the use of non-wood jute derived natural fibre polymer composite reinforcement with PLA [104,105]. The work of Jiang et al. [104] on jute/PLA composite hydrothermal ageing and structural damage found by X-ray tomography. The study claimed that biodegradable PLA composites reinforced with natural fibre could replace conventional composites reinforced with synthetic fibre, but they might experience hygrothermal ageing from the combined effects of heat and moisture. The effect of 50 °C liquid ageing on water absorption, chemical fibre, matrix degradation, and the tensile properties of the composite jute fibre/PLA were associated with internal structure modifications of the composite jute fibre/PLA observed on X-ray computed tomography. The result showed that the hygrothermal ageing caused the jute fibre/PLA composite to reduce in strength, tensile elastic modulus, ductility, and mechanical properties. In conclusion, X-ray tomography will determine the effects of hygrothermal degradation of a composite jute fibre/PLA.

A variety of definitions of the term PLA have been suggested. This paper will use the definition suggested by Yu et al. [105], who saw it as PLA is a linear aliphatic thermoplastic polyester manufactured with high stiffness and UV stability from renewable resources. The research reported that PLA is produced by two methods; ring-opening polymerisation of lactide and polycondensation of lactic acid, while the combination PLA/jute fibre material indicated that total tensile strength was slightly above that of the PLA. Interestingly, the PLA was observed to have the lowest tensile strength than PLA based composites. The results of this study revealed that with the incorporation of the fibre, the mechanical properties of materials dependent on PLA will increase and then decline until the fibre content is above 30%.

Oksman and Selin [106] discovered that plastics and composites from PLA showed that in a composite system, where natural fibres are used as reinforcements, PLA may be used as the matrix. Flax-reinforced PLA composites that can easily be extruded and compressed were found to be 50% stronger than many other thermoplastic composites reinforced with flax that are currently used in automotive panels. With the addition of 30% flax fibres, the PLA stiffness was increased from 3.4 GPa to 8.4 GPa. The results of this research indicated that the possibility of using typical production methods is an important element in the industrial use of renewable resources. There were no problems with extrusion and compression moulding in PLA/flax composites in this situation, and they can be easily processed.

In addition, an informative study by Omar et al. [107] involved the application of kenaf fibre reinforced composite in the automotive industry. The study reviewed the latest findings in kenaf fibre reinforced composite. In contrast to synthetic fibres, natural fibre materials in the automotive sector could possess several benefits, for example, weight and price reduction, recyclability, renewability and being environmentally friendly. The results suggested that kenaf fibres have fewer environmental impacts from the processing of natural fibres compared to the manufacturing of glass fibres, the higher natural fibre content in the composite replaces the matrix (synthetic polymer), higher fuel efficiency and lower phase emissions due to weight reduction, and energy and carbon offsets from end-of-life natural fibre incineration [108]. The research concluded that better kenaf plantation procedures are also needed to be practised upstream in order to guarantee a consistent availability of good kenaf fibres and to satisfy the growing demand for kenaf biocomposites.

In an investigation conducted by Zhang et al. [109] on flame retardancy of flax/PLA biocomposites, it was discovered that the bio-inspired approach to fibre surface modification on flax/PLA composites could enhance flame retardancy of the material. Raw flax fibre was first coated with a thin adhesive polydopamine (PDA) film in an aqueous dopamine solution followed by iron phosphonate growth on the fibre surface in situ. In order to prepare a flame retardant biocomposite, the altered flax fibre was added to PLA. The improved composite showed reductions of 16% and 21% in peak heat release rate and total smoke generation, respectively, with a limited amount of fibre surface flame retardant. In summary, this research revealed a novel approach to high-performing reinforced polymer composite preparation of flame retardant material.

For a comprehensive understanding, detailed research on flame retardancy of flax/PLA composites was performed by Pornwannachai et al. [110]. Their study on fire-resistant natural fibre reinforced flame retardant textiles discovered that by treating fabrics with common flame retardants, composites made from natural thermoplastic blended materials were often effectively flame retarded before melting. The melting of intermingled flax thermoplastic fibre was used to prepare thermoplastic composites. Before composite preparation with different common flame retardants (FRs) for textiles, fabrics were prepared. UL-94 flame output tests showed that only the organophosphate FR passed the test while the flax/PLA control and the rest of the flame-retardant composite specimens failed. On the opposite hand, the entire flame retarded flax/PLA samples reached a V0 rating which represented improved flame retardant properties of the composites. However, the study concluded that each fire retardant had also decreased the mechanical properties of the laminates, although giving the composite laminate high fire-retardant properties.

Seed/fruit

Previous research on natural fibres was focused on reinforcing natural fibre with PLA composite. In a study conducted by Jang et al. [111] on the flammability of reinforced PLA composites with coconut fibre, the mechanical and thermophysical behaviour of composite treated with plasma therapy was established. The coconut fibre interfacial adhesion strength was increased by the means of plasma treatment. In most cases, the coconut fibre/PLA will shrink in size with an increase in fibre weight after surface treatment using plasma. To be precise, the longitudinal direction of composites has higher shrinkage than the transverse direction. In this research, it was stated that plasma treatment could be important in optimizing the mechanical and thermophysical properties of natural fibre reinforced composites.

Graupner [10] proposed the usage of lignin as a promoter to natural adhesion in composites of cotton fibre-reinforced PLA. This research examined the method of lignin reinforcement with biodegradable thermoplastic cotton as a natural adhesion promoter that influenced the composites' mechanical properties. Further data collection was required to determine whether compression moulding affected fibre mass proportions by 40%. The findings will be of interest to the composites' fractured surfaces when examined using electron scanning microscopy (SEM). The findings of the composite studies revealed that cellulose affected the cotton/PLA composite characteristics. From SEM images, it was found that by adding lignin, the correlation between fibre and matrix will be strengthened. It was possible to enhance tensile characteristics by reducing the impact properties.

Grass

In this section, several studies that investigated the reinforcement of grass originated natural fibre with PLA composite will be discussed, which will include bamboo fibre [112,113], elephant grass [114], and switch grass. Previous research by Gunti et al. [114] on mechanical and degradation properties for PLA reinforced jute, sisal and elephant grass-based composites showed that the mechanical properties of elephant grass composite were superior compared to jute and sisal. The composites were synthesized using an injection moulding method with various percentages of untreated and treated fibres. Furthermore, the tensile strengths of the PLA composite with treated elephant grass were 18.14% and 24% higher at 20% fibre loading than that of the treated jute/PLA composite and plain PLA. This research showed that the integration of fibres into the PLA matrix had greatly increased the strength and strengthened the modulus.

A study by Sukmawan et al. [112] on strength assessment of bamboo fibre-reinforced cross-ply green composite laminates had revealed the impacts of the fibre material on the mechanical properties and failure characteristics. Laminate PLA/bamboo fibre cross-ply composite (0/90) was fabricated by using the hand-layup method before hot pressing using dispersion type biodegradable PLA. The outcome showed that the PLA/bamboo fibre composite was like that of conventional glass fibre reinforced laminate plastic and the basic strength was three times greater than mild steel. The research concluded that PLA/bamboo fibre laminate cross-ply (0/90) might be a substitute for glass fibre-reinforced composites with the potential to be used in sandwich structures as a skin material.

A study of bamboo fibre reinforced biocomposites was conducted by Abdul Khalil et al. [113]. In the last few years, the usage of bamboo fibres as insulation in composite materials has increased tremendously and has experienced a large development in response to the increasing need for the production of biodegradable and renewable materials. This result is somewhat counterintuitive; the rapid development in technology for producing goods makes it possible for customers to make a suitable choice and to have their own desirable preferences. This study has extended their quality skills through the use of raw materials such as bamboo fibre, which is stronger and can be used to produce sustainable high-end manufacturing products of high quality.

Straw fibre

Crop straw is a large-scale, low-price and sustainable farm waste. For the reuse of agricultural waste, research into straw fibre composite materials is very important. According to Elmessiry and Deeb [115], one of the difficulties of modern intensive farming is agricultural waste, which requires novel solutions for useful commercial use. The production of composites from sustainable raw materials has increased considerably over the last years as they are environmentally friendly materials. Thus, in recent years, many researchers had put more focused on straw fibre-reinforced composites including wheat straw [115,116], rice straw [117,118], corn straw [119], soy straw [120] and abutilon fibre [121].

A heated two roll-mill was used to prepare the poly (lactic acid)/rice straw (RS) composites by Mat Zubir et al. [117] using different ratios of RS. They investigated the mechanical performance of prepared PLA/RS composites. The tensile strength and elongation of the composite (E_b) decreases with the rice straw fibre content increasing from 5% to 25% with the Young's modulus increasing. These improvements also reported by Nyambo et al. [116]. They had discovered that enhanced tensile strength (20%) and flexural strength (14%) of the composites had increased considerably in line with that of the neat polymer compared to the 3 and 5 phr PLA-g-MA addition to composites. Plus, the strong interfacial adhesion between the fibre and matrix was due to the observed enhanced strength.

Pradhan et al. [120] had studied the compostability and biodegradation degree of soy straw/PLA and also wheat straw biocomposites. From their discoveries, it is found that PLA composites have shown to be clearly compostable materials for untreated soy and wheat straw. The deterioration of the PLA component is improved due to the existence of the natural biomass, indicating that modified/treated components can be used in composites. This discovery raises the chance of adopting a changed or handled biomass in the composites as any lack of deterioration that might occur due to the biomass alteration may be leveraged, due to the inclusion of readily degradable biomass components with the priming or favourable results.

On the other hand, Ding [119] had prepared corn straw fibre/PLA composite via hot pressing process, and its effect on the mechanical properties and degradation performance had been studied. The findings show that the mechanical properties of the composites first increased and then decreased with the rise of fibre content of the cornstalk (tensile strength and elongation during breakage). At 10%, the elongation ratio of break was 20.3%. The ratio of breakage is 10%. When the corn stem fibre content was 13%, the composite tensile strength hit 24.38 MPa. The intensity of corn straw fibre and rate of degradation of polylactic acid composites increased after 120 days of degradation. At same time, the mass loss of composites increased and the PLA molecular weight decreased more rapidly with the rise in corn straw fibre.

Abutilon fibre-reinforced PLA natural composites had been characterized and studied by Wang et al. [121]. This research uses a melting mix and an extruder to prepare biocomposites of PLA and abutilon fibres The DSC findings show that the fibres have acted as a central agent, resulting in an improvement in PLA crystallization. The findings also show that abutilon fibres have increased the thermal stability of PLA. In addition, higher storage modulus values due to heavy interfacial adhesion are observed. Thetan delta is also diminished by fibre material being added to the PLA matrix, which reduces the mobility of PLA polymer molecules in the presence of fibres. The enhancement of the properties and energy absorption of such biocomposites indicates that abutilon fibres have a tremendous potential as a refinement of green composites.

5.2.3. Wood Natural Fibre/PLA Composite

A number of scientists have begun to examine the reinforcement of natural fibre with PLA composite. A study by Ozyhar et al. [122] on the effect of the functional mineral additive on the properties of the material and the processability of PLA reinforced wood fibre (WF) composites had analysed the use of alkenyl succinic anhydride (ASA) combined with calcium carbonate to act as the practical mineral supplement for WF reinforced PLA composites. With additional mineral amounts of 10, 20, and 30 wt.%, respectively, the effect of the number of minerals on the material properties of 40% fibre reinforcement PLA composites was investigated. The findings indicated that up to 20 wt.% of PLA can be substituted with ASA while preserving material properties. The research concluded that the addition of ASA-treated calcium carbonate had improved fibre adhesion with the PLA, enabling the composite formulation to substantially reduce the content of PLA while maintaining the properties of the material.

In a comprehensive study on PLA/natural fibre composites by Zhang et al. [123], the analysis was conducted as an essential first step towards the use of a mixture of modified natural fibres such as modified bamboo fibre, coconut fibre, and WF as reinforcing materials in the PLA matrix. Via the casting process, PLA composites with three sorts of natural fibres, like bamboo fibres, wood fibres, and coconut fibres were prepared. The results showed that adding three sorts of natural fibres could both strengthen the mechanical and thermal properties of composites, which the durability of composites might be further improved by natural fibres after modification. Additionally, as compared, the composite of the PLA/coconut fibre had the most excellent lastingness and thermal properties. This study presented a viable approach for the PLA industry to be introduced. In conclusion, this study offered a feasible method for the implementation of the PLA industry.

Du et al. [124] discovered that the inclusion of pulp fibres had improved the crystallisation of PLA and its tensile strength. Polymer composites were produced with natural fibres of PLA and cellulose, incorporating the process of shaping wet-layered fibre sheets with conventional composite manufacturing methods. The highest fibre amounts for maximum composite strength were 40% for high yield hardwood pulp fibre and 50% for high yield softwood and kraft pulp fibre. The highest tensile strength achieved was 121 MPa, almost double the natural PLA's tensile strength. The results indicated that the inclusion of pulp fibres had successfully improved the composite storage modulus, elasticity, and even enhanced PLA crystallization. However, there were no major changes in the transition temperature of the composite glass and PLA crystallinity observed.

5.2.4. Natural Mineral Fibre/PLA Biocomposites

One such material of interest currently compared to other natural fibres is natural mineral fibre, which is already being extensively used due to its cost effective, strong mechanical, physical as well as its chemical properties and biodegradable. Mineral fibres used as reinforcement material for composites including basalt [125] and asbestos fibre, via specific treatments. Recent research by Sang et al. [126] on the fabrication of PLA composites strengthened by short basalt fibre and their feasible assessment for 3D-printing applications evaluated the performance of KH550-treated composite reinforced PLA/basalt fibre (KBF) as a possible 3D-printed feedstock. PLA/KBF feedstock filaments were successfully manufactured and printed using the fused-deposition modelling (FDM) technique to variable shape and size pieces. The results indicated that PLA/KBF exhibited similar tensile properties and better flexural properties than PLA/carbon fibre (CF), which can be attributed to the highly complex PLA/CF viscosity which affected the interlayer adhesion. PLA/KBF was shown by the latest research as a technologically enhanced and cheap feedstock for uses in complex design and adjustable scale 3D-printing.

A study by Kurniawan et al. [127] was conducted on the effects of plasma polymerization on silane treated basalt fibre in order to evaluate the mechanical and thermal properties of the basalt fibre reinforced PLA composites. The findings revealed that the composite's mechanical properties were enhanced 45% and 18% higher than the untreated properties respectively. This enhancement also related to the optimized time of 4.5 min irradiating the basalt fibres.

5.3. Processing Method Developments of Natural Fibre/PLA Composite

Natural fibre reinforcement with PLA composite has been the subject of a great deal of previous natural fibre studies. In a study conducted by Khan et al. [128], the mechanical properties of reinforced PLA composites for woven jute fabrics, an eco-friendly bio-composite was introduced as an alternative to the non-biodegradable synthetic fibre composite. The hot press moulding process was used to prepare the PLA reinforced plain-woven jute fabric (WJF). The average values of tensile strength, tensile modulus, flexural strength, flexural modulus, and impact strength of raw warp directed woven jute composite were increased by approximately 103%, 211%, 95.2%, 42.4%, and 85.9%, respectively, and the strain at maximum tensile stress was increased by 11.7% after the reinforcement. It was deduced that jute fabric composites based on PLA can be a suitable synthetic fibre composite alternative even for high load-bearing applications.

The value of natural fibre reinforced polymer composites has been illustrated by several studies. In an engaging study by Ogin et al. [89] on the components, design, and generic degradation of composite materials, the fundamental components of composite materials and the generic defects arising from the manufacturing process and the external loading of the material were described. Due to resin shrinkage, common process-related defects include porosity, shrinkage cracking, and fibre matrix debonding were observed. The curing kinetics, e.g., temperature cycles, gelation mechanisms, and resin shrinkage, resulted in internal curing and temperature stresses owing to the difference in thermal expansion coefficients (CTE) between the various basic components, which could lead to material defects too. This study also emphasised the essential component of material architecture optimization that was the avoidance of excess material use (weight reduction) and unnecessary strength, given that structural performance is only required in the parts and directions of the structure that bear stress.

Many studies have investigated natural fibres systematically reinforced PLA composites. There is a growing body of literature recognized by Bergeret and Benezet [129] about natural fibre-reinforced bio foams. Previous studies investigated their criteria for selection on starches and PLA as major bio-based and biodegradable polymers for biofoam applications. A number of techniques were developed such as blowing agent to melt extrusion to produce starch foams. The starch foam of natural fibres resulted in a decrease in density by up to 33% and void content of 48% for cellulose to increase mechanical properties of fibre-reinforced PLA foams.

Jauhari et al. [130] studied natural fibre reinforced composite laminates. By using this approach, researchers have been able to investigate natural fibre reinforcement by assembling the long or short bundles of natural fibres. In polymeric composite terminology, it created a flat sheet between one to tens layers of fibres. Different methods were proposed to classify the interlock of the fibres themselves with a binder in mechanical properties. To understand the conduct of FRPs under axial load, a PLA drop-off laminate was also modelled and analysed using ANSYS Software. The results of this investigation showed that the layers were held together to keep these materials together.

A study conducted by Lim et al. [73] on PLA processing technology, which could be considered as an option for mitigating waste management problem and reducing dependency on synthetic plastic for food packaging. This was attributed to the biodegradability and eco-friendly properties of PLA-based composite. It was stated that there are various PLA processing approaches which include methods of injection moulding, extrusion, and thermoforming fibre spinning. Demand for agricultural feedstock to produce PLA will also rise as the use of PLA continues to develop. Overall, the existing data highlighted the importance of the developments of sustainable natural sources in order to address possible alternative raw materials in the food supply chains.

In one study by Nechwatal et al. [131], the characterisation and use of natural fibre properties for composites were studied. The reinforcement of natural fibre had led to increasing composites' tensile strength and Young's modulus. Moreover, a new system for the processing of long fibre reinforced thermoplastic granules using regular plastic equipment was also proposed. In order to test a thread-like structure, the single fibre and fibre bundle tests were created. The fibre bundle test was known for its speed, whereas the precision of the single fibre test was established. The emphasis of this experiment was more on precision and hence, the single fibre test was used. A significant alternative to the current processes, for example, pultrusion or extrusion are the processes that involve long fibre granules. The new long fibre granules have a helical structure that can permit large fibre length for the composite.

Many studies have investigated natural fibres reinforced PLA composites in systematic ways. There is a growing body of literature recognized by Van De Velde and Kiekens [132] on thermoplastic pultrusion of natural fibre reinforced composites. As a reinforcement material for composites, natural flax fibre offered good opportunities in terms of mechanical characteristics and environmental advantage. One method to satisfy continuous demand for composites is pultrusion. As mentioned in the study, the capacity to become a major new force in the pultrusion industry was demonstrated by thermoplastic pultrusion and showed that the growth of thermoplastic pultruded composites reinforced with flax fibre might respond to the need for constantly reinforced profiles that are environmentally friendly.

Table 2. Reported work on natural fibre reinforced PLA composites.

Polymer	Natural Fibre	Effect of Reinforcement	Mechanical Strength			Reference
			Tensile (MPa)	Flexural (MPa)	Impact (kJ/m^2)	
PLA	Bamboo fibre	-Increased mechanical properties -Bamboo fibre was the most effective in increasing tensile strength	54	-	-	[133]
PLA	Wood fibre	-Improved the fibre adhesion with the PLA,	73.8	-	21.5	[122]
PLA	Hard wood high yield pulp	-The incorporation of pulp fibres significantly increased the composite storage moduli and elasticity	90	-	-	[124]
PLA	Soft wood high yield pulp	-The incorporation of pulp fibres significantly increased the composite storage moduli and elasticity	110	-	-	[124]
PLA	Kraft	-Suggesting the fibre–fibre bond also positively contributed to the composites' strengths	121	-	-	[124]
PLA	Wood fibre	-Improved mechanical and thermal properties of composites	23.0	-	-	[123]
PLA	Vetiver fibre	-Increased mechanical properties -Bamboo fibre was the most effective in increasing tensile strength	48	-	-	[133]

Table 2. *Cont.*

Polymer	Natural Fibre	Effect of Reinforcement	Mechanical Strength			Reference
			Tensile (MPa)	Flexural (MPa)	Impact (kJ/m^2)	
PLA	Coconut fibre	-Increased mechanical properties -Bamboo fibre was the most effective in increasing tensile strength	50	-	-	[133]
PLA	Kenaf fibre	-Comparable mechanical and physical properties to glass-fibre composite -Safe for environment	-	-	-	[107]
PLA	Jute fibre	-Elephant grass fibre had higher tensile strength and flexural strength than treated Jute/PLA composite and plain PLA -Untreated composite with elephant grass, jute, and sisal fibre had higher impact strength than plain PLA	65	112	5.3	[114]
PLA	Sisal fibre	-Elephant grass fibre had higher tensile strength and flexural strength than treated Jute/PLA composite and plain PLA -Untreated composite with elephant grass, jute, and sisal fibre had higher impact strength than plain PLA	62	110	5.3	[114]
PLA	Elephant grass	-Elephant grass fibre had higher tensile strength and flexural strength than treated Jute/PLA composite and plain PLA -Untreated composite with elephant grass, jute, and sisal fibre had higher impact strength than plain PLA	55	86	3.0	[114]
PLA	Flax fibre	-Showed high Limiting Oxygen Index (LOI) of 26.1% -Achieved V-2 rating	59.1	-	-	[109]
PLA	Wheat straw	-Excellent tensile mechanical properties	62	98	27	[116]
PLA	Rice straw fibre	-Increased the shear modulus and impact strength	50	-	-	[117]
PLA	Corn straw fibre	-Shorter curing time and enhanced mechanical properties.	24.38	-	-	[119]
PLA	Abutilon fibre	-Increased the shear modulus and impact strength	53	-	1.7	[121]

Table 2. *Cont.*

Polymer	Natural Fibre	Effect of Reinforcement	Mechanical Strength			Reference
			Tensile (MPa)	Flexural (MPa)	Impact (kJ/m^2)	
PLA	Woven jute fibre	-Increased mechanical properties	81	82	16.4	[128]
PLA	Hemp fibre	-Improved mechanical properties	54.6	112.7	-	[134]
PLA	Jute fibre	-Water absorption during ageing caused fibre/matrix bonding failure	60	-	-	[104]
PLA	Jute fibre	-Increased the shear modulus and impact strength	52	100	9	[105]
PLA	Ramie fibre	-Shorter curing time and enhanced mechanical properties.	48	105	10	[105]
PLA	Flax fibre	-Optimised mechanical properties of composites for automotive applications	53	-	12	[18]
PLA	Ramie fibre	-The composites had better mechanical properties than pure PLA	72	-	11	[135]
PLA	Hemp fibre	-Improved tensile, impact, and flexural properties	45	60	9.7	[136]
PLA	Hemp-lyocell fibre	-Improved tensile, impact, and flexural properties	60	102	21.7	[136]
PLA	Lyocell fibre	-Improved tensile, impact, and flexural properties	80	121	26	[136]
PLA	Flax fibre	-Flame retardant flax/PLA samples achieved V0 rating -All flame retardant reduced the mechanical properties	80	40	-	[110]
PLA	Basalt fibre	-PLA/KBF exhibited comparable tensile properties and superior flexural properties	79.5	128	5.6	[126]
PLA	Grewia optiva fibre	-Tensile strength of the composite increased by 75% of the neat polymer -Microwave joining was better than adhesive bonding	-	-	-	[137]
PLA	Palm fibre	-Increased mechanical strength	35	-	-	[138]
PLA	Kenaf fibre	-Increased mechanical strength	27	-	-	[138]
PLA	Manicaria Saccifera palm fibre	-Tensile strength, elastic modulus, and impact resistance were improved by 26%, 51% and 56%, respectively	68.45	133.12	26.62	[139]
PLA	Cordenka rayon fibres	–Mechanical properties improved	58	-	72	[140]
PLA	Kenaf fibre	-Very high tensile strength and Young's modulus values	52.9	-	9	[103]

Table 2. *Cont.*

Polymer	Natural Fibre	Effect of Reinforcement	Mechanical Strength			Reference
			Tensile (MPa)	Flexural (MPa)	Impact (kJ/m^2)	
PLA	Jute fibres	-Increased tensile stiffness and strength significantly -Different improvements of the mechanical parameters	81.9	-	3	[141]
PLA	Keratin-based fibre	-Green and safe environment -Joints and bone fixtures to alleviate pain for patients	-	-	-	[142]
PLA	Flax fibre	-Mechanical properties improved -Tensile properties were in the same range	253.7	-	-	[143]
PLA	Flax fibre	-Tensile properties improved -Tensile strength and modulus of flax fibre/PLLA composite close to glass fibres	100	-	-	[144]
PLA	Ramie, flax and cotton fibres	-Difference intrinsic viscosity and melt flow index	-	-	-	[145]
PLA	Kenaf fibre	-Excellent tensile mechanical properties	52.88	-	8.97	[10]
PLA	Kenaf fibres	-Perfectly improved features of the polymers -Suitable for construction materials application	-	-	-	[146]
PLA	Hemp fibres	-Increased PLA transcrystallinity -Improved chemical bonding	-	-	-	[147]
PLA	Plant fibres	-Susceptibility to chemical degradation	-	-	-	[148]
PLA	Woven flax and jute fabrics	-Flax fibre resulted in composites with better mechanical strength than the woven jute fibre composites	-	-	-	[149]
PLA	Abaca fibre	-No weight loss observed for neat PLA and PLA/AA-abaca composite, meanwhile the PLA/untreated abaca composite showed ca. 10% weight loss	-	-	-	[150]
PLA	Banana/sisal fibre	-Mechanical properties were enhanced	80	125	48	[101]
PLA	Lyocell fibre	-Very high tensile strength and Young's modulus values -Good impact characteristics	81.8	-	39.7	[103]
PLA	Abaca fibres	-Increased tensile stiffness and strength significantly -Different improvements of the mechanical parameters	74	-	5	[141]

Table 2. *Cont.*

Polymer	Natural Fibre	Effect of Reinforcement	Mechanical Strength			Reference
			Tensile (MPa)	Flexural (MPa)	Impact (kJ/m^2)	
PLA	Coir, sisal and jutes fibres	-Can be used to replace petroleum-based polymer -High specific strength	16.17	29.26	46.17	[151]
PLA	Coconut fibre	-Improved mechanical and thermal properties of composites	23.5	-	-	[123]
PLA	Bamboo fibre	-Improved mechanical and thermal properties of composites	22.5	-	-	[123]
PLA	Bamboo fibre	-Tensile strength was comparable with ordinary glass fibre-reinforced plastics laminate -Three times higher specific strength than mild steel	210	-	-	[112]
PLA	Coconut fibre	-Excellent tensile mechanical properties	65	-	-	[111]
PLA	Bamboo fibre	-Thermal conductivity of the natural composite was lower than that of synthetic composites	-	-	-	[152]
PLA	Cotton fibre	-Very high tensile strength and Young's modulus values -Good impact characteristics	41.2	-	28.7	[103]
PLA	Cotton fibre	-Increased stiffness, tensile strength, elongation at break, and impact strength	41.20	-	28.71	[10]
PLA	Bamboo fibre, vetiver grass fibre and coconut fibre	-Impact of strength decreased	-	-	1.8	[153]

6. 3D and 4D Printings of PLA Biocomposite

Polylactic acid, a thermoplastic aliphatic polyester is the main nature-based raw material for 3D printing. It is a wholly biodegradable thermoplastic polymer synthesised from renewable raw materials [8,154]. Among the entire 3D printing materials, PLA is one of the most common feedstocks used for additive manufacturing. Ease of processing is a key advantage of PLA, being one of the easiest materials to print, despite tend to slightly shrink after the 3D printing process [155]. The main advantage of PLA compared to acrylonitrile butadiene styrene (ABS) material is it does not need a heated platform during printing, hence PLA can be printed at low temperatures between 190–230 °C [156]. PLA also does not require complex post-processing since it can be treated with acetone or sanded when necessary and the supports are usually very easily removable. Many manufacturers fabricated PLA filament to be used with 3D and 4D printing such as Prusament, Amolen, Polymaker PolyMax, Proto Pasta, MatterHackers, Fillamentum, Colorfabb, Sunlu, and Paramount 3D. One of the most prominent PLA filament manufacturers is an Austrian company WeforYou. In other side, the German company named Evonik focuses on PLA development for medical sector. Whereas the American company NaturaWorks is another large producer of biopolymers, and the company Corbion, based in the Netherlands, centralises the development of high-performance resins with PLA.

Since the material is suitable for interaction with foods, this material is used as a replacement for petroleum-based plastics for packaging application, especially in the food industry [157]. PLA can be utilized in 3D printing using the FDM (fused deposition modelling) technology, which manufactures parts through the extrusion of thermoplastic filaments, and PLA is one of the commonly used materials for this technology.

Composites are exceptionally advantageous in manufacturing lightweight with strong mechanical properties parts. The fibres function to contribute mechanical strength to parts without compromising weight, which is the primary factor of recognising composites as fibre reinforced materials [21,158]. There are two classes of reinforcements, short fibre (Figure 6) or continuous fibre (Figure 7). In short fibre case, chopped fibres consisting segments of less than a millimetre in length were mixed with thermoplastics such as PLA, ABS, or nylon into 3D printing plastics to raise the stiffness and to a lesser extent the strength of components. On the other hand, continuous addition of the fibres to the thermoplastics could be performed to produce stronger parts. Commonly, the most utilised fibre in 3D printing is carbon fibre along with glass, Kevlar or natural fibres. Figure 8 shows poplar as biofibre reinforcement in composites for large-scale 3D printing. Besides that, there are various hybrid materials that combine plastics with powders to give new colour, finish, or extra material properties. Usually for PLA, these materials are typically fabricated from 70% PLA biopolymer and another 30% hybrid natural material. For instance, wood-based filaments such as poplar, and non-wood filaments such as bamboo, jute, cork, wood dust kenaf, and so forth are combined with PLA to provide a more organic texture to the hybrid filament which in line with Sustainable Development Goals (SDGs) that are to (I) encourage inclusive and sustainable industrialisation and foster innovation, and (II) safeguard sustainable consumption and production patterns.

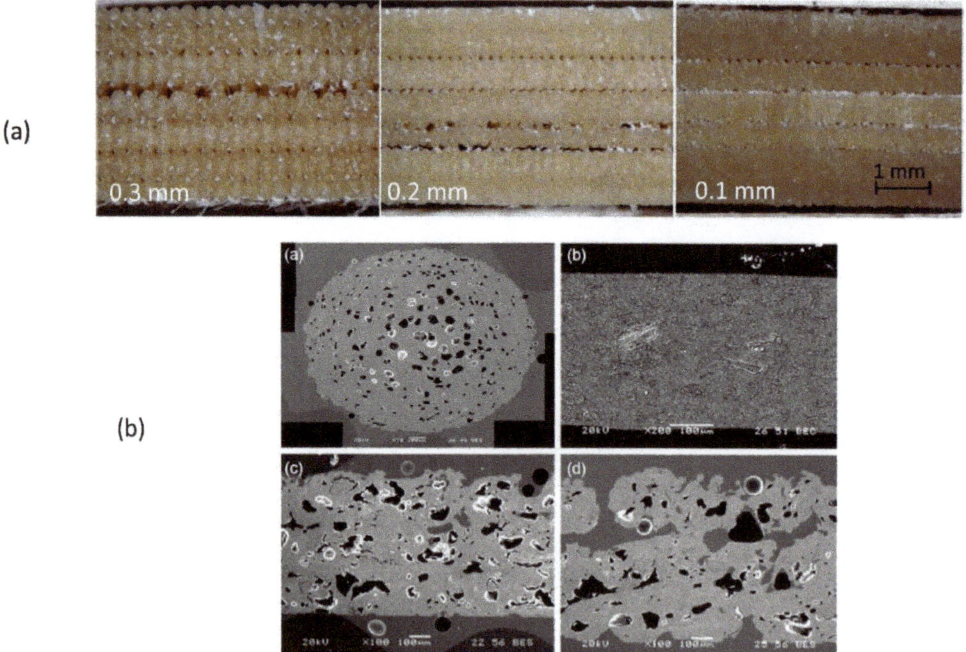

Figure 6. Short natural fibre reinforced PLA. (**a**) Effect of layer height on wood, 0.3, 0.2, and 0.1 mm [126]. Higher porosity content was observed with higher layer height. (**b**) Scanning electron micrograph of printed wood/PLA biocomposites with raw filament cross-section. Adapted from Le Duigou et al. [159] and Ayrilmis et al. [160] with copyright permission of Elsevier.

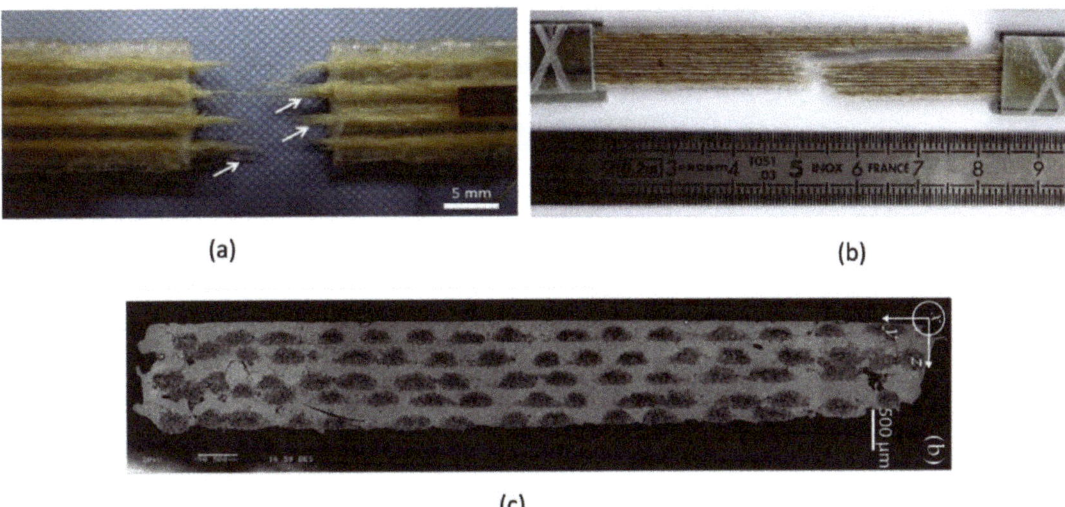

(a)

(b)

(c)

Figure 7. Continuous natural fibre (**a**) Fractured continuous unidirectional jute fibres/PLA sample printed with in-nozzle reinforced with various fibre debonding. (**b**) Fractured continuous unidirectional flax fibre/PLA sample printed with pre-impregnated filaments with a transverse crack followed by propagation along the tensile axis. (**c**) SEM micrograph of a cross-section of Flax/PLA composite microstructure. Adapted with copyright permission from Le Duigou et al. [143] and Matsuzaki et al. [161].

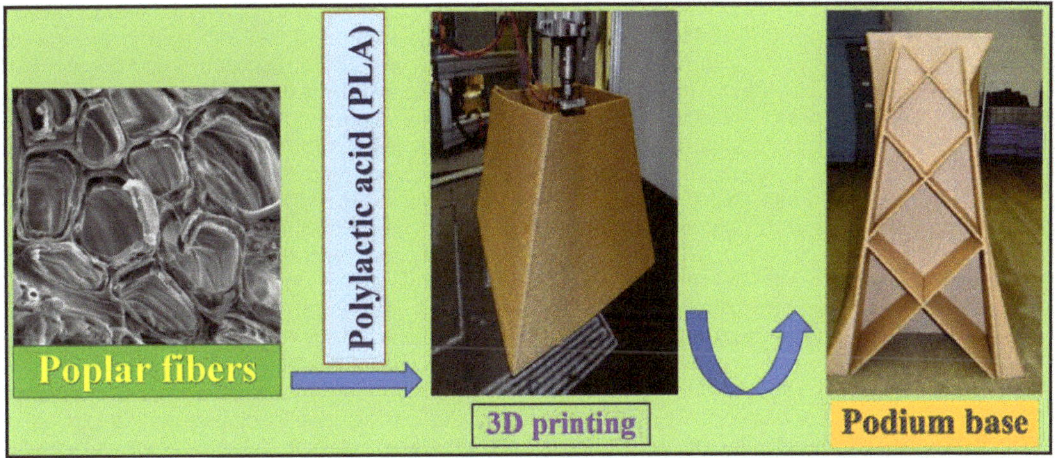

Figure 8. Large scale printing process with poplar/PLA composites. Adapted from Zhao X, Tekinalp H, Meng X, Ker D, Benson B, Pu Y, et al. [162]. ACS Appl Bio Mater 2019. Copyright (2019) American Chemical Society.

Four dimensional printing is a relatively recent trend to develop 3D printed structures that can change their shape or properties over time [163–165]. The difference is that 4D printed objects can transform themselves over time, while 3D printed objects maintain fixed shape like any plastic or metal parts. The fourth dimension of 4D is the transformation over time, where 4D printing technology offers an output of smart structures by using new manufacturing techniques of 3D printing, advanced materials, and customized design. 4D printed objects need a stimulus to begin the deformation phase; the trigger can be an exposure to water, heat, light, or magnetic field [166–169]. Four dimensional printing technology is a combination of 3D printing, smart materials, and customized design for object transformation. The self-transformation of the structure is also called self-assembly as the structure can be designed to assemble itself. The concept of 4D printing is a smart structure that consists of rigid materials connected with expandable elements, or it can also be a whole structure made from expandable materials depending on what materials properties are needed and what are the applications [170,171]. The expandable elements can change their shape when exposed to certain stimuli, and this causes the hard parts to move or rotate, resulting in the whole structure transformation. The expandable element of smart materials can be hydrogel and elements with shape memory. A hydrogel is a polymeric material that is capable of absorbing a large amount of water [172,173]. It can be programmed to expand or shrink when there are changes in the external environmental conditions [163,174]. Hydrogels are biocompatible and easy to be modified. Elements with shape memory can be considered as smart materials due to their capability to return to their original shape from a deformed shape when stimuli are applied. Figure 9 shows the moisture-induced deployable structure based on curved-line folding inspired by Aldrovanda [175]. Hygromorph biocomposite (HBC) actuators made use of the transport properties of plant fibres to generate an out-of-plane displacement when a moisture gradient was present. Le Duigou et al. [175] developed a theoretical actuating response (curvature) formulation of maleic anhydride polypropylene (MAPP)/plant fibres (i.e., flax, jute, kenaf, and coir) based on bimetallic actuators theory (Figure 9). The result showed that the actuation was directly related to the fibres biochemical composition and microstructure, in which flax and jute fibres were observed and found to be the best candidates to be used in HBCs. Thus, PLA biopolymer might have huge potential to be used for hygromorph biocomposite (HBC) actuators. Simon Poppinga et al. [176] researched plant movements as concept generators for the development of biomimetic compliant mechanisms. Lilium 'Casa Blanca' or lily flower model was used in this experiment. The structure of the hygroscopic made up of wood-based hygromorph composite was translated into a 4D printed mechanism or printed under warm and dry conditions at a temperature of 21 °C and relative humidity of 18%. The composite then underwent deformation when submerged in water at the temperature of 19 °C. The submerged composite resulted to edge growth-driven actuation as identified from the petals of the lily flower (Figure 10). Their novel biomimetic compliant mechanisms highlighted the feasibility of modern printing techniques for designing and developing versatile tailored motion responses for technical applications. Alief et al. [177] conducted experiment on modelling the shape memory properties of 4D printed polylactic acid (PLA) for application of disk spacer in minimally invasive spinal fusion. Figure 11 shows the deformation of PLA model. They observed that specific PLA structure possessed thermal shape memory behaviour that can be thermo-mechanically trained into temporary shape and returned to its permanent shape when heated. Besides that, based on the simulation result, non-uniformed hollow spaces pattern displayed favourable result that could be a reference for future research in order to design the suitable pattern for the 4D PLA model. The possible advanced applications of 4D printing are medical devices for stents placed into blood vessels, drug capsules that release medicine, home appliances for control and that adjust according to humidity and heat, footwear and clothes, implants for humans and animals made from biocompatible materials, soft robots that can be activated without reliance on an electric device, smart valves and sensors for infrastructure lines [178–183].

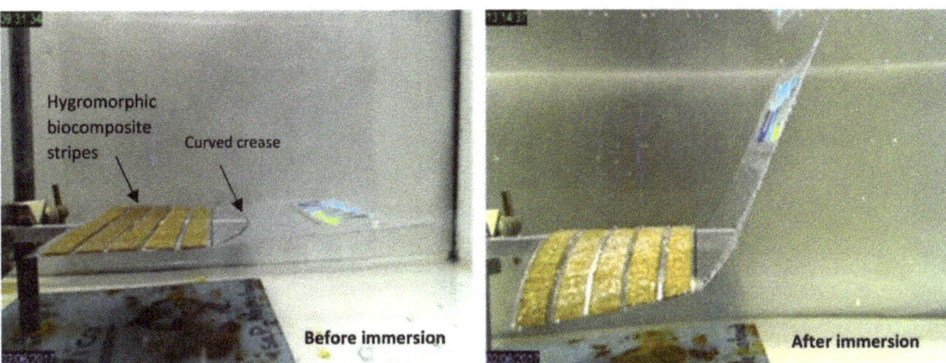

Figure 9. Moisture-induced deployable structure based on curved-line folding inspired by Aldrovanda, adapted from Le Duigou [175] with copyright permission of Elsevier.

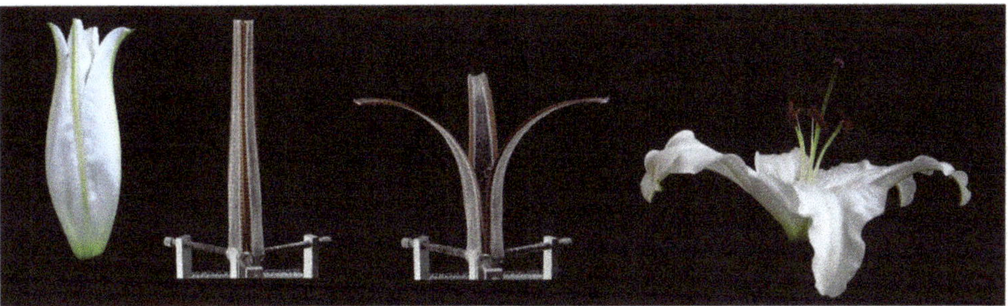

Figure 10. Edge growth causing a strain gradient in the Lilium 'Casa Blanca' drives the flower opening mechanism, a principle that is translated into a 4D printed mechanism with a wood-based hygromorph composite edge. Adapted from Poppinga et al. [176] with copyright permission of Oxford University Press.

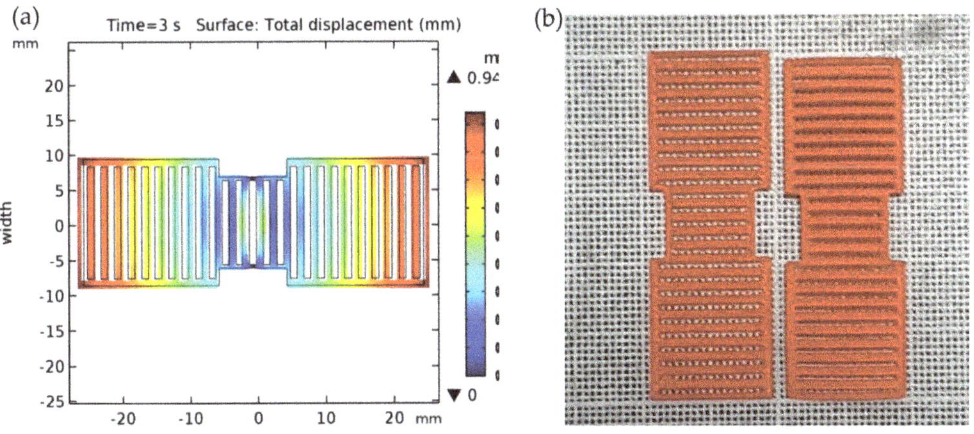

Figure 11. (**a**) PLA model deformation after 3 s. Uniformed vertical hollow spaces in 1 mm width. (**b**) Side by side comparison before and after adding stimulus. Left: a PLA model before adding stimulus and right: after adding the stimulus. Adapted from Alief et al. [177] with copyright permission of AIP Conference Proceedings.

7. Potential Applications of Natural Fibre Reinforced PLA Composite

Natural fibre reinforced PLA composite has a lot of potential usages. Possessing high mechanical strength similar to conventional glass fibre, it is seen as a potential substitution of glass fibre. Glass fibres are hard to biodegrade and is detrimental to the environment. Natural fibre reinforced PLA composite has high biodegradability and recyclability which is environmentally friendly. Among possible alternatives to the traditional fibres and synthetic polymers are PLA and natural fibres due to their renewability and ease of recycling [94]. As they have almost similar mechanical and physical properties, kenaf biocomposites have a promising ability to replace petrol-based composites such as glass-fibre composites [107] and used as packaging material. An increasing number of automotive parts and packaging materials using natural fibre composites is observed in India [184]. Thus, natural fibre reinforced PLA composite can even replace the conventional petroleum-based products with well-known environmental impacts.

Furthermore, due to the excellent mechanical strength of natural fibre reinforced PLA composites, they are favoured to be used in the automotive sector. A study reported that natural fibre reinforced PLA had comparable mechanical strength to glass fibre. The strength of the composites was comparable to that of normal laminated glass fibre-reinforced plastics, and the specific strength was 3 times higher than mild steel's [112]. Jute fibre composite was found to have a higher damping behaviour, so jute fibre composites can be practical options as automotive parts since they possessed low vibration and noise [185]. Due to its relatively cheap cost, good properties, environmental friendliness and ease of manufacturing, natural fibre reinforced PLA composite can be applied in a wide range of applications, including the aerospace and automotive industries [100]. In the automotive industry, coconut fibre is used to make furniture in cars, while cotton is used for noise cancellation. Sometimes, wood fibre is also used as furniture and accessory in vehicles while flax, sisal, and hemp are applied in the refining of seatback linings and floor panels [96].

Besides the automotive industry, the construction sector also offers broad potential usages of natural fibre composite. The evaluation of building materials based on renewable resources such as natural fibres and their reinforcement in cement-based materials are being researched [98]. Structural beams and panels using bio-based composite materials were developed, manufactured and tested, particularly on plant oil-based resins along with natural fibre composite and composite building materials made from straw bales in the United States [184].

Natural fibre reinforced PLA composite has huge potential usage in the industry but its flammability might be a main concern. The flame retardant test on natural fibre reinforced PLA exhibited promising results and thus, this problem should not be worrying. The altered flax fibre reinforced PLA had a Limiting Oxygen Index (LOI) value of 26.1 with a UL-94 V-2 rating and the release of flammable gaseous during thermal decomposition of the PLA composite was suppressed after the addition of flame retardant, which enabled the modified composite to display high resistance to minor flammable sources of ignition [109]. The water absorption rate for all composites was observed to increase gently over the first 24 h before levelling off, and the rate of water absorption was increased in all composites as the fibre content was increased. The absorption rate was found to decrease with successive alkali treatment of the fibres [186].

8. Conclusions

The goal of this study is to investigate the potential of polylactic acid (PLA) composite reinforced with natural fibres to enhance the quality of the produced composite. The studies have proven that coconut and kenaf fibres exhibited fewer environmental effects from the processing of natural fibres, whereas elephant grass fibre reinforced PLA composite was stronger than the composites of jute and sisal. Essentially, the findings of the PLA matrix showed significant improvement in the smooth properties of polymer wear. Natural fibre reinforced PLA composites demonstrated their possibility to provide a possible backup to natural fabrics and polymers that are impossible and expensive to reuse. Numerous approaches have been employed to study the composites of natural fibre reinforced PLA, including flexural testing experiment, single-fibre test, fibre-bundle test, chemical treatments and plasma treatment. For example, tensile and flexural testing experiments on composites were created by strengthening a matrix of polyester resin lower than a new natural fibre. In the extrusion and compression moulding processes, PLA/flax composites did not demonstrate any difficulties and can be processed similarly to PP-based composites. In particular, for the mechanical properties of a composite, the quality of the fibre and matrix relationship is crucial. Fibre/matrix adhesion is a dynamic mechanism that deals with several variables. Generally, PLA natural fibre composites have durable mechanical properties. The beginning of 3D and 4D printings represents a great chance for PLA biocomposites to progress for the first time on the same time scale as their synthetic counterparts. Natural fibre reinforced PLA biocomposites have huge potential to be a new class of smart materials, namely hygromorph biocomposites, in which they are utilisable as raw materials in 4D printing to develop specialized shape-changing mechanisms and structures.

Author Contributions: Conceptualization, R.A.I. and S.M.S.; formal analysis, M.Y.A.Y.H. and M.Z.M.H.; investigation, M.Y.A.Y.H. and M.Z.M.H.; resources, S.M.S.; writing—original draft preparation, M.Y.A.Y.H., M.Z.M.H. and M.M.H.; writing—review and editing, M.R.M.A, M.R.I., M.R.R., R.A.I., S.M.S., N.M.N., M.N.F.N., H.A., M.A. and M.S.N.A.; supervision, R.A.I. and S.M.S.; project administration, R.A.I. and S.M.S.; funding acquisition, M.R.R. All authors have read and agreed to the published version of the manuscript.

Funding: This research was supported by Universiti Kebangsaan Malaysia, PP/LESTARI/2021 and XX-2018-008.

Institutional Review Board Statement: Not applicable.

Informed Consent Statement: Not applicable.

Data Availability Statement: No new data were created or analyzed in this study. Data sharing is not applicable to this article.

Acknowledgments: The authors would like to express gratitude for the financial support received from PP/LESTARI/2021 and XX-2018-008 by Universiti Kebangsaan Malaysia, Malaysia.

Conflicts of Interest: The authors declare no conflict of interest.

References

1. Plackett, D.; Andersen, T.L.; Pedersen, W.B.; Nielsen, L. Biodegradable composites based on L-polylactide and jute fibres. *Compos. Sci. Technol.* **2003**, *63*, 1287–1296. [CrossRef]
2. Harussani, M.M.; Sapuan, S.M.; Khalina, A.; Ilyas, R.A.; Hazrol, M.D. Review on Green Technology Pyrolysis for Plastic Wastes. In *Proceedings of the 7th Postgraduate Seminar on Natural Fibre Reinforced Polymer Composites 2020*; Sapuan, S.M., Ilyas, R.A., Eds.; Institute of Tropical Forestry and Forest Products (INTROP), Universiti Putra Malaysia: Serdang, Selangor, 2020; pp. 50–53.
3. Syafiq, R.; Sapuan, S.M.; Zuhri, M.Y.M.; Ilyas, R.A.; Nazrin, A.; Sherwani, S.F.K.; Khalina, A. Antimicrobial activities of starch-based biopolymers and biocomposites incorporated with plant essential oils: A review. *Polymers* **2020**, *12*, 2403. [CrossRef] [PubMed]
4. Atikah, M.S.N.; Ilyas, R.A.; Sapuan, S.M.; Ishak, M.R.; Zainudin, E.S.; Ibrahim, R.; Atiqah, A.; Ansari, M.N.M.; Jumaidin, R. Degradation and physical properties of sugar palm starch/sugar palm nanofibrillated cellulose bionanocomposite. *Polimery* **2019**, *64*, 27–36. [CrossRef]
5. Drumright, R.E.; Gruber, P.R.; Henton, D.E. Polylactic acid technology. *Adv. Mater.* **2000**, *12*, 1841–1846. [CrossRef]
6. Nurazzi, N.M.; Harussani, M.M.; Zulaikha, N.D.S.; Norhana, A.H.; Syakir, M.I.; Norli, A. Composites based on conductive polymer with carbon nanotubes in DMMP gas sensors—An overview. *Polimery* **2021**, *66*, 85–97. [CrossRef]
7. Ali, S.S.S.; Razman, M.R.; Awang, A. The nexus of population, GDP growth, electricity generation, electricity consumption and carbon emissions output in Malaysia. *Int. J. Energy Econ. Policy* **2020**, *10*, 84–89. [CrossRef]
8. Nazrin, A.; Sapuan, S.M.; Zuhri, M.Y.M.; Ilyas, R.A.; Syafiq, R.; Sherwani, S.F.K. Nanocellulose Reinforced Thermoplastic Starch (TPS), Polylactic Acid (PLA), and Polybutylene Succinate (PBS) for Food Packaging Applications. *Front. Chem.* **2020**, *8*, 1–12. [CrossRef]

9. Oksman, K.; Skrifvars, M.; Selin, J.-F. Natural fibres as reinforcement in polylactic acid (PLA) composites. *Compos. Sci. Technol.* **2003**, *63*, 1317–1324. [CrossRef]

10. Graupner, N. Application of lignin as natural adhesion promoter in cotton fibre-reinforced poly(lactic acid) (PLA) composites. *J. Mater. Sci.* **2008**, *43*, 5222–5229. [CrossRef]

11. Ilyas, R.A.; Sapuan, S.M.; Atikah, M.S.N.; Asyraf, M.R.M.; Rafiqah, S.A.; Aisyah, H.A.; Nurazzi, N.M.; Norrrahim, M.N.F. Effect of hydrolysis time on the morphological, physical, chemical, and thermal behavior of sugar palm nanocrystalline cellulose (*Arenga pinnata* (Wurmb.) Merr). *Text. Res. J.* **2020**, 004051752093239. [CrossRef]

12. Sabaruddin, F.A.; Paridah, M.T.; Sapuan, S.M.; Ilyas, R.A.; Lee, S.H.; Abdan, K.; Mazlan, N.; Roseley, A.S.M.; Abdul Khalil, H.P.S. The effects of unbleached and bleached nanocellulose on the thermal and flammability of polypropylene-reinforced kenaf core hybrid polymer bionanocomposites. *Polymers* **2020**, *13*, 116. [CrossRef]

13. Ayu, R.S.; Khalina, A.; Harmaen, A.S.; Zaman, K.; Isma, T.; Liu, Q.; Ilyas, R.A.; Lee, C.H. Characterization Study of Empty Fruit Bunch (EFB) Fibers Reinforcement in Poly(Butylene) Succinate (PBS)/Starch/Glycerol Composite Sheet. *Polymers* **2020**, *12*, 1571. [CrossRef]

14. Wambua, P.; Ivens, J.; Verpoest, I. Natural fibres: Can they replace glass in fibre reinforced plastics? *Compos. Sci. Technol.* **2003**, *63*, 1259–1264. [CrossRef]

15. Asyraf, M.R.M.; Ishak, M.R.; Sapuan, S.M.; Yidris, N. Comparison of Static and Long-term Creep Behaviors between Balau Wood and Glass Fiber Reinforced Polymer Composite for Cross-arm Application. *Fibers Polym.* **2021**, *22*. [CrossRef]

16. Asyraf, M.R.M.; Ishak, M.R.; Sapuan, S.M.; Yidris, N. Influence of Additional Bracing Arms as Reinforcement Members in Wooden Timber Cross-Arms on Their Long-Term Creep Responses and Properties. *Appl. Sci.* **2021**, *11*, 2061. [CrossRef]

17. Asyraf, M.R.M.; Ishak, M.R.; Sapuan, S.M.; Yidris, N.; Ilyas, R.A. Woods and composites cantilever beam: A comprehensive review of experimental and numerical creep methodologies. *J. Mater. Res. Technol.* **2020**. [CrossRef]

18. Suriani, M.J.; Radzi, F.S.M.; Ilyas, R.A.; Petrů, M.; Sapuan, S.M.; Ruzaidi, C.M. Flammability, Tensile, and Morphological Properties of Oil Palm Empty Fruit Bunches Fiber/Pet Yarn-Reinforced Epoxy Fire Retardant Hybrid Polymer Composites. *Polymers* **2021**, *13*, 1282. [CrossRef]

19. Asyraf, M.R.M.; Rafidah, M.; Azrina, A.; Razman, M.R. Dynamic mechanical behaviour of kenaf cellulosic fibre biocomposites: A comprehensive review on chemical treatments. *Cellulose* **2021**. [CrossRef]

20. Jumaidin, R.; Ilyas, R.A.; Saiful, M.; Hussin, F.; Mastura, M.T. Water Transport and Physical Properties of Sugarcane Bagasse Fibre Reinforced Thermoplastic Potato Starch Biocomposite. *J. Adv. Res. Fluid Mech. Therm. Sci.* **2019**, *61*, 273–281.

21. Ilyas, R.A.; Sapuan, S.M. The Preparation Methods and Processing of Natural Fibre Bio-polymer Composites. *Curr. Org. Synth.* **2020**, *16*, 1068–1070. [CrossRef]

22. Nurazzi, N.M.; Asyraf, M.R.M.; Khalina, A.; Abdullah, N.; Aisyah, H.A.; Rafiqah, S.A.; Sabaruddin, F.A.; Kamarudin, S.H.; Norrrahim, M.N.F.; Ilyas, R.A.; et al. A Review on Natural Fiber Reinforced Polymer Composite for Bullet Proof and Ballistic Applications. *Polymers* **2021**, *13*, 646. [CrossRef]

23. Asyraf, M.R.M.; Rafidah, M.; Ishak, M.R.; Sapuan, S.M.; Ilyas, R.A.; Razman, M.R. Integration of TRIZ, Morphological Chart and ANP method for development of FRP composite portable fire extinguisher. *Polym. Compos.* **2020**, 1–6. [CrossRef]

24. Nurazzi, N.M.; Khalina, A.; Sapuan, S.M.; Ilyas, R.A.; Rafiqah, S.A.; Hanafee, Z.M. Thermal properties of treated sugar palm yarn/glass fiber reinforced unsaturated polyester hybrid composites. *J. Mater. Res. Technol.* **2020**, *9*, 1606–1618. [CrossRef]

25. Aisyah, H.A.; Paridah, M.T.; Sapuan, S.M.; Khalina, A.; Berkalp, O.B.; Lee, S.H.; Lee, C.H.; Nurazzi, N.M.; Ramli, N.; Wahab, M.S.; et al. Thermal Properties of Woven Kenaf/Carbon Fibre-Reinforced Epoxy Hybrid Composite Panels. *Int. J. Polym. Sci.* **2019**, *2019*, 5258621. [CrossRef]

26. Norizan, M.N.; Abdan, K.; Ilyas, R.A.; Biofibers, S.P. Effect of fiber orientation and fiber loading on the mechanical and thermal properties of sugar palm yarn fiber reinforced unsaturated polyester resin composites. *Polimery* **2020**, *65*, 34–43. [CrossRef]

27. Sapuan, S.M.; Aulia, H.S.; Ilyas, R.A.; Atiqah, A.; Dele-Afolabi, T.T.; Nurazzi, M.N.; Supian, A.B.M.; Atikah, M.S.N. Mechanical properties of longitudinal basalt/woven-glass-fiber-reinforced unsaturated polyester-resin hybrid composites. *Polymers* **2020**, *12*, 2211. [CrossRef]

28. Asyraf, M.R.M.; Ishak, M.R.; Sapuan, S.M.; Yidris, N.; Ilyas, R.A.; Rafidah, M.; Razman, M.R. Evaluation of Design and Simulation of Creep Test Rig for Full-Scale Crossarm Structure. *Adv. Civ. Eng.* **2020**, *2020*. [CrossRef]

29. Asyraf, M.R.M.; Ishak, M.R.; Sapuan, S.M.; Yidris, N. Conceptual design of multi-operation outdoor flexural creep test rig using hybrid concurrent engineering approach. *J. Mater. Res. Technol.* **2020**, *9*, 2357–2368. [CrossRef]

30. Syafri, E.; Yulianti, E.; Asrofi, M.; Abral, H.; Sapuan, S.; Ilyas, R.; Fudholi, A. Effect of sonication time on the thermal stability, moisture absorption, and biodegradation of water hyacinth (*Eichhornia crassipes*) nanocellulose-filled bengkuang (*Pachyrhizus erosus*) starch biocomposites. *J. Mater. Res. Technol.* **2019**, *8*, 6223–6231. [CrossRef]

31. Abral, H.; Atmajaya, A.; Mahardika, M.; Hafizulhaq, F.; Kadriadi; Handayani, D.; Sapuan, S.M.; Ilyas, R.A. Effect of ultrasonication duration of polyvinyl alcohol (PVA) gel on characterizations of PVA film. *J. Mater. Res. Technol.* **2020**, *9*, 2477–2486. [CrossRef]

32. Jumaidin, R.; Saidi, Z.A.S.; Ilyas, R.A.; Ahmad, M.N.; Wahid, M.K.; Yaakob, M.Y.; Maidin, N.A.; Rahman, M.H.A.; Osman, M.H. Characteristics of Cogon Grass Fibre Reinforced Thermoplastic Cassava Starch Biocomposite: Water Absorption and Physical Properties. *J. Adv. Res. Fluid Mech. Therm. Sci.* **2019**, *62*, 43–52.

33. Rozilah, A.; Jaafar, C.N.A.; Sapuan, S.M.; Zainol, I.; Ilyas, R.A. The Effects of Silver Nanoparticles Compositions on the Mechanical, Physiochemical, Antibacterial, and Morphology Properties of Sugar Palm Starch Biocomposites for Antibacterial Coating. *Polymers* **2020**, *12*, 2605. [CrossRef] [PubMed]
34. Mazani, N.; Sapuan, S.M.; Sanyang, M.L.; Atiqah, A.; Ilyas, R.A. Design and Fabrication of a Shoe Shelf From Kenaf Fiber Reinforced Unsaturated Polyester Composites. In *Lignocellulose for Future Bioeconomy*; Ariffin, H., Sapuan, S.M., Hassan, M.A., Eds.; Elsevier: Amsterdam, The Netherlands, 2019; pp. 315–332. ISBN 9780128163542.
35. Asyraf, M.R.M.; Ishak, M.R.; Sapuan, S.M.; Yidris, N.; Ilyas, R.A.; Rafidah, M.; Razman, M.R. Potential Application of Green Composites for Cross Arm Component in Transmission Tower: A Brief Review. *Int. J. Polym. Sci.* **2020**, *2020*, 8878300. [CrossRef]
36. Asyraf, M.R.M.; Ishak, M.R.; Sapuan, S.M.; Yidris, N. Conceptual design of creep testing rig for full-scale cross arm using TRIZ-Morphological chart-analytic network process technique. *J. Mater. Res. Technol.* **2019**, *8*, 5647–5658. [CrossRef]
37. Alam, M.M.; Maniruzzaman, M.; Morshed, M.M. Application and Advances in Microprocessing of Natural Fiber (Jute)–Based Composites. In *Comprehensive Materials Processing*; Hashmi, S., Batalha, G.F., Van Tyne, C.J., Eds.; Elsevier: London, UK, 2014; pp. 243–260.
38. Chemiefaser, I. Worldwide Production Volume of Chemical and Textile Fibers from 1975 to 2018. Available online: https://www.statista.com/statistics/263154/worldwide-production-volume-of-textile-fibers-since-1975/ (accessed on 2 December 2019).
39. Migneault, S.; Koubaa, A.; Erchiqui, F.; Chaala, A.; Englund, K.; Krause, C.; Wolcott, M. Effect of fiber length on processing and properties of extruded wood-fiber/HDPE composites. *J. Appl. Polym. Sci.* **2008**, *110*, 1085–1092. [CrossRef]
40. Mukherjee, T.; Kao, N. PLA Based Biopolymer Reinforced with Natural Fibre: A Review. *J. Polym. Environ.* **2011**, *19*, 714–725. [CrossRef]
41. Cosgrove, D.J. Growth of the plant cell wall. *Nat. Rev. Mol. Cell Biol.* **2005**, *6*, 850–861. [CrossRef]
42. Martins, M.A.; Kiyohara, P.K.; Joekes, I. Scanning electron microscopy study of raw and chemically modified sisal fibers. *J. Appl. Polym. Sci.* **2004**, *94*, 2333–2340. [CrossRef]
43. Ilyas, R.A.; Sapuan, S.M.; Sanyang, M.L.; Ishak, M.R.; Zainudin, E.S. Nanocrystalline cellulose as reinforcement for polymeric matrix nanocomposites and its potential applications: A Review. *Curr. Anal. Chem.* **2018**, *14*, 203–225. [CrossRef]
44. Ferreira, F.V.; Mariano, M.; Rabelo, S.C.; Gouveia, R.F.; Lona, L.M.F. Isolation and surface modification of cellulose nanocrystals from sugarcane bagasse waste: From a micro- to a nano-scale view. *Appl. Surf. Sci.* **2018**, *436*, 1113–1122. [CrossRef]
45. Abral, H.; Ariksa, J.; Mahardika, M.; Handayani, D.; Aminah, I.; Sandrawati, N.; Pratama, A.B.; Fajri, N.; Sapuan, S.M.; Ilyas, R.A. Transparent and antimicrobial cellulose film from ginger nanofiber. *Food Hydrocoll.* **2020**, *98*, 105266. [CrossRef]
46. Ilyas, R.A.; Sapuan, S.M.; Ishak, M.R. Isolation and characterization of nanocrystalline cellulose from sugar palm fibres (Arenga Pinnata). *Carbohydr. Polym.* **2018**, *181*, 1038–1051. [CrossRef] [PubMed]
47. Ilyas, R.A.; Sapuan, S.M.; Ishak, M.R.; Zainudin, E.S. Sugar palm nanofibrillated cellulose (*Arenga pinnata* (Wurmb.) Merr): Effect of cycles on their yield, physic-chemical, morphological and thermal behavior. *Int. J. Biol. Macromol.* **2019**, *123*, 379–388. [CrossRef] [PubMed]
48. Ilyas, R.A.; Sapuan, S.M.; Ishak, M.R.; Zainudin, E.S. Effect of delignification on the physical, thermal, chemical, and structural properties of sugar palm fibre. *BioResources* **2017**, *12*, 8734–8754. [CrossRef]
49. Ahmad Ilyas, R.; Mohd Sapuan, S.; Ibrahim, R.; Abral, H.; Ishak, M.R.; Zainudin, E.S.; Asrofi, M.; Siti Nur Atikah, M.; Muhammad Huzaifah, M.R.; Radzi, M.A.; et al. Sugar palm (*Arenga pinnata* (Wurmb.) Merr) cellulosic fibre hierarchy: A comprehensiveapproach from macro to nano scale. *J. Mater. Res. Technol.* **2019**, *8*, 2753–2766. [CrossRef]
50. Alemdar, A.; Sain, M. Isolation and characterization of nanofibers from agricultural residues—Wheat straw and soy hulls. *Bioresour. Technol.* **2008**, *99*, 1664–1671. [CrossRef] [PubMed]
51. Julie Chandra, C.S.; George, N.; Narayanankutty, S.K. Isolation and characterization of cellulose nanofibrils from arecanut husk fibre. *Carbohydr. Polym.* **2016**, *142*, 158–166. [CrossRef]
52. Chirayil, C.J.; Joy, J.; Mathew, L.; Mozetic, M.; Koetz, J.; Thomas, S. Isolation and characterization of cellulose nanofibrils from Helicteres isora plant. *Ind. Crops Prod.* **2014**, *59*, 27–34. [CrossRef]
53. Cherian, B.M.; Leão, A.L.; de Souza, S.F.; Thomas, S.; Pothan, L.A.; Kottaisamy, M. Isolation of nanocellulose from pineapple leaf fibres by steam explosion. *Carbohydr. Polym.* **2010**, *81*, 720–725. [CrossRef]
54. Syafri, E.; Kasim, A.; Abral, H.; Asben, A. Cellulose nanofibers isolation and characterization from ramie using a chemical-ultrasonic treatment. *J. Nat. Fibers* **2018**, 1145–1155. [CrossRef]
55. Megashah, L.N.; Ariffin, H.; Zakaria, M.R.; Hassan, M.A. Properties of Cellulose Extract from Different Types of Oil Palm Biomass. *IOP Conf. Ser. Mater. Sci. Eng.* **2018**, *368*. [CrossRef]
56. Jonoobi, M.; Khazaeian, A.; Tahir, P.M.; Azry, S.S.; Oksman, K. Characteristics of cellulose nanofibers isolated from rubberwood and empty fruit bunches of oil palm using chemo-mechanical process. *Cellulose* **2011**, *18*, 1085–1095. [CrossRef]
57. Corrêa, A.C.; de Morais Teixeira, E.; Pessan, L.A.; Mattoso, L.H.C. Cellulose nanofibers from curaua fibers. *Cellulose* **2010**, *17*, 1183–1192. [CrossRef]
58. Tibolla, H.; Pelissari, F.M.; Menegalli, F.C. Cellulose nanofibers produced from banana peel by chemical and enzymatic treatment. *LWT Food Sci. Technol.* **2014**, *59*, 1311–1318. [CrossRef]
59. de Morais Teixeira, E.; Bondancia, T.J.; Teodoro, K.B.R.; Corrêa, A.C.; Marconcini, J.M.; Mattoso, L.H.C. Sugarcane bagasse whiskers: Extraction and characterizations. *Ind. Crops Prod.* **2011**, *33*, 63–66. [CrossRef]

60. Jonoobi, M.; Harun, J.; Shakeri, A.; Misra, M.; Oksmand, K. Chemical composition, crystallinity, and thermal degradation of bleached and unbleached kenaf bast (*Hibiscus cannabinus*) pulp and nanofibers. *BioResources* **2009**, *4*, 626–639. [CrossRef]
61. Bendahou, A.; Habibi, Y.; Kaddami, H.; Dufresne, A. Physico-chemical characterization of palm from Phoenix Dactylifera-L, preparation of cellulose whiskers and natural rubber-based nanocomposites. *J. Biobased Mater. Bioenergy* **2009**, *3*, 81–90. [CrossRef]
62. Chan, C.H.; Chia, C.H.; Zakaria, S.; Ahmad, I.; Dufresne, A. Production and characterisation of cellulose and nano-crystalline cellulose from kenaf core wood. *BioResources* **2013**, *8*, 785–794. [CrossRef]
63. Abral, H.; Dalimunthe, M.H.; Hartono, J.; Efendi, R.P.; Asrofi, M.; Sugiarti, E.; Sapuan, S.M.; Park, J.W.; Kim, H.J. Characterization of Tapioca Starch Biopolymer Composites Reinforced with Micro Scale Water Hyacinth Fibers. *Starch Staerke* **2018**, *70*, 1–8. [CrossRef]
64. Alemdar, A.; Sain, M. Biocomposites from wheat straw nanofibers: Morphology, thermal and mechanical properties. *Compos. Sci. Technol.* **2008**, *68*, 557–565. [CrossRef]
65. Li, M.; Wang, L.J.; Li, D.; Cheng, Y.L.; Adhikari, B. Preparation and characterization of cellulose nanofibers from de-pectinated sugar beet pulp. *Carbohydr. Polym.* **2014**, *102*, 136–143. [CrossRef] [PubMed]
66. Sheltami, R.M.; Abdullah, I.; Ahmad, I.; Dufresne, A.; Kargarzadeh, H. Extraction of cellulose nanocrystals from mengkuang leaves (*Pandanus tectorius*). *Carbohydr. Polym.* **2012**, *88*, 772–779. [CrossRef]
67. Mohanty, A.K.; Misra, M.; Dreal, L.T. Surface modifications of natural fibres and peformance of the resulting biocomposite. *Compos. Interfaces* **2001**, *8*, 313–343. [CrossRef]
68. Sawpan, M.A.; Pickering, K.L.; Fernyhough, A. Improvement of mechanical performance of industrial hemp fibre reinforced polylactide biocomposites. *Compos. Part A Appl. Sci. Manuf.* **2011**. [CrossRef]
69. Li, X.; Tabil, L.G.; Panigrahi, S. Chemical treatments of natural fiber for use in natural fiber-reinforced composites: A review. *J. Polym. Environ.* **2007**, *15*, 25–33. [CrossRef]
70. Hazrol, M.D.; Sapuan, S.M.; Zuhri, M.Y.M.; Zainudin, E.S.; Wahab, N.I.A.; Ilyas, R.A.; Harussani, M.M.; Jamal, T.; Nazrin, A.; Syafiq, R. Effect of sorbitol and glycerol plasticizer and concentration on physical properties of corn starch (Zea mays) biodegradable films. In *Proceedings of the 7th Postgraduate Seminar on Natural Fibre Reinforced Polymer Composites 2020*; Sapuan, S.M., Ilyas, R.A., Eds.; Institute of Tropical Forestry and Forest Products (INTROP), Universiti Putra Malaysia: Serdang, Selangor, 2020; pp. 1–10.
71. Kandemir, N.; Yemeniciogwlu, A.; Mecitogwlu, Ç.; Elmaci, Z.S.; Arslanogwlu, A.; Göksungur, Y.; Baysal, T. Production of antimicrobial films by incorporation of partially purified lysozyme into biodegradable films of crude exopolysaccharides obtained from Aureobasidium pullulans fermentation. *Food Technol. Biotechnol.* **2005**, *43*, 343–350.
72. Södergård, A.; Stolt, M. Industrial Production of High Molecular Weight Poly(Lactic Acid). In *Poly(Lactic Acid)*; Auras, R., Lim, L., Selke, S.E.M., Tsuji, H., Eds.; John Wiley & Sons, Inc.: Hoboken, NJ, USA, 2010; pp. 27–41. ISBN 9780470293669.
73. Lim, L.T.; Auras, R.; Rubino, M. Processing technologies for poly(lactic acid). *Prog. Polym. Sci.* **2008**, *33*, 820–852. [CrossRef]
74. Rasal, R.M.; Janorkar, A.V.; Hirt, D.E. Poly(lactic acid) modifications. *Prog. Polym. Sci.* **2010**, *35*, 338–356. [CrossRef]
75. Arrieta, M.P.; Samper, M.D.; Aldas, M.; López, J. On the use of PLA-PHB blends for sustainable food packaging applications. *Materials* **2017**, *10*, 1008. [CrossRef]
76. Devaux, E.; Aubry, C.; Campagne, C.; Rochery, M. PLA/carbon nanotubes multifilament yarns for relative humidity textile sensor. *J. Eng. Fiber. Fabr.* **2011**, *6*, 155892501100600300. [CrossRef]
77. Tanase, C.E.; Spiridon, I. PLA/chitosan/keratin composites for biomedical applications. *Mater. Sci. Eng. C* **2014**, *40*, 242–247. [CrossRef]
78. Pozo Morales, A.; Güemes, A.; Fernandez-Lopez, A.; Carcelen Valero, V.; De La Rosa Llano, S. Bamboo–polylactic acid (PLA) composite material for structural applications. *Materials* **2017**, *10*, 1286. [CrossRef]
79. Pickering, K.L.; Efendy, M.G.A. Preparation and mechanical properties of novel bio-composite made of dynamically sheet formed discontinuous harakeke and hemp fibre mat reinforced PLA composites for structural applications. *Ind. Crops Prod.* **2016**, *84*, 139–150. [CrossRef]
80. Kumar, N.; Das, D. Fibrous biocomposites from nettle (*Girardinia diversifolia*) and poly (lactic acid) fibers for automotive dashboard panel application. *Compos. Part B Eng.* **2017**, *130*, 54–63. [CrossRef]
81. Mohanty, A.K.; Misra, M.; Hinrichsen, G. Biofibres, biodegradable polymers and biocomposites: An overview. *Macromol. Mater. Eng.* **2000**, *24*, 1–24. [CrossRef]
82. Nurazzi, N.M.; Khalina, A.; Sapuan, S.M.; Ilyas, R.A. Mechanical properties of sugar palm yarn/woven glass fiber reinforced unsaturated polyester composites: Effect of fiber loadings and alkaline treatment. *Polimery* **2019**, *64*, 12–22. [CrossRef]
83. Abral, H.; Ariksa, J.; Mahardika, M.; Handayani, D.; Aminah, I.; Sandrawati, N.; Sapuan, S.M.; Ilyas, R.A. Highly transparent and antimicrobial PVA based bionanocomposites reinforced by ginger nanofiber. *Polym. Test.* **2019**, 106186. [CrossRef]
84. Jumaidin, R.; Khiruddin, M.A.A.; Asyul Sutan Saidi, Z.; Salit, M.S.; Ilyas, R.A. Effect of cogon grass fibre on the thermal, mechanical and biodegradation properties of thermoplastic cassava starch biocomposite. *Int. J. Biol. Macromol.* **2020**, *146*, 746–755. [CrossRef]
85. İşmal, Ö.E.; Paul, R. Composite textiles used in high-performance apparel. In *High-Performance Apparel: Materials, Development, and Applications*; McLoughlin, J., Sabir, T., Eds.; Woodhead Publishing: Cambridge, UK, 2017; pp. 377–420. ISBN 9780081009048.

86. Mohd Nurazzi, N.; Muhammad Asyraf, M.R.; Khalina, A.; Abdullah, N.; Sabaruddin, F.A.; Kamarudin, S.H.; Ahmad, S.; Mahat, A.M.; Lee, C.L.; Aisyah, H.A.; et al. Fabrication, Functionalization, and Application of Carbon Nanotube-Reinforced Polymer Composite: An Overview. *Polymers* **2021**, *13*, 1047. [CrossRef]

87. Sari, N.H.; Pruncu, C.I.; Sapuan, S.M.; Ilyas, R.A.; Catur, A.D.; Suteja, S.; Sutaryono, Y.A.; Pullen, G. The effect of water immersion and fibre content on properties of corn husk fibres reinforced thermoset polyester composite. *Polym. Test.* **2020**, *91*, 106751. [CrossRef]

88. Amir, S.M.M.; Sultan, M.T.H.; Jawaid, M.; Ariffin, A.H.; Mohd, S.; Salleh, K.A.M.; Ishak, M.R.; Md Shah, A.U. Nondestructive testing method for Kevlar and natural fiber and their hybrid composites. *Durab. Life Predict. Biocompos. Fibre Reinf. Compos. Hybrid Compos.* **2018**, 367–388. [CrossRef]

89. Ogin, S.L.; Brøndsted, P.; Zangenberg, J. Composite materials: Constituents, architecture, and generic damage. In *Modeling Damage, Fatigue and Failure of Composite Materials*; Talreja, R., Varna, J., Eds.; Elsevier Ltd.: London, UK, 2016; pp. 3–23. ISBN 9781782422860.

90. Ku, H.; Wang, H.; Pattarachaiyakoop, N.; Trada, M. A review on the tensile properties of natural fiber reinforced polymer composites. *Compos. Part B Eng.* **2011**, *42*, 856–873. [CrossRef]

91. Jem, K.J.; Tan, B. The development and challenges of poly (lactic acid) and poly (glycolic acid). *Adv. Ind. Eng. Polym. Res.* **2020**, *3*, 60–70. [CrossRef]

92. Jiménez-Gómez, C.P.; Cecilia, J.A. Chitosan: A Natural Biopolymer with a Wide and Varied Range of Applications. *Molecules* **2020**, *25*, 3981. [CrossRef] [PubMed]

93. Saratale, R.G.; Cho, S.-K.; Saratale, G.D.; Kadam, A.A.; Ghodake, G.S.; Kumar, M.; Bharagava, R.N.; Kumar, G.; Kim, D.S.; Mulla, S.I. A comprehensive overview and recent advances on polyhydroxyalkanoates (PHA) production using various organic waste streams. *Bioresour. Technol.* **2021**, *325*, 124685. [CrossRef] [PubMed]

94. Siakeng, R.; Jawaid, M.; Ariffin, H.; Sapuan, S.M.; Asim, M.; Saba, N. Natural fiber reinforced polylactic acid composites: A review. *Polym. Compos.* **2019**, *40*, 446–463. [CrossRef]

95. Hinchcliffe, S.A.; Hess, K.M.; Srubar, W.V. Experimental and theoretical investigation of prestressed natural fiber-reinforced polylactic acid (PLA) composite materials. *Compos. Part B Eng.* **2016**, *95*, 346–354. [CrossRef]

96. Holbery, J.; Houston, D. Natural-fiber-reinforced polymer composites in automotive applications. *Jom* **2006**, *58*, 80–86. [CrossRef]

97. Dunne, R.; Desai, D.; Sadiku, R.; Jayaramudu, J. A review of natural fibres, their sustainability and automotive applications. *J. Reinf. Plast. Compos.* **2016**, *35*, 1041–1050. [CrossRef]

98. Faruk, O.; Bledzki, A.K.; Fink, H.-P.; Sain, M. Biocomposites reinforced with natural fibers: 2000–2010. *Prog. Polym. Sci.* **2012**, *37*, 1552–1596. [CrossRef]

99. Herrera-Franco, P.J.; Valadez-González, A. Mechanical properties of continuous natural fibre-reinforced polymer composites. *Compos. Part A Appl. Sci. Manuf.* **2004**, *35*, 339–345. [CrossRef]

100. Alkbir, M.F.M.; Sapuan, S.M.; Nuraini, A.A.; Ishak, M.R. Fibre properties and crashworthiness parameters of natural fibre-reinforced composite structure: A literature review. *Compos. Struct.* **2016**, *148*, 59–73. [CrossRef]

101. Asaithambi, B.; Ganesan, G.; Ananda Kumar, S. Bio-composites: Development and mechanical characterization of banana/sisal fibre reinforced poly lactic acid (PLA) hybrid composites. *Fibers Polym.* **2014**, *15*, 847–854. [CrossRef]

102. Bajpai, P.K.; Singh, I.; Madaan, J. Tribological behavior of natural fiber reinforced PLA composites. *Wear* **2013**, *297*, 829–840. [CrossRef]

103. Graupner, N.; Herrmann, A.S.; Müssig, J. Natural and man-made cellulose fibre-reinforced poly(lactic acid) (PLA) composites: An overview about mechanical characteristics and application areas. *Compos. Part A Appl. Sci. Manuf.* **2009**, *40*, 810–821. [CrossRef]

104. Jiang, N.; Yu, T.; Li, Y.; Pirzada, T.J.; Marrow, T.J. Hygrothermal aging and structural damage of a jute/poly (lactic acid) (PLA) composite observed by X-ray tomography. *Compos. Sci. Technol.* **2019**. [CrossRef]

105. Yu, T.; Li, Y.; Ren, J. Preparation and properties of short natural fiber reinforced poly(lactic acid) composites. *Trans. Nonferrous Met. Soc. China Engl. Ed.* **2009**, *19*, s651–s655. [CrossRef]

106. Oksman, K.; Selin, J.-F. Plastics and Composites from Polylactic Acid. In *Natural Fibers, Plastics and Composites*; Wallenberger, F.T., Weston, N.E., Eds.; Springer: Boston, MA, USA, 2004; pp. 149–165.

107. Omar, M.F.; Jaya, H.; Zulkepli, N.N. Kenaf Fiber Reinforced Composite in the Automotive Industry. *Encycl. Renew. Sustain. Mater.* **2020**, *5*, 95–101. [CrossRef]

108. Ilyas, R.A.; Sapuan, S.M.; Atikah, M.S.N.; Ibrahim, R.; Hazrol, M.D.; Sherwani, S.F.K.; Harussani, M.M.; Jamal, T.; Nazrin, A.; Syafiq, R. Natural fibre: A promising source for the production of nanocellulose. In *Proceedings of the 7th Postgraduate Seminar on Natural Fibre Reinforced Polymer Composites 2020*; Sapuan, S.M., Ilyas, R.A., Eds.; Institute of Tropical Forestry and Forest Products (INTROP), Universiti Putra Malaysia: Serdang, Selangor, 2020; pp. 2–9.

109. Zhang, L.; Li, Z.; Pan, Y.T.; Yáñez, A.P.; Hu, S.; Zhang, X.Q.; Wang, R.; Wang, D.Y. Polydopamine induced natural fiber surface functionalization: A way towards flame retardancy of flax/poly(lactic acid) biocomposites. *Compos. Part B Eng.* **2018**, *154*, 56–63. [CrossRef]

110. Pornwannachai, W.; Ebdon, J.R.; Kandola, B.K. Fire-resistant natural fibre-reinforced composites from flame retarded textiles. *Polym. Degrad. Stab.* **2018**, *154*, 115–123. [CrossRef]

111. Jang, J.Y.; Jeong, T.K.; Oh, H.J.; Youn, J.R.; Song, Y.S. Thermal stability and flammability of coconut fiber reinforced poly(lactic acid) composites. *Compos. Part B Eng.* **2012**, *43*, 2434–2438. [CrossRef]

112. Sukmawan, R.; Takagi, H.; Nakagaito, A.N. Strength evaluation of cross-ply green composite laminates reinforced by bamboo fiber. *Compos. Part B Eng.* **2016**, *84*, 9–16. [CrossRef]
113. Abdul Khalil, H.P.S.; Bhat, I.U.H.; Jawaid, M.; Zaidon, A.; Hermawan, D.; Hadi, Y.S. Bamboo fibre reinforced biocomposites: A review. *Mater. Des.* **2012**, *42*, 353–368. [CrossRef]
114. Gunti, R.; Ratna Prasad, A.V.; Gupta, A.V.S.S.K.S. Mechanical and degradation properties of natural fiber-reinforced PLA composites: Jute, sisal, and elephant grass. *Polym. Compos.* **2018**, *39*, 1125–1136. [CrossRef]
115. Elmessiry, M.; Deeb, E. Analysis of the wheat straw/flax fiber reinforced polymer hybrid composites. *J. App. Mech. Eng* **2016**, *5*, 1–5.
116. Nyambo, C.; Mohanty, A.K.; Misra, M. Effect of maleated compatibilizer on performance of PLA/wheat Straw-Based green composites. *Macromol. Mater. Eng.* **2011**, *296*, 710–718. [CrossRef]
117. Mat Zubir, N.H.; Sam, S.T.; Santiagoo, R.; Noimam, N.Z.; Wang, J. Tensile properties of rice straw fiber reinforced poly (lactic acid) biocomposites. In *Advanced Materials Research*; Trans Tech Publications Ltd.: Bäch, Switzerland, 2016; Volume 1133, pp. 598–602.
118. Ismail, M.R.; Yassen, A.A.M.; Afify, M.S. Mechanical properties of rice straw fiber-reinforced polymer composites. *Fibers Polym.* **2011**, *12*, 648. [CrossRef]
119. Ding, F. Mechanical properties and degradation characteristics of corn straw fibers/polylactic acid composite materials. *J. Agric. Resour. Environ.* **2018**, *35*, 455–458.
120. Pradhan, R.; Misra, M.; Erickson, L.; Mohanty, A. Compostability and biodegradation study of PLA–wheat straw and PLA–soy straw based green composites in simulated composting bioreactor. *Bioresour. Technol.* **2010**, *101*, 8489–8491. [CrossRef]
121. Wang, H.; Hassan, E.; Memon, H.; Elagib, T.; Abad AllaIdris, F. Characterization of natural composites fabricated from Abutilon-fiber-reinforced Poly (Lactic Acid). *Processes* **2019**, *7*, 583. [CrossRef]
122. Ozyhar, T.; Baradel, F.; Zoppe, J. Effect of functional mineral additive on processability and material properties of wood-fiber reinforced poly(lactic acid) (PLA) composites. *Compos. Part A Appl. Sci. Manuf.* **2020**, *132*, 105827. [CrossRef]
123. Zhang, Q.; Shi, L.; Nie, J.; Wang, H.; Yang, D. Study on poly(lactic acid)/natural fibers composites. *J. Appl. Polym. Sci.* **2012**, *125*, E526–E533. [CrossRef]
124. Du, Y.; Wu, T.; Yan, N.; Kortschot, M.T.; Farnood, R. Fabrication and characterization of fully biodegradable natural fiber-reinforced poly(lactic acid) composites. *Compos. Part B Eng.* **2014**, *56*, 717–723. [CrossRef]
125. Dhand, V.; Mittal, G.; Rhee, K.Y.; Park, S.-J.; Hui, D. A short review on basalt fiber reinforced polymer composites. *Compos. Part B Eng.* **2015**, *73*, 166–180. [CrossRef]
126. Sang, L.; Han, S.; Li, Z.; Yang, X.; Hou, W. Development of short basalt fiber reinforced polylactide composites and their feasible evaluation for 3D printing applications. *Compos. Part B Eng.* **2019**, *164*, 629–639. [CrossRef]
127. Kurniawan, D.; Kim, B.S.; Lee, H.Y.; Lim, J.Y. Atmospheric pressure glow discharge plasma polymerization for surface treatment on sized basalt fiber/polylactic acid composites. *Compos. Part B Eng.* **2012**, *43*, 1010–1014. [CrossRef]
128. Khan, G.M.A.; Terano, M.; Gafur, M.A.; Alam, M.S. Studies on the mechanical properties of woven jute fabric reinforced poly(L-lactic acid) composites. *J. King Saud Univ. Eng. Sci.* **2016**, *28*, 69–74. [CrossRef]
129. Bergeret, A.; Benezet, J.C. Natural fibre-reinforced biofoams. *Int. J. Polym. Sci.* **2011**, *2011*, 569871. [CrossRef]
130. Jauhari, N.; Mishra, R.; Thakur, H. Natural Fibre Reinforced Composite Laminates—A Review. *Mater. Today Proc.* **2015**, *2*, 2868–2877. [CrossRef]
131. Nechwatal, A.; Mieck, K.-P.; Reußmann, T. Developments in the characterization of natural fibre properties and in the use of natural fibres for composites. *Compos. Sci. Technol.* **2003**, *63*, 1273–1279. [CrossRef]
132. Van De Velde, K.; Kiekens, P. Thermoplastic pultrusion of natural fibre reinforced composites. *Compos. Struct.* **2001**, *54*, 355–360. [CrossRef]
133. Sujaritjun, W.; Uawongsuwan, P.; Pivsa-Art, W.; Hamada, H. Mechanical property of surface modified natural fiber reinforced PLA biocomposites. *Energy Procedia* **2013**, *34*, 664–672. [CrossRef]
134. Hu, R.; Lim, J.K. Fabrication and mechanical properties of completely biodegradable hemp fiber reinforced polylactic acid composites. *J. Compos. Mater.* **2007**, *41*, 1655–1669. [CrossRef]
135. Choi, H.Y.; Lee, J.S. Effects of surface treatment of ramie fibers in a ramie/poly(lactic acid) composite. *Fibers Polym.* **2012**, *13*, 217–223. [CrossRef]
136. Baghaei, B.; Skrifvars, M.; Rissanen, M.; Ramamoorthy, S.K. Mechanical and thermal characterization of compression moulded polylactic acid natural fiber composites reinforced with hemp and lyocell fibers. *J. Appl. Polym. Sci.* **2014**, *131*. [CrossRef]
137. Singh, I.; Bajpai, P.K.; Malik, D.; Madaan, J.; Bhatnagar, N. Microwave Joining of Natural Fiber Reinforced Green Composites. *Adv. Mater. Res.* **2011**, *410*, 102–105. [CrossRef]
138. Neoh, K.W.; Tshai, K.Y.; Khiew, P.S.; Chia, C.H. Micro Palm and Kenaf Fibers Reinforced PLA Composite: Effect of Volume Fraction on Tensile Strength. *Appl. Mech. Mater.* **2011**, *145*, 1–5. [CrossRef]
139. Porras, A.; Maranon, A.; Ashcroft, I.A. Thermo-mechanical characterization of Manicaria Saccifera natural fabric reinforced poly-lactic acid composite lamina. *Compos. Part A Appl. Sci. Manuf.* **2016**, *81*, 105–110. [CrossRef]
140. Bax, B.; Müssig, J. Impact and tensile properties of PLA/Cordenka and PLA/flax composites. *Compos. Sci. Technol.* **2008**, *68*, 1601–1607. [CrossRef]
141. Bledzki, A.K.; Jaszkiewicz, A. Mechanical performance of biocomposites based on PLA and PHBV reinforced with natural fibres—A comparative study to PP. *Compos. Sci. Technol.* **2010**, *70*, 1687–1696. [CrossRef]

142. Cheung, H.; Ho, M.; Lau, K.; Cardona, F.; Hui, D. Natural fibre-reinforced composites for bioengineering and environmental engineering applications. *Compos. Part B Eng.* **2009**, *40*, 655–663. [CrossRef]
143. Le Duigou, A.; Barbé, A.; Guillou, E.; Castro, M. 3D printing of continuous flax fibre reinforced biocomposites for structural applications. *Mater. Des.* **2019**, *180*, 107884. [CrossRef]
144. Bodros, E.; Pillin, I.; Montrelay, N.; Baley, C. Could biopolymers reinforced by randomly scattered flax fibre be used in structural applications? *Compos. Sci. Technol.* **2007**, *67*, 462–470. [CrossRef]
145. Van Den Oever, M.J.A.; Beck, B.; Müssig, J. Agrofibre reinforced poly(lactic acid) composites: Effect of moisture on degradation and mechanical properties. *Compos. Part A Appl. Sci. Manuf.* **2010**, *41*, 1628–1635. [CrossRef]
146. Saba, N.; Paridah, M.T.; Jawaid, M. Mechanical properties of kenaf fibre reinforced polymer composite: A review. *Constr. Build. Mater.* **2015**, *76*, 87–96. [CrossRef]
147. Sawpan, M.A.; Pickering, K.L.; Fernyhough, A. Effect of fibre treatments on interfacial shear strength of hemp fibre reinforced polylactide and unsaturated polyester composites. *Compos. Part A Appl. Sci. Manuf.* **2011**, *42*, 1189–1196. [CrossRef]
148. Hughes, M. Defects in natural fibres: Their origin, characteristics and implications for natural fibre-reinforced composites. *J. Mater. Sci.* **2012**, *47*, 599–609. [CrossRef]
149. Bledzki, A.K.; Zhang, W.; Chate, A. Natural-fibre-reinforced polyurethane microfoams. *Compos. Sci. Technol.* **2001**, *61*, 2405–2411. [CrossRef]
150. Teramoto, N.; Urata, K.; Ozawa, K.; Shibata, M. Biodegradation of aliphatic polyester composites reinforced by abaca fiber. *Polym. Degrad. Stab.* **2004**, *86*, 401–409. [CrossRef]
151. Verma, D.; Gope, P.C.; Shandilya, A.; Gupta, A.; Maheshwari, M.K. Coir fibre reinforcement and application in polymer composites: A review. *J. Mater. Environ. Sci.* **2013**, *4*, 263–276.
152. Takagi, H.; Kako, S.; Kusano, K.; Ousaka, A. Thermal conductivity of PLA-bamboo fiber composites. *Adv. Compos. Mater. Off. J. Japan Soc. Compos. Mater.* **2007**, *16*, 377–384. [CrossRef]
153. Nuthong, W.; Uawongsuwan, P.; Pivsa-Art, W.; Hamada, H. Impact property of flexible epoxy treated natural fiber reinforced PLA composites. *Energy Procedia* **2013**, *34*, 839–847. [CrossRef]
154. Azlin, M.N.M.; Sapuan, S.M.; Zainudin, E.S.; Zuhri, M.Y.M.; Ilyas, R.A. Natural Polylactic Acid-Based Fiber Composites: A Review. In *Advanced Processing, Properties, and Applications of Starch and Other Bio-Based Polymers*; Al-Oqla, F.M., Sapuan, S.M., Eds.; Elsevier: Oxford, UK, 2020; pp. 21–34. ISBN 9780128196618.
155. Chia, H.N.; Wu, B.M. High-resolution direct 3D printed PLGA scaffolds: Print and shrink. *Biofabrication* **2014**, *7*, 015002. [CrossRef] [PubMed]
156. Mazzanti, V.; Malagutti, L.; Mollica, F. FDM 3D Printing of Polymers Containing Natural Fillers: A Review of their Mechanical Properties. *Polymers* **2019**, *11*, 1094. [CrossRef] [PubMed]
157. Jamshidian, M.; Tehrany, E.A.; Imran, M.; Jacquot, M.; Desobry, S. Poly-Lactic Acid: Production, Applications, Nanocomposites, and Release Studies. *Compr. Rev. Food Sci. Food Saf.* **2010**, *9*, 552–571. [CrossRef]
158. Ilyas, R.A.; Sapuan, S.M. Biopolymers and biocomposites: Chemistry and technology. *Curr. Anal. Chem.* **2020**, *16*, 500–503. [CrossRef]
159. Le Duigou, A.; Castro, M.; Bevan, R.; Martin, N. 3D printing of wood fibre biocomposites: From mechanical to actuation functionality. *Mater. Des.* **2016**, *96*, 106–114. [CrossRef]
160. Ayrilmis, N.; Kariz, M.; Kwon, J.H.; Kitek Kuzman, M. Effect of printing layer thickness on water absorption and mechanical properties of 3D-printed wood/PLA composite materials. *Int. J. Adv. Manuf. Technol.* **2019**, *102*, 2195–2200. [CrossRef]
161. Matsuzaki, R.; Ueda, M.; Namiki, M.; Jeong, T.-K.; Asahara, H.; Horiguchi, K.; Nakamura, T.; Todoroki, A.; Hirano, Y. Three-dimensional printing of continuous-fiber composites by in-nozzle impregnation. *Sci. Rep.* **2016**, *6*, 23058. [CrossRef]
162. Zhao, X.; Tekinalp, H.; Meng, X.; Ker, D.; Benson, B.; Pu, Y.; Ragauskas, A.J.; Wang, Y.; Li, K.; Webb, E.; et al. Poplar as Biofiber Reinforcement in Composites for Large-Scale 3D Printing. *ACS Appl. Bio Mater.* **2019**, *2*, 4557–4570. [CrossRef]
163. Champeau, M.; Heinze, D.A.; Viana, T.N.; de Souza, E.R.; Chinellato, A.C.; Titotto, S. 4D Printing of Hydrogels: A Review. *Adv. Funct. Mater.* **2020**, *30*, 1910606. [CrossRef]
164. Spiegel, C.A.; Hippler, M.; Münchinger, A.; Bastmeyer, M.; Barner-Kowollik, C.; Wegener, M.; Blasco, E. 4D Printing at the Microscale. *Adv. Funct. Mater.* **2020**, *30*, 1907615. [CrossRef]
165. Zolfagharian, A.; Kaynak, A.; Kouzani, A. Closed-loop 4D-printed soft robots. *Mater. Des.* **2020**, *188*, 108411. [CrossRef]
166. Quanjin, M.; Rejab, M.R.M.; Idris, M.S.; Kumar, N.M.; Abdullah, M.H.; Reddy, G.R. Recent 3D and 4D intelligent printing technologies: A comparative review and future perspective. *Procedia Comput. Sci.* **2020**, *167*, 1210–1219. [CrossRef]
167. Ma, S.; Zhang, Y.; Wang, M.; Liang, Y.; Ren, L.; Ren, L. Recent progress in 4D printing of stimuli-responsive polymeric materials. *Sci. China Technol. Sci.* **2020**, *63*, 532–544. [CrossRef]
168. Subash, A.; Kandasubramanian, B. 4D printing of shape memory polymers. *Eur. Polym. J.* **2020**, *134*, 109771. [CrossRef]
169. Zafar, M.Q.; Zhao, H. 4D Printing: Future Insight in Additive Manufacturing. *Met. Mater. Int.* **2020**, *26*, 564–585. [CrossRef]
170. Chu, H.; Yang, W.; Sun, L.; Cai, S.; Yang, R.; Liang, W.; Yu, H.; Liu, L. 4D Printing: A Review on Recent Progresses. *Micromachines* **2020**, *11*, 796. [CrossRef]
171. Zolfagharian, A.; Kaynak, A.; Bodaghi, M.; Kouzani, A.Z.; Gharaie, S.; Nahavandi, S. Control-Based 4D Printing: Adaptive 4D-Printed Systems. *Appl. Sci.* **2020**, *10*, 3020. [CrossRef]

172. Yang, J.; Bai, R.; Chen, B.; Suo, Z. Hydrogel Adhesion: A Supramolecular Synergy of Chemistry, Topology, and Mechanics. *Adv. Funct. Mater.* **2020**, *30*, 1901693. [CrossRef]
173. Nautiyal, U.; Sahu, N.; Gupta, D. Hydrogel: Preparation, Characterization and Applications. *Asian Pacific J. Nurs. Health Sci.* **2020**, *3*. [CrossRef]
174. Zheng, Y.; Chen, Z.; Jiang, Q.; Feng, J.; Wu, S.; del Campo, A. Near-infrared-light regulated angiogenesis in a 4D hydrogel. *Nanoscale* **2020**, *12*, 13654–13661. [CrossRef] [PubMed]
175. Le Duigou, A.; Requile, S.; Beaugrand, J.; Scarpa, F.; Castro, M. Natural fibres actuators for smart bio-inspired hygromorph biocomposites. *Smart Mater. Struct.* **2017**, *26*, 125009. [CrossRef]
176. Poppinga, S.; Correa, D.; Bruchmann, B.; Menges, A.; Speck, T. Plant Movements as Concept Generators for the Development of Biomimetic Compliant Mechanisms. *Integr. Comp. Biol.* **2020**, *60*, 886–895. [CrossRef]
177. Alief, N.A.; Supriadi, S.; Whulanza, Y. Modelling the shape memory properties of 4D printed polylactic acid (PLA) for application of disk spacer in minimally invasive spinal fusion. *AIP Conf. Proc.* **2019**, *2092*, 020005. [CrossRef]
178. Martins, S.S.; Evangelista, A.C.J.; Hammad, A.W.A.; Tam, V.W.Y.; Haddad, A. Evaluation of 4D BIM tools applicability in construction planning efficiency. *Int. J. Constr. Manag.* **2020**, 1–14. [CrossRef]
179. Zolfagharian, A.; Mahmud, M.A.P.; Gharaie, S.; Bodaghi, M.; Kouzani, A.Z.; Kaynak, A. 3D/4D-printed bending-type soft pneumatic actuators: Fabrication, modelling, and control. *Virtual Phys. Prototyp.* **2020**, *15*, 373–402. [CrossRef]
180. Miao, S.; Cui, H.; Esworthy, T.; Mahadik, B.; Lee, S.; Zhou, X.; Hann, S.Y.; Fisher, J.P.; Zhang, L.G. 4D Self-Morphing Culture Substrate for Modulating Cell Differentiation. *Adv. Sci.* **2020**, *7*, 1902403. [CrossRef]
181. Le Fer, G.; Becker, M.L. 4D Printing of Resorbable Complex Shape-Memory Poly(propylene fumarate) Star Scaffolds. *ACS Appl. Mater. Interfaces* **2020**, *12*, 22444–22452. [CrossRef]
182. Durga Prasad Reddy, R.; Sharma, V. Additive manufacturing in drug delivery applications: A review. *Int. J. Pharm.* **2020**, *589*, 119820. [CrossRef] [PubMed]
183. Wang, L.; Leng, J.; Du, S. 4D printing of shape memory polymers and their composites: Research status and application progress. *Harbin Gongye Daxue Xuebao J. Harbin Inst. Technol.* **2020**. [CrossRef]
184. Saravana Bavan, D.; Mohan Kumar, G.C. Potential use of natural fiber composite materials in India. *J. Reinf. Plast. Compos.* **2010**, *29*, 3600–3613. [CrossRef]
185. Furtado, S.C.R.; Araújo, A.L.; Silva, A.; Alves, C.; Ribeiro, A.M.R. Natural fibre-reinforced composite parts for automotive applications. *Int. J. Automot. Compos.* **2014**, *1*, 18. [CrossRef]
186. Barkoula, N.M.; Alcock, B.; Cabrera, N.O.; Peijs, T. Flame-Retardancy Properties of Intumescent Ammonium Poly(Phosphate) and Mineral Filler Magnesium Hydroxide in Combination with Graphene. *Polym. Polym. Compos.* **2008**, *16*, 101–113. [CrossRef]

Review

Embracing Additive Manufacturing Technology through Fused Filament Fabrication for Antimicrobial with Enhanced Formulated Materials

Waleed Ahmed [1,*], Sidra Siraj [2] and Ali H. Al-Marzouqi [2]

1 Engineering Requirements Unit, College of Engineering, United Arab Emirates University, Al Ain 15551, United Arab Emirates
2 Chemical Engineering Department, COE, United Arab Emirates University, Al Ain 15551, United Arab Emirates; sidra.siraj@uaeu.ac.ae (S.S.); hassana@uaeu.ac.ae (A.H.A.-M.)
* Correspondence: w.ahmed@uaeu.ac.ae

Abstract: Antimicrobial materials produced by 3D Printing technology are very beneficial, especially for biomedical applications. Antimicrobial surfaces specifically with enhanced antibacterial property have been prepared using several quaternary salt-based agents, such as quaternary ammonium salts and metallic nanoparticles (NPs), such as copper and zinc, which are incorporated into a polymeric matrix mainly through copolymerization grafting and ionic exchange. This review compared different materials for their effectiveness in providing antimicrobial properties on surfaces. This study will help researchers choose the most suitable method of developing antimicrobial surfaces with the highest efficiency, which can be applied to develop products compatible with 3D Printing Technology.

Keywords: antimicrobial; antibacterial; 3D printing; fused filament fabrication; composite material

Citation: Ahmed, W.; Siraj, S.; Al-Marzouqi, A.H. Embracing Additive Manufacturing Technology through Fused Filament Fabrication for Antimicrobial with Enhanced Formulated Materials. *Polymers* **2021**, *13*, 1523. https://doi.org/10.3390/polym13091523

Academic Editor: Emin Bayraktar

Received: 9 April 2021
Accepted: 3 May 2021
Published: 9 May 2021

Publisher's Note: MDPI stays neutral with regard to jurisdictional claims in published maps and institutional affiliations.

1. Introduction

3D Printing, also known as fused filament fabrication (FFF), continues to open new routes to the production of high-performance and complex structures with enhanced properties and dynamic shapes that are unattainable via conventional fabrication methods. Developing new material for different applications of the 3D Printing technology [1] usually involves many challenges, such as the mechanical properties [2] as well as the failure mechanism [3–5]. The FFF technology opened new horizons for a wide range of applications [1,2] but limitations still face this fast-moving sector [6–8]. Although the 3D Printing technology contributes to the increasing levels of wasted material of polymeric base [9,10], on the contrary, there are unlimited benefits that are continuously increasing day by day. Owing to its innovation-driven approach, 3D Printing has been expanding to several fields, such as analytical chemistry for chromatography [3] and fluorescence techniques [4], optic fibers for increased optical and mechanical performance [5,6], membrane technology for increased adsorption of chemicals [7], space technology to develop tools on-site [8], and medicine for the development of implants and medical devices [3,5,7,9–13].

The ease of modeling and experimentation helps with progressive growth and expansion of knowledge in these fields, which can be of significant value to clinical areas. 3D Printing has been extensively employed in the biomedical field, especially in biomodelling, for the fabrication of scaffolds and implants, such as for dental restorations and tissue engineering, bone and fracture healing [1,14] and even for drug delivery because it is highly flexible and faster than current methods such as machining [15,16]. Moreover, the positive effects on the mechanical and biological properties of designed microporous scaffolds manufactured by commercial 3D Printing technology promote the use of 3D printed scaffolds as viable candidates for further research for clinical applications [16,17]. 3D Printing is also a highly cost-effective approach, cutting costs and improving economic efficiency.

Several techniques that have been commonly applied worldwide for 3D Printing include FFF, selective laser sintering (SLS), digital light processing (DLP), and stereolithography (SLA) [13,18–20]. Of these, FFF has been extensively used to develop medical devices. In FFF, the feedstock to the 3D printer is a simple filament made from a thermoplastic material that is heated to soften and extrude through a printer head or nozzle, which builds layers driven in an X–Y orientation. FFF has been reported to be an excellent technique to fabricate small mechanical parts, providing sufficient precision, which allows rapid modifications to be made during the process itself. A simple FFF schematic is shown in Figure 1 [18].

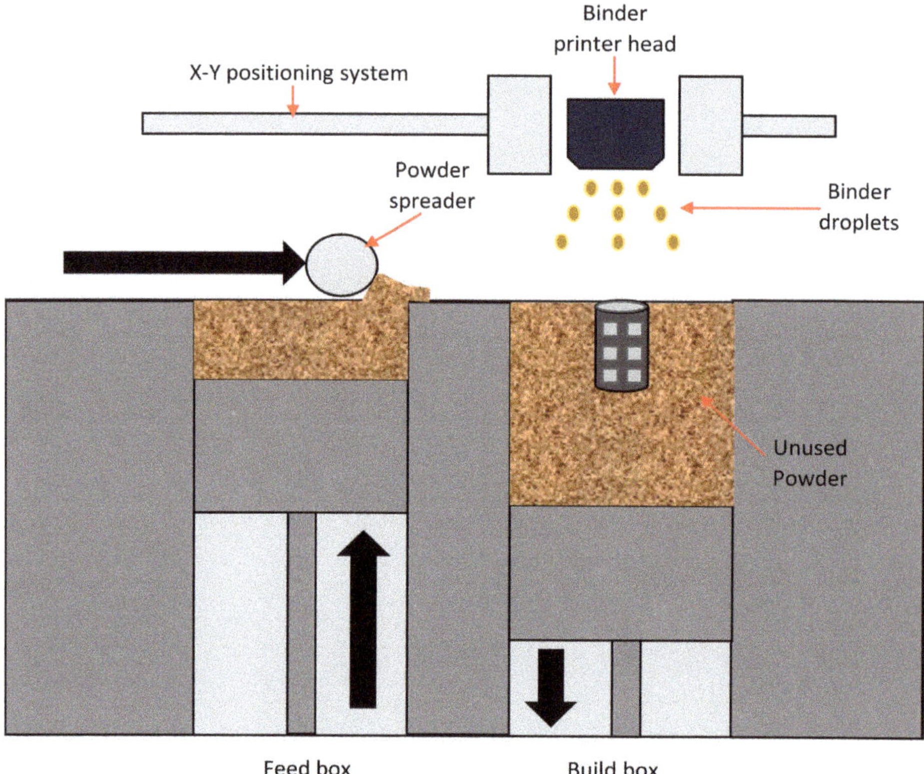

Figure 1. Schematic representation of FFF.

With the use of 3D Printing increasing in biomedical applications, the surface properties of the developed 3D printed material are essential because this material will be in direct contact with the human body. Common microorganisms that are highly associated with infections include Staphylococcus spp. strains, such as Staphylococcus epidermidis [21], and Gram-negative Enterobacteriaceae strains, such as *Escherichia coli* (*E. coli*) [22]. The adhesion of bacteria to the surface is facilitated by several factors, such as hydrophobicity and surface tension, which is followed by accumulation resulting in a biofilm [23]; however, it may not always be harmful [24]. Thus, having an active antibacterial surface on the 3D-printed material is of vital necessity to prevent any microorganisms from developing on human tissues and infecting the human body. Even though conventional approaches involve the release of metal ions, such as fluoride, silver, or zinc, to prevent and fight bacterial infections, this method still results in decreased mechanical properties, causing toxicity to surround tissues. Additionally, the fabrication of antimicrobial materials is preferred for coating the surfaces of materials, even though this adds an additional process, costing time and money in any clinical field [25].

Polymer composite materials have gained the attention of researchers worldwide in a variety of fields, from printing and packing applications to several other applications [17,26–28]. Due to filler addition, significantly enhanced properties, such as mechanical and thermal properties and conductivity, are attained; thus, increasingly, applications prefer the use of polymer composite materials instead of pure polymers. One of the most promising techniques that has been used is the addition of metallic particles as fillers to provide enhanced antimicrobial and antiviral properties to the commonly developed polymer composite materials being used in several applications. Currently, viral transmission through either touch or contaminated liquids, such as blood, poses a massive threat to populations worldwide. The issue of the inactivation of viruses in media requires an upfront solution, and the development of novel means to inactivate viruses is highly desirable. One of the most novel developed strategies is to incorporate copper nanoparticles (NPs) into the matrix. Copper has potent virucidal properties, and copper's neutralization of infectious bronchitis virus, poliovirus, human immunodeficiency virus type 1 (HIV-1), and other enveloped or nonenveloped single- or double-stranded DNA or RNA viruses has been well reported [29]. The ability of free copper ions in copper NPs to break the membranes of microorganisms alters their DNA, thus eliminating them, which confirms their great potential for keeping material surfaces free of pathogens. Based on the broad-spectrum antiviral effectiveness of copper, the incorporation of copper NPs in any intended hybrid composite material could inherently add significant value to the final product, especially with regard to the COVID-19 pandemic situation [30].

Another metal that has caught the attention of researchers worldwide is silica. Due to its high chemical and thermal stability and excellent antimicrobial properties, silica has been used in several applications to provide enhanced features to composites. For instance, for the treatment of bacteria, silver NPs are widely used in several industries. The main challenges in the use of silver NPs is to provide an antimicrobial specifically for antibacterial surface suitable for a wide range of substrates with mechanical and thermal resistance with prolonged antibacterial properties and that is suitable for wide temperature ranges. Moreover, the agglomeration of silver NPs has always remained a challenge because it decreases their antiviral effects. A study reported excellent antiviral capabilities for a hybrid material against two model viruses for different water conditions. silver NPs incorporated with SiO_2 showed minimized agglomeration and minimal particle release in water environments [31].

Another study reported the production of a series of mesoporous silica nanoparticles (MSNPs) using room-temperature ionic liquids (RTILs) to study the mass-transport properties by investigating the controlled release profiles of these materials and utilizing the RTIL templates as antibacterial agents. The study showed that the antibacterial activity was dependent on the rate of diffusional release of the pore-encapsulated RTIL, which was in turn governed by the particle and pore morphology of the MSN materials, thus showing the good potential for silica-based mesoporous materials to be used in several applications that require controlled release delivery [32]. The literature also supports the idea of the production of antibacterial silver nanocluster–silica composite coatings by radiofrequency co-sputtering techniques, which can result in peculiar antibacterial properties [33]. Additionally, another method by which silica is gaining popularity is allowing monomers to be attached to its surface, providing enhanced properties. For instance, silica particles were used to treat water, and the results supported excellent biocidal efficacy against bacteria, thus showing their good potential to combat microorganisms in water [34,35]. Thus, the use of silica-based composite materials with added antimicrobial and antiviral properties enhanced with copper NPs can open pathways for its significant use in fields that require high-purity surfaces, such as medicine, food packing, and space industries [2,10,23].

Additionally, polysaccharides are now also considered to develop material for antimicrobial materials. Bacteria is a large reservoir for the development of polysaccharides.

These materials have exhibited antibacterial property by formulating films which have shown broad anti-bacterial activity [36]. Since these bacterial polysaccharides can follow an ordered structure, it is seen as an excellent bioprinting material to provide antibacterial effect. Commonly used bionics are alginate, bacterial cellulose and hyaluronic acids [37].

Antimicrobial mechanisms in 3D printed material simply by exploiting the antimicrobial property of the incorporated NPs. NPs are usually synthesized following up physio-chemical processes such as using solvents and through reactions such as reduction and oxidation. However, a lot of green/bio NPs are also considered to reduce the adverse effects of toxic solvents and harmful by-products produced [38].

Furthermore, to fully understand the mechanism of how metal ions produce this antimicrobial effect for instance against bacteria, a simple representation of how these metallic particles at a micro or nanoscale are illustrated in Figure 2 for a polymer/composite material. Initially, the adsorption of bacteria on the surface of the polymer triggers the diffusion of water through the matrix. Next, the water, along with dissolved oxygen, reaches the surface of the metal particles, which allows the dissolution process. This leads to the formation of metal ions. Consequently, the metal ions deposit on the surface of the composite and damage the membrane of the bacteria, which results in the metal ions diffusing into the bacteria. The same can be applied for any polymer/metal material with a biocide agent that is embedded in the matrix [39].

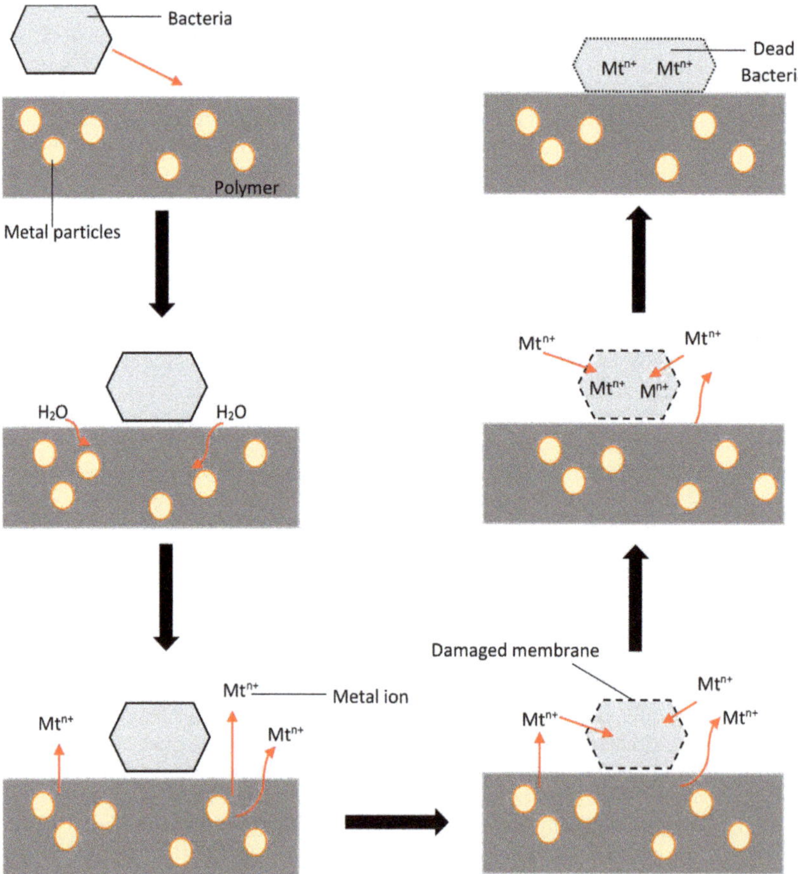

Figure 2. Mechanism of the antibacterial behavior of polymer composites with metal ions [39].

Since metallic ions have been extensively used to provide antipathogenic effects, the combination of 3D Printing alongside metallic particles can result in successful developments of antimicrobial materials, tools, and surfaces [40]. Li et al. [19] successfully prepared an antimicrobial photosensitive resin using the DLP technique with active bacterial resistance. Yue et al. [41] also successfully developed an antimicrobial composite FFF that was positively grafted with metal ions capable of killing any bacteria contacting the surface using SLA. Another study conducted by Muwaffak et al. [42] used hot-melt extrusion to integrate metal ions into a polymer-based filament, which was then used to 3D print a wound dressing. Gutierrez et al. [15] followed a more natural approach by developing antimicrobial alginate hydrogels that were used to 3D print. Ben et al. [29] used surface-modified β-tri-calcium phosphate (β-TCP) granules for 3D Printing for bone tissue engineering. In another study conducted by Huang et al., polymer/ceramic blends prepared with the FFF technique to develop bone scaffolds were characterized [43].

In particular, the use of antimicrobial composite material along with FFF specifically for bio-applications has been widely studied [44,45]. For instance, biodegradable polymers such polylactic acid- (PLA) based polymers are often used to develop filament that possess antibacterial properties using FFF as discussed by Bronstein et al. [46]. Moreover, the antibacterial property is added with the help of processing techniques such as co-extrusion with particles for instance using chitosan as studied by Mania et al. [18] or using metallic nanoparticles as done by Rezić et al. [47]. Biomedical applications such as dental resins as studied by Bayarsaikhan et al. [48] and drug delivery as highlighted by Shaqour et al. [49,50] highly promote a practical approach of using 3D Printing along with compatible antibacterial material. Bartolomé et al. [51] and Navaruckiene et al. [52] both emphasize the use of FFF as a promising approach for bio fabrication of antibacterial material and suggest that using metallic particles alongside biodegradable polymers show great compatibility to the human body and is expected to be considered as the more superior and dynamic approach of development in the field in the upcoming years.

Thus, several metal ions have been incorporated to prepare antibacterial materials compatible with 3D Printing technology, which can serve as a cost-effective approach, especially in the field of biomedicine, which requires high purity and highly anti-pathogen surface materials. 3D Printing can offer a higher precision compared to conventional methods that can cater to more customizable solutions that are increasingly being desired. This article provides a comprehensive review of the development of new antibacterial material and their use in 3D Printing technology. The potential application of using metallic fillers as antibacterial materials in 3D Printing technology in several other fields will also be highlighted.

2. Techniques and Effectiveness of Developed Antibacterial 3D-Printed Materials

To obtain a clear understanding of whether antibacterial materials are providing good effectiveness against bacteria development on surfaces, selected recent studies (2015–2020) on developing antibacterial materials for 3D Printing were compared. Standard laboratory procedures for ISO 22,196 were followed to test the antibacterial properties in all studies. The tests were mainly conducted against methicillin-resistant *Staphylococcus aureus* (*S. aureus*), standard *Staphylococcus aureus* (*S. aureus*), and *Escherichia coli* (*E. coli*) because these are familiar strains known to cause infections at homes and hospitals.

One of the biggest challenges of medical implants is the ability to consistently inhibit bacterial adhesion and infection. Alternate materials to antibiotics are now being explored, particularly carbon-derived materials, such as graphene-based materials (GBMs), and they have started to receive increased attention owing to their potential antibacterial properties. Melo et al. [53] incorporated graphene oxide (GO) to develop a polycaprolactone (PCL) fibrous-based scaffold because smaller oxidized versions of GBM's have been reported to have a higher biocompatibility with cells. GO was dispersed in PCL using suitable solvents. After attaining optimum concentrations, these formulations were extruded through a needle for scaffold plotting. The time-dependent bacterial effect was studied, and a significant increase in the bacterial death rate from <20% in neat PCL up to 80% in the GO/PCL composite material was observed. The results also revealed that the composite material allowed human cell activity to show fibroblast adhesion, spreading, and colonization. The results showed the potential of GO concerning its antibacterial properties in scaffolds that could be used in medical implants and devices.

Photosensitive resins with antibacterial properties are vital, especially for biomedical applications. Studies have reported that methyl acrylates are widely used in DLP 3D Printing because they cure fast and result in a rich variety of products. Moreover, thiol-ene-based systems can accelerate the printing speed and use a lower amount of photoinitiator, resulting in reduced costs and lower yellowing of the resin. They are also employed in DLP 3D Printing techniques that require increased resolution and higher accuracy. In a study conducted by Li et al. [19], a ternary, thiol-ene-acrylate, which has a high photopolymerization rate, was selected as the 3D Printing matrix material owing to its reduced shrinkage property, which can result in increased 3D Printing accuracy. The study incorporated two quaternary ammonium salt-based antibacterial agents, QAC and SH-QAC (ammonium salt with added thiol group), which were prepared via dissolution in solvents, to provide the desired antibacterial effect. The QAC group was added to the matrix via in situ copolymerization, and SH-QAC was added to further increase the compatibility between the matrix and the QAC group. It have been reported that a 100% antibacterial rate was achieved with 4 wt% QAC and 10 wt% SH-QAC. However, the antibacterial effect of the QAC was superior to that of SH-QAC with the same amount. This could be because SH-QAC has long chain segments and the antibacterial group is possibly immobilized inside the resin after the photocuring process. Both antibacterial photosensitive resins were used to develop a 3D-printed tooth model, which showed high precision, suggesting that this technique is feasible for biomedical applications.

Antibacterial resins are commonly used in dental applications to deposit antibacterial agents, such as fluoride, zinc ions, and antimicrobial peptides into resins [54]. Additionally, silver ions have been reported to be able to kill bacteria, then dissociate from dead bacteria, and effectively repeat the sterilization process. Silver-based salts or layered inorganic fillers, such as zeolites, are commonly used to develop silver-based inorganic antibacterial agents via ion-exchange or chemical deposition techniques. Moreover, nanomaterials, such as halloysite nanotubes, which have a high aspect ratio and large volume, are often used with silver ions to provide increased antibacterial effects for a long duration. In a study conducted by Sa et al. [55], DLP-3D-printed stereolithographic resins (SLR) with silver halloysite (Ag-HNT) were prepared to study the antibacterial effect against Streptococcus mutans, which is a main reason for dental caries. Urethane dimethacrylate (UDMA) was used as the main component, and triethylene glycol dimethacrylate (TEGDMA) was used as a diluent to decrease the viscosity of the resin fluid; a photosensitizer was also added. The material exhibited strong antibacterial property up to 99% even after repeated extraction for over 1 week. This prolonged antibacterial effect is highly desired in dental applications because saliva cannot be easily replaced in the retention area, promoting Ag-HNT/SLR antibacterial material as having good potential for use in dental applications.

Biomedical applications have also began utilizing biopolymer-based hydrogels or salt-hydrogels owing to their excellent biocompatibility and relatively low cytotoxicity [56,57]. This is a result of their structural similarity to the soft tissues and surrounding cell matrix and due to their high water content in their structure. Moreover, because hydrogels provide a moist environment, they are believed to promote wound healing. Alginate-based hydrogels are widely accepted as good biopolymer hydrogels. They can be easily ionically cross-linked with divalent ions owing to the presence of the negatively charged carboxylic acid group. The ionic cross-linking is a relatively fast process that specifically results in the incorporation of alginates in the 3D Printing technique for artificial implants and biomedical devices. In a study conducted by Guiterrez et al. [15], alginate beads were developed by ionic cross-linking an aqueous alginate solution at 1 wt%. Once the beads were prepared, an infusion pump was used to drop a copper nitrate solution into the bead solution for the ion-exchange process. Bacterial cellulose was used as a rheological modifier in this case with the fermentation of Gluconacetobacter xylinus. During extrusion, the alginate was ionically cross-linked by an ionotropic process with 0.1 M $CaCl_2$ or $Cu(NO_3)_2$ solutions. Then, the 3D-printed scaffolds with a 4 wt% alginate concentration were tested against *S. aureus* and *E. coli*. The tests showed a distinct inhibition halo independent of the bacteria strain with a 50-mM copper salt concentration. The results were similar to those in the literature in terms of the obtained biocidal effect but without a specific effectiveness percent reported.

A study led by Zuniga J. [58] focused on developing a low-cost upper limb prosthesis for the human finger, and shoulder prosthesis [59] using 3D Printing technology because using a customized prosthesis can help improve bi-manual activities and unilateral activities in terms of the functions of grasping and holding. For the developed shoulder prosthesis, manually adjustable options increased the wrist movements, elbow extensions, and shoulder rotations. For this study, PLACTIVE™ with 1% antibacterial nanoparticle additive was purchased as the ready-made 3D Printing material. PLACTIVE™ is a pure high-grade type of polylactic acid (PLA), which has copper NPs incorporated as the key material that provides the antibacterial effect. Copper NPs are highly effective at killing bacteria as well as other pathogens, including fungi and viruses. The customized prosthesis was designed and developed using extrusion and FFF. The results showed that the surface of the developed prosthesis was effective against 99.99% to *S. aureus* and *E. coli*. The study also stated that the antibacterial properties of the 3D Printing PLA filament were not altered after addition of the NPs or after the extrusion, which supports during the positive post-processing modifications needed for customization of any type of prostheses. From this study, they found that using PLACTIVE™ in 3D Printing technology could provide a simple inexpensive method to develop customizable patient-specific prostheses and potentially medical devices.

As discussed earlier, metals, such as zinc, copper, and silver, are extensively used to provide antibacterial properties to pure polymeric matrices. In a study conducted by Muwaffak et al. [42], polycaprolactone (PCL) was used to produce filaments for 3D Printing to develop patient-specific wound healing dressings. Wound dressings are used to provide protection from contamination and can be used to deliver topical bioactive agents in the form of ointments and creams to aid faster healing. In that study, silver-loaded filaments, copper-loaded filaments, and zinc-loaded filaments were prepared using dissolution in suitable solvents and PCL. The obtained filaments were chopped and sent into a hot-melt extruder to increase the homogeneity between the polymer matrix and filler content. The study reported Ag-PCL and Cu-PCL having the highest bactericidal effects against the *S. aureus* bacteria strain with increasing amounts resulting in a more potent inhibition compared to Zn-PCL. As 3D scanning was used to develop the patient-specific 3D model for wound dressing, this approach is a novel technique to tailor patient-specific antibacterial wound dressings.

Dental restorations with potential antibacterial properties have been one of the leading branches of biomedicine that has incorporated 3D Printing technology. In a study conducted by Yue et al. [41], positively charged quaternary ammonium-based salts were grafted onto the surface of a material to kill bacteria on contact. These charged groups interact with the cell wall of the bacteria and break their lipid membrane, releasing cytoplasmic elements. The frame component was chosen to be diurethane metacrylate (UDMA) because it can be rapidly cross-linked, along with camphorquinone (CQ) as the photosensitizer, ethyl 4-(dimethylamino)benzoate (EDMAB) as the co-initiator, and glycerol dimethacrylate (GDMA) as a diluent to decrease the viscosity. The antibacterial agent, which was the quaternary ammonium-based salt, was varied according to the alkyl chains from $n = 2$ to $n = 12$ (QA_Cn). The quaternary ammoniums groups were covalently added into the matrix by in situ copolymerization with the resin. FFF was used to 3D print a tooth model wherein all the layers fused well during photocuring. The antibacterial efficacy of the surface of the tooth model was observed before and after coating with a salivary conditioning film (pQA_C12). The results showed almost no presence of bacteria with increasing chain length. Moreover, incorporation of the QA_C12 within the polymer resulted in less bacteria contact killing than the copolymerized QA_C12 that was directly present in the resin. However, the use of antimicrobial resins is a promising approach for several dental applications, such as adhesives, cement, or dental restorations, with the use of high alkyl chain ammonium salts.

Table 1 shows a comprehensive comparison of selected parameters of the above-mentioned studies in terms of polymers chosen, particles, and chemical agents used to provide the antibacterial effect. Table 2 compares the 3D Printing parameters and tests done to develop the antibacterial materials. Additionally, a summary of the selected studies is shown in Table 3 along with the particles used, their antibacterial efficacy, and the chosen application.

Table 1. Parameters used in selected studies in the 3D printing of antimicrobial material.

Polymer Name	Particle Name	Particle Size	Weight/Volume %	Chemical Agent	Ref
PCL	Graphene oxide	Not specified	0%, 5%, and 7.5%,	Not specified	[53]
Thiol-ene-acrylate	QUA, SH-QUA	Not specified	2 wt%, 4 wt%, 6, wt%, 8 wt%, 10 wt%	phenylbis(2,4,6-trimethylbenzoyl), phosphineoxide photoinitiator, 1,2,3-benzenetriol	[19]
stereolithographic resins- UDMA	Ag-HNT	Nanoparticles	1%, 2%, 5%	triethylene glycol dimethacrylate (TEGDMA), photosensitizer	[55]
alginate beads	Copper nanoparticles	Nanoparticles	4 wt% alginate concentration; 50-mM copper salt concentration	Bacterial cellulose	[15]
PLA	Copper nanoparticles	Nanoparticles	1% antibacterial nanoparticle additive	-	[58]
PCL	Metal ions (Ag, Cu, Zn)	Not specified	Ag-10% w/w Cu-10% and 25% w/w Zn-10% w/w	Tetrahydrofuran (THF) dichloromethane (DCM) -	[42]
UDMA	QA_C$_{n\ (n=2\ to\ n=16)}$pQA_C$_{12}$	Not specified	Nitrogen (N%) linearly related to alkyl chain pQA-C$_{12}$–25 wt%	camphorquinone (CQ); ethyl 4-(dimethylamino)benzoate (EDMAB); glycerol dimethacrylate (GDMA); 2-hydroxyethyl methacrylate (HEMA) for pQA_Cn$_{12}$	2015 [41]

Table 2. Printing parameters used in selected studies.

Polymer Name	Nozzle Diameter	Filament Diameter/ Printer Settings	Printing Temperature	Filament Process	Tests Done	Reference
PCL	xy distances ranged from 200 μm to 400 μm, z-steps from 20 μm to 80 μm, and staggering between layers from 50 μm to 200 μm	Not specified	Not specified	internal diameter of 184 μm (28 G) for scaffold plotting	1,4,6,12,16	[53]
Thiol-ene-acrylate	Not specified	Not specified	Not specified	Not specified	1,2,9,14	[19]
stereolithographic resins	Not specified	Resolution of the device = 50 lm in the Z-direction; exposure time of each layer = 12 s	Not specified	Not specified	1,2,8,5,14,15,18	[55]

Table 2. *Cont.*

Polymer Name	Nozzle Diameter	Filament Diameter/ Printer Settings	Printing Temperature	Filament Process	Tests Done	Reference
alginate beads	1.75 mm	3D structures of $30 \times 30 \times 1$ mm^3 (length × width × height) 1.5 mm of thread spacing; dispensing head temperature of 25 °C; ink extruded with 23-G needle tip at 25 °C; printing speed of 50 mm s^{-1}; extrusion pressure of 1 bar	25 °C	Not specified	1,5,18	[15]
PLA	Not specified	40% infill (hexagon pattern); 50-mm/s print speed; 150–200-mm/s travel speed; 50 °C heated bed; 0.15-mm layer height; 1-mm shell thickness	200 °C	Not specified	1,13	[58]
PCL	1.75 mm	- square dressings (20 _ 20-1 mm) for antimicrobial studies and circular dressings (10-mm diameter; 1-mm thickness) −0.1 (mm layer height), with two shells, 100% infill and speed while extruding and while traveling was set to 50 mm/s	170 °C	Extruded/ Single screw Ag-80 °C Cu-60 °C Zn-75 °C	1,2,3,4,7,11,17	[41]
UDMA	Not specified	z-stage with the substrate moved upward by 200 μm; resolution of the device ~300 μm in the XY-plane and 25 μm in the Z-direction	Not specified	Not specified	1,7,12,14,15,19	[41]

1-Antibacterial testing, 2-TGA, 3-DSC, 4-SEM, 5-TEM, 6-Optical Microscopy, 7-FTIR, 8-EDS, 9-NMR, 10-3D Scanning, 11-3D Scanning, 12-XPS, 13-Box and block test, 14-Mechanical test, 15-Cytotoxicity test, 16-In vivo test, 17-Dissolution test, 18-UV, 19-Contact angle measurements.

Table 3. Comparison of antibacterial efficacies from the abovementioned recent studies.

Reinforcement Material	Antimicrobial Synthesis Method	Printing Method	Antimicrobial Activity (%)	Application	Reference
Graphene oxide	Dissolution	Wet spinning + AM	80%	Fibrous scaffolds	[53]
QUA, SH-QUA	Dissolution (Copolymerization)	DLP	100%	Dental tooth	[19]
Ag-HNT	Dissolution in SLR	DLP	99%	Dental composite resin	[55]
Copper nanoparticles	Ionic cross-linking	FFF	Not specified	Composite hydrogel	[15]
Copper nanoparticles	Purchased PLACTIVE©	Extrusion + FFF	99%	Finger prosthesis	[58]
Metal ions (Ag, Cu, Zn)	Hot-melt extrusion	FFF	Not specified	Wound dressing	[41]
QA_C$_n$ (n=2 to n=16) pQA_C$_{12}$	Dissolution (Copolymerization)	SLA	99%	Dental composite resin	[41]

3. Applications Using 3D-Printed Antibacterial Materials

3.1. Biomedical Field

Over the last 20 years, the drawbacks of current materials used in the medical field, mainly for bone applications, have led advances in the area, particularly toward developing a synthetic bone material alternative. Owing to the ease of experimentation, prototyping, and customized tailored output of 3D Printing, its use has significantly increased in bone, dental, and tissue restorations. 3D Printing has resulted in materials that are biocompatible, osteoconductive, and can result in the desired porous structure that is fully compatible with the growth of the surrounding capillaries and promotes drug delivery. The aim of any alternative bone material or scaffold is to provide mechanical support needed and over a prolonged time to become resorbed as new bone and aid the growth of surrounding tissues [16].

In a study by Roopavath et al. [60], hydroxyapatite (HA) scaffolds were printed using extrusion-based 3D Printing for bone grafting applications. The results indicated that the porosity and mechanical properties of the printed scaffolds could be controlled using 3D Printing technology. As the scaffolds can be printed on models based on computed tomography, the accuracy and precision of this approach inevitably depict it as a promising technique for bone implant developments. Recently, the FFF approach was tested with an applied ceramic coating of silicon nitride (Si_3N_4) on biomedical grade commercially available pure titanium (cp-Ti) via the laser sintering process. The Si_3N_4 coating was able to provide both the antibacterial and osteogenic properties of the bulk material to the cp-Ti substrate, resulting in properties that match current biomedical implants [61].

As discussed earlier, studies have reported copper compounds that have high antipathogenic properties; thus, their use in the medical field is quite extensive. As they can aid in developing low-cost medical devices with high antibacterial properties, copper particles are excellent alternatives to other compounds, such as silver, which in some cases has been reported to cause skin irritations [42]. Several studies on 3D-printed medical prostheses for the upper limbs, including arms, shoulders, and fingers, have stated the capability of 3D Printing to provide low-cost, customizable solutions with the possibility of incorporating the required antibacterial surfaces [59,62,63].

3D Printing is widely used in dental applications. In a study conducted by Yamada et al. [64], silver compounds were incorporated to develop dental prosthesis to provide an antibacterial effect to the surroundings, mainly preventing caries. Silver compounds were also studied by Sa et al. for the preparation of dental restorations with excellent antibacterial effects [55].

Wound dressing applications have also resulted from the 3D Printing approach. The use of a wound dressing will help isolate the injury and external surroundings and aid in faster healing by reducing tissue edema and promoting hemostasis. Topical agents combined with wound dressings are used to provide increased wound healing and to treat any buildup of local infections by minimizing their residence times [65,66]. A 3D-printed chitosan-pectin (CS-PEC) biopolymeric hydrogel wound dressing that incorporated lidocaine hydrochloride (LDC) has been studied [67]. These hydrogels were developed using cross-linking of CS and PEC, and the scaffolds were 3D printed using extrusion-based 3D Printing. The wound dressing exhibited good physical integrity, flexibility, and skin adhesion. Constant drug release was also observed, supporting the applicability of 3D Printing for use in wound dressing applications. Another study conducted by Muwaffak et al. involved 3D scanning to develop 3D models to customize the shape and size of a wound dressing. The extrusion of PCL pellets combined with zinc, copper, and silver metal particles of different loadings was developed. The results of the antibacterial property against the common *S. aureus* strain of bacteria of the developed wound dressing suggested that silver and copper wound dressings resulted in the most potent antibacterial properties. This simple 3D Printing customizable approach has excellent potential for developing wound dressings with added antimicrobial properties.

3.2. Use of Composite Materials in 3D Printing for Material Chemistry

As mentioned earlier, AM can deliver high-performance and multifaceted structures with improved properties that are unattainable by current conventional fabrication methods. The use of ceramics and fillers, such as silica, aluminum, and zirconium, are leading to rising developments in the field of advanced materials owing to their high thermal stability and high stability. Powdered ceramics have been widely incorporated into 3D Printing to provide enhanced properties, such as increased strength and porosity to the developed material. Several examples of 3D Printing technology in the field of material science and chemistry have been studied, such as in mass spectrophotometric techniques and fluorescence techniques [4,68]. The results support FFF, providing an excellent method to fabricate small mechanical parts with sufficient precision, which allows rapid modifications to be made during the process itself. A study focused on the modification of a low-cost open-source 3D printer into a 3D silica gel layer printer, which can be applied to thin layer chromatography. Using this technique, the potential of 3D Printing of adsorbents for analytical chemistry was studied. The results supported the fast, inexpensive production of the layers, opening new pathways for novel planar chromatographic separations [3]. A study was conducted to develop a novel silica–titania-based 3D-printed material, resulting in increased use of optical performance materials [69].

Another study also reported the production of high-quality fused silica glass by 3D Printing using a silica nanocomposite [70]. In that study, the ability to develop fused silica glass using silica nanocomposites showed the ability to produce complex shapes and structures on a macro and micro scale using 3D Printing technology. 3D-printed ceramic-based scaffolds have also gained wide interest in the medical industry [16,71]. For instance, the effects of the mechanical and biological properties of designed microporous scaffolds manufactured by commercial 3D Printing technology have been studied. The results showed the use of silica-based 3D-printed scaffolds as viable candidates for further research for clinical applications [16,17].

Another study by Li et al. reported the effects of Al_2O_3 dispersed in a UV-curable acrylic-based resin, which may be compatible with SLA-3D Printing of micro components with complex shapes [46]. Yang et al. reported a 3D Printing gel printing process that was based on a water-based gelation system that incorporates zirconia, which can be applied to develop complex parts for magnetic applications [72]. Rapid advancements in this field are expected owing to the ease of incorporating ceramic powders with polymers and their ability to control and vary the desired properties of the developed material.

3.3. Use of Metallic Fillers in Composites in 3D Printing Membrane Technology

The increased use of metallic fillers in composite materials has recently gained huge interest in the water treatment sectors to remove waste compounds or heavy metals owing to their increased chemical and thermal stability, higher permeation rate, and increased selectivity compared to polymeric sheet membranes. However, owing to the high costs involved in developing larger quantities of ceramic–polymer composite materials, there is a need to shift toward a cost-effective approach to develop them commercially that can be easily applied industrially, which could potentially be accomplished via 3D Printing to provide good antipathogenic surfaces.

Several techniques have been utilized to develop polymer composite materials such as spray-coating, spin-coating, and dip-coating via nano-filtration and ultra-filtration (UF). UF has also been extensively studied to treat wastewater because it can remove heavy colloidal substances; however, studies have shown that it does not have the ability to remove 1% oil by volume because the sizes of the oil droplets are below 10 μm [73,74]. Moreover, polyvinyl alcohol (PVA) and cellulose acetate as membranes have also been studied owing to their anti-fouling properties and biodegradability [75].

Recently, 3D Printing technology using ceramics and clays has also gained wide interest. Metal–organic frameworks (MOFs) are commonly used as adsorbing materials. SLS to print highly porous flow-through filters containing the MOF copper (II) benzene-

1,3,5-tricarboxylate (HKUST-1) has been studied. The printing did not have any negative impact on the structure of the MOF, thus supporting the use of 3D Printing technology for filter use [13]. Inject 3D Printing has also been employed in another case of ceramic filters for water treatment using clay as the raw material. That study reported that inkjet 3D Printing is an effective technology for forming clay ceramics that have the potential for use as microfiltration membranes, for example, as pre-filtration media to improve water quality. Another study also reported the use of clay to develop filter material using 3D Printing because it is a cost-effective approach for treating waste from water. The study showed that excellent filtration was achieved, thus increasing the water quality [7]. Additionally, as metallic particles are proven to provide high antibacterial efficacy, there may be the added benefit of killing bacteria and treating water.

In another study, microporous silica-based composite material was developed to study the effective removal of Sr-90 and Cs-137 (both are high heat emitting nuclides) from acidic high-level liquid waste, which is produced in the reprocessing of nuclear spent fuel. As these ions can be used for energy generation, it is vital to study the effective removal of such nuclides [76]. Despite the challenges involved in the removal of such highly beneficial ions, silica has been used as a desirable filler in a polymer composite material owing to its high thermal stability and selectivity to Sr-90, thus showing the versatility of its application for treating liquid wastes. Another study highlighted the ability of a silica-based composite material to recover gold and palladium from $AuCl_3$ and $AgCl_2$ in acidic aqueous solutions [77]. Moreover, polymer/silica/amine fibers have been reported to show improved adsorption of CO_2 from flue gas compared with currently used functional fibers [78]. Thus, silica-based polymer composites have been of significant interest because they have the capability to recover precious metals, remove toxic materials, and be incorporated with 3D Printing. Silica has also been proven to provide added antibacterial properties, which can be very useful for water treatment.

3.4. Space Applications

3D Printing has recently also been accepted for use in space applications. 3D Printing in the consistent microgravity environment of the International Space Station (ISS) can be used to develop parts made from particular materials needed in space. For example, the first 3D-printed tool on the ISS was a wrench built using FFF by depositing 104 layers of plastic [79]. The ability to manufacture 3D-printed parts and tools on demand will dramatically reduce the time it takes to get parts to orbit and increase the reliability and safety of space missions while reducing costs. Current space missions take months to years to get parts to orbit [8]. Moreover, FFF with extrusion-based machines functions similarly in microgravity as it does on the ground, allowing for a fool-proof concept of 3D Printing in space [10].

The extrusion-based fused filament fabrication method using thermoplastics represents a robust and straightforward methodology that is applicable to printing parts for both current and future human spaceflight exploration missions [80]. Literature reports have indicated that polymeric-based materials have potential as space-capable materials because they can produce extravehicular activity, repair tools, and even satellite structures that can be created on-site and on-demand. This will enable safer, less mass-intensive missions and scientific experiments, provide an advantage to FFF [81]. Among the most commonly used thermoplastic polymer materials are acrylonitrile butadiene styrene and PLA because these materials have been reported to maintain their shape at cabin temperatures [80].

According to the database maintained by NASA on the ISS, 17% of the total spare parts are made of metallic materials. It has been reported that powder bed fusion and, in particular, laser beam melting (LBM) are the preferred FFF technologies for commercially producing ready-to-use high-performance metallic parts with desired size specifications [82]. A stainless steel metal powder was successfully processed in the LBM process under microgravity (μ-g) conditions in a parabolic flight campaign, which showed no significant deviations from a part manufactured at 1 g, thus supporting the feasibility of the LBM process for 3D Printing ready-to-use metal parts in space.

3D Printing systems have been advancing in the medical field as well. A bioprinting system for the ISS has already been developed. This technique allows 3D Printing of thick tissue and organs using adult stem cells, supporting that 3D bioprinting can result in exact movements and control of the final dispensation [83]. Advances in the medical field could be a great breakthrough for the space industry. The literature also shows that NPs of various metals and metal oxides, such as zinc oxide nanoparticles (ZnONPs), cuprous oxide nanoparticles (CuONPs), silver nanoparticles (AgNPs), copper (I) iodide nanoparticles (CuINPs), gold nanoparticles on silica nanoparticles (Au-SiO2NPs) and some quaternary ammonium cations, which are commonly called QUATs, are very promising for inactivating viruses. The increasing use of polymer composite materials worldwide in a variety of fields promotes filler addition (whether metallic or non-metallic), which significantly enhances the properties such as mechanical, thermal, and conductivity properties [17,26–28]. One of the most novel strategies developed is the incorporation of copper NPs into the polymer matrix. Copper has strong virucidal properties, and copper's neutralization of infectious bronchitis virus, poliovirus, human immunodeficiency virus type 1 (HIV-1), and other enveloped or nonenveloped single- or double-stranded DNA or RNA viruses has been well reported [29]. Based on the wide range antiviral potency of copper, the incorporation of copper NPs in an intended composite material could inherently add significant value to the final 3D-printed product, especially with regard to providing antipathogenic surfaces, such as medical equipment that requires high purity for use by astronauts in a safer and more efficient way. Additionally, another mineral that has caught the attention of researchers worldwide is silica. Owing to its high chemical and thermal stability and good antimicrobial properties, silica has been used in several applications to provide enhanced properties to composite materials, thus making it a good candidate for the final 3D-printed product.

Investigations into the properties of regolith materials [84] (material deposited on the soils of planets) are also being conducted via powder bed fusion-based 3D Printing, which show the recent advancements in this field. Nevertheless, developing such particulate matter is not easy or efficient. Perhaps already developed materials, such as composite materials, would provide a better alternative to investigate further. However, owing to the high costs involved to develop larger quantities of polymer composite material, there is a need to shift toward a cost-effective approach to develop them commercially, which can be easily applied industrially and could potentially be advanced using 3D Printing.

Moreover, because thermoplastics have the ability to be recycled as feedstock and then printed again, using recycled materials as printer feedstocks could be a viable approach. One goal is to utilize non-metallic composite materials (recycled fiber-based composite materials) that could provide an effective long-term solution to the buildup of trash in space and its effective disposal [85]. Currently, the in-space manufacturing program conducted by NASA is focused on evolving manufacturing technologies from Earth-reliant to Earth-independent technologies. This is directed toward future hardware fabrication via FFF, which includes feedstock recycling, the development of parts, printable electronics, and investigations into additive manufacturing of metallics and external repairs [80,86].

Furthermore, it is noteworthy to state that due to the sudden rise in the COVID-19 pandemic, extensive researches worldwide have been conducted focusing on the production and development of antimicrobial material just in the span of a year [87–90]. More and more focus has been provided on the fabrication of antimicrobial surfaces with the incorporation of polymeric base along with additives [18,46,91–93]. The positive results from such antimicrobial studies have been promising and are providing hope to fight against such potent diseases.

4. Gaps, Challenges, Opportunities and Future Trends

The ease of customization of devices which can be developed with the minimum processing time makes this technology the preferred alternative for tailoring specific shapes and tools [54]. Moreover, since polymer science is bound to advance in the medical field, the rapid prototyping approach allowed by the 3D Printing technology results in control over the developed product with the desired shape and properties [94]. However, one of the critical challenges of 3D Printing is the alignment of the composite fibers used. Research is underway to improve this. One of the solutions developed to tackle this issue is using continuous fiber-reinforced with thermoplastics, which can be implemented for large-scale industrial independent processes [95,96]. Another challenge is that the influence of parameters can be material dependent, and this interconnection requires further investigation to fully develop complex materials via 3D Printing. 3D Printing of antibacterial substances is still a relatively new field, and there is still a lack of a database on modeling of this method. However, this technique has considerable potential to analyze composite structures and further improve enhancements of 3D Printing technology with slight modifications. Complex fields, such as membrane technology and space technology, that incorporate 3D Printing with the addition of antibacterial materials to provide high-purity tools and devices promote the use of 3D Printing and its increasing potential and reliability for such fields in the future. Moreover, adopting antimicrobial material through 3D Printing technology opens the doors for a wide range of applications, especially the medical sector, such as producing assistive devices [97,98]. In addition, due to the COVID-19 pandemic crisis, the antimicrobial material can be used to fight the spread of the virus of the possible contaminated buttons instead of replacing the exist system [99]. Moreover, the FFF is an excellent, affordable, traditional technology [100] that can be incorporated with antimicrobial material with different natural material to produce different engineering components [101,102]. Eventually, using antimicrobial material through the 3D printing technology must be evaluated [103] to predict the final product properties [104] as well as to explore the possibilities of the failure while being in service [105] through experimental investigation, and even could be achieved theoretically through simulation [106], especially we are dealing with composite reinforced with different types and material of particulates. Eventually, using antimicrobial material could be not feasible from the economic view point especially if used for bulky objects or parts, but antimicrobial material could be efficiently used by using the FFF technology to apply it on the top surface of the object that is potentially subjected to the contamination and this could be optimized using multilayered sandwich components [107,108] to reduce the quantity consumed.

5. Conclusions

3D Printing is one the most promising technologies in the field of fabrication and synthesis and has received considerable attention owing to its ability to incorporate antibacterial agents within the structure. Metallic ions, such as copper and silver, are mainly being considered to provide high antibacterial effects in several 3D-developed products. Quaternary ammonium-based salts have also shown high antibacterial potency. The utilization of 3D Printing technology results in faster, compact materials with increased accuracy and enhanced properties as desired. With the increasing rise of 3D Printing in several fields, the potential of is use as an alternative approach is expected. However, because this approach is relatively new, its growth is progressing slowly, and questions remain regarding the reliability of the technique compared with existing well-developed methods that are currently being used.

Author Contributions: Conceptualization, software, methodology, project administration, funding acquisition; validation; writing—review and editing, W.A.; formal analysis, S.S.; investigation, S.S.; resources, A.H.A.-M., S.S., W.A.; data curation, S.S.; writing—original draft preparation, S.S.; visualization; supervision, A.H.A.-M.; All authors have read and agreed to the published version of the manuscript.

Funding: This research was funded by UAEU, grant number G00003334.

Institutional Review Board Statement: Not applicable.

Informed Consent Statement: Not applicable.

Data Availability Statement: The data presented in this study are available on request from the corresponding author.

Conflicts of Interest: The authors declare no conflict of interest.

References

1. Kudzin, M.H.; Mrozińska, Z. Biofunctionalization of Textile Materials. 2. Antimicrobial Modification of Poly(Lactide) (PLA) Nonwoven Fabricsby Fosfomycin. *Polymers* **2020**, *12*, 768. [CrossRef]
2. Biniaś, D.; Biniaś, W.; Machnicka, A.; Hanus, M. Preparation of Antimicrobial Fibres from the EVOH/EPC Blend Containing Silver Nanoparticles. *Polymers* **2020**, *12*, 1827. [CrossRef]
3. Fichou, D.; Morlock, G.E. Open-Source-Based 3D Printing of Thin Silica Gel Layers in Planar Chromatography. *Anal. Chem.* **2017**, *89*, 2116–2122. [CrossRef] [PubMed]
4. Prikryl, J.; Foret, F. Fluorescence Detector for Capillary Separations Fabricated by 3D Printing. *Anal. Chem.* **2014**, *86*, 11951–11956. [CrossRef]
5. Chauhan, I.; Chattopadhyay, S.; Mohanty, P. Fabrication of titania nanowires incorporated paper sheets and study of their optical properties. *Mater. Express* **2013**, *3*, 343–349. [CrossRef]
6. Silva, C.; Bobillier, F.; Canales, D.; Sepúlveda, F.A.; Cament, A.; Amigo, N.; Rivas, L.M.; Ulloa, M.T.; Reyes, P.; Ortiz, J.A.; et al. Mechanical and Antimicrobial Polyethylene Composites with CaO Nanoparticles. *Polymers* **2020**, *12*, 2132. [CrossRef] [PubMed]
7. Hwa, L.C.; Uday, M.; Ahmad, N.; Noor, A.M.; Rajoo, S.; Bin Zakaria, K. Integration and fabrication of the cheap ceramic membrane through 3D printing technology. *Mater. Today Commun.* **2018**, *15*, 134–142. [CrossRef]
8. NASA. Space Tools on Demand NASA OK. Available online: https://www.nasa.gov/sites/default/files/files/3D_Printing-v3.pdf (accessed on 12 May 2020).
9. Chou, Y.-C.; Lee, D.; Chang, T.-M.; Hsu, Y.-H.; Yu, Y.-H.; Chan, E.-C.; Liu, S.-J. Combination of a Biodegradable Three-Dimensional (3D)—Printed Cage for Mechanical Support and Nanofibrous Membranes for Sustainable Release of Antimicrobial Agents for Treating the Femoral Metaphyseal Comminuted Fracture. *J. Mech. Behav. Biomed. Mater.* **2017**, *72*, 209–218. [CrossRef] [PubMed]
10. Snyder, M.; Dunn, J.; Gonzalez, E. The Effects of Microgravity on Extrusion Based Additive Manufacturing. In Proceedings of the Conference and Exposition, American Institute of Aeronautics and Astronautics, San Diego, CA, USA, 10 September 2013.
11. Werkheiser, M.J.; Dunn, J.; Snyder, M.P.; Edmunson, J.; Cooper, K.; Johnston, M.M. 3D Printing in Zero-G ISS Technology Demonstration. In Proceedings of the Conference and Exposition, American Institute of Aeronautics and Astronautics, San Diego, CA, USA, 31 August 2014.
12. Mills, D.K.; Jammalamadaka, U.; Tappa, K.; Weisman, J. Studies on the Cytocompatibility, Mechanical and Antimicrobial Properties of 3D Printed Poly(Methyl Methacrylate) Beads. *Bioact. Mater.* **2018**, *3*, 157–166. [CrossRef]
13. Lahtinen, E.; Precker, R.L.M.; Lahtinen, M.; Hey-Hawkins, E.; Haukka, M. Selective Laser Sintering of Metal-Organic Frameworks: Production of Highly Porous Filters by 3D Printing onto a Polymeric Matrix. *ChemPlusChem* **2019**, *84*, 222–225. [CrossRef]
14. Ahmed, W.; Siraj, S.; Alnajjar, F.; Al Marzouqi, A.H. 3D Printed Implants for Joint Replacement. In *Applications of 3D Printing in Biomedical Engineering*; Sharma, N.R., Subburaj, K., Sandhu, K., Sharma, V., Eds.; Springer: Singapore, 2021; pp. 97–119.
15. Gutierrez, E.; Burdiles, P.A.; Quero, F.; Palma, P.; Olate-Moya, F.; Palza, H. 3D Printing of Antimicrobial Alginate/Bacterial-Cellulose Composite Hydrogels by Incorporating Copper Nanostructures. *ACS Biomater. Sci. Eng.* **2019**, *5*, 6290–6299. [CrossRef] [PubMed]
16. Fielding, G.A.; Bandyopadhyay, A.; Bose, S. Effects of silica and zinc oxide doping on mechanical and biological properties of 3D printed tricalcium phosphate tissue engineering scaffolds. *Dent. Mater.* **2012**, *28*, 113–122. [CrossRef]
17. Attaran, S.A.; Hassan, A.; Wahit, M.U. Materials for food packaging applications based on bio-based polymer nanocomposites. *J. Thermoplast. Compos. Mater.* **2017**. [CrossRef]
18. Mania, S.; Ryl, J.; Jinn, J.-R.; Wang, Y.-J.; Michałowska, A.; Tylingo, R. The Production Possibility of the Antimicrobial Filaments by Co-Extrusion of the PLA Pellet with Chitosan Powder for FDM 3D Printing Technology. *Polymers* **2019**, *11*, 1893. [CrossRef] [PubMed]
19. Li, Z.; Wang, C.; Qiu, W.; Liu, R. Antimicrobial Thiol–ene–acrylate Photosensitive Resins for DLP 3D Printing. *Photochem. Photobiol.* **2019**, *95*, 1219–1229. [CrossRef] [PubMed]
20. Musić, S.; Filipović-Vinceković, N.; Sekovanić, L. Precipitation of amorphous SiO2 particles and their properties. *Braz. J. Chem. Eng.* **2011**, *28*, 89–94. [CrossRef]
21. Falagas, M.; Kasiakou, S. Mesh-related infections after hernia repair surgery. *Clin. Microbiol. Infect.* **2005**, *11*, 3–8. [CrossRef] [PubMed]
22. Sharma, G.; Sharma, S.; Sharma, P.; Chandola, D.; Dang, S.; Gupta, S.; Gabrani, R. Escherichia coli biofilm: Development and therapeutic strategies. *J. Appl. Microbiol.* **2016**, *121*, 309–319. [CrossRef]
23. Galdbart, J.O.; Allignet, J.; Tung, H.S.; Rydèn, C.; El Solh, N. Screening for Staphylococcus Epidermidis Markers Discriminating between Skin-Flora Strains and Those Responsible for Infections of Joint Prostheses. *J. Infect. Dis.* **2000**, *182*, 351–355. [CrossRef]
24. Aljohani, W.; Ullah, M.W.; Li, W.; Shi, L.; Zhang, X.; Yang, G. Three-dimensional printing of alginate-gelatin-agar scaffolds using free-form motor assisted microsyringe extrusion system. *J. Polym. Res.* **2018**, *25*, 62. [CrossRef]
25. González-Henríquez, C.M.; Sarabia-Vallejos, M.A.; Hernandez, J.R. Antimicrobial Polymers for Additive Manufacturing. *Int. J. Mol. Sci.* **2019**, *20*, 1210. [CrossRef] [PubMed]

26. Corrêa, A.C.; De Santi, C.R.; Manrich, S. Synthetic Paper from Plastic Waste: The Effect of $CaCO_3$ on Physical, Surface Properties and Printability. *Macromol. Symp.* **2006**, *245–246*, 611–620. [CrossRef]

27. Krupa, I.; Cecen, V.; Boudenne, A.; Prokeš, J.; Novák, I. The mechanical and adhesive properties of electrically and thermally conductive polymeric composites based on high density polyethylene filled with nickel powder. *Mater. Des.* **2013**, *51*, 620–628. [CrossRef]

28. Lee, D.W.; Yoo, B.R. Advanced silica/polymer composites: Materials and applications. *J. Ind. Eng. Chem.* **2016**, *38*, 1–12. [CrossRef]

29. Borkow, G.; Sidwell, R.W.; Smee, D.F.; Barnard, D.L.; Morrey, J.D.; Lara-Villegas, H.H.; Shemer-Avni, Y.; Gabbay, J. Neutralizing Viruses in Suspensions by Copper Oxide-Based Filters. *Antimicrob. Agents Chemother.* **2007**, *51*, 2605–2607. [CrossRef] [PubMed]

30. Cevik, M.; Bamford, C.; Ho, A. COVID-19 pandemic—A focused review for clinicians. *Clin. Microbiol. Infect.* **2020**, *26*, 842–847. [CrossRef] [PubMed]

31. Park, S.; Ko, Y.-S.; Jung, H.; Lee, C.; Woo, K.; Ko, G. Disinfection of Waterborne Viruses Using Silver Nanoparticle-Decorated Silica Hybrid Composites in Water Environments. *Sci. Total Environ.* **2018**, *625*, 477–485. [CrossRef] [PubMed]

32. Trewyn, B.G.; Whitman, C.M.; Lin, V.S.-Y. Morphological Control of Room-Temperature Ionic Liquid Templated Mesoporous Silica Nanoparticles for Controlled Release of Antibacterial Agents. *Nano Lett.* **2004**, *4*, 2139–2143. [CrossRef]

33. Ferraris, M.; Perero, S.; Miola, M.; Ferraris, S.; Verné, E.; Morgiel, J. Silver nanocluster–silica composite coatings with antibacterial properties. *Mater. Chem. Phys.* **2010**, *120*, 123–126. [CrossRef]

34. Jiang, Z.; Demir, B.; Broughton, R.M.; Ren, X.; Huang, T.S.; Worley, S.D. Antimicrobial silica and sand particles functionalized with anN-halamine acrylamidesiloxane copolymer. *J. Appl. Polym. Sci.* **2016**, *133*. [CrossRef]

35. Zuniga, J.M.; Thompson, M. Applications of antimicrobial 3D printing materials in space. *J. 3D Print. Med.* **2019**, *3*, 5–9. [CrossRef]

36. McCarthy, R.R.; Ullah, M.W.; Pei, E.; Yang, G. Antimicrobial Inks: The Anti-Infective Applications of Bioprinted Bacterial Polysaccharides. *Trends Biotechnol.* **2019**, *37*, 1155–1159. [CrossRef] [PubMed]

37. McCarthy, R.R.; Ullah, M.W.; Booth, P.; Pei, E.; Yang, G. The use of bacterial polysaccharides in bioprinting. *Biotechnol. Adv.* **2019**, *37*, 107448. [CrossRef]

38. Ullah, M.W.; Manan, S.; Khattak, W.A.; Shahzad, A.; Ul-Islam, M.; Yang, G. Biotemplate-Mediated Green Synthesis and Applications of Nanomaterials. *Curr. Pharm. Des.* **2020**, *26*, 5819–5836. [CrossRef]

39. Palza, H. Antimicrobial Polymers with Metal Nanoparticles. *Int. J. Mol. Sci.* **2015**, *16*, 2099–2116. [CrossRef]

40. Kirsh, I.; Frolova, Y.; Beznaeva, O.; Bannikova, O.; Gubanova, M.; Tveritnikova, I.; Romanova, V.; Filinskaya, Y. Influence of the Ultrasonic Treatment on the Properties of Polybutylene Adipate Terephthalate, Modified by Antimicrobial Additive. *Polymers* **2020**, *12*, 2412. [CrossRef] [PubMed]

41. Yue, J.; Zhao, P.; Gerasimov, J.Y.; Van De Lagemaat, M.; Grotenhuis, A.; Rustema-Abbing, M.; Van Der Mei, H.C.; Busscher, H.J.; Herrmann, A.; Ren, Y. 3D-Printable Antimicrobial Composite Resins. *Adv. Funct. Mater.* **2015**, *25*, 6756–6767. [CrossRef]

42. Muwaffak, Z.; Goyanes, A.; Clark, V.; Basit, A.W.; Hilton, S.T.; Gaisford, S. Patient-specific 3D scanned and 3D printed antimicrobial polycaprolactone wound dressings. *Int. J. Pharm.* **2017**, *527*, 161–170. [CrossRef] [PubMed]

43. Huang, B.; Bártolo, P.J. Rheological characterization of polymer/ceramic blends for 3D printing of bone scaffolds. *Polym. Test.* **2018**, *68*, 365–378. [CrossRef]

44. Wang, Y.; Wang, S.; Zhang, Y.; Mi, J.; Ding, X. Synthesis of Dimethyl Octyl Aminoethyl Ammonium Bromide and Preparation of Antibacterial ABS Composites for Fused Deposition Modeling. *Polymers* **2020**, *12*, 2229. [CrossRef]

45. Yang, F.; Zeng, J.; Long, H.; Xiao, J.; Luo, Y.; Gu, J.; Zhou, W.; Wei, Y.; Dong, X. Micrometer Copper-Zinc Alloy Particles-Reinforced Wood Plastic Composites with High Gloss and Antibacterial Properties for 3D Printing. *Polymers* **2020**, *12*, 621. [CrossRef] [PubMed]

46. Brounstein, Z.; Yeager, C.M.; Labouriau, A. Development of Antimicrobial PLA Composites for Fused Filament Fabrication. *Polymers* **2021**, *13*, 580. [CrossRef] [PubMed]

47. Rezić, I.; Majdak, M.; Bilić, V.L.; Pokrovac, I.; Martinaga, L.; Škoc, M.S.; Kosalec, I. Development of Antibacterial Protective Coatings Active against MSSA and MRSA on Biodegradable Polymers. *Polymers* **2021**, *13*, 659. [CrossRef] [PubMed]

48. Bayarsaikhan, E.; Lim, J.-H.; Shin, S.-H.; Park, K.-H.; Park, Y.-B.; Lee, J.-H.; Kim, J.-E. Effects of Postcuring Temperature on the Mechanical Properties and Biocompatibility of Three-Dimensional Printed Dental Resin Material. *Polymers* **2021**, *13*, 1180. [CrossRef] [PubMed]

49. Shaqour, B.; Reigada, I.; Górecka, Ż.; Choińska, E.; Verleije, B.; Beyers, K.; Święszkowski, W.; Fallarero, A.; Cos, P. 3D-Printed Drug Delivery Systems: The Effects of Drug Incorporation Methods on Their Release and Antibacterial Efficiency. *Materials* **2020**, *13*, 3364. [CrossRef]

50. Ramos, M.; Fortunati, E.; Beltrán, A.; Peltzer, M.; Cristofaro, F.; Visai, L.; Valente, A.J.M.; Jiménez, A.; Kenny, J.M.; Garrigós, M.C. Controlled Release, Disintegration, Antioxidant, and Antimicrobial Properties of Poly (Lactic Acid)/Thymol/Nanoclay Composites. *Polymers* **2020**, *12*, 1878. [CrossRef] [PubMed]

51. Puertas-Bartolomé, M.; Mora-Boza, A.; García-Fernández, L. Emerging Biofabrication Techniques: A Review on Natural Polymers for Biomedical Applications. *Polymers* **2021**, *13*, 1209. [CrossRef] [PubMed]

52. Navaruckiene, A.; Bridziuviene, D.; Raudoniene, V.; Rainosalo, E.; Ostrauskaite, J. Influence of Vanillin Acrylate-Based Resin Composition on Resin Photocuring Kinetics and Antimicrobial Properties of the Resulting Polymers. *Materials* **2021**, *14*, 653. [CrossRef]

53. Melo, S.F.; Neves, S.C.; Pereira, A.T.; Borges, I.; Granja, P.L.; Magalhães, F.D.; Gonçalves, I.C. Incorporation of Graphene Oxide into Poly(ε-Caprolactone) 3D Printed Fibrous Scaffolds Improves Their Antimicrobial Properties. *Mater. Sci. Eng. C* **2020**, *109*, 110537. [CrossRef] [PubMed]

54. Guerra, A.J.; Lammel-Lindemann, J.; Katko, A.; Kleinfehn, A.; Rodriguez, C.A.; Catalani, L.H.; Becker, M.L.; Ciurana, J.; Dean, D. Optimization of Photocrosslinkable Resin Components and 3D Printing Process Parameters. *Acta Biomater.* **2019**, *97*, 154–161. [CrossRef]

55. Sa, L.; Kaiwu, L.; Shenggui, C.; Junzhong, Y.; Yongguang, J.; Lin, W.; Li, R. 3D printing dental composite resins with sustaining antibacterial ability. *J. Mater. Sci.* **2019**, *54*, 3309–3318. [CrossRef]

56. Liang, J.; Li, J.; Zhou, C.; Jia, W.; Song, H.; Zhang, L.; Zhao, F.; Lee, B.P.; Liu, B. In situ synthesis of biocompatible imidazolium salt hydrogels with antimicrobial activity. *Acta Biomater.* **2019**, *99*, 133–140. [CrossRef] [PubMed]

57. Xu, C.; Dai, G.; Hong, Y. Recent advances in high-strength and elastic hydrogels for 3d printing in biomedical applications. *Acta Biomater.* **2019**, *95*, 50–59. [CrossRef] [PubMed]

58. Zuniga, J.M. 3D Printed Antibacterial Prostheses. *Appl. Sci.* **2018**, *8*, 1651. [CrossRef]

59. Zuniga, J.M.; Carson, A.M.; Peck, J.M.; Kalina, T.; Srivastava, R.M.; Peck, K. The development of a low-cost three-dimensional printed shoulder, arm, and hand prostheses for children. *Prosthetics Orthot. Int.* **2017**, *41*, 205–209. [CrossRef]

60. Ahmed, W.; Al-Douri, Y. Chapter 17: Three-Dimensional Printing of Ceramic Powder Technology. In *Metal Oxide Powder Technologies: Fundamentals, Processing Methods and Applications*; Al-Douri, Y., Ed.; Elsevier: Amsterdam, The Netherlands, 2020; pp. 351–383.

61. Zanocco, M.; Boschetto, F.; Zhu, W.; Marin, E.; McEntire, B.J.; Bal, B.S.; Adachi, T.; Yamamoto, T.; Kanamura, N.; Ohgitani, E.; et al. 3D-additive deposition of an antibacterial and osteogenic silicon nitride coating on orthopaedic titanium substrate. *J. Mech. Behav. Biomed. Mater.* **2020**, *103*, 103557. [CrossRef]

62. Zuniga, J.; Katsavelis, D.; Peck, J.; Stollberg, J.; Petrykowski, M.; Carson, A.; Fernandez, C. Cyborg beast: A low-cost 3d-printed prosthetic hand for children with upper-limb differences. *BMC Res. Notes* **2015**, *8*, 1–9. [CrossRef] [PubMed]

63. Young, K.J.; Pierce, J.E.; Zuniga, J.M. Assessment of body-powered 3D printed partial finger prostheses: A case study. *3D Print. Med.* **2019**, *5*, 7–8. [CrossRef]

64. Yamada, R.; Nozaki, K.; Horiuchi, N.; Yamashita, K.; Nemoto, R.; Miura, H.; Nagai, A. Ag Nanoparticle–Coated Zirconia for Antibacterial Prosthesis. *Mater. Sci. Eng. C* **2017**, *78*, 1054–1060. [CrossRef] [PubMed]

65. Zahedi, P.; Rezaeian, I.; Ranaei-Siadat, S.-O.; Jafari, S.-H.; Supaphol, P. A review on wound dressings with an emphasis on electrospun nanofibrous polymeric bandages. *Polym. Adv. Technol.* **2009**, *21*, 77–95. [CrossRef]

66. Thapa, R.K.; Diep, D.B.; Tønnesen, H.H. Topical antimicrobial peptide formulations for wound healing: Current developments and future prospects. *Acta Biomater.* **2020**, *103*, 52–67. [CrossRef]

67. Long, J.; Etxeberria, A.E.; Nand, A.V.; Bunt, C.R.; Ray, S.; Seyfoddin, A. A 3D printed chitosan-pectin hydrogel wound dressing for lidocaine hydrochloride delivery. *Mater. Sci. Eng. C* **2019**, *104*, 109873. [CrossRef]

68. Martínez-Jarquín, S.; Moreno-Pedraza, A.; Alonso, H.G.; Winkler, R. Template for 3D Printing a Low-Temperature Plasma Probe. *Anal. Chem.* **2016**, *88*, 6976–6980. [CrossRef] [PubMed]

69. Dudukovic, N.A.; Wong, L.L.; Nguyen, D.T.; Destino, J.F.; Yee, T.D.; Ryerson, F.J.; Suratwala, T.; Duoss, E.B.; Dylla-Spears, R. Predicting Nanoparticle Suspension Viscoelasticity for Multimaterial 3D Printing of Silica–Titania Glass. ACS Appl. *Nano Mater.* **2018**, *1*, 4038–4044. [CrossRef]

70. Kotz, F.; Arnold, K.; Bauer, W.; Schild, D.; Keller, N.; Sachsenheimer, K.; Nargang, T.M.; Richter, C.; Helmer, D.; Rapp, B.E. Three-dimensional printing of transparent fused silica glass. *Nat. Cell Biol.* **2017**, *544*, 337–339. [CrossRef] [PubMed]

71. Arango, M.A.T.; Kwakye-Ackah, D.; Agarwal, S.; Gupta, R.K.; Sierros, K.A. Environmentally Friendly Engineering and Three-Dimensional Printing of TiO2 Hierarchical Mesoporous Cellular Architectures. *ACS Sustain. Chem. Eng.* **2017**, *5*, 10421–10429. [CrossRef]

72. Yang, F.; Zhang, X.; Guo, Z.; Volinsky, A.A. 3D gel-printing of Sr ferrite parts. *Ceram. Int.* **2018**, *44*, 22370–22377. [CrossRef]

73. Li, S.; Zhang, Y.; Zhao, T.; Han, W.; Duan, W.; Wang, L.; Dou, R.; Wang, G. Additive manufacturing of SiBCN/Si3N4w composites from preceramic polymers by digital light processing. *RSC Adv.* **2020**, *10*, 5681–5689. [CrossRef]

74. Benito, J.; Ríos, G.; Ortea, E.; Fernández, E.; Cambiella, A.; Pazos, C.; Coca, J. Design and construction of a modular pilot plant for the treatment of oil-containing wastewaters. *Desalination* **2002**, *147*, 5–10. [CrossRef]

75. Chen, W.; Su, Y.; Zheng, L.; Wang, L.; Jiang, Z. The Improved Oil/Water Separation Performance of Cellulose Acetate-Graft-Polyacrylonitrile Membranes. *J. Membr. Sci.* **2009**, *337*, 98–105. [CrossRef]

76. Sciencedirect. Modification of a Novel Macroporous Silica-Based Crown Ether Impregnated Polymeric Composite with 1-Dodecanol and Its Adsorption for Some Fission and Non-Fission Products Contained in High Level Liquid Waste-Science Direct. Available online: https://www.sciencedirect.com/science/article/abs/pii/S0014305708003625 (accessed on 1 April 2020).

77. Neoh, K.; Tan, K.; Goh, P.; Huang, S.; Kang, E. Electroactive polymer–SiO 2 nanocomposites for metal uptake. *Polymers* **1999**, *40*, 887–893. [CrossRef]

78. Rezaei, F.; Lively, R.P.; Labreche, Y.; Chen, G.; Fan, Y.; Koros, W.J.; Jones, C.W. Aminosilane-Grafted Polymer/Silica Hollow Fiber Adsorbents for CO2Capture from Flue Gas. *ACS Appl. Mater. Interfaces* **2013**, *5*, 3921–3931. [CrossRef] [PubMed]

79. Harbaugh, J. Space Station 3-D Printer Builds Ratchet Wrench to Complete First Phase. Available online: http://www.nasa.gov/mission_pages/station/research/news/3Dratchet_wrench (accessed on 26 May 2020).

80. Cowley, A.; Perrin, J.; Meurisse, A.; Micallef, A.; Fateri, M.; Rinaldo, L.; Bamsey, N.; Sperl, M. Effects of variable gravity conditions on additive manufacture by fused filament fabrication using polylactic acid thermoplastic filament. *Addit. Manuf.* **2019**, *28*, 814–820. [CrossRef]
81. Space Assembly Pushes 3D Printing Boundaries—ProQuest. Available online: https://search-proquest-com.uaeu.idm.oclc.org/docview/2102339029?accountid=62373 (accessed on 17 May 2020).
82. Zocca, A.; Lüchtenborg, J.; Mühler, T.; Wilbig, J.; Mohr, G.; Villatte, T.; Léonard, F.; Nolze, G.; Sparenberg, M.; Melcher, J.; et al. Enabling the 3D Printing of Metal Components in M-Gravity. *Adv. Mater. Technol.* **2019**, *4*, 1900506. [CrossRef]
83. Michelle, J. Zero Gravity Bioprinter: Ready for the International Space Station. Available online: https://www.3dnatives.com/en/zero-gravity-bioprinter220820184/ (accessed on 17 May 2020).
84. Goulas, A.; Binner, J.G.; Harris, R.A.; Friel, R.J. Assessing extraterrestrial regolith material simulants for in-situ resource utilisation based 3D printing. *Appl. Mater. Today* **2017**, *6*, 54–61. [CrossRef]
85. Johnson, M. Solving the Challenges of Long Duration Space Flight with 3D Printing. Available online: http://www.nasa.gov/mission_pages/station/research/news/3d-printing-in-space-long-duration-spaceflight-applications (accessed on 18 May 2020).
86. Prater, T.J.; Bean, Q.A.; Beshears, R.D.; Rolin, T.D.; Werkheiser, N.J.; Ordonez, E.A.; Ryan, R.M.; Iii, F.E.L. *Summary Report on Phase I Results From the 3D Printing in Zero-G Technology Demonstration Mission*; NASA: Washington, WA, USA, 2016.
87. Mathew, E.; Gilmore, B.F.; Larrañeta, E.; Lamprou, D.A. Antimicrobial 3D Printed Objects in the Fight Against Pandemics. *3D Print. Addit. Manuf.* **2021**, *8*, 79–86. [CrossRef]
88. Čech Barabaszová, K.; Holešová, S.; Hundáková, M.; Kalendová, A. Tribo-Mechanical Properties of the Antimicrobial Low-Density Polyethylene (LDPE) Nanocomposite with Hybrid ZnO–Vermiculite–Chlorhexidine Nanofillers. *Polymers* **2020**, *12*, 2811. [CrossRef]
89. Zhang, Z.; Li, J.; Ma, L.; Yang, X.; Fei, B.; Leung, P.H.M.; Tao, X. Mechanistic Study of Synergistic Antimicrobial Effects between Poly (3-hydroxybutyrate) Oligomer and Polyethylene Glycol. *Polymers* **2020**, *12*, 2735. [CrossRef]
90. Strasakova, M.; Pummerova, M.; Filatova, K.; Sedlarik, V. Immobilization of Caraway Essential Oil in a Polypropylene Matrix for Antimicrobial Modification of a Polymeric Surface. *Polymers* **2021**, *13*, 906. [CrossRef]
91. Spirescu, V.; Chircov, C.; Grumezescu, A.; Andronescu, E. Polymeric Nanoparticles for Antimicrobial Therapies: An up-to-date Overview. *Polymers* **2021**, *13*, 724. [CrossRef] [PubMed]
92. Kupnik, K.; Primožič, M.; Kokol, V.; Leitgeb, M. Nanocellulose in Drug Delivery and Antimicrobially Active Materials. *Polymers* **2020**, *12*, 2825. [CrossRef]
93. Dong, P.; Feng, J.; Li, S.; Sun, T.; Shi, Q.; Xie, X. Synthesis, Characterization, and Antimicrobial Evaluation of Random Poly(ester-Carbonate)s Bearing Pendant Primary Amine in the Main Chain. *Polymers* **2020**, *12*, 2640. [CrossRef]
94. Culbreath, C.J.; Gaerke, B.; Taylor, M.S.; McCullen, S.D.; Mefford, O.T. Effect of Infill on Resulting Mechanical Properties of Additive Manufactured Bioresorbable Polymers for Medical Devices. *Materialia* **2020**, *12*, 100732. [CrossRef]
95. Mazzanti, V.; Malagutti, L.; Mollica, F. FDM 3D Printing of Polymers Containing Natural Fillers: A Review of their Mechanical Properties. *Polymers* **2019**, *11*, 1094. [CrossRef] [PubMed]
96. Parandoush, P.; Lin, D. A review on additive manufacturing of polymer-fiber composites. *Compos. Struct.* **2017**, *182*, 36–53. [CrossRef]
97. Alnajjar, F.; Alsaedi, N.F.N.; Ahmed, W.K. Robotic Gripping Assist. U.S. Patent Application No. 1049677B1, October 2019.
98. Alnajjar, F.; Umari, H.; Ahmed, W.K.; Gochoo, M.; Vogan, A.A.; Aljumaily, A.; Mohamad, P.; Shimoda, S. CHAD: Compact Hand-Assistive Device for enhancement of function in hand impairments. *Robot. Auton. Syst.* **2021**, *142*, 103784. [CrossRef]
99. Alnajjar, F.; Ahmed, W.; Gochoo, M. Touchless Elevator Keyboard System. U.S. Patent Application No. 10968073B1, April 2021.
100. Ahmed, W.; Alabdouli, H.; Alqaydi, H.; Mansour, A.; Al, K.H.; Al, J.H. 3D printer: A case study. In Proceedings of the 10th Annual International IEOM Conference, Dubai, United Arab Emirates, 10–12 March 2020.
101. Ahmed, W.; Alnajjar, F.; Zaneldin, E.; Al-Marzouqi, A.H.; Gochoo, M.; Khalid, S. Implementing FDM 3D Printing Strategies Using Natural Fibers to Produce Biomass Composite. *Materials* **2020**, *13*, 4065. [CrossRef] [PubMed]
102. Ahmed, W.; Siraj, S.; Al-Marzouqi, A.H. 3D Printing PLA Waste to Produce Ceramic Based Particulate Reinforced Composite Using Abundant Silica-Sand: Mechanical Properties Characterization. *Polymers* **2020**, *12*, 2579. [CrossRef] [PubMed]
103. Mansour, A.; Alabdouli, H.; Alqaydi, H.; Al, K.H.; Ahmed, W.; Al, J.H. Evaluating the 3D printing capabilities. In Proceedings of the 10th Annual International IEOM Conference, Dubai, United Arab Emirates, 10–12 March 2020.
104. Al Khawaja, H.; Alabdouli, H.; Alqaydi, H.; Mansour, A.; Ahmed, W.; Al Jassmi, H. Investigating the Mechanical Properties of 3D Printed Components. In Proceedings of the 2020 Advances in Science and Engineering Technology International Conferences (ASET), Dubai, United Arab Emirates, 11–26 March 2020; pp. 1–7.
105. Ahmed, W.; Zaneldin, E.; Kabbani, S. *Fracture Mechanics Performance of Through-Thickness Crack of Polymeric 3D Printed Components*; Metzler, J.B., Ed.; Springer: Berlin/Heidelberg, Germany, 2021; pp. 269–279.
106. Ahmed, W.; Al-Rifaie, W.; Zaneldin, E. Chapter 4—Mathematical modeling and simulation of interfaces between fiber and its matrix. In *Micro and Nano Technologies, Fiber-Reinforced Nanocomposites: Fundamentals and Applications*; Elsevier: Amsterdam, The Netherlands, 2020; pp. 91–99.

107. Ahmed, W.; Ahmed, S.; Alnajjar, F.; Zaneldin, E. Mechanical performance of three-dimensional printed sandwich composite with a high-flexible core. *Proc. Inst. Mech. Eng. Part L J. Mater. Des. Appl.* **2021**, 14644207211011729. [CrossRef]
108. Ahmed, W. Product and method to manufacture multi-layered, multi-material composite sandwich structure with hyper elasticity rubber like core made by fusion deposition modeling. U.S. Patent Application No. US10974444B1, April 2021.

polymers

Article

Effect of Chemically Treated Kenaf Fibre on Mechanical and Thermal Properties of PLA Composites Prepared through Fused Deposition Modeling (FDM)

Aida Haryati Jamadi [1], Nadlene Razali [1,2,*], Michal Petrů [3,*], Mastura Mohammad Taha [2,4], Noryani Muhammad [1,2] and Rushdan Ahmad Ilyas [5,6]

[1] Fakulti Kejuruteraan Mekanikal, Universiti Teknikal Malaysia Melaka, Melaka 76100, Malaysia; aidaaaharyati@gmail.com (A.H.J.); noryani@utem.edu.my (N.M.)

[2] Centre for Advanced Research on Energy, Universiti Teknikal Malaysia Melaka, Melaka 76100, Malaysia; mastura.taha@utem.edu.my

[3] Faculty of Mechanical Engineering, Technical University of Liberec, Studentská 2, 46117 Liberec, Czech Republic

[4] Fakulti Teknologi Kejuruteraan Mekanikal dan Pembuatan, Universiti Teknikal Malaysia Melaka, Melaka 76100, Malaysia

[5] School of Chemical and Energy Engineering, Faculty of Engineering, Universiti Teknologi Malaysia, Johor Bahru 81310, Malaysia; ahmadilyas@utm.my

[6] Centre for Advanced Composite Materials (CACM), Universiti Teknologi Malaysia, Johor Bahru 81310, Malaysia

* Correspondence: nadlene@utem.edu.my (N.R.); michal.petru@tul.cz (M.P.)

check for **updates**

Citation: Jamadi, A.H.; Razali, N.; Petrů, M.; Taha, M.M.; Muhammad, N.; Ilyas, R.A. Effect of Chemically Treated Kenaf Fibre on Mechanical and Thermal Properties of PLA Composites Prepared through Fused Deposition Modeling (FDM). *Polymers* **2021**, *13*, 3299. https:// doi.org/10.3390/polym13193299

Academic Editors: Emin Bayraktar and Markus Gahleitner

Received: 8 September 2021
Accepted: 22 September 2021
Published: 27 September 2021

Publisher's Note: MDPI stays neutral with regard to jurisdictional claims in published maps and institutional affil-iations.

Abstract: Natural fibre as a reinforcing agent has been widely used in many industries in this era. However, the reinforcing agent devotes a better strength when embedded with a polymer matrix. Nevertheless, the characteristic of natural fibre and polymer matrix are in contrast, as natural fibre is hydrophilic, while polymer is hydrophobic in nature. Natural fibre is highly hydrophilic due to the presence of a hydroxyl group (-OH), while polymer matrix has an inherent hydrophobic characteristic which repels water. This issue has been fixed by modifying the natural fibre's surface using a chemical treatment combining an alkaline treatment and a silane coupling agent. This modifying process of natural fibre might reduce the attraction of water and moisture content and increase natural fibre surface roughness, which improves the interfacial bonding between these two phases. In this paper, the effect of alkaline and silane treatment has been proven by performing the mechanical test, Scanning Electron Micrograph (SEM), and Fourier Transform Infrared spectrometry (FTIR) to observe the surface structure. The chemical compositions and thermal properties of the composites have been obtained by performing Differential Scanning Calorimetry (DSC) and Thermogravimetric Analysis (TGA) tests. 1.0% silane treatment displayed better strength performance as compared to other composites, which was proven by performing Scanning Electron Micrograph (SEM). The assumption is that by enduring chemical treatment, kenaf fibre composites could develop high performance in industry applications.

Keywords: kenaf fibre; fibre treatment; mechanical properties; thermal properties; Fused Deposition Modelling (FDM)

1. Introduction

Nowadays, attention of engineers and professionals has been triggered regarding the increased consumption of petroleum and the depletion of these sources. In addition, the emission of harmful gas into the environment and the greenhouse effect during incineration produced an alternative in the development and sustainability of natural polymer compos-ites [1,2]. Aerospace, automotive, and construction industries have widely used advanced polymer composites, which contain carbon and glass fibre as the primary materials [3]. It was found that these primary materials are hardly reusable and reutilised [3]. Therefore, natural fibre has been introduced to replace the consumption of petroleum-based and synthetic fibres. Other than that, the characteristics between natural fibres and synthetic fibres are quite similar, such as low density, high stiffness, and good mechanical properties [1]. In comparison to the characteristics of other fibres such as synthetic, glass, and carbon, natural fibre [4] shows an advantage in biodegradability, renewability, non-toxicity, CO_2 neutral life cycle, degradablity, sustainability, and environmentally friendliness [2,3,5–10]. Table 1 shows the properties of natural fibre compared to other synthetic fibre.

Table 1. Fibre characteristic values for tensile strength (MPa), Young's Modulus (GPa), elongation (%) and density (g/cm^3) [11–15].

Fibre	Tensile Strength (MPa)	Young's Modulus (GPa)	Elongation (%)	Density (g/cm^3)
Cotton	287–800	5.5–12.6	3.0–10.0	1.5–1.6
Jute	393–800	10.0–30.0	1.16–1.8	1.3–1.6
Flax	345–1500	27.6	1.2–3.2	1.4–1.5
Hemp	550–900	70.0	1.6–4.0	1.47–1.48
Sisal	400–700	9.0–38.0	2.0–14	1.33–1.5
E-glass	2000–3500	70.0–73.0	2.5–3.4	2.50–2.55
Carbon (standard)	3400–4800	230–425	1.4–1.8	1.4–1.78
Kenaf	930	53.0	1.6	1.2–1.45
PALF	170–1627	60.0–82.5	1.6–2.4	1.56

Next, besides greenhouse protection, advantages of natural fibre properties include less machine wear during processing, no health hazards, a high degree of flexibility, low cost, its light weight, ease of separation, fracture resistance, high sound absorption, less equipment abrasion, low respiratory irritation, vibration damping, enhanced energy recovery, and good thermal insulation [1,3,5–7,9,16,17], all of which encourage researchers and industrial engineers to use natural fibre as a main material in development [4]. Mechanical characteristics of natural fibre have high specific benefits such as stiffness, impact resistance, modulus, strength, and durability [9,16,18]. These characteristics of natural fibres are well utilised in industrial product applications such as windows, frames, door panels, railroad sleepers, furniture, automotive dashboards and brake linings, shelves, egg-boxes, electronics packaging, textiles, and also building material applications [2,7]. Plant, animal, and mineral fibres are classified under the natural fibres division [5]. The most famous natural fibres used in the industrial platform is plant fibre, also known as lignocellulosic or cellulosic fibre. Pineapple (PALF), kenaf, coir, sisal, jute, hemp, flax, ramie and wood are examples of natural fibres that are popular in the applications platform and are usually used as reinforcing agents when combined with a biodegradable or non-biodegradable polymer matrix [2,3,7,19]. Kenaf fibre (KF), also known as *Hibiscus cannibus* L. [2,10], is one of the excellent substitutes for synthetic fibres [7]; it is also bio-based and available in tropical countries such as Thailand and Malaysia [20]. The kenaf plant demand is high because it grows rapidly which sustains its availability, and it is also low in cost [2,7,21]. Kenaf bast is a popular fibre used by researchers and scientists due to its outstanding mechanical properties such as flexural and tensile strength [10]. In addition to kenaf bast, flax and hemp are also popular natural fibre choices. [16]. As the pollution rate and environmental conditions consistently worsen, natural fibre, as a common material used in biodegradable products has been proposed by researchers, scientists, and engineers for application in industrial platforms such as the automotive, aerospace, aircraft, marine, and packaging industries [5,8,17]. Natural fibres are lightweight and have good mechanical properties [22]. Currently, most industries have resorted to the use of plastics [23,24]. In advanced applications, thermoplastic polymers are widely used, but due to their disadvantages which are lower in thermal stability and strength, some applications might not be applicable [24]. One of the renewable and biodegradable base polymers in the polyester group is Polylactic Acid (PLA) [21,22,25,26], which emits less CO_2 gas and shows that this material is not harmful to the greenhouse, humans, and animals [22]. A number of reactive groups, which mostly contain biopolymers offer excellent composite blends between natural fibres and matrix polymers [26].

The hydrophilic property is innate in natural fibres such as kenaf. This property states that due to the presence of a hydroxyl group (-OH) in their cellulose structure as (-OH) exists in natural fibre's structure, the moisture content might increase gradually [1,4,7,16,27]. This moisture content effect can cause swelling in structure and instability of dimension and also can lead to cracking [10]. This characteristic also has a disadvantage, which is that it could affect the adhesion bonding between the fibre and a polymer matrix, thereby producing unsatisfactory test results [16,19]. It could, in fact, contribute to low mechanical properties, low strength and short life span due to low interphase bonding between the fibre and polymer matrix [5,6]. Regarding these hydrophilic and hydrophobic issues, researchers state that one solution that might work is enhancement adhesion bonding between two phases and also improving the mechanical properties of the biocomposites by applying a surface modification [28]. M. Shirazi et al. (2019) stated that alkaline treatment was the most popular treatment in the surface modification process, and required the immersion of natural fibre in a sodium hydroxide solution within a certain timeframe [20,21]. This process could also remove impurities such as wax and oil from natural fibres and increase its surface roughness [2,9,21,24,26]. Some research experiments suggest surface modification will endure the alkaline treatment and silane coupling agent. In order to improve the wettability of natural fibres by polymer matrix and promote interfacial bonding, methods such as applying a coupling agent are used [8]. Silane is an example of a coupling agent in surface modification that shows excellent treatment, and importantly can improve interlocking adhesion between the fibre and polymer matrix better than other treatments [1,29,30]. It also interacts with chemical bonds of natural fibre and polymer matrix.

M. Asim et. al. (2016) conducted an experiment regarding surface treatment between kenaf fibre and PALF composites. In this experiment, data were collected from four different parameters: untreated fibre, alkaline-treated fibre, alkaline–silane-treated fibre, and silane-treated fibre. From the researchers' observation, by enduring alkaline treatment, all the impurities in fibre can be removed completely, depending on alkali concentration and soaking time. By performing surface treatment, enhancement of strength in composites could occur [7].

Overall, the production of natural fibre is a new issue that has been introduced by many researchers. Natural fibre-reinforced polymer biocomposites using environmentally friendly FDM technology has attracted many industries and researchers. The implementation of natural fibres in the filament of FDM to replace the current fillers has attracted many competitors and market platforms [31]. The most popular polymer that acts as the main material in FDM is acrylonitrile butadiene styrene (ABS). However, the use of a thermoplastics polymer as the main material for FDM is still not recommended. The important elements of a polymer are its mechanical properties, which are strength and stiffness. As previously stated, the mechanical aspects of many bio-based polymers have been investigated to enhance the technology of FDM. Acrylonitrile butadiene styrene (ABS) and polylactic acid (PLA) are popular because they are stable. The most frequent thermoplastic that had been produced in this technology is PLA. The advantages of using PLA are that it is recyclable, biodegradable and has a temperature of 145–160 °C [32]. PLA is one of the biopolymers that is obtained from the fermentation of the recyclable product and has good mechanical properties such as tensile strength and low thermal stability that prevents crystallisation [33]. As previously stated, PLA is getting attention as a biodegradable and renewable plastic. It is also environmentally friendly, and the study of natural fibres such as hemp and kenaf as reinforcement combined with PLA using a standard method has also been done [25]. The fibre loading optimisation and also the chemical treatment of the reinforcement can affect the mechanical properties of the product. Therefore, the natural fibre that combines with the PLA is firm and requires dried feedstock and storage [25].

In this paper, the authors studied the treatment of kenaf fibre with a NaOH concentration of 6% for 24 h, followed by the chemical treatment of a silane coupling agent with three different concentrations, 0.5%, 1%, and 2%, respectively for 3 h to modify the surface characterisation of the natural fibre. This paper aims to investigate the effect of chemically treated kenaf fibres on mechanical and thermal properties of kenaf fibre-reinforced PLA composites. The effect of alkaline and silane treatment for surface modification has also been studied.

2. Materials and Methods

2.1. Materials

Kenaf fibre powder (unsieved) was supplied locally from Lembaga Kenaf dan Tembakau Negara (LKTN) before being treated and mixed with Poly Lactic Acid (PLA) pellets. Poly Lactic Acid (PLA), Silane (Aminopropyltriethoxysilane Agent) was obtained from Mecha Solve Engineering (Selangor, Malaysia).

2.2. Methodology

2.2.1. Alkaline Treatment

In this experiment, kenaf fibre powder with random size (unsieved) within 100–650 µ were treated with alkaline treatment. The kenaf fibres were immersed in sodium hydroxide solution with a fixed concentration of 6% for 24 h [20]. After alkaline treatment, the kenaf fibres were washed thoroughly with running water and dried in an oven at a temperature of 110 °C for 24 h.

2.2.2. Silane Treatment

Surface treatment is then followed with silane coupling agent method. In this treatment, 0.5%, 1%, and 2%, respectively of APS (aminopropyltriethoxy silane) was dissolved in a solution which contained 70% methanol and 30% water. Next, the solution was stirred for 30 min. Then, the kenaf fibre which had already endured the alkaline treatment soaked in silane solution for 3 h and dried in an oven at a temperature of 110 °C for 24 h to remove all the fibre's moisture content.

Samples have been classified depending on three different types of silane concentration, neat polymer, and untreated kenaf fibre as tabulated in Table 2.

Table 2. Sample classification.

Parameter	Explanation
PLA	Neat polymer
Untreated	Untreated kenaf fibre composites
0.5% silane	0.5 wt % silane concentration + 6% alkali concentration kenaf fibre composites
1.0% silane	1.0 wt % silane concentration + 6% alkali concentration kenaf fibre composites
2.0% silane	2.0 wt % silane concentration + 6% alkali concentration kenaf fibre composites

2.2.3. Composite Mixture

The kenaf fibre and polymer matrix were prepared using the law of mixture formula as per in Table 3. To obtain the composition of composites, weight of elements has been calculated using Equation (1).

$$\text{Weight Percentage of Element, we} \times \text{Weight of composites} = \text{Weight of elements} \quad (1)$$

Table 3. Composition of composites.

Samples	Weight of Composites (g)	Weight of Fibre (g) 2.5 wt %	Weight of Matrix (g) 97.5 wt %
All samples	500	12.5	487.5

2.2.4. Extrusion of Filament

A twin screw extruder has been used to produce filament composites with different parameters as shown in Table 4.

Table 4. Parameter of extrusion.

Samples	Melting Temperature (°C)	Screw Speed (rpm)
PLA	210	25
Untreated fibre	190	29
0.5% silane	204	25
1.0% silane	204	25
2.0% silane	204	25

With a constant pulling speed of 25.5 rpm and constant filament size of 1.75 mm [34], the twin screw extruder is shown in Figure 1 while Figure 2 shows the biodegradable filament of kenaf fibre and neat PLA.

Figure 1. Twin screw extruder.

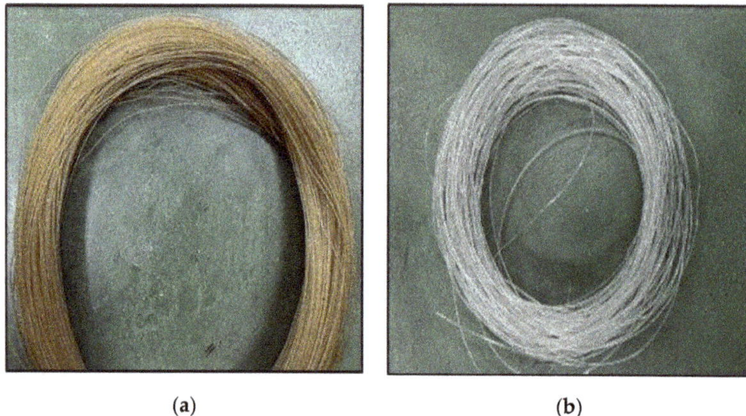

(a)   (b)

Figure 2. Kenaf fibre reinforced PLA composites: (**a**) Kenaf fibre PLA composites filament; (**b**) Neat polymer filament.

2.2.5. Sample Extrusion

The sample had been extruded using Flashforge 3D printing as illustrated in Figure 3. Next, there were several parameters that needed to be considered such as the temperature of the nozzle, the temperature of the bed, and the percentage of infill. The parameter that had been set up where the solid infill is set to 100% in-line shape. The shell's parameter is two layers, while the upper and bottom layer is three repeated numerical layers. Layer height was 0.18 mm, while first layer height is 0.27 mm. Next, nozzle temperature had been set up to 210 °C and bed temperature at 60 °C as PLA polymer does not require a high temperature. The printing speed also affected the performance of printed samples. In this printing process, the speed of the nozzle was 60 m/s while travel printing was 80 m/s. Figure 4 depicts the tensile and flexural specimen via 3D printing.

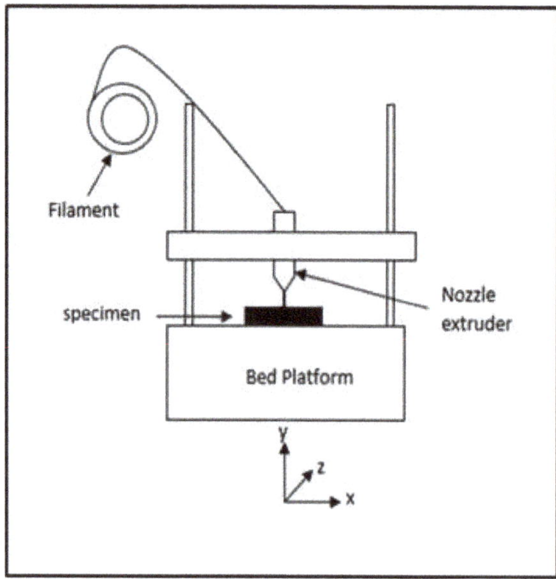

Figure 3. Schematic 3D printing process.

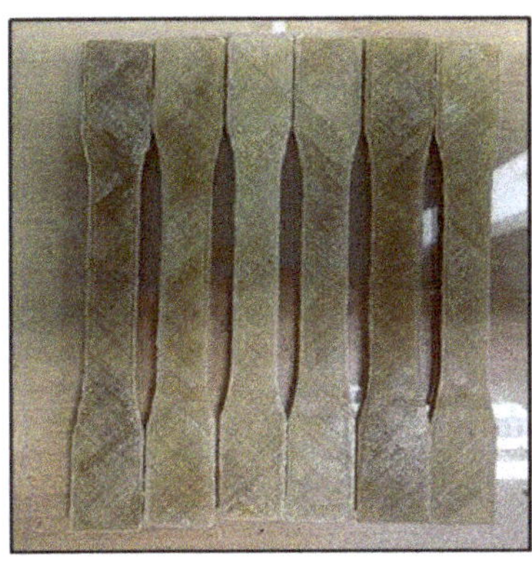
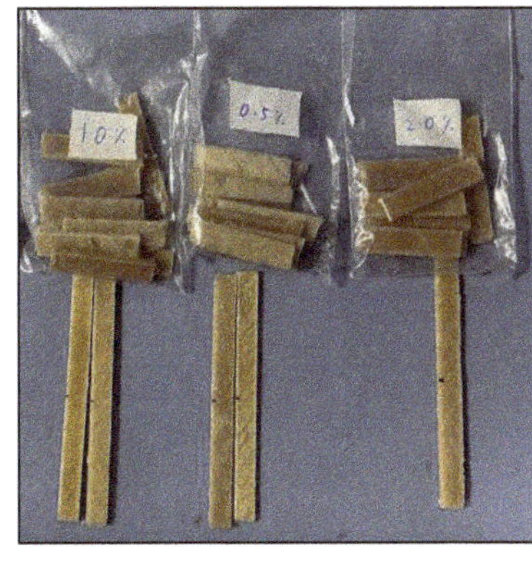

(a) (b)

Figure 4. (a) Tensile specimens (b) Flexural specimens.

3. Sample Characterisation

3.1. Mechanical Test

In order to evaluate the mechanical properties of the biodegradable composites, the tensile test was applied. Some of the properties that can be obtained after performing the tensile test include Young's Modulus, maximum elongation, tensile strain, and yield stress. The sample size is in a "dog bone shape", which is type 1 of the three listed types. For this research, the testing was carried out by following the ASTM D638 standard. By using this standard testing, the crosshead speed is 1 mm/min with a load cell of 5 kN with a span length of 50 mm. The tensile properties of composites were determined using the Universal Testing Machine model Instron 887, manufactured in Norwood, Massachusetts, United States.

The tensile strength of the single fibre can be calculated using Equation (2).

$$\sigma = \frac{F}{A} \tag{2}$$

where, σ is the tensile strength of the fibre (Pa), F is the maximum force at break (N), and A is the area of the cross section (m^2).

Using a three-point bending set up by following the ASTM D790 standard, the flexural test was conducted. Using this standard testing, the crosshead speed is 1 mm/min with a load cell of 5 kN. About five samples each from samples A, B, C, D, E, F, G, and H were taken and tested using the Universal Testing Machine model Instron 5585 manufactured in Norwood, Massachusetts, United States. The sample size is $100 \times 10 \times 3$ mm following the ASTM standard with a span length of 50 mm.

The flexural test of the single fibre can be calculated using Equation (3).

$$\sigma = \frac{3PL}{2bd^2} \tag{3}$$

where, σ is flexural strength of the fibre (Pa), P is maximum force at break (N), L is support span (mm), b is the width of the beam tested (mm), and d is the depth of the beam tested (mm).

3.2. Thermogravimetric Analysis (TGA)

The thermogravimetric analysis (TGA) was performed in order to obtain the degradation of the kenaf fibre under a high temperature before forming into composites. This analysis is conducted by using a machine from TA instruments and a filament specimen following the ASTM D3850 standard. The temperature rate was between 10 °C and 900 °C with a heating rate of 10 °C/min. The TGA was obtained using a Thermogravimetric Analyser located at Mettler-Toledo (M) Sdn. Bhd., Selangor, Malaysia.

3.3. Fourier Transform Infrared Spectrometry (FTIR)

The Fourier Transform Infrared Spectrometry (FTIR) was conducted using Jasco FT/IR-6100 (manufactured in the United States) on five different samples in an untreated powder state, at 0.5% silane, 1.0% silane, and 2.0% silane, in order to obtain the functional group for each different surface treatment. All spectra were recorded in the range of 4000 cm^{-1} to 400 cm^{-1}.

3.4. Differential Scanning Calorimetry (DSC)

Differential Scanning Calorimetry (DSC) was performed on five sample filaments in a nitrogen atmosphere; neat polymer, untreated, 0.5% silane, 1.0% silane and 2.0% silane. The temperature range is 10 °C up to 300 °C, with a heating rate of 25 °C. The Differential Scanning Calorimetry (DSC) was obtained using DSC Q20 V24.11 Build 124.

3.5. Morphological Analysis

For this research, morphological studies were performed in detail on the fractured surface of the tensile test sample using a Scanning Electron Microscope (SEM). The five different samples taken from the tensile specimen were tested; neat polymer with untreated composites at 0.5% silane, 1.0% silane and 2.0% silane were taken in 2.5 wt % of fibre loading. The samples were coated with platinum to get a better result of resolution as it offers good electrical conductivity. The micrograph was obtained by using a JSM-6010PLUS/LV Scanning Electron Microscope (Jeol Ltd., Tokyo, Japan).

4. Results and Discussion

4.1. Mechanical Test

Tensile and flexural strength were performed by mechanical testing and measured their strength and Young's Modulus [30].

Strength of composites might be influenced by several factors such as interfacial bonding. Good stress distribution could obtain good results [35].

Ververis et al. (2012) stated that tensile could be represented as one type of stress working in one direction (1-D). This testing could indicate whether the sample has good or bad in interfacial bonding. Maximum tensile strength, elastic modulus, and strain to failure can be obtained by enduring the tensile test [36].

In this experiment, the data in Figure 5 was obtained with three different classes which were neat polymer (PLA), untreated kenaf fibre, and treated kenaf fibre. Treated kenaf fibre used the process of immersion in alkali and silane. Three different concentrations of silane solution were used: concentrations 0.5%, 1.0% and 2.0%. Asim et al. (2016) and Oushabi et al. (2017) stated that alkaline treatment could remove the impurity, lignin, and hemicellulose of kenaf fibre and enhance the interfacial bonding between two phases (fibre and polymer), but the parameters that need to be considered are the concentration of alkali itself and the time of immersion. This paper also claimed that by enduring alkali and silane treatment, a good result in tensile strength could be obtained as compared to the untreated fibre [7].

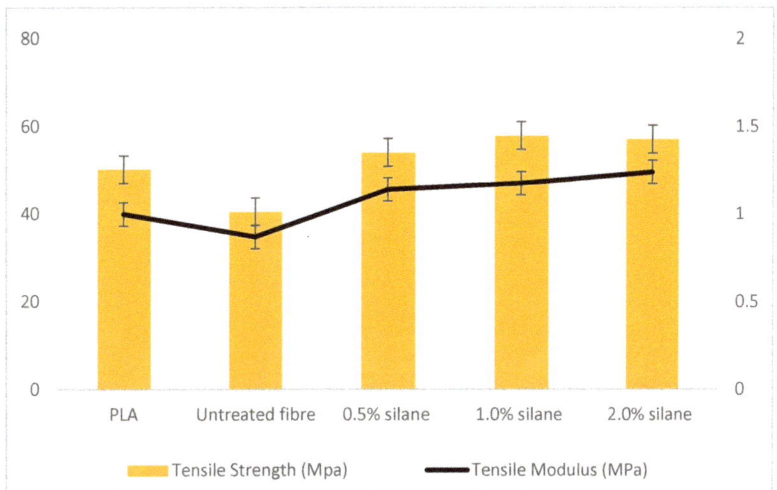

Figure 5. Results of composites tensile strength, MPa, and tensile modulus, MPa.

The tensile test was carried out following all the specific guidelines from ASTM D638 Standard. The graph bar above shows that the treated fibre indicates good strength as compared to the neat polymer and untreated fibre. It can be concluded that good mechanical properties were obtained due to surface treatment as compared to the untreated fibre [37]. Fibre orientation is not applicable in this experiment because the fibre used is in a powder state, known as isotropic, which means it doesn't have a specific orientation involved.

Next, the treated group using different types of silane concentration obviously shows that 1.0% (57.85 MPa) of silane treatment is the strongest as compared to 0.5% (54.01 MPa) and 2.0% (56.99 MPa) silane concentration, respectively. The obtained results showed that a 1% concentration of silane gave the optimum tensile strength as compared to the others. It also showed that introducing the fibres to PLA plastic had improved the mechanical properties. These results revealed that the removal of lignin and hemicellulose by enduring silane treatment showed good interfacial bonding between the matrix and fibres. The efficient removal of impurities might be done by using a higher concentration of silane but might degrade the tensile strength due to rupture of the surface fibre and degradation of the fibre chemical content [7]. Nevertheless, if the concentration is lower, the impurities might not be removed perfectly, and the strength of the composites was also affected due to the hydrophilic property of the fibre. In previous research in which optimum concentration was studied, it was reported that lower concentrations might not work efficiently [6].

Lee (2009) experimented by varying the concentration of silane and concluded that approximately 1% created optimal strength and binding of composite as compared to 3% and 5% [3]. Yucheng Liu [5] investigated corn stalk fibre-reinforced polymer composites by using four different types of silane concentration. From all the testing performed, this paper concludes that 1% is the most optimal as compared to other silane concentrations, as a higher concentration might affect the surface and reduce the special characteristic of the fibre itself. Another paper by Yucheng Liu [1] also observed the different type of concentrations on the results of a mechanical test involving the tensile, as well as an impact test. 1% silane concentration treatment is optimal according to the mechanical test, as compared to other concentrations. This is because silane is an acidic liquid, and if the usage is high, it will corrode the original structure and strength of the fibre [1]. The variation of concentration, time and effect of the surface fibre has been discussed by R. Mahjoub [12]. Mahjoub et. al. (2014) also concluded in their paper that the higher the concentration and immersion time, the greater the decrement in the breaking strength of fibre.

In addition, by referring to the literature review, one might say that the condition of composites might also be affected during the manufacturing process. The difference between treated and untreated fibre composites might be due to the surface cleaning, as treated fibre promotes better adhesion bonding between the two phases and increases the strength of composites [2].

Flexural testing was done in order to determine the strength and the ability of the material to resist the deformation under loading before reaching the break point [38]. This technique to evaluate and obtain modulus elasticity in bending and flexural stress as the material was set up as supported beam under two supports with load applied at a point [39].

Flexural testing observes whether the composites could withstand bending load and deformation before they fail [36]. A flexural test was carried out following all the specific guidelines from the ASTM D790 standard. Three categories have been set up in this flexural test which are neat polymer, untreated, and treated fibre, using silane in different concentrations. The data in Figure 6 shows that 1.0% silane obtained high flexural strength (84.22 MPa), while the flexural modulus was the second highest as compared to 2.0% silane (82.74 MPa). Whereas the untreated composites achieved low strength (59.30 MPa) as compared to treated composites, but overall, PLA achieved the lowest strength(50.93 MPa). The treated fibre obtained good results, presumably due to the silane treatment which enhanced the interfacial bonding and had good dispersion stress while applying force.

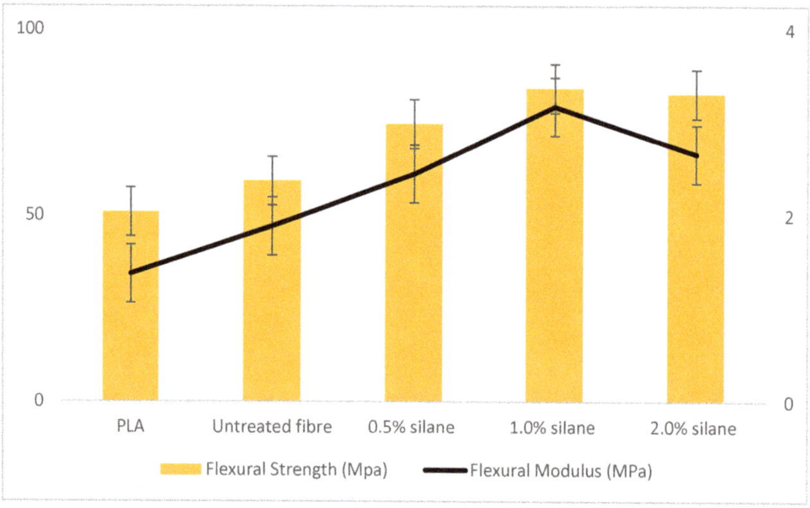

Figure 6. Results of composite flexural strength (MPa) and flexural modulus (MPa).

Regarding the flexural modulus, 1.0% silane indicated a high modulus (3174.76 Pa) as compared to other composites and neat polymers. This proved that 1.0% silane treatment is the optimal concentration for kenaf fibre composites. The untreated polymer had low data, which may be due to low interaction linkage between the fibre and polymer, or poor dispersion of fibre towards matrix which leads to weak load transfer and may be due to voids during manufacturing of the composites as compared to treated composites [30,38]. It can be concluded that the higher the strength of composites, the stronger the bonding between two phases (reinforced and polymer matrix). The good interlocking composites can also be achieved by enduring the chemical treatment with the optimal concentration [36]. This concluded that a good surface treatment parameter process can lead to good strength and elasticity while performing any test.

From the results obtained, it was found that the treatment enhanced the tensile properties of the composite samples due to the interfacial adhesion between the fibres and matrix. Figure 7a indicates the molecular structure of trialkoxysilane, which acts as a silane coupling agent for the kenaf fibre. In the presence of water (H_2O), the trialkoxysilane would develop the active agent of silanol in the reaction with the kenaf substrate. This silanol structure underwent a condensation process and naturally deposited on the kenaf surface to form a siloxane bond between the kenaf and silane coupling agent. This can result in a functional kenaf surface where the organofunctional groups can react with the PLA resin and produce better adhesion between the fibre and the matrix. Figure 7b illustrates how alkaline and silane treatment react towards natural fibre surfaces.

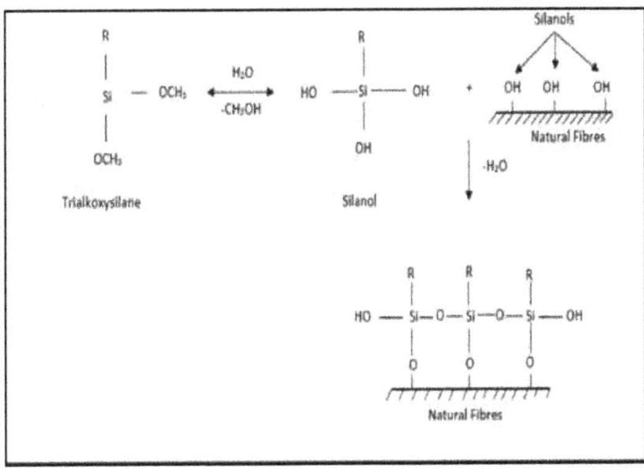

(a)

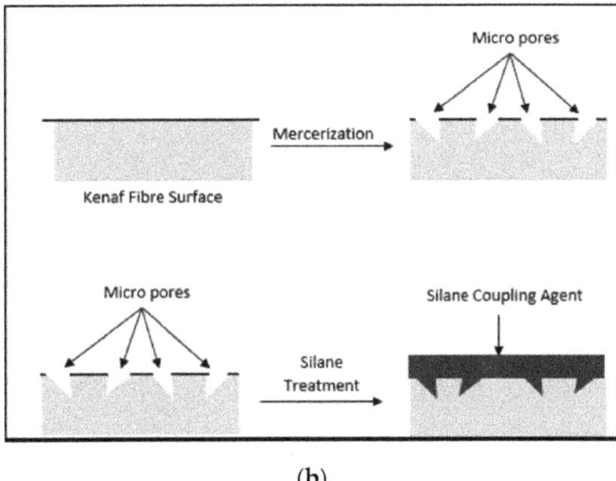

(b)

Figure 7. The overview reaction of surface modification (**a**) Silane reaction; (**b**) Mercerisation and silane treatment process.

4.2. Thermogravimetric Analysis (TGA)

Thermogravimetric analysis (TGA) and difference thermogravimetry (DTG) illustrated in Figure 8a,b in a nitrogen atmosphere is used to measure the thermal stability, thermal decomposition, and mass changes of PLA, untreated, and treated kenaf fibre composites. Due to properties of thermoplastics that allow them to be recycled and reused, thermal degradation is required to observe the degradation of composites at certain temperatures. This procedure can provide information on the composites capability to withstand high temperatures. TGA and DTG show the characteristic of composite degradation under nitrogen air. Five samples were conducted and started to degrade at certain temperatures.

Three phases of degradation had been stated by K. Krishna [40]. First, the degradation phase where the moisture content started to evaporate, followed by a second phase in which a high temperature was applied, the chemical content such as hemicellulose, cellulose, pectin and lignin was removed, and in the last stage, the final residue was below 10 wt %.

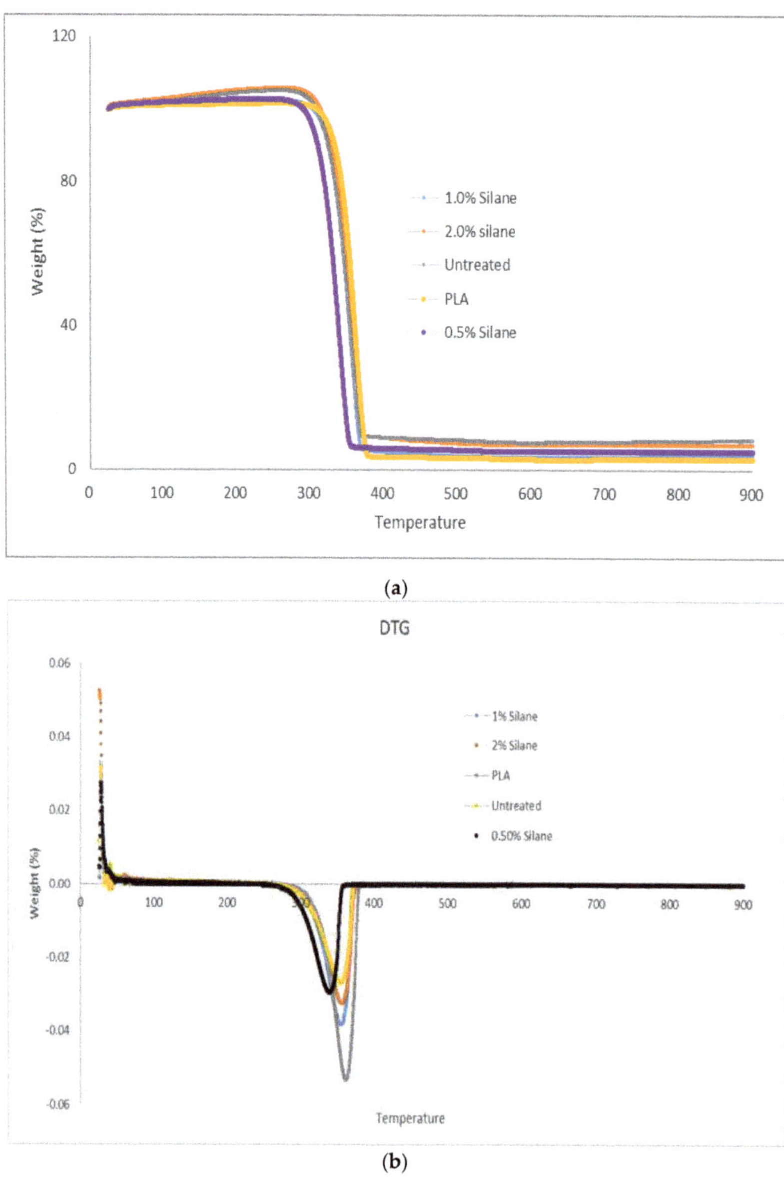

Figure 8. The results of composites (**a**) Thermogravimetric analysis (TGA); (**b**) Difference thermogravimetry (DTG).

In TGA analysis for kenaf fibre-reinforced PLA composites, the first phase degradation occurs at 10–300 °C. In this phase, fibres start to lose moisture as it is evaporated [41] and the weight loss in this stage is currently below 9 wt % [42]. At temperatures of 300 °C up to 400 °C is where the chemical content such as cellulose, hemicellulose, pectin, and lignin in fibre slowly degraded [40–42]. This is due to the high temperature applied towards kenaf fibre. Hemicellulose is the functional group that degraded first, followed by cellulose and lignin; cellulose is more stable than hemicellulose [40]. From the observation, 1.0% silane composites needed a high temperature to break the functional group. Lastly, the final phase was the remaining of composites after applying the maximum temperature.

The maximum temperature applied was 90 °C, leaving the weight residue below 10 wt %. From the obtained results, the main degradation temperature of the structural component of kenaf fibre was 300–400 °C. This means that it is suitable for the 3D printing application as the process temperature is around 160 °C to 210 °C. The process temperature depends on the types of filament. It is recommended to set the melting temperature of the material a bit higher to ensure all the filament is fully melted during the extrusion process [34].

Char residue is produced when cellulose is decomposed at a high temperature. This is evident in Figure 8a,b, where the percentage of char residue in treated fibre is lower in comparison to untreated fibre because the presence of lignin and cellulose which is not removed from untreated fibre. The result of TGA parallels previous researchers' findings which concluded that thermal stability may be improved through alkali treatment. In addition to that, the reduction in char residue content is a result of decrement in the formation of carbonaceous char.

4.3. Chemical Analysis by Using Fourier Infrared Spectrometry (FTIR)

Figure 9 and Table 5 show the FTIR spectra for neat polymer, untreated, and treated kenaf fibre-reinforced PLA composites. Cellulose, hemicellulose, and lignin are components that were detected in the FTIR spectrum [43]. For example, the group of C–O stretching from lignin indicated clearly at peaks of 1000–1300 cm^{-1} [7,43]. From the data, the untreated fibre shows fibre lignin at a peak of 1035 cm^{-1}, and 6% NaOH at a peak of 1030 cm^{-1}; for fibre that endured silane treatment (0.5%, 1.0%, 2.0%), the peak was at a range of 1029–1030 cm^{-1}. The decreasing trend of the wave number shows that the lignin was removed from the fibre.

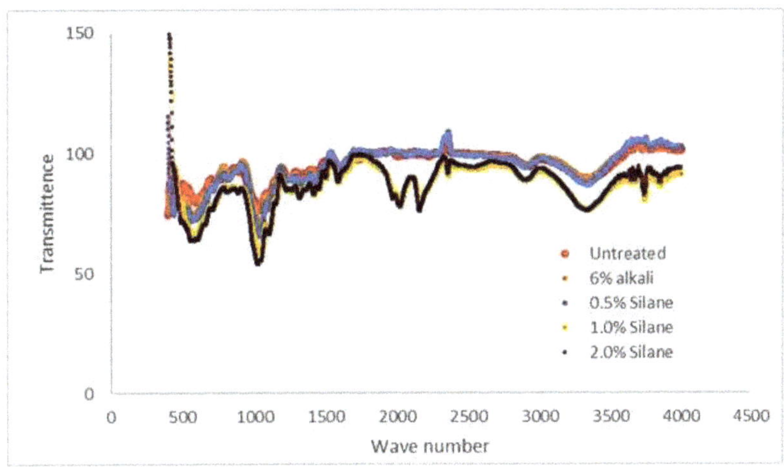

Figure 9. Fourier Transform Infrared Spectrometry (FTIR).

Table 5. Parameter of extrusion.

	Untreated	6% NaOH	0.5% Silane	1.0% Silane	2.0% Silane
Lignin	1035	1030	1029	1030	1029
Hemicellulose CH	2924 (2936–2916)	2899	2899	2902	2902
Cellulose CH$_2$	3308	3334	3333	3334	3335
Absorption of H$_2$O	1597	1499	1420	1421	1421
Hydroxyl Group -OH	3400–3200	3400–3200	3400–3200	3400–3200	3400–3200
Si-C Stretching Bond Silane	8958	826.3	813.8	900–700 (no peak)	900–700 (no peak)
Stretching N-H Vibration	3328–3250 (SYM_STR)	3400–3332 (SYM_STR)	-	-	-
Ester Carbonyl Group C=O	1727	1592	1593	1593	1593

Based on theoretical peak data, CH and -CH$_2$, which are hemicellulose and cellulose, exist at a range of 2858 cm^{-1} to 2926 cm^{-1}. Untreated fibre has shown that hemicellulose and cellulose exist at peaks of 2924 cm^{-1} and 3308 cm^{-1}, respectively. From the data we can see that fibres that treated with the alkaline treatment and 0.5% silane have no marked difference for hemicellulose, which is 2899 cm^{-1}, but for cellulose, NaOH and 1.0% silane indicated no difference where the peak is at 3334 cm^{-1}.

4.4. Differential Scanning Calorimetry (DSC)

Figure 10 and Table 6 compare the temperature value of neat polymer, untreated, and treated kenaf fibre-reinforced PLA composites. The values have been tabulated into Table 6. The terms exothermic and endothermic are the main keys in graph reading. Exothermic peak for the PLA polymer is the crystallisation temperature, while endothermic peak is at the melting temperature of 151.23 °C and the degradation temperature of 298.75 °C [44].

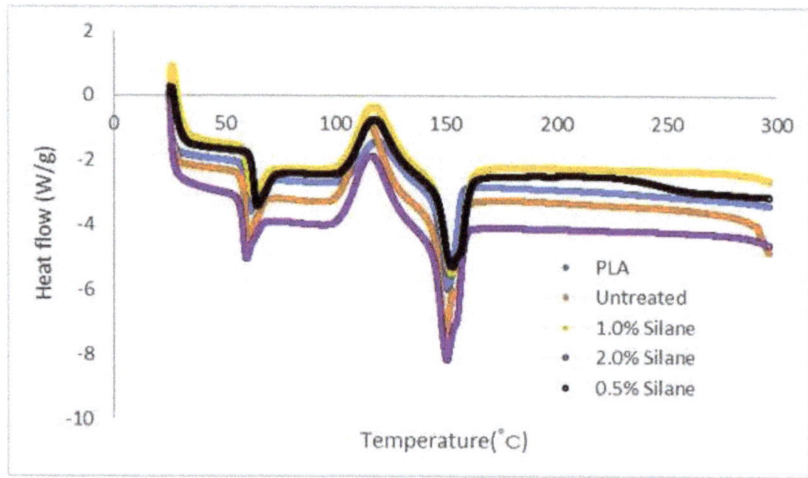

Figure 10. Differential Scanning Calorimetry (DSC).

Table 6. Differential Scanning Calorimetry (DSC) data.

	Thermal Properties		
Parameter	Tg (°C)	Tcc (°C)	T$_m$ (°C)
PLA	58.69	120.12	151.23
Untreated kenaf/PLA	58.06	115.36	149.61
0.5% silane kenaf/PLA	61.89	118.67	152.35
1.0% silane kenaf/PLA	59.32	118.29	152.87
2.0% silane kenaf/PLA	57.04	116.88	150.94

The neat PLA DSC curves and those of PLA composites with untreated, 0.5 % silane, 1.0% silane, and 2.0% silane of treated kenaf fibre composites indicate the glass transition temperature of PLA (58.69 °C), untreated (58.06 °C), at 0.5% silane (61.89 °C), 1.0% (59.32 °C), and 2.0% silane (57.04 °C). By setting the neat polymer crystal temperature benchmark (120.12 °C), the thermography shows that in untreated 0.5% silane, 1.0% silane, and 2.0% silane fibre composites, the PLA polymer chains did not crystallise completely, as illustrated in the temperature peaks of kenaf fibre composites. The crystallisation temperatures that had been achieved from the data are 115.36 °C, 118.67 °C, 118.29 °C, and 116.88 °C, respectively. The melting temperature is significantly different between composites and neat PLA. By referring to the table, the melting temperature of untreated (149.61 °C), 0.5% silane (152.35 °C), 1.0% silane (152.87 °C) and 2.0% silane (150.94 °C). The difference between each data for melting temperature of PLA composites is in the range of 1 °C, indicating that kenaf fibre does not interfere with the processing temperature. The temperature of degradation for each of the parameters circulated between 291.75 °C and 29.75 °C. This occurred because the PLA itself degraded the polymer chain and there was a loss of hydrogen elements after the rupture happened. The thermal properties of the filament are one of the important factors to be determined before the printing process. This is due to the need to add thermal properties of the filament as an input parameter during the printing process. If the thermal energy is not adequate during the process, it will affect the quality of the samples and lower the mechanical properties. The exact melting temperature will also help the uniform distribution between fibre and polymer in the extrusion process and avoid the problem of a clogged nozzle [34].

4.5. Morphological Analysis

Scanning Electron Microscope (SEM) is a method to review the detail of a kenaf fibre-reinforced PLA surface. This test was performed to check the adhesion bonding between treated and untreated fibre composites after enduring the tensile test. Figure 11 shows the SEM micrographs ($\times$500) of the kenaf fibre composites.

Figure 11a indicates the smooth neat polymer surface defects. From the observations, it has been proven that PLA resins have a ductile manner as compared to untreated and treated fibres, which are brittle. Figure 11b of untreated kenaf clearly shows that fibre pull-out occurred, and as proven, a hole exists on the surface of the composites. This case happened due to the impurity of untreated fibre that get affected because of weak interfacial bonding between natural fibres and polymers (2). As a result, when load was applied, the fibre could not withstand and pulled off from the grip of the matrix. As a comparison to the treated fibre shown in Figure 11c–e, with three different concentrations of silane, it can be seen that there is still fibre left. Figure 11c, which was treated with 0.5% of silane shows impurities still exist on the composite surface. In conclusion, 0.5% silane cannot remove impurities as well as 1.0% and 2.0% silane. The fibre left on the surface proves that the bonding between the fibre and the polymer occurs perfectly; thereby the dispersion of load towards composites is distributed equally. Therefore, 1.0% silane and 2.0% silane of treated fibre achieved a high result in strength as compared to other composites, but results show that 1.0% silane is the most optimal concentration for silane.

Torrado et al. (2014) stated that the use of silane in addition to surface modification after the alkaline treatment improved some of the minor factors such as dispersion and adhesion of the reinforcement and polymer matrix [45], and the SEM micrograph in Figure 12 has been compared with Figure 11 in silane treatment towards the fibre surface. In past experiments, Petchwattana et al. [32] stated that a silane coupling agent enhanced the bonding interaction between hydrophilic wood flour and hydrophobic PLA polymers [46,47]. Figure 12 shows the bonding between fibre and PLA polymer, untreated and treated with a silane coupling agent. Figure 12a,c show poor interfacial adhesion between the untreated fibre and polymer, while Figure 12b,d show good interfacial bonding with treated fibre.

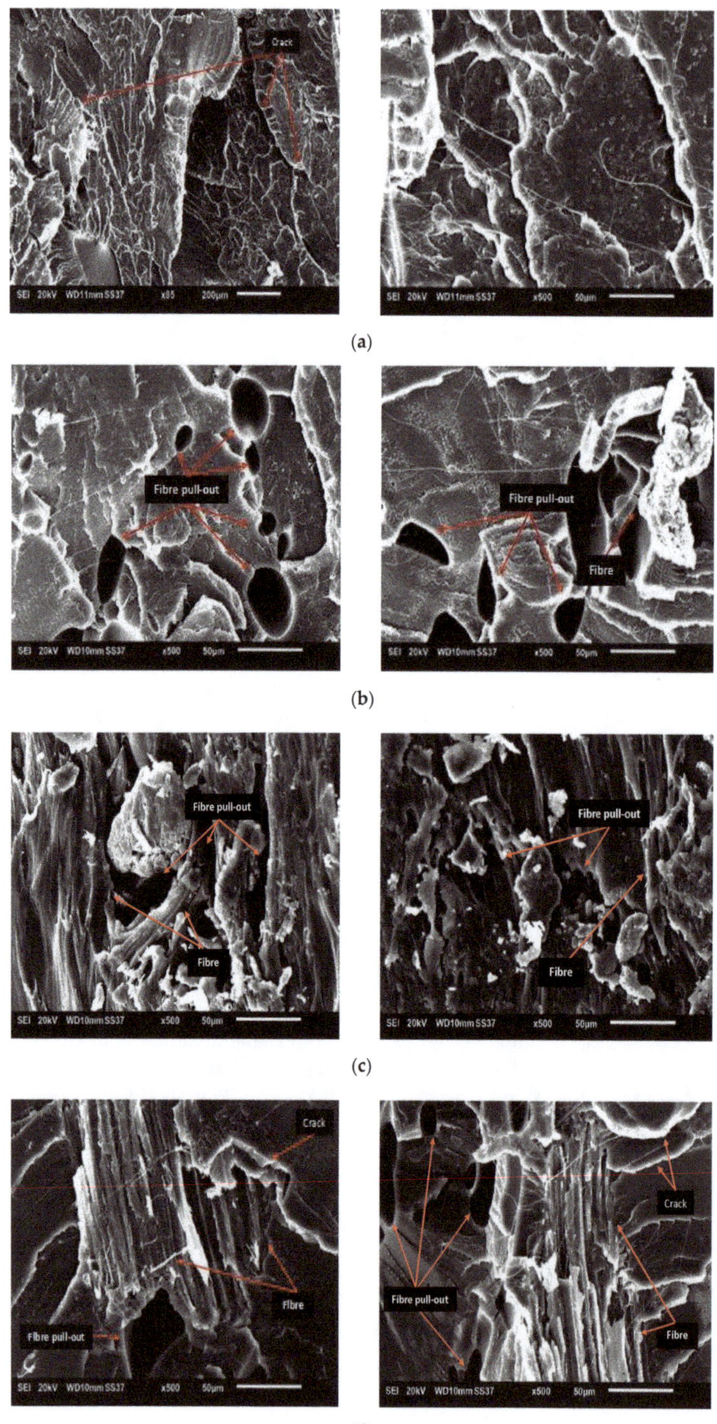

Figure 11. *Cont.*

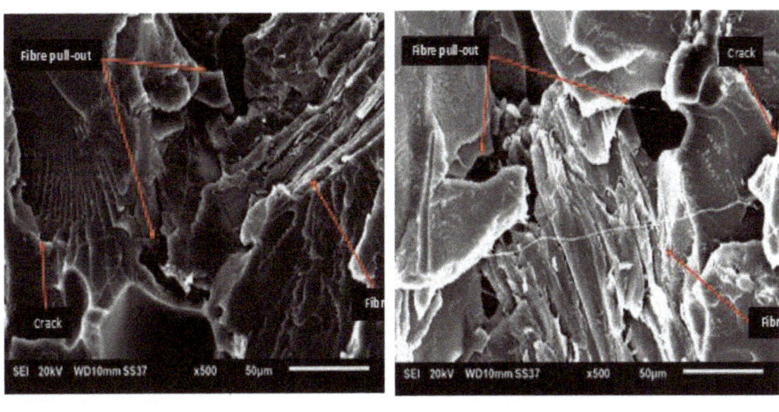

(e)

Figure 11. Scanning Electron micrograph (**a**) PLA; (**b**) Untreated fibre; (**c**) 0.5% silane; (**d**) 1.0% silane; (**e**) 2.0% silane.

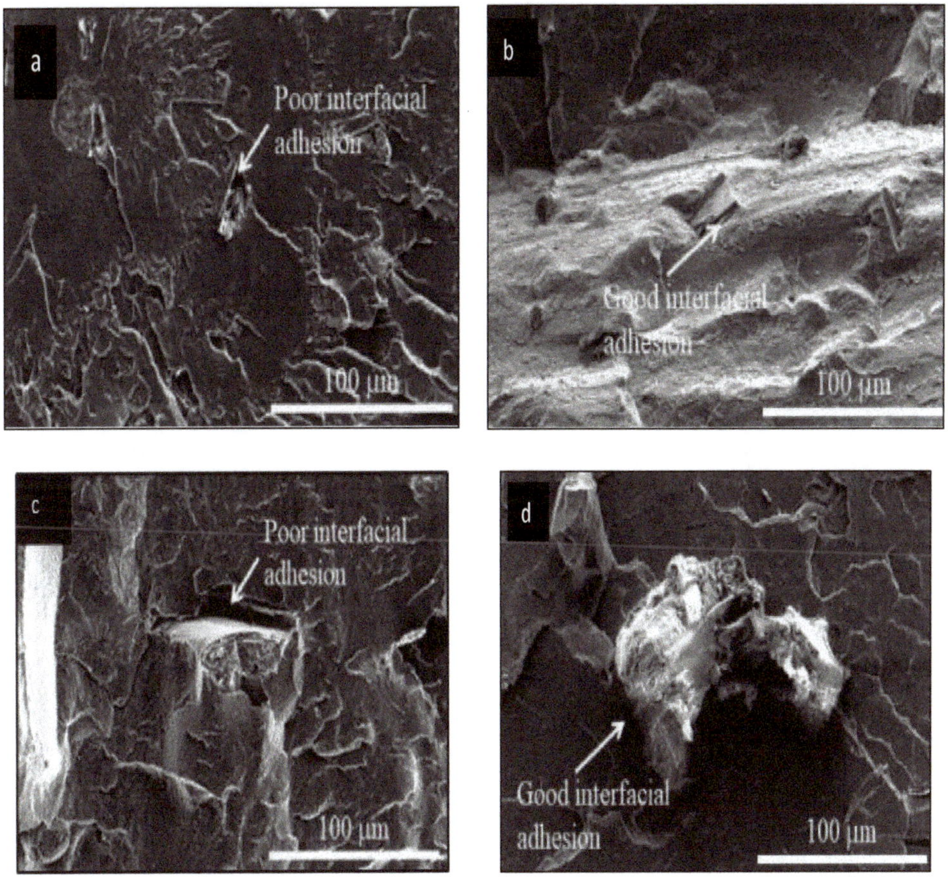

Figure 12. Scanning Electron Micrograph (**a**) Untreated wood PLA (**b**) Treated wood PLA (**c**) Untreated wood PLA (**d**) Treated wood PLA [32].

5. Conclusions

For many years, researchers have tried to determine the correct method to produce high-performance materials by using natural fibres as reinforcement agents. Many methods have been discovered in combining reinforcing agents with a polymer matrix, and this paper is focuses on mixing kenaf fibre with thermoplastic PLA, and observing the mechanical and physical properties of the composites. FDM is one of the additive technologies that has many advantages, such as rapid production, good finishing, ability to generate many shapes with complicated geometries and dimensions, and low cost. This research focuses on the effect of using chemical treatment on kenaf fibre and analysis thereof by performing mechanical and physical tests. In this paper, 2.5 wt % of kenaf fibre was mixed with a PLA polymer by using a twin screw extruder, and a 3D printer filament was extruded. Five different parameters were produced, including neat polymer (PLA with 0 wt % of fibre), untreated kenaf fibre composites, and three different silane treatment parameters. Samples that were printed following ASTM standards were used to performed the tests, and data has been collected. Based on data observation, 1.0% silane concentration after being treated with a 6% alkali solution indicated that this parameter can enhance the interfacial bonding between two phases, which occur due to removal of chemical content in the natural fibre itself such as cellulose, hemicellulose, and lignin. This experiment also proves that applying a higher silane concentration can lead to fibre damage. For example, 2.0% silane is the highest concentration of silane, and the strength data shows that 2.0% silane is lower than a concentration of 1.0 %. Meanwhile, untreated natural fibre composites obtain the lowest strength due to poor interfacial bonding because the stress cannot distribute equally on the surface, unlike treated composites. In a nutshell, for natural fibre modifying composites, the most crucial element that needs to be considered is the optimal concentration of silane for natural fibre surface treatment, as it creates appropriate bonding to achieve a high strength for application development.

Author Contributions: Research idea, N.R. and M.M.T.; methodology, R.A.I.; validation, N.M. and N.R.; formal analysis, A.H.J. and M.M.T.; investigation, A.H.J.; resources, M.P.; data curation, A.H.J., N.M. and N.R.; writing—original draft preparation, A.H.J.; writing—review and editing, N.R. and R.A.I.; supervision, N.R.; funding acquisition, N.R. and M.P. All authors have read and agreed to the published version of the manuscript.

Funding: The result was obtained through the financial support of the Ministry of Education Malaysia under grant number RACER/2019/FKM-CARE/F00408, and publishing of the results was financially supported by the Ministry of Education, Youth and Sports of the Czech Republic, and the European Union (European Structural and Investment Funds—Operational Programme Research, Development, and Education) within the scope of the project "International Research Laboratories", Reg. No. CZ.02.2.69/0.0/0.0/18_054/0014685.

Institutional Review Board Statement: Not applicable.

Informed Consent Statement: Not applicable.

Data Availability Statement: The data presented in this study are available on request from the corresponding author.

Acknowledgments: The author would like to thank Universiti Teknikal Malaysia, Melaka, for providing the facility support to carry out the experiments.

Conflicts of Interest: The authors declare no conflict of interest.

References

1. Liu, Y.; Lv, X.; Bao, J.; Xie, J.; Tang, X.; Che, J.; Ma, Y.; Tong, J. Characterization of silane treated and untreated natural cellulosic fibre from corn stalk waste as potential reinforcement in polymer composites. *Carbohydr. Polym.* **2019**, *218*, 179–187. [CrossRef]
2. Fiore, V.; Di Bella, G.; Valenza, A. The effect of alkaline treatment on mechanical properties of kenaf fibers and their epoxy composites. *Compos. Part B Eng.* **2015**, *68*, 14–21. [CrossRef]
3. Lee, B.H.; Kim, H.S.; Lee, S.; Kim, H.J.; Dorgan, J.R. Bio-composites of kenaf fibers in polylactide: Role of improved interfacial adhesion in the carding process. *Compos. Sci. Technol.* **2009**, *69*, 2573–2579. [CrossRef]

4. Van de Weyenberg, I.; Ivens, J.; De Coster, A.; Kino, B.; Baetens, E.; Verpoest, I. Influence of processing and chemical treatment of flax fibres on their composites. *Compos. Sci. Technol.* **2003**, *63*, 1241–1246. [CrossRef]

5. Liu, Y.; Xie, J.; Wu, N.; Wang, L.; Ma, Y.; Tong, J. Influence of silane treatment on the mechanical, tribological and morphological properties of corn stalk fiber reinforced polymer composites. *Tribol. Int.* **2019**, *131*, 398–405. [CrossRef]

6. Oushabi, A.; Sair, S.; Hassani, F.O.; Abboud, Y.; Tanane, O.; el Bouari, A. The effect of alkali treatment on mechanical, morphological and thermal properties of date palm fibers (DPFs): Study of the interface of DPF–Polyurethane composite. *S. Afr. J. Chem. Eng.* **2017**, *23*, 116–123. [CrossRef]

7. Asim, M.; Jawaid, M.; Abdan, K.; Ishak, M.R. Effect of Alkali and Silane Treatments on Mechanical and Fibre-matrix Bond Strength of Kenaf and Pineapple Leaf Fibres. *J. Bionic Eng.* **2016**, *13*, 426–435. [CrossRef]

8. Sreenivasan, D.P.; Sujith, A.; Rajesh, C. Cure characteristics and mechanical properties of biocomposites of natural rubber reinforced with chicken feather fibre: Effect of fibre loading, alkali treatment, bonding and vulcanizing systems. *Mater. Today Commun.* **2019**, *20*, 100555. [CrossRef]

9. Sreenivasan, V.S.; Ravindran, D.; Manikandan, V.; Narayanasamy, R. Influence of fibre treatments on mechanical properties of short Sansevieria cylindrica/polyester composites. *J. Mater.* **2012**, *37*, 111–121. [CrossRef]

10. Edeerozey, A.M.; Akil, H.M.; Azhar, A.; Ariffin, M.Z. Chemical modification of kenaf fibers. *Mater. Lett.* **2007**, *61*, 2023–2025. [CrossRef]

11. Siakeng, R.; Jawaid, M.; Ariffin, H.; Sapuan, S.M.; Asim, M.; Saba, N. Natural fiber reinforced polylactic acid composites: A review. *Polym. Compos.* **2019**, *40*, 446–463. [CrossRef]

12. Mahjoub, R.; Yatim, J.M.; Sam, A.R.M.; Hashemi, S.H. Tensile properties of kenaf fiber due to various conditions of chemical fiber surface modifications. *Constr. Build. Mater.* **2014**, *55*, 103–113. [CrossRef]

13. Akil, H.M.; Omar, M.F.; Mazuki, A.A.M.; Safiee, S.; Ishak, Z.; Abu Bakar, A. Kenaf fiber reinforced composites: A review. *Mater. Des.* **2011**, *32*, 4107–4121. [CrossRef]

14. Saba, N.; Paridah, M.; Jawaid, M. Mechanical properties of kenaf fibre reinforced polymer composite: A review. *Constr. Build. Mater.* **2015**, *76*, 87–96. [CrossRef]

15. Sreenivasan, S.; Sulaiman, S.; Baharudin, B.T.H.T.; Ariffin, M.K.A.; Abdan, K. Recent developments of kenaf fibre reinforced thermoset composites. *Review. Mater. Res. Innov.* **2013**, *17*, s2–s11. [CrossRef]

16. Sgriccia, N.; Hawley, M.; Misra, M. Characterization of natural fiber surfaces and natural fiber composites. *Compos. Part A Appl. Sci. Manuf.* **2008**, *39*, 1632–1637. [CrossRef]

17. Khan, A.; Vijay, R.; Singaravelu, D.L.; Arpitha, G.R.; Sanjay, M.R.; Siengchin, S.; Jawaid, M.; Alamry, K.; Asiri, A.M. Extraction and characterization of vetiver grass (*Chrysopogon zizanioides*) and kenaf fiber (*Hibiscus cannabinus*) as reinforcement materials for epoxy based composite structures. *J. Mater. Res. Technol.* **2020**, *9*, 773–778. [CrossRef]

18. Xing, X.; Liu, T.; Pei, J.; Huang, J.; Li, R.; Zhang, J.; Tian, Y. Effect of fiber length and surface treatment on the performance of fiber-modified binder. *Constr. Build. Mater.* **2020**, *248*, 118702. [CrossRef]

19. Herrera-Franco, P.J.; Valadez-González, A. A study of the mechanical properties of short natural-fiber reinforced composites. *Compos. Part B Eng.* **2005**, *36*, 597–608. [CrossRef]

20. Shirazi, M.G.; Rashid, A.S.A.; Bin Nazir, R.; Rashid, A.H.A.; Kassim, A.; Horpibulsuk, S. Investigation of tensile strength on alkaline treated and untreated kenaf geotextile under dry and wet conditions. *Geotext. Geomembr.* **2019**, *47*, 522–529. [CrossRef]

21. Asumani, O.; Reid, R.G.; Paskaramoorthy, R. The effects of alkali–silane treatment on the tensile and flexural properties of short fibre non-woven kenaf reinforced polypropylene composites. *Compos. Part A Appl. Sci. Manuf.* **2012**, *43*, 1431–1440. [CrossRef]

22. Huda, M.S.; Drzal, L.T.; Mohanty, A.K.; Misra, M. Effect of fiber surface-treatments on the properties of laminated bio-composites from poly(lactic acid) (PLA) and kenaf fibers. *Compos. Sci. Technol.* **2008**, *68*, 424–432. [CrossRef]

23. Reich, S.; Elsabbagh, A.; Steuernagel, L. Improvement of Fibre-Matrix-Adhesion of Natural Fibres by Chemical Treatment. *Macromol. Symp.* **2008**, *262*, 170–181. [CrossRef]

24. Manral, A.; Bajpai, P.K. Analysis of properties on chemical treatment of kenaf fibers. *Mater. Today Proc.* **2020**, *40*, S35–S38. [CrossRef]

25. Mazzanti, V.; de Luna, M.S.; Pariante, R.; Mollica, F.; Filippone, G. Natural fiber-induced degradation in PLA-hemp biocomposites in the molten state. *Compos. Part A Appl. Sci. Manuf.* **2020**, *137*, 105990. [CrossRef]

26. Ghaffar, S.H.; Madyan, O.A.; Fan, M.; Corker, J. The Influence of Additives on the Interfacial Bonding Mechanisms between Natural Fibre and Biopolymer Composites. *Macromol. Res.* **2018**, *26*, 851–863. [CrossRef]

27. Sawpan, M.A.; Pickering, K.; Fernyhough, A. Effect of various chemical treatments on the fibre structure and tensile properties of industrial hemp fibres. *Compos. Part A Appl. Sci. Manuf.* **2011**, *42*, 888–895. [CrossRef]

28. Najeeb, M.; Sultan, M.; Andou, Y.; Shah, A.; Eksiler, K.; Jawaid, M.; Ariffin, A. Characterization of silane treated Malaysian Yankee Pineapple AC6 leaf fiber (PALF) towards industrial applications. *J. Mater. Res. Technol.* **2020**, *9*, 3128–3139. [CrossRef]

29. Liu, Y.; Guo, L.; Wang, W.; Sun, Y.; Wang, H. Modifying wood veneer with silane coupling agent for decorating wood fiber/high-density polyethylene composite. *Constr. Build. Mater.* **2019**, *224*, 691–699. [CrossRef]

30. Arslan, C.; Dogan, M. The effects of silane coupling agents on the mechanical properties of basalt fiber reinforced poly(butylene terephthalate) composites. *Compos. Part B Eng.* **2018**, *146*, 145–154. [CrossRef]

31. Stoof, D.; Pickering, K.; Zhang, Y. Fused Deposition Modelling of Natural Fibre/Polylactic Acid Composites. *J. Compos. Sci.* **2017**, *1*, 8. [CrossRef]

32. Petchwattana, N.; Channuan, W.; Naknaen, P.; Narupai, B. 3D Printing Filaments Prepared from Modified Poly(Lactic Acid)/Teak Wood Flour Composites: An Investigation on the Particle Size Effects and Silane Coupling Agent Compatibilisation. *J. Phys. Sci.* **2019**, *30*, 169–188. [CrossRef]
33. Coppola, B.; Garofalo, E.; Di Maio, L.; Scarfato, P.; Incarnato, L. Investigation on the use of PLA/hemp composites for the Fused Deposition Modelling (FDM) 3D printing. *AIP Conf. Proc.* **2018**, *1981*, 020086.
34. Rahim, T.N.A.T.; Abdullah, A.M.; Akil, H.M. Recent Developments in Fused Deposition Modeling-Based 3D Printing of Polymers and Their Composites. *Polym. Rev.* **2019**, *59*, 589–624. [CrossRef]
35. El-Shekeil, Y.A.; Sapuan, S.M.; Abdan, K.; Zainudin, E.S. Influence of fiber content on the mechanical and thermal properties of Kenaf fiber reinforced thermoplastic polyurethane composites. *Compos. Part B Eng.* **2012**, *43*, 245–254. [CrossRef]
36. Maslinda, A.B.; Majid, M.S.A.; Ridzuan, M.J.M.; Afendi, M.; Gibson, A.G. Effect of water absorption on the mechanical properties of hybrid interwoven cellulosic-cellulosic fibre reinforced epoxy composites. *Compos. Struct.* **2017**, *167*, 227–237. [CrossRef]
37. Singh, J.I.P.; Dhawan, V.; Singh, S.; Jangid, K. Study of Effect of Surface Treatment on Mechanical Properties of Natural Fiber Reinforced Composites. *Mater. Today Proc.* **2017**, *4*, 2793–2799. [CrossRef]
38. Razali, N.; Sapuan, S.M.; Razali, N. *Mechanical Properties and Morphological Analysis of Roselle/Sugar Palm Fiber Reinforced Vinyl Ester Hybrid Composites*; Elsevier Ltd.: Amsterdam, The Netherlands, 2018.
39. Sood, M.; Dwivedi, G. Effect of fiber treatment on flexural properties of natural fiber reinforced composites: A review. *Egypt. J. Pet.* **2018**, *27*, 775–783. [CrossRef]
40. Krishna, K.V.; Kanny, K. The effect of treatment on kenaf fiber using green approach and their reinforced epoxy composites. *Compos. Part B Eng.* **2016**, *104*, 111–117. [CrossRef]
41. Thomason, J.L.; Rudeiros-Fernández, J.L. Thermal degradation behaviour of natural fibres at thermoplastic composite processing temperatures. *Polym. Degrad. Stab.* **2021**, *188*, 109594. [CrossRef]
42. Kumar, N.; Grewal, J.S.; Singh, T.; Kumar, N. Mechanical and thermal properties of chemically treated Kenaf natural fiber reinforced polymer composites. *Mater. Today Proc.* **2021**, in presss. [CrossRef]
43. Nadlene, R.; Sapuan, S.; Jawaid, M.; Ishak, M.; Yusriah, L. The effects of chemical treatment on the structural and thermal, physical, and mechanical and morphological properties of roselle fiber-reinforced vinyl ester composites. *Polym. Compos.* **2016**, *39*, 274–287. [CrossRef]
44. Cobos, C.M.; Salesiana, U.P.; de València, U.P.; Garz, L. Study of thermal and rheological properties of PLA loaded with carbon and halloysite nanotubes for additive manufacturing. *Rapid Prototyp. J.* **2019**, *25*, 738–743. [CrossRef]
45. Perez, A.R.T.; Roberson, D.A.; Wicker, R.B. Fracture Surface Analysis of 3D-Printed Tensile Specimens of Novel ABS-Based Materials. *J. Fail. Anal. Prev.* **2014**, *14*, 343–353. [CrossRef]
46. Ilyas, R.; Sapuan, S.; Harussani, M.; Hakimi, M.; Haziq, M.; Atikah, M.; Asyraf, M.; Ishak, M.; Razman, M.; Nurazzi, N.; et al. Polylactic Acid (PLA) Biocomposite: Processing, Additive Manufacturing and Advanced Applications. *Polymers* **2021**, *13*, 1326. [CrossRef] [PubMed]
47. Aisyah, H.A.; Paridah, M.T.; Sapuan, S.M.; Ilyas, R.A.; Khalina, A.; Nurazzi, N.M.; Lee, S.H.; Lee, C.H. A Comprehensive Review on Advanced Sustainable Woven Natural Fibre Polymer Composites. *Polymers* **2021**, *13*, 471. [CrossRef] [PubMed]

polymers

Article

Determination of Fire Parameters of Polyamide 12 Powder for Additive Technologies

Richard Kuracina [1,*], Zuzana Szabová [1,*], Eva Buranská [1], Alica Pastierová [1], Peter Gogola [2] and Ivan Buranský [2]

1   Institute of Integral Safety, Faculty of Materials Science and Technology in Trnava, Slovak University of Technology in Bratislava, Ul. Jána Bottu 2781/25, SK-917 24 Trnava, Slovakia; eva.buranska@stuba.sk (E.B.); alica.pastierova@stuba.sk (A.P.)
2   Institute of Materials, Faculty of Materials Science and Technology in Trnava, Slovak University of Technology in Bratislava, Ul. Jána Bottu 2781/25, SK-917 24 Trnava, Slovakia; peter.gogola@stuba.sk (P.G.); ivan.buransky@stuba.sk (I.B.)
*   Correspondence: richard.kuracina@stuba.sk (R.K.); zuzana.szabova@stuba.sk (Z.S.)

Abstract: The use of additive technologies keeps growing. Increasingly, flammable powder materials are also used in additive technologies, and there is a risk of explosion or fire when using them. The current article deals with the determination of fire parameters of a powder sample of polyamide Sinterit PA12 Smoth in accordance with the EN 14034 and EN ISO/IEC 80079-20-2 standards. For that purpose, a sample at a median size of 27.5 μm and a humidity of 0% wt. was used. The measurements showed that the maximum explosion pressure of the PA12 polyamide sample was 6.78 bar and the value of the explosion constant K_{st} was 112.2 bar·m·s^{-1}. It was not possible to determine the MIT value of the settled dust, since the melting point of polyamide sample is low. The MIT of the dispersed dust was 450 °C. Based on the measured results, it can be stated that the powdered polyamide PA12 poses a risk in terms of explosions and fires. Therefore, when using polyamide PA12 in additive technologies, it is necessary to ensure an effective explosion prevention.

Keywords: minimum ignition temperature of dispersed dust; dust explosion; dust cloud; polyamide 12; additive technologies

check for **updates**

Citation: Kuracina, R.; Szabová, Z.; Buranská, E.; Pastierová, A.; Gogola, P.; Buranský, I. Determination of Fire Parameters of Polyamide 12 Powder for Additive Technologies. *Polymers* **2021**, *13*, 3014. https://doi.org/10.3390/polym13173014

Academic Editors: Emin Bayraktar, S. M. Sapuan and R. A. Ilyas

Received: 31 July 2021
Accepted: 2 September 2021
Published: 6 September 2021

Publisher's Note: MDPI stays neutral with regard to jurisdictional claims in published maps and institutional affiliations.

1. Introduction

Additive manufacturing (AM) is gaining increasing importance in industry, not just as a technology for prototyping, but also, and in most cases, for production of functional parts in various fields. Polyamides, or nylons, were the first materials to be recognized as engineering thermoplastics, owing to their superior mechanical properties (especially when exposed to elevated temperatures or solvents) [1–5].

Additive manufacturing is defined by ISO 17296 and ASTM F2792 as "the process of joining materials to make parts or objects from 3D model data, usually layer upon layer, as opposed to subtractive manufacturing methodologies" [6]. Three of the basic types of AM used laser for building up the parts, wherein the starting materials used in the AM laser processes may be in the form of a powder [7].

Dust explosion poses a risk for the processes where dispersed dust occurs. It is therefore necessary to apply the principles of effective explosion protection [8–11].

In the case of powdered polymers, it is possible to design suitable explosion prevention measures based on the measurement of fire characteristics (minimum ignition temperature of settled and agitated dust (MIT), maximum pressure rise rate, explosion constant (Kst) and lower explosion limit).

There are several databases and literature sources dealing with the explosiveness of dispersed dusts [12–16]. In the field of polyamide polymer powders, only information on nylon 6 and 6,6 fibrous materials can be found. Although the amount of polyamide PA12 used has increased significantly, information on its hazards is quite limited (MIE in [17]), and we have not encountered the research into the explosion parameters of polyamide PA12 so far.

Polyamide PA12 is being increasingly used in industry in additive technologies. It is therefore necessary to know its fire characteristics of MIT [18] and explosion parameters. The PA12 powder is handled in relatively large quantities (additive technologies), and the initiating source for the explosion of dispersed dust is almost always present in these technologies (laser, high temperature...).

2. Materials and Methods

2.1. Sinterit PA12 Smooth Powder

Sinterit PA12 Smooth polymer polyamide powder was used in this study. According to MSDS, the melting point of PA12 is 182 °C [19]. The moisture content of the PA 12 sample is 0% wt. The proportion of dimensional fractions of the PA12 polyamide sample was determined by sieve analysis. The analysis procedure was performed according to the ISO 3310-1:2016 (25 9610) Standard. The analysis was conducted on a Retsch AS 200 sieving machine with a sieving time of 15 min and an amplitude of 2 mm/G. The results of the PA12 sample sieve analysis are given in Table 1. The median value of the sample was 27.5 µm. The particle shape of the PA12 powder is shown in Figure 1.

Table 1. Proportion of particle sizes in the sample.

Mesh Size [µm]	Proportion of Particles Sizes in the Sample [%]
500	0.02
250	0.08
200	0.02
150	0.06
90	0.88
71	1.78
56	3.36
45	15.49
32	74.92
20	3.39
0	0.00
median:	**27.5 µm**

Topography of powder particles was documented using a ZEISS LSM700 scanning confocal microscope. 405 nm light source was used which in combination with a Epiplan-Apochromat 100x/0.95 objective enabled to reach step sizes of 110 nm on the X and Y axis as well as 60 nm in the Z axis, Figure 1.

The X-ray diffraction measurements were performed using a Panalytical Empyrean diffractometer. The Cu–Kα nickel filtered radiation was detected in the range of 10–140° 2Theta. Figure 2 shows the XRD pattern for the investigated PA-12 powder. Good agreement with related publications was found. In the current powder both crystalline α and γ phases were observed with their corresponding peaks. Peaks at 20° and 23° 2theta correspond to the α phase. Peaks at 11.2° and 21.5° 2theta correspond to the γ phase [20–24].

For FTIR (FT-IR Spectrometer 660, Varian, Palo Alto, CA, USA) analysis, samples of sheaths were directly applied to a diamante crystal of ATR (GladiATR, Pike Technology, Fitchburg, WI, USA). The spectra were recorded using a Varian Resolutions Pro, and the samples were measured in the region of 400–4000 cm^{-1}; each spectrum was measured 48 times, at the resolution of 4.

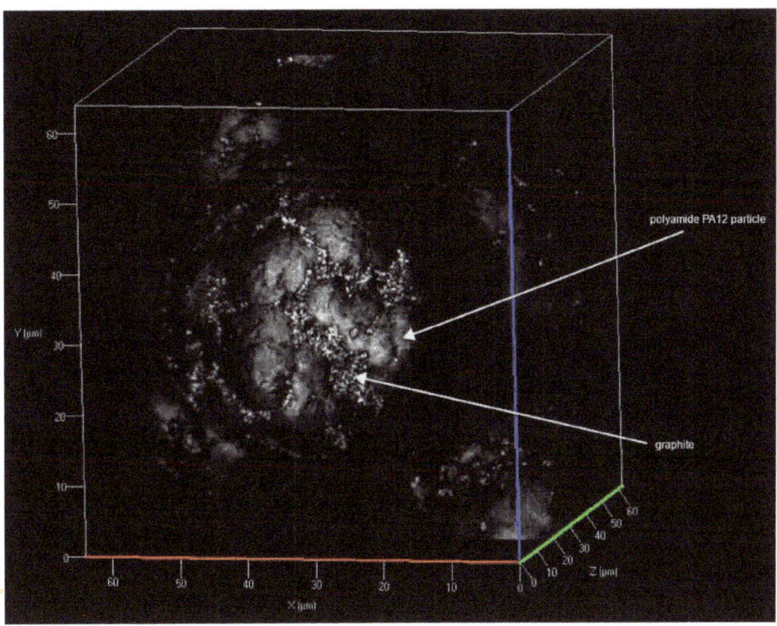

Figure 1. Particles of "Sinterit PA 12 smooth" for laser sintering (confocal laser scanning microscope).

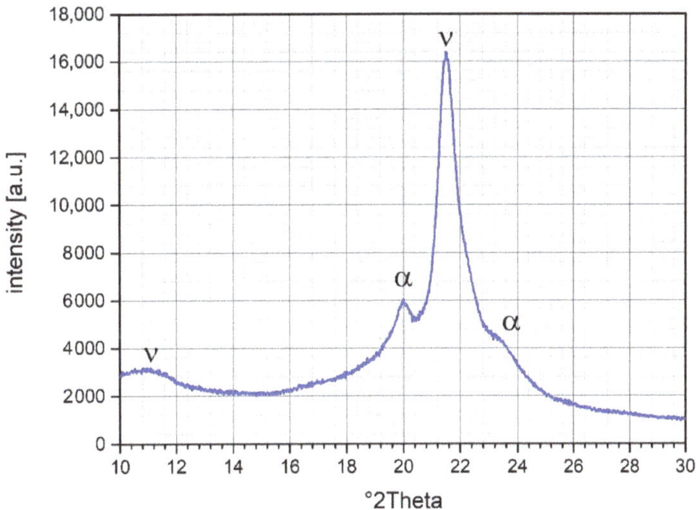

Figure 2. X-ray diffraction patterns of PA12 sample PA 12.

Figure 3 shows the infrared spectra of PA12. The infrared spectrum of the sample is typical for polyamide. The characteristic bands at a wavelength of 3278 cm^{-1} and 1645 cm^{-1} are attributed to the bending and stretching –NH$_2$ bond. In addition, the vibration at 1650 cm^{-1} corresponds to the –CO, and the band at 1540 cm^{-1} is due to the -NH deformation and -CN of the secondary amides. The peaks between 2800 and 2900 cm^{-1} owing to the presence of -CH$_2$, -CH$_3$ and NH; the bands at 720, 1470 and 1539–1645 cm^{-1} are assigned to the bonds N–H, C–CO–NH$_2$ and C=O of the primary amide.

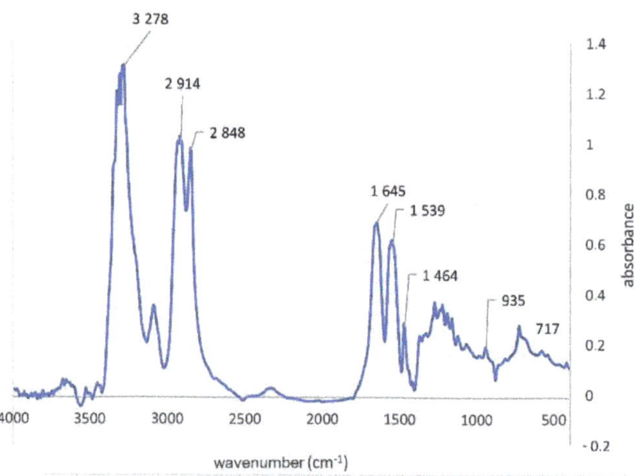

Figure 3. Infrared spectra of PA12 sample.

2.2. MIT of Dispersed Dust

The MIT measurements of dispersed dust were performed on a standardized equipment, Godbert-Greenwald furnace, Figures 4 and 5 [25].

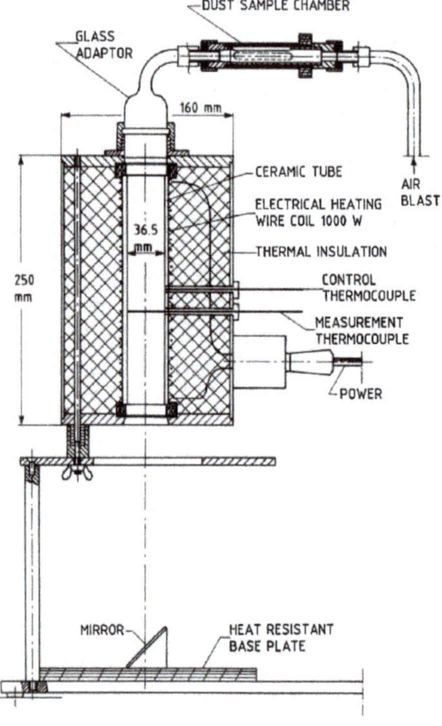

Figure 4. Cross-section of Godbert-Greenwald furnace [25].

Figure 5. G-G furnace for determination of MIT of dispersed dust.

Small quantities of dust are blown vertically downward through a heated furnace, and ignition is detected by visual inspection. The test material is dispersed in the furnace by air blast. Since most of the sample consists of two fractions (>32 μm, >45 μm), the resulting MIT value depends mainly on these two fractions. Owing to their percentage in the sample, other fractions have a negligible effect on the MIT value of the dispersed dust from the hot surface.

The dust quantity is 0.15 g (corresponds to the concentration with the highest P_{max} value) and the dust is dispersed at the air pressure 20 kPa and 50 kPa. If a burst of flame is seen below the end of the furnace tube, this shall be considered as an ignition. For each combination of temperature and pressure, five measurements were performed. The measurements were assessed "YES" if at least one test was performed with positive results. MIT of the dispersed dust is recorded as the lowest temperature of the furnace at which ignition was obtained, minus 20 K [26].

Owing to the dust characteristics according to MSDS (melting point 182 °C), the MIT determination of settled dust was not performed.

2.3. Explosion Parameters

The explosion parameters of PA12 polyamide were determined in the KV 150M2 explosion chamber, Figures 6 and 7.

A KV 150-M2 explosion chamber was used to determine the explosion parameters of PA12 polyamide. Compressed air for dispersing the dust is supplied from a compressed air vessel (6.5 L at 10 bar) through a fast-opening valve into the chamber. The volume of the chamber is 365 L. The sample is placed on a disperser plate and is dispersed by a stream of compressed air. The sample is then ignited by an igniter with an energy of 2×5 kJ. The igniter is located in the middle of the explosion chamber according to the EN 14034 Standard [27].

The time between opening the dispersing valve and activation of the igniter is 350 ms. Pressure changes inside the chamber are recorded by pressure transducers. The Keller pressure transmitter has a response of 2000/s, the Kulite pressure transmitter has a response of 410,000/s. The values are recorded at a speed of 50,000/s. Pressure changes during the explosion of dust clouds were measured at the following concentrations: 15 g·m^{-3}, 30 g·m^{-3}, 60 g·m^{-3}, 125 g·m^{-3}, 250 g·m^{-3}, 500 g·m^{-3}, 750 g·m^{-3} and 1000 g·m^{-3}.

The measurement was performed three times at each concentration. The P_{max} value is the highest value obtained during the measurements. dP/dt values were obtained by deriving a smoothed P-t curve (FFT filter, 200 Hz = 125 points of window).

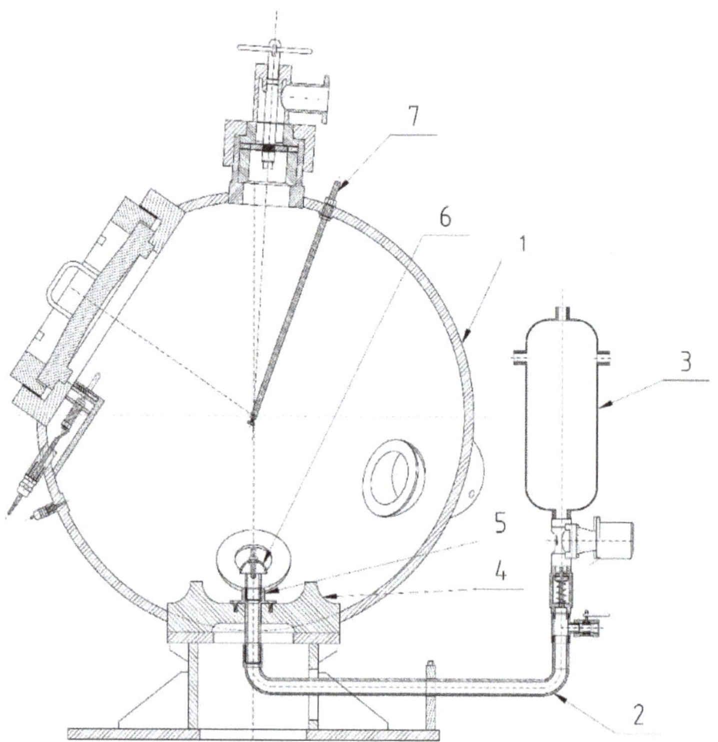

Figure 6. Cross-section of the KV 150M2 explosion chamber (1, chamber; 2, disperser tube; 3, air pressure vessel; 4, dispersing plate; 5, disperser; 6, air flow reverser; 7, igniter rod).

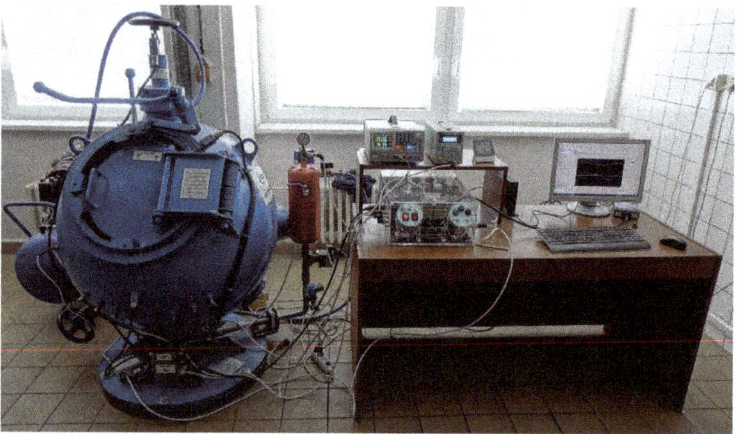

Figure 7. KV 150M2 explosion chamber.

3. Results

The minimum hot surface temperature (MIT) causing ignition of dispersed dust was determined in a G-G furnace. The measured values are listed in Table 2.

Table 2. Measured values of MIT (sample weight 0.15 g).

Air Pressure [kPa]	Temperature [°C]	Result
20	400	NO
20	410	NO
50	420	NO
50	430	NO
50	440	NO
50	450	NO
50	460	**YES**
50	470	**YES**
50	465	**YES**

The measurement of the explosion parameters of dispersed dust sample of PA 12 polyamide was performed in the KV 150M2 explosion chamber. The pressure record of the measurement for the concentration of 500 g·m^{-3}, where the highest value of dP/dt was reached, is shown in Figure 8. The pressure record for the concentration of 750 g·m^{-3}, where the highest value of Pmax was reached, is given in Figure 9. The pressure records important for determining LEL are recorded in Figure 10. All measured values are listed in Table 3.

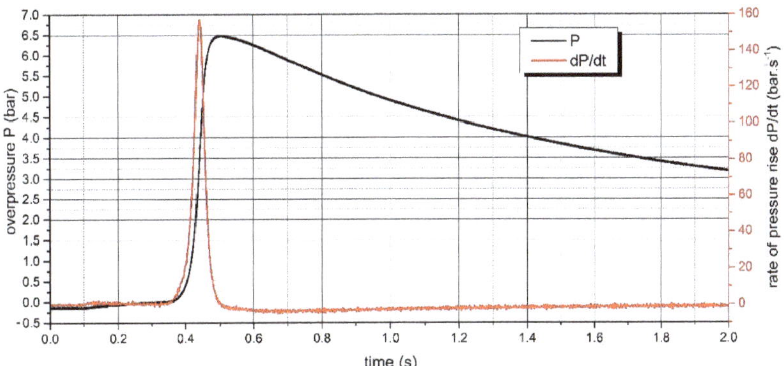

Figure 8. Pressure record of PA12 polyamide explosion at concentration 500 g·m^{-3}.

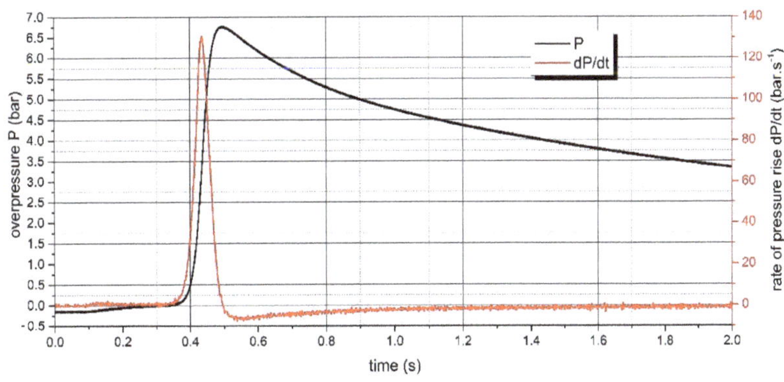

Figure 9. Pressure record of PA12 polyamide explosion at concentration 750 g·m^{-3}.

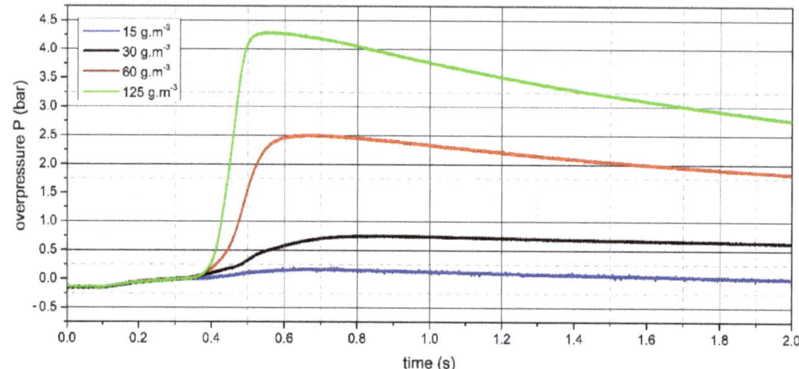

Figure 10. Pressure records for the LEL determination of PA12 polyamide sample.

Table 3. Explosion parameters of PA12 polyamide sample.

Concentration $[g \cdot m^{-3}]$	P_{max} [bar]	dP/dt $[bar \cdot s^{-1}]$
30	0.75	4.4
60	2.52	23.6
125	4.31	54.4
250	5.67	109.0
500	6.49	**157.0**
750	**6.78**	130.7
1000	6.33	107.2

The value of constant explosion K_{st} of the sample Sinterit PA12 Smooth was calculated as follows:

$$K_{st} = \sqrt[3]{V} \times (dP/dt)_{max} = \sqrt[3]{0.365} \times 157.0 = 112.2 \text{ bar} \cdot s^{-1} \cdot m \tag{1}$$

4. Discussion

In our research, the explosion parameters and MIT of dispersed dust of PA12 polyamide were measured. The highest value of explosion overpressure of 6.78 bar for the sample was measured at a concentration of 750 $g \cdot m^{-3}$. The value of the explosion constant was 112.2 $bar \cdot m \cdot s^{-1}$. The MIT value for dispersed PA12 polyamide dust is 450 °C. As shown in Figure 10, the LEL of the Sinterit PA12 Smooth sample is 15 $g \cdot m^{-3}$.

It is not possible to compare the measured results of this particular sample with the measurement results of identical sample attained by other researchers; measurements of explosion parameters of this type of explosive sample (Sinterit PA 12 Smooth) have not been published yet. Generally, only parameters of PA12 polyamide fibers and flocks from the textile industry are available. For the sample "Polyamide flock" of the mean particle size 37 μm, Field [13] states the value of MIT (dispersed dust) 520 °C, the value P_{max} = 9.8 bar and the value K_{st} = 93 $bar \cdot m \cdot s^{-1}$ and LEL = 30 $g \cdot m^{-3}$. Iarossi et al. [28] determined P_{max} = 6.6 bar, K_{st} = 50 $bar \cdot m \cdot s^{-1}$ and MIT (dispersed dust) 485 °C for the fibers with a length of 0.5 mm and dtex 1.7. For polyamide (7367) with a sample median size of 30 μm, the Gestis database [29] determined LEL = 30 $g \cdot m^{-3}$, P_{max} = 7.3 bar, explosion constant K_{st} = 117 $bar \cdot m \cdot s^{-1}$ and MIT (dispersed dust 480 °C). It can therefore be stated that the measured values that Sinterit PA12 Smooth fire properties are comparable with the results indicated in the scientific literature and databases.

5. Conclusions

Frequently utilized for the advantages they provide, additive technologies widely use powder materials, including PA12 polyamide (Sinterit PA12 Smooth). It is therefore necessary to deal which safety when using polyamide PA12 in additive technologies. One of the PA12 polyamide is its explosiveness. Explosion and ignition of dispersed dust can cause damage to property and lives.

Based on these values, it can be concluded that the use of PA12 in additive technologies can pose a significant risk of explosion or fire. When using PA12 polyamide in additive technologies, it is therefore necessary to ensure effective explosion prevention.

Author Contributions: R.K., Z.S., and E.B. conceived and designed the experiments; R.K., Z.S., A.P., I.B., and P.G. performed the experiments and analysed the data; and R.K. and Z.S. managed all the experimental and writing process as the corresponding authors. All authors discussed the results and commented on the manuscript. All authors have read and agreed to the published version of the manuscript.

Funding: This research was supported by the Cultural and Educational Grant Agency of the Ministry of Education, Science, Research and Sport of the Slovak Republic under the Contract No. 020STU-4/2021, by the Science Grant Agency of the Ministry of Education, Science, Research and Sport of the Slovak Republic under the Contract No. 1/0747/19 and by the Slovak Research and Development Agency under the Contract No. APVV-16-0223.

Institutional Review Board Statement: Not applicable.

Informed Consent Statement: Not applicable.

Data Availability Statement: The data presented in this study are available on request from the corresponding author.

Conflicts of Interest: The authors declare no conflict of interest.

References

1. Cicala, G.; Latteri, A.; Del Curto, B.; Russo, A.L.; Recca, G.; Farè, S. Engineering Thermoplastics for Additive Manufacturing: A Critical Perspective with Experimental Evidence to Support Functional Applications. *J. Appl. Biomater. Funct. Mater.* **2017**, *15*, 10–18. [CrossRef] [PubMed]
2. Peters, E.N. 1—Engineering Thermoplastics—Materials, Properties, Trends. In *Applied Plastics Engineering Handbook*, 2nd ed.; Kutz, M., Ed.; William Andrew Publishing: Norwich, NY, USA, 2017; pp. 3–26. ISBN 780323390408. [CrossRef]
3. Gibson, I.; Rosen, D.; Stucker, B.; Mahyar, K. *Additive Manufacturing Technologies*, 3rd ed.; Springer: Berlin/Heidelberg, Germany, 2021; 472p, ISBN 978-1-4419-1119-3.
4. Martynková, G.S.; Slíva, A.; Kratošová, G.; Barabaszová, K.; Študentová, S.; Klusák, J.; Brožová, S.; Dokoupil, T.; Holešová, S. Polyamide 12 Materials Study of Morpho-Structural Changes during Laser Sintering of 3D Printing. *Polymers* **2021**, *13*, 810. [CrossRef] [PubMed]
5. Schneider, K.; Wudy, K.; Drummer, D. Flame-Retardant Polyamide Powder for Laser Sintering: Powder Characterization, Processing Behavior and Component Properties. *Polymers* **2020**, *12*, 1697. [CrossRef]
6. ASTM F2792-12a. *Standard Terminology for Additive Manufacturing Technologies*; ASTM International: West Conshohocken, PA, USA, 2013; pp. 10–12.
7. Schmidt, M.; Merklein, M.; Bourell, D.; Dimitrov, D.; Hausotte, T.; Wegener, K.; Overmeyer, L.; Vollertsen, F.; Levy, G.N. Laser based additive manufacturing in industry and academia. *CIRP Ann.* **2017**, *66*, 561–583. [CrossRef]
8. Hassan, J.; Khan, F.; Amyotte, P.; Ferdous, R. Industry specific dust explosion likelihood assessment model with case studies. *J. Chem. Health Saf.* **2014**, *21*, 13–27. [CrossRef]
9. Dobashi, R. Studies on accidental gas and dust explosions. *Fire Saf. J.* **2017**, *91*, 21–27. [CrossRef]
10. Eckhoff, R. Prevention and mitigation of dust explosions in the process industries: A survey of recent research and development. *J. Loss Prev. Process. Ind.* **1996**, *9*, 3–20. [CrossRef]
11. Fumagalli, A.; Derudi, M.; Rota, R.; Copelli, S. Estimation of the deflagration index K St for dust explosions: A review. *J. Loss Prev. Process. Ind.* **2016**, *44*, 311–322. [CrossRef]
12. Amyotte, P.R.; Pegg, M.J.; Khan, F.I.; Nifuku, M.; Yingxin, T. Moderation of dust explosions. *J. Loss Prev. Process. Ind.* **2007**, *20*, 675–687. [CrossRef]
13. Field, P. *Dust Explosions*; Elsevier: Amsterdam, The Netherlands, 1982; Volume 4, p. 212, ISBN 0-444-40746-4.
14. Amyotte, P.R. Some myths and realities about dust explosions. *Process. Saf. Environ. Prot.* **2014**, *92*, 292–299. [CrossRef]

15. Amyotte, P.R.; Cloney, C.T.; Khan, F.I.; Ripley, R.C. Dust explosion risk moderation for flocculent dusts. *J. Loss Prev. Process. Ind.* **2012**, *25*, 862–869. [CrossRef]
16. Addo, A.; Dastidar, A.G.; Taveau, J.R.; Morrison, L.S.; Khan, F.I.; Amyotte, P.R. Niacin, lycopodium and polyethylene powder explosibility in 20-L and 1-m3 test chambers. *J. Loss Prev. Process. Ind.* **2019**, *62*, 103937. [CrossRef]
17. Bernard, S.; Youinou, L.; Gillard, P. MIE determination and thermal degradation study of PA12 polymer powder used for laser sintering. *J. Loss Prev. Process. Ind.* **2013**, *26*, 1493–1500. [CrossRef]
18. Półka, M.; Salamonowicz, Z.; Wolinski, M.; Kukfisz, B. Experimental Analysis of Minimal Ignition Temperatures of a Dust Layer and Clouds on a Heated Surface of Selected Flammable Dusts. *Procedia Eng.* **2012**, *45*, 414–423. [CrossRef]
19. Sinterit, PA12 Smooth—Technical Datasheet. 2021, p. 2. Available online: https://seeda.nl/wp-content/uploads/2018/10/PA1 2_SMOOTH.pdf (accessed on 27 August 2021).
20. Salmoria, G.V.; Paggi, R.A.; Lago, A.; Beal, V.E. Microstructural and mechanical characterization of PA12/MWCNTs nanocomposite manufactured by selective laser sintering. *Polym. Test.* **2011**, *30*, 611–615. [CrossRef]
21. Ishikawa, T.; Nagai, S.; Kasai, N. Effect of casting conditions on polymorphism of nylon-12. *J. Polym. Sci.* **1980**, 291–299. [CrossRef]
22. Liu, Y.; Zhu, L.; Zhou, L.; Li, Y. Microstructure and mechanical properties of reinforced polyamide 12 composites prepared by laser additive manufacturing. *Rapid Prototyp. J.* **2019**, *25*, 1127–1134. [CrossRef]
23. Androsch, R.; Stolp, M.; Radusch, H.-J. Simultaneous X-ray diffraction and differential thermal analysis of polymers. *Thermochim. Acta* **1996**, *271*, 1–8. [CrossRef]
24. Schmid, M.; Kleijnen, R.; Vetterli, M.; Wegener, K. Influence of the Origin of Polyamide 12 Powder on the Laser Sintering Process and Laser Sintered Parts. *Appl. Sci.* **2017**, *7*, 462. [CrossRef]
25. Eckhoff, R.K. Origin and development of the Godbert-Greenwald furnace for measuring minimum ignition temperatures of dust clouds. *Process. Saf. Environ. Prot.* **2019**, *129*, 17–24. [CrossRef]
26. EN ISO/IEC 80079-20-2. In *Explosive Atmospheres—Part 20—2: Material Characteristics—Combustible Dusts Test Methods*; British Standards Institution: London, UK, 2016.
27. STN EN 14034+A1. In *Determination of Explosion Characteristics of Dust Clouds*; Slovak Standards Institute: Bratislava, Slovakia, 2011.
28. Iarossi, I.; Amyotte, P.R.; Khan, F.I.; Marmo, L.; Dastidar, A.G.; Eckhoff, R.K. Explosibility of polyamide and polyester fibers. *J. Loss Prev. Process. Ind.* **2013**, *26*, 1627–1633. [CrossRef]
29. GESTIS Dust Ex Database. Available online: https://staubex.ifa.dguv.de/explokomp.aspx?nr=7367&lang=e (accessed on 27 August 2021).

Article

Mechanical Properties of Flexible TPU-Based 3D Printed Lattice Structures: Role of Lattice Cut Direction and Architecture

Victor Beloshenko [1], Yan Beygelzimer [1], Vyacheslav Chishko [1], Bogdan Savchenko [2], Nadiya Sova [2], Dmytro Verbylo [3], Andrei Voznyak [4] and Iurii Vozniak [5,*]

[1] Donetsk Institute for Physics and Engineering Named after O.O. Galkin, National Academy of Sciences of Ukraine, pr. Nauki, 46, 03028 Kyiv, Ukraine; biloshenko.va@gmail.com (V.B.); yanbeygel@gmail.com (Y.B.); chishko@ukr.net (V.C.)

[2] Department of Applied Ecology, Technology of Polymers and Chemical Fibers, Kyiv National University of Technologies and Design, Nemirovicha Danchenko Str., 2, 03056 Kyiv, Ukraine; 1079@ukr.net (B.S.); djanc@ukr.net (N.S.)

[3] Institute for Problems in Materials Science I.M. Frantsevich, National Academy of Sciences of Ukraine, Krzhizhanovsky Str., 3, 03142 Kyiv, Ukraine; ver@ipms.kiev.ua

[4] Department of General Technical, Disciplines and Vocational Training, Kryvyi Rih State Pedagogical University, Gagarin Av. 54, 50086 Kryvyi Rih, Ukraine; avvoznyak76@gmail.com

[5] Centre of Molecular and Macromolecular Studies, Polish Academy of Sciences, Sienkiewicza Str., 112, 90363 Lodz, Poland

* Correspondence: wozniak@cbmm.lodz.pl; Tel.: +48-42-6803237

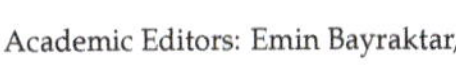

Citation: Beloshenko, V.; Beygelzimer, Y.; Chishko, V.; Savchenko, B.; Sova, N.; Verbylo, D.; Voznyak, A.; Vozniak, I. Mechanical Properties of Flexible TPU-Based 3D Printed Lattice Structures: Role of Lattice Cut Direction and Architecture. *Polymers* **2021**, *13*, 2986. https://doi.org/10.3390/polym13172986

Academic Editors: Emin Bayraktar, Mohd Sapuan Salit and R. A. Ilyas

Received: 25 July 2021
Accepted: 1 September 2021
Published: 3 September 2021

Publisher's Note: MDPI stays neutral with regard to jurisdictional claims in published maps and institutional affiliations.

Abstract: This study addresses the mechanical behavior of lattice materials based on flexible thermoplastic polyurethane (TPU) with honeycomb and gyroid architecture fabricated by 3D printing. Tensile, compression, and three-point bending tests were chosen as mechanical testing methods. The honeycomb architecture was found to provide higher values of rigidity (by 30%), strength (by 25%), plasticity (by 18%), and energy absorption (by 42%) of the flexible TPU lattice compared to the gyroid architecture. The strain recovery is better in the case of gyroid architecture (residual strain of 46% vs. 31%). TPUs with honeycomb architecture are characterized by anisotropy of mechanical properties in tensile and three-point bending tests. The obtained results are explained by the peculiarities of the lattice structure at meso- and macroscopic level and by the role of the pore space.

Keywords: lattice material; flexible TPU; 3D printing; internal architecture; mechanical properties

1. Introduction

Currently, much attention is paid to the fabrication and study of lattice (cellular) structures because of their excellent functionality due to their superior mechanical properties, large surface area, and open pores [1]. They can be used in various technological fields, including structural lightweight design [2,3], acoustic and thermal insulation [4–6], shock absorption [7], and as biomaterials for implants and scaffolds for tissue engineering [8,9]. Conventional technologies for the fabrication of lattice materials severely limit the possibility of forming structures with complex architecture characterized by a strictly defined shape and size of cells [10]. Many of these limitations can be overcome by using additive technologies [11–13].

The mechanical behavior of lattice materials, their crash resistance, and energy absorption capacity are important issues in additive manufacturing of regular cell structures. Rigid thermoplastic polymers such as ABS or PLA are mainly used to fabricate lattice polymer materials with different architectures by 3D printing [14–16]. The use of flexible thermoplastics for these purposes is not so widespread. This is because the 3D printing process with flexible material is more demanding and cannot be realized with most 3D printers available on the market. Besides, the determination of suitable technological parameters for flexible polymers requires additional, time-consuming optimization studies [17–19].

Thermoplastic polyurethane (TPU) is one of the promising polymers that can find wide practical application as a lattice material. In [20,21], it was shown that 3D printing of TPU provides a unique opportunity to fabricate customized flexible cellular structures that can be designed and optimized for specific energy absorbing applications. TPU lattice materials can be fabricated from both rigid TPU and flexible TPU and can also differ in architecture (i.e., type of cell structure). For example, for TPU honeycombs (hexagonal cells), it was shown that the energy absorption properties under compression depend on the orientation of the cells as well as the strain rate [21]. In [22], the effect of different gradient gradings produced by TPU materials on the energy absorption of honeycomb structures was demonstrated. The results showed that the energy absorption curves of these structures can be controlled by grading the density of the structures. In [23], the effects of cyclic loading and impact loading on the gradient energy absorption of honeycomb structures were investigated. The authors [17] analyzed the deformation process of regular cell structures based on the flexible TPU under quasi-static loading conditions. As shown in [24], TPU-based honeycomb structures have the potential for repeatable and high specific energy absorption, with absorption efficiency not worse than rigid polyurethane foams. Varying the cell size and wall thickness of TPU honeycombs facilitated the change in stiffness but provides only a modest opportunity to change the shape of the stress–strain curve [24].

The mechanical properties of lattice materials are also largely determined by the parameters of 3D printing (layer thickness, printing speed, raster angle, raster width, air gap etc.). The build orientation and the loading direction during the experiment also have a significant influence on the mechanical properties of additively manufactured components. For example, it was shown in [25] that TPU parts printed flat and edge oriented have better tensile strength and deformability than those printed on edge. It should be noted that the TPU samples studied did not have a lattice structure but consisted of a solid material. The authors of [20] also stated the importance and need for further research to understand the difference in elasticity and energy absorption capacity between the axial printing and transverse directions to determine whether or not 3D printing orientation has an effect on the mechanical properties of TPU-based lattice structures.

In our previous work [26], we investigated the influence of build direction and loading direction on the mechanical properties in three-point bending test of 3D-printed rigid TPU-based lattice structures. For lattice structures with square cells, it was shown that the investigated specimens are characterized by a strong anisotropy of the mechanical properties, which depends on the build direction and the loading direction. Thereby, the influence of the loading direction is significantly stronger for the samples printed vertically or at an angle of 45°, while the properties of the horizontally printed lattice structures are almost isotropic.

The main characteristic of flexible TPU is high deformability. This, combined with the right architecture of the lattice material, can lead to high efficiency in energy absorption and crashworthiness, for example, 3D-printed flexible TPU-based lattice materials also have great potential in the fabrication of lightweight, custom-shaped structures and functional parts for applications in various fields such as aerospace engineering, medical devices and sports equipment [27–31]. The aim of this work is to present results regarding the mechanical response of 3D printed lattice structures made of flexible TPU subjected to static (tension, three-point bending, compression) and dynamic (dynamic mechanical thermal analysis) tests. The novelty of this work lies in the complex investigation of the effects of architecture and lattice cut direction on the mechanical properties of 3D printed flexible TPU under different types of loading.

2. Experimental Section

2.1. Materials and Processing

TPU Desmopan 3690 AU (Covestro AG, Leverkusen, Germany) was used as raw material. Two types of lattice architectures were investigated: honeycomb and gyroid (Figure 1).

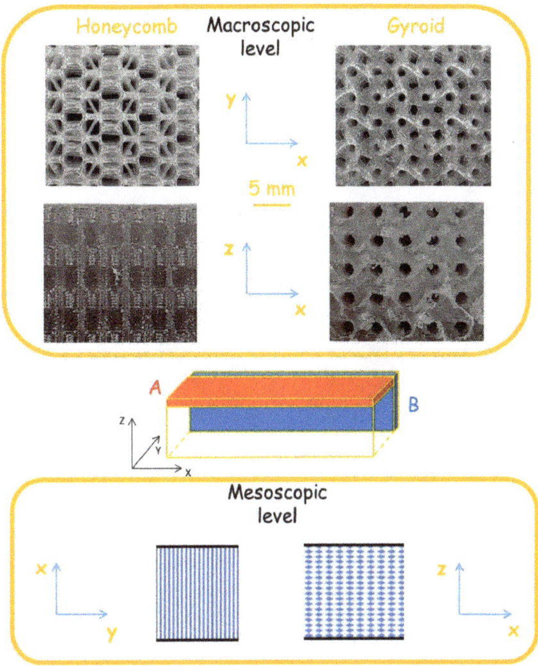

Figure 1. The peculiarities of the multiscale structure of flexible TPU lattice materials with honeycomb and gyroid architecture as well as the arrangement of A- and B-type test specimens.

Models for the creation of honeycomb and gyroid architecture were created by using a program generated infill Simplify3d and Ultimaker Cura. The honeycomb structure was interpenetrating open hexagonal cells with sizes of 3 and 6 mm and a wall thickness of 0.36 mm. The corresponding gyroid structure was described by the formula:

$$\sin(1.5x)\cdot\cos(1.5y) + \sin(1.5y)\cdot\cos(1.5z) + \sin(1.5z)\cdot\cos(1.5x) = 0 \tag{1}$$

They were later converted into rectangular bars by linear array duplication (Figure 1). The resulting models were saved in .stl format and sliced with the Simplify 3D package. Isometric views of individual cells of lattice structures are presented in Figure 2.

The 3D printing was done with the printer Flashforge Creator pro (Zhejiang Flashforge 3D Technology Co., Ltd., Jinhua City, China). The parameters of 3D printing were as follows: the nozzle temperature was 245 °C, the temperature of the building platform was 90 °C, the printing speed was 2000 mm/min. The filament was produced by processing of TPU in a single-screw extruder (Polymer Mash LTD, Kyiv, Ukraine; D = 25 mm, L/D = 16 and 50 rpm), followed by two-stage cooling (in water at temperatures of 60 °C and 15 °C) and drying in a heating chamber with air circulation (8 h, 50 °C). The temperatures within the zones were 190–200–210 °C. Filament production speed was 16 m/min. The diameter of the monofilament is 1.75 ± 0.05 mm.

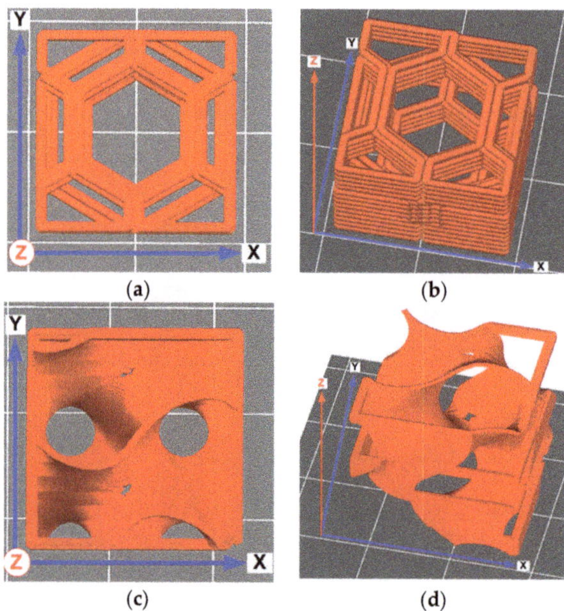

Figure 2. Isometric views of individual cells of honeycomb (**a**,**b**) and gyroid (**c**,**d**) architectures.

The samples of the lattice TPU with honeycomb and gyroid architecture were printed in horizontal directions, which, as we showed in [26], provides the least anisotropy of properties. In all the cases, the structure used is self-supporting due to the small size of the open area and the sufficient bridging properties of the material and settings used for 3D printing. The thickness of the melt layers was 0.2 mm. The TPU content in the sample volume was 29.7% and 37.0% for lattice material with honeycomb and gyroid architecture, respectively. The samples were cut in two mutually perpendicular planes: perpendicular and parallel to the print direction and are designated hereinafter as Sample, type A and Sample, type B, respectively. The large side of the specimens was oriented in the x direction (Figure 1). The mechanical characteristics of materials obtained by testing samples A and B are labeled below as (z, x) and (y, x), where the first letter in parentheses indicates the normal to the sample plane, and the second indicates the direction of tension-compression.

2.2. Characterization

The tensile and static three-point bending tests were performed at room temperature using the universal electromechanical device UTM-100 (G.S. Pisarenko Institute for Problems of Strength of the National Academy of Sciences of Ukraine, Kyiv, Ukraine) with a speed of the crosshead movement of 10 mm/min. The distance between the outer rollers in three-point bending test was 100 mm. The samples were parallelepipeds $20 \times 13 \times 125$ mm^3 in size. To measure the strains, a specialized sensor for measuring large deformations (up to 500 mm) was used, the probes of which were fixed directly on the samples with the help of springs. In the experiment, the difference in the stroke of the probes was recorded. The measurement error of the sample elongation was not worse than $\pm 0.5\%$. The standard deviation of the observation results was taken as a measure of measurement error. An uniaxial compression test was performed using the loading frame of an universal tensile testing machine (Instron, Model 5582, High Wycombe, UK) controlled by the Bluehill® II software and a compression fixture equipped with LVDT transducer, mounted close to the specimen for precise determination of the strain. It is known that boundary effects are considered negligible when the ratio of specimen diameter to unit cell size is sufficient to represent the compressive strength of the "bulk" porous material. According to [32–34] this ratio should be around 3, which was chosen in the compression test. The samples were cylindrical in shape with 10 mm diameter and 15 mm thickness. The temperature of the test was 25 °C. The compression rate was 5% of the initial thickness per minute. The mean values of the mechanical characteristic were calculated from the results of testing at least 5 specimens.

The viscoelastic properties were tested on rectangular specimens, 24 mm × 10 mm × 3 mm, by dynamic mechanical thermal analysis (DMA) using the DMAQ800 (TA instrument, New Castle, DE, USA). The measurement of the tangent of the loss angle tgδ, the storage modulus E', and the loss modulus E'' was performed in the deformation mode of single cantilevered bending. The frequency of forced sine oscillations was 1 Hz, the temperature varied from −100 to 150 °C at the heating rate of 3 °C/min.

Differential scanning calorimetry (DSC) was performed with a device DSC Q20 (TA Instruments, New Castle, DE, USA) under heating from −60 to 140 °C at the rate of 10 °C/min. The samples with a mass of 4–5 mg were cut from lattice TPU and placed in standard aluminum pans. During the measurement the DSC cell was blown off with dry nitrogen (20 mL/min).

3. Results and Discussion

3.1. Tensile and Static Three-Point Bending Tests

Figure 3 shows the stress–strain curves of lattice flexible TPU with honeycomb and gyroid architecture during tension (loading and unloading). The average values of modulus of elasticity E, maximum tensile strength σ_T, strain at break ε_b, and residual strain ε_r (after 24 h) are summarized in Table 1.

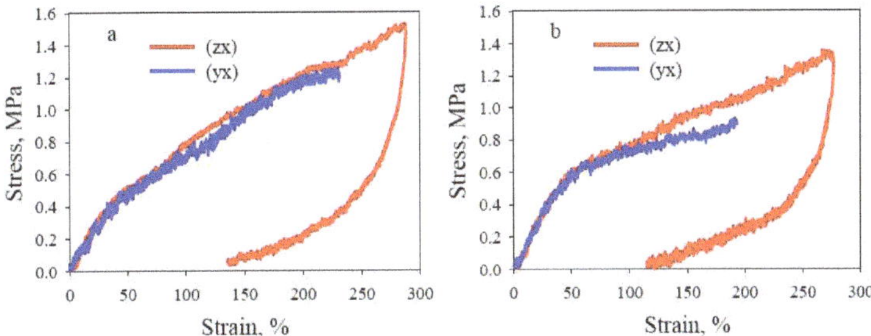

Figure 3. Representative stress–strain curves of lattice flexible TPU under tension. (**a**) honeycomb, (**b**) gyroid architecture.

Table 1. Mechanical properties of lattice flexible TPU. Tensile test.

Internal Architecture	Sample Orientation	E, MPa	σ_T, MPa	ε_b, %	ε_r, %
Honeycomb	(zx)	2.1 ± 0.1	Not break	>280	46 ± 2
	(yx)	1.6 ± 0.1	1.26 ± 0.05	230 ± 5	21 ± 2
Gyroid	(zx)	1.5 ± 0.1	Not break	>280	31 ± 2
	(yx)	1.4 ± 0.1	0.94 ± 0.05	190 ± 6	12 ± 1

It can be seen that all stress–strain curves exhibit nonlinear behavior, regardless of the internal architecture of the sample. An anisotropy of the mechanical properties of the studied specimens is observed as a function of their orientation relative to the compression direction (Table 1). For the same orientation, the values of E, σ_T and ε_b are higher for lattice TPU with honeycomb architecture. In turn, the degree of anisotropy of these parameters is much lower for lattice TPU with gyroid architecture. This fact is in agreement with the results of the static three-point bending test of lattice TPU. Indeed, the modulus of elasticity for a sample with honeycomb architecture is 1.5 MPa (sample, type A) and 0.66 MPa (sample, type B), and for a sample with gyroid architecture, it is 0.72 and

0.71 MPa, respectively. It should be noted that the specimens of type A do not fail within the travel of the testing machine and show higher deformability and higher maximum tensile strength. The unloading paths show a large hysteresis loop with residual strain. After unloading, a greater strain recovery occurs with time. The residual strains measured 24 h after the tests (Table 1) also show anisotropy as a function of the lattice cut direction, indicating a more complete strain recovery for lattice TPU with a gyroid architecture.

The obtained results can be explained by the peculiarities of the multiscale structure of A- and B-type test specimens for TPUs with honeycomb and gyroid architecture. Let us consider them at the mesoscopic and macroscopic scales. The meso-scale refers to the scale where the location of each structural element is visible, which is determined by the 3D printing technology and does not depend on the architecture of the material. The macro scale refers to the scale on which the architecture of the material manifests itself.

Figure 1 shows that in the A- and B-type samples the mesostructure is different: in the first case the polymer filaments are interconnected along the width of the sample, in the second-along the thickness, which is much smaller than the width. This explains the lower strength of the B-type samples for the two studied architectures. At the macro level, the structural features of the A- and B-type samples are fundamentally different for the honeycomb and gyroid architectures. In the case of honeycomb architecture, the structures of the images of the two types are completely different, because in the XY plane this structure looks like hexagons, and in the ZX plane-layers (Figure 1). This explains the large difference in their mechanical properties (Table 1). At the same time, for a gyroid architecture at this scale, the structures of the samples of types A and B are the same, which explains the closeness of their mechanical properties. After making the change of variables in relation (1):

$$x^* = y, \ y^* = z, \ z^* = x \tag{2}$$

the following Equation (3) of the gyroid in new coordinates is obtained:

$$\sin(1.5y^*) \cdot \cos(1.5z^*) + \sin(1.5z^*) \cdot \cos(1.5x^*) + \sin(1.5x^*) \cdot \cos(1.5y^*) = 0 \tag{3}$$

A comparison of Equations (1) and (3) shows that these relations agree. It follows that the sample of the first type, cut from the gyroid in coordinates $(x^*, \ y^*, \ z^*)$, has the same architecture as the sample of the first type, cut in coordinates (x, y, z). However, the plane $(x^*, \ y^*)$ corresponds to the plane (y, z), i.e., the sample of the first type in the new coordinates is the sample of the second type in the old coordinates. This shows that in the case of a gyroid, the samples of the first and second types have the same structure at the macroscopic level.

3.2. DMA and DSC Analyzes

Figure 4, as an example, shows the results obtained from DMA tests of 3D printed lattice TPU with honeycomb architecture. DMA curves for the lattice TPU with gyroid architecture are qualitatively the same. In Table 2, E', E'' are storage modulus and loss modulus at room temperature, respectively; T_E, $T_{tg\delta}$ are the temperatures of the peaks of the maxima of E'' and tgδ, respectively; tgδ is the tangent of loss angle at room temperature; tgδ_m is the maximum of tgδ.

The $E'(T)$, $E''(T)$ and tg$\delta(T)$ curves of lattice TPU show a shape typical of semicrystalline thermoplastic elastomers throughout the temperature range below the flow region [35]. The temperature-independent rubber plateau corresponds to rigid TPU segments that form stable crystals softening in a relatively narrow temperature range, which is evident from the increase in tg$\delta(T)$ at the end of the DMA curve. The temperature location of the maximum of the temperature dependences of E'' and tgδ, which determine the glass transition of soft TPU segments, as well as the height of the maximum of tgδ, which characterizes the energy dissipation of a material (mechanical losses), depend weakly on the architecture and the lattice intersection direction. A similar conclusion was drawn from the DSC data (shown in the supplementary materials, Figure S1). For all tested lattice flexible TPU samples, a T_g associated with the glass transition temperatures of the soft TPU segments was observed at the temperatures around $-22\ ^\circ C$. At the same time, the storage modulus and the loss modulus are higher in the case of lattice TPU with honeycomb architecture in the same lattice cut direction.

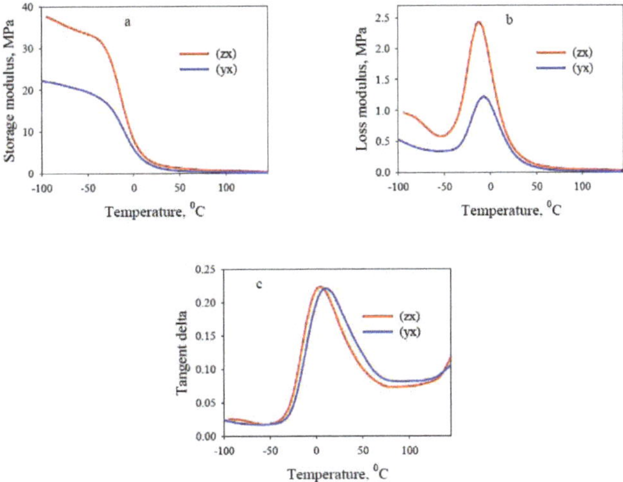

Figure 4. DMA curves for lattice flexible TPU with honeycomb architecture. (**a**)–$E'(T)$, (**b**)–$E''(T)$, (**c**)–tg$\delta(T)$ dependencies.

Table 2. Viscoelastic properties of lattice flexible TPU.

Internal Architecture	Sample Orientation	E', MPa	E'', MPa	E'', MPa	T_E, $^\circ$C	$T_{tg\delta}$, $^\circ$C	tgδ	tgδ_m
Honeycomb	(zx)	2.5	0.40	2.6	-13	7	0.18	0.23
	(yx)	1.7	0.35	1.4	-11	6	0.19	0.23
Gyroid	(zx)	0.7	0.16	1.1	-12	6	0.2	0.27
	(yx)	0.8	0.15	0.85	-11	5	0.2	0.25

In general, the dynamic mechanical analysis data (Table 2) are in good agreement with the results of the tensile and static three-point bending tests, which show similar regularities. The comparison of viscoelastic properties of honeycomb architecture lattice rigid TPU [26] and lattice flexible TPU showed that a decrease in matrix rigidity naturally leads to a significant (up to 10-fold) decrease in E' and E'' values.

3.3. Compression Test

Theoretically, anisotropic structures are not conducive to energy absorption because under uncertain loading conditions, the energy absorption in the weak direction of the structure will be extremely low. Figure 5 shows the stress-strain curves of lattice TPU specimens of different architectures under compression. In both cases, three typical stages of compression of such materials are observed: the linear-elastic stage, the plastic plateau stage, and the densification stage, when all cells collapse, and the material deforms like a continuous medium. It is important to note that the studied samples were not destroyed-they were completely compressed, and over time they almost completely restore their shape (shown in the supplementary materials, Figure S2). The modulus of elasticity, plateau stress σ_L, and densification strain ε_D of the lattice TPU structures are given in Table 3.

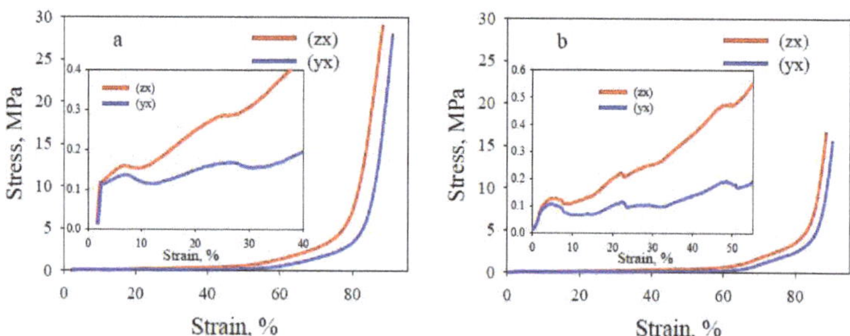

Figure 5. Representative stress–strain curves of lattice flexible TPU under compression. (**a**)–honeycomb, (**b**)–gyroid architecture. The inset shows the enlarged initial part of the same curves.

Table 3. Mechanical properties of lattice flexible TPU. Compression test.

Internal Architecture	Sample Orientation	E, MPa	σ_L, MPa	ε_y, %	ε_D, %
Honeycomb	(zx)	10.8 ± 0.3	4.5 ± 0.2	1.6 ± 0.1	79 ± 3
	(yx)	10.7 ± 0.2	4.3 ± 0.1	1.4 ± 0.1	84 ± 4
Gyroid	(zx)	7.5 ± 0.3	4.1 ± 0.1	1.3 ± 0.1	83 ± 3
	(yx)	7.3 ± 0.3	4.0 ± 0.1	1.1 ± 0.2	86 ± 3

In contrast to the above results of the tensile and three-point bending tests, the compression test shows an almost complete absence of anisotropy of the mechanical properties (modulus of elasticity) for both the lattice TPU with a honeycomb and gyroid architecture (Table 3). This behavior of the materials under compression, in contrast to bending and tension, can be explained by the fact that in this case the properties are largely determined by a reduction in the volume of the pore space. In terms of its influence on the properties of the materials, this factor outweighs the role of the architecture, which explains the closeness of the mechanical properties of the type A and B specimens.

Regardless of the lattice section direction, the TPU lattice with a honeycomb architecture is characterized by a higher strain hardening rate in both small and large deformations (Figure 5), higher values of modulus of elasticity, plateau stress (Table 3) and energy absorption (Figure 6) compared to the TPU lattice with a gyroid architecture. With the increase of compressive strain, several plateaus appeared corresponding to a gradual layer-by-layer collapse (Figure 5, inset). The value of onset densification strain ε_D of the lattice structures, where the compaction region begins (which is the practical limit for energy absorption applications), is slightly lower for samples with a honeycomb architecture than for samples with a gyroid architecture (Table 3).

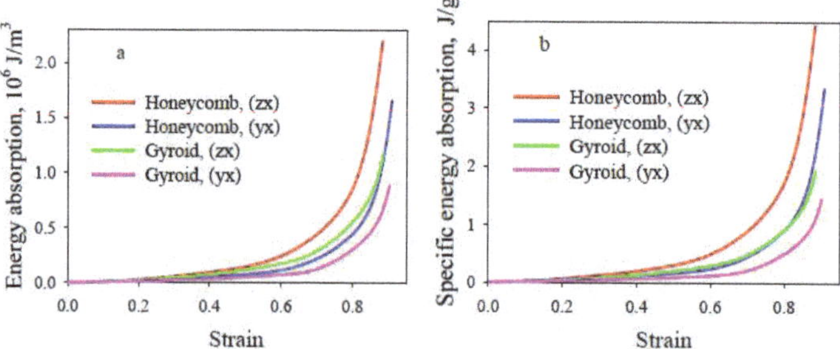

Figure 6. Energy absorption (**a**) and specific energy absorption (**b**) of the lattice flexible TPU structures.

The energy absorption (*EA*) capacity of lattice materials can be calculated from the area under the stress–strain curve:

$$EA = \int_0^{\varepsilon_D} \sigma(\varepsilon)d\varepsilon \tag{4}$$

where σ is stress and ε is strain. The specific energy absorption (*SEA*) is the energy absorption per unit mass which is used to evaluate the energy-absorbing efficiency. The *SEA* is defined as:

$$SEA = \frac{EA}{m} \tag{5}$$

where m is the mass of the lattice material, *EA*-is the absorbed deformation energy per unit volume.

By integrating $\sigma(\varepsilon)$ at different strains, the energy absorption of the lattice structure under different strains can be obtained, as shown in Figure 6a. It can be seen that at the initial stage, the energy absorption capacity of the studied samples is practically the same. However, when the strain exceeds 0.3, the energy absorption of the lattice TPU with honeycomb architecture is higher than that of the gyroid TPU. The differences in energy absorption between the two lattice cut directions correlate with those mentioned earlier and amount to about 25% for both lattice TPU with honeycomb and gyroid architectures. The data from SEA show that the energy absorption performance is significantly higher for lattice TPU with honeycomb architecture, regardless of the lattice cut direction (Figure 6b).

4. Conclusions

The presented study considers the possibility of 3D printing flexible TPU lattice materials with two types of lattice architectures: honeycombs and gyroids. The effects of lattice cut direction and architecture on the mechanical response under various types of loading are systematically investigated. The main findings from this study are as follows.

It is shown that compared to the gyroid architecture, the honeycomb architecture determines the TPU lattice with 30%, 25% and 18% higher values of E, σ_T, ε_b, respectively, measured in both tension and compression.

The presence of the anisotropy of the mechanical properties of the TPU lattice in the case of tension and three-point bending and its absence during compression are noted. The effect of the lattice cut direction is much stronger for samples with honeycomb architecture, while the properties of gyroid-based lattice structures are almost isotropic. The characteristics of the mechanical behavior of the studied TPU lattices are associated with differences in their structure at the meso and macro levels, as well as with the leading role of the pore space in compression tests.

The energy absorption capacity of a lattice structures depends on its architecture and lattice cut direction. The energy absorption performance (by 42%) is significantly higher for lattice TPU with honeycomb architecture, regardless of the lattice cut direction.

Flexible TPU-based 3d printed lattice materials have broad prospects for use in many areas of everyday life, including in medicine, furniture, automotive, civil engineering, etc. Their undoubted advantage is lightness, high mechanical, energy absorption and elastic recovery properties. The results from this study provide experimental data for analyzing and optimizing products made of 3D printed flexible TPU.

Supplementary Materials: The following are available online at https://www.mdpi.com/article/10.3390/polym13172986/s1, Figure S1: DSC curves of lattice flexible TPU with honeycomb and gyroid architecture. Figure S2: The recovery degree of lattice flexible TPU with honeycomb and gyroid architecture over time after the compression tests.

Author Contributions: Conceptualization, V.B. and Y.B.; methodology, V.C., B.S., N.S., D.V., A.V.; software, B.S., I.V.; validation, V.C., A.V. and I.V.; formal analysis, Y.B., V.C., I.V.; investigation, B.S., N.S., D.V., A.V., I.V.; resources, N.S., B.S.; data curation, N.S., D.V.; writing—original draft preparation, V.C.; writing—review and editing, V.B., Y.B., I.V.; visualization, B.S., N.S., A.V.; supervision, V.B.;

project administration, V.B.; funding acquisition, V.B. All authors have read and agreed to the published version of the manuscript.

Funding: The project was financed from funds of the National Academy of Sciences of Ukraine, the project "Formation of spatially structured polymer composites for functional purposes using additive technologies", contract N 15–21 from 9 March 2021.

Data Availability Statement: All the data will be available to the readers.

Acknowledgments: Statutory funds of Donetsk Institute for Physics and Engineering named after O.O. Galkin and the Centre of Molecular and Macromolecular Studies, Polish Academy of Sciences are acknowledged. The personal exchange was financed by the Polish Academy of Sciences and National Academy of Sciences of Ukraine under the agreement for the years 2018–2020.

Conflicts of Interest: The authors declare no conflict of interest.

References

1. Bai, L.; Gong, C.; Chen, X.; Sun, Y.; Xin, L.; Pu, H.; Peng, Y.; Jun Luo, J. Mechanical properties and energy absorption capabilities of functionally graded lattice structures: Experiments and simulations. *Int. J. Mech. Sci.* **2000**, *182*, 105735. [CrossRef]
2. Yan, C.; Hao, L.; Hussein, A.; Bubb, S.L.; Young, P.; Raymont, D. Evaluation of light-weight AlSi10Mg periodic cellular lattice structures fabricated via direct metal laser sintering. *J. Mater. Proc. Techn.* **2014**, *214*, 856–864. [CrossRef]
3. Yan, C.; Hao, L.; Hussein, A.; Young, P.; Raymont, D. Advanced lightweight 316L stainless steel cellular lattice structures fabricated via selective laser melting. *Mater. Des.* **2014**, *55*, 533–541. [CrossRef]
4. Li, Y.; Gu, D. Parametric analysis of thermal behavior during selective laser melting additive manufacturing of aluminum alloy powder. *Mater. Des.* **2014**, *63*, 856–867. [CrossRef]
5. Hakamada, M.; Kuromura, T.; Chen, Y.; Kusuda, H.; Mabuchi, M. High sound absorption of porous aluminum fabricated by spacer method. *Appl. Phys. Lett.* **2006**, *88*, 254106. [CrossRef]
6. Hakamada, M.; Kuromura, T.; Chen, Y.; Kusuda, H.; Mabuchi, M. Sound absorption characteristics of porous aluminum fabricated by spacer method. *J. Appl. Phys.* **2006**, *100*, 114908. [CrossRef]
7. Levy, D.; Shirizly, A.; Rittel, D. Static and dynamic compressive response of additively manufactured discrete patterns of Ti6Al4V. *Int. J. Impact Eng.* **2018**, *122*, 1225. [CrossRef]
8. Bidan, C.M.; Wang, F.M.; Dunlop, J.W.C. A three-dimensional model for tissue deposition on complex surfaces. *Comp. Meth. Biomech. Biomed. Eng.* **2013**, *16*, 1056–1070. [CrossRef]
9. Bidan, C.M.; Kommareddy, K.P.; Rumpler, M.; Kollmannsberger, P.; Brechet, Y.J.M.; Fratzl, P.; Dunlop, J.W.C. How linear tension converts to curvature: Geometric control of bone tissue growth. *PLoS ONE* **2012**, *7*, 36336. [CrossRef] [PubMed]
10. Markkula, S.; Storck, S.; Burns, D.; Zupan, M. Compressive Behavior of pyramidal, tetrahedral, and strut-reinforced tetrahedral ABS and electroplated cellular solids. *Adv. Eng. Mater.* **2009**, *11*, 56–62. [CrossRef]
11. Jiang, J.; Xiong, Y.; Zhang, Z.; Rosen, D. Machine learning integrated design for additive manufacturing. *J. Intell. Manuf.* **2020**, *31*, 865–884.
12. Silva, R.G.; Estay, C.S.; Pavez, G.M.; Viñuela, J.Z.; Torres, M.J. Influence of Geometric and Manufacturing Parameters on the Compressive Behavior of 3D Printed Polymer Lattice Structures. *Materials* **2021**, *14*, 1462. [CrossRef]
13. Nazir, A.; Abate, K.M.; Kumar, A.; Jeng, J.Y. A state-of-the-art review on types, design, optimization, and additive manufacturing of cellular structures. *Int. J. Adv. Manuf. Technol.* **2019**, *104*, 3489–3510. [CrossRef]
14. Zaharia, S.M.; Enescu, L.A.; Pop, M.A. Mechanical Performances of lightweight sandwich structures produced by material extrusion-based additive manufacturing. *Polymers* **2020**, *12*, 1740. [CrossRef]
15. Gautam, R.; Idapalapati, S.; Feih, S. Printing and characterization of kagome lattice structures by fused deposition modelling. *Mater. Des.* **2018**, *137*, 266–275. [CrossRef]
16. Ye, G.; Bi., H.; Chen, L.; Hu, Y. Compression and energy absorption performances of 3D printed polylactic acid lattice core sandwich structures. *3D Print. Additive Manuf.* **2019**, *6*, 333–343. [CrossRef]
17. Płatek, P.; Rajkowski, K.; Cieplak, K.; Sarzyński, M.; Małachowski, J.; Woźniak, R.; Janiszewski, J. Deformation process of 3D printed structures made from flexible material with different values of relative density. *Polymers* **2020**, *12*, 2120. [CrossRef]
18. Bakradze, G.; Arajs, E.; Gaidukovs, S.; Vijay Kumar Thakur, V.K. On the heuristic procedure to determine processing parameters in additive manufacturing based on materials extrusion. *Polymers* **2020**, *12*, 3009. [CrossRef] [PubMed]
19. Zhang, J.Z.; Peng, X.Y.; Liu, S.; Jiang, B.P.; Ji, S.C.; Shen, X.C. The Persistence Length of Semiflexible Polymers in Lattice Monte Carlo Simulations. *Polymers* **2019**, *11*, 295. [CrossRef] [PubMed]
20. Ge, C.; Priyadarshini, L.; Cormier, D.; Pan, L.; Tuber, J. A preliminary study of cushion properties of a 3D printed thermoplastic polyurethane Kelvin foam. *Packag. Technol. Sci.* **2018**, *31*, 361–368. [CrossRef]
21. Bates, S.R.G.; Farrow, I.R.; Trask, R.S. 3D printed polyurethane honeycombs for repeated tailored energy absorption. *Mater. Des.* **2016**, *112*, 172–183. [CrossRef]

22. Bates, S.R.; Farrow, I.R.; Trask, R.S. 3D printed elastic honeycombs with graded density for tailorable energy absorption. Active and Passive Smart Structures and Integrated Systems 2016. In Proceedings of the 2016 International Society for Optics and Photonics; Society of Photo-Optical Instrumentation Engineers (SPIE), San Diego, CA, USA, 28 August–1 September 2016.

23. Bates, S.R.; Farrow, I.R.; Trask, R.S. Compressive behavior of 3D printed thermoplastic polyurethane honeycombs with graded densities. *Mater. Des.* **2019**, *162*, 130–142. [CrossRef]

24. Townsend, S.; Adams, R.; Robinson, M.; Hanna, B.; Theobald, P. 3D printed origami honeycombs with tailored out-of-plane energy absorption behavior. *Mater. Des.* **2020**, *195*, 108930. [CrossRef]

25. Xu, T.; Shen, W.; Lin, X.; Xie, Y.M. Mechanical properties of additively manufactured thermoplastic polyurethane (TPU) material affected by various processing parameters. *Polymers* **2020**, *12*, 3010. [CrossRef]

26. Beloshenko, V.; Beygelzimer, Y.; Chishko, V.; Savchenko, B.; Sova, N.; Verbylo, D.; Vozniak, I. Mechanical Properties of Thermoplastic Polyurethane-Based Three-Dimensional-Printed Lattice Structures: Role of Build Orientation, Loading Direction, and Filler. Available online: https://www.liebertpub.com/doi/10.1089/3dp.2021.0031 (accessed on 14 May 2021).

27. Rodríguez-Parada, L.; Rosa, S.; Mayuet, P.F. Influence of 3D-printed TPU properties for the design of elastic products. *Polymers* **2021**, *13*, 2519. [CrossRef] [PubMed]

28. Vu, C.C.; Nguyen, T.T.; Kim, S.; Kim, J. Effects of 3D printing-line directions for stretchable sensor performances. *Materials* **2021**, *14*, 1791. [CrossRef] [PubMed]

29. Kwon, J.; Ock, J.; Kim, N. Mimicking the mechanical properties of aortic tissue with pattern-embedded 3D printing for a realistic phantom. *Materials* **2020**, *13*, 5042. [CrossRef] [PubMed]

30. Wang, J.; Yang, B.; Lin, X.; Gao, L.; Liu, T.; Lu, Y.; Wang, R. Research of TPU materials for 3D printing aiming at non-pneumatic tires by FDM method. *Polymers* **2020**, *12*, 2492. [CrossRef]

31. Haryńska, A.; Carayon, I.; Kosmela, P.; Brillowska-Dąbrowska, A.; Łapiński, M.; Kucińska-Lipka, J.; Janik, H. Processing of polyester-urethane filament and characterization of FFF 3D printed elastic porous structures with potential in cancellous bone tissue engineering. *Materials* **2020**, *13*, 4457. [CrossRef]

32. Andrews, E.W.; Gioux, G.; Onck, P.; Gibson, L.J. Size effects in ductile cellular solids. Part II: Experimental results. *Int. J. Mech. Sci.* **2001**, *43*, 701–713. [CrossRef]

33. Pham, A.; Kelly, C.; Gall, K. Free boundary effects and representative volume elements in 3D printed Ti–6Al–4V gyroid structures. *J. Mater. Res.* **2020**, *35*, 2547–2555. [CrossRef]

34. Yoder, M.; Thompson, L.; Summers, J. Size effects in lattice structures and a comparison to micropolar elasticity. *Inter. J. Solid. Struct.* **2018**, *143*, 245–261. [CrossRef]

35. Kopal, I.; Harničárová, M.; Valíček, J.; Kušnerová, M. Modeling the temperature dependence of dynamic mechanical properties and visco-elastic behavior of thermoplastic polyurethane using artificial neural network. *Polymers* **2017**, *9*, 519. [CrossRef] [PubMed]

 *polymers*

Article

Effects of Printing Parameters on the Fatigue Behaviour of 3D-Printed ABS under Dynamic Thermo-Mechanical Loads

Feiyang He [1],* and Muhammad Khan [2]

[1] School of Aerospace, Transport and Manufacturing, Cranfield University, College Road, Cranfield MK43 0AL, UK

[2] Centre for Life-Cycle Engineering and Management, Cranfield University, College Road, Cranfield MK43 0AL, UK; Muhammad.A.Khan@cranfield.ac.uk

* Correspondence: feiyang.he@cranfield.ac.uk; Tel.: +44-7851975879

Abstract: Fused deposition modelling (FDM) is the most widely used additive manufacturing process in customised and low-volume production industries due to its safe, fast, effective operation, freedom of customisation, and cost-effectiveness. Many different thermoplastic polymer materials are used in FDM. Acrylonitrile butadiene styrene (ABS) is one of the most commonly used plastics owing to its low cost, high strength and temperature resistance. The fabricated FDM ABS parts commonly work under thermo-mechanical loads in actual practice. For producing FDM ABS components that show high fatigue performance, the 3D printing parameters must be effectively optimized. Hence, this study evaluated the bending fatigue performance for FDM ABS beams under different thermo-mechanical loading conditions with varying printing parameters, including building orientations, nozzle size, and layer thickness. The combination of three building orientations (0°, ±45°, and 90°), three nozzle sizes (0.4, 0.6, and 0.8 mm) and three-layer thicknesses (0.05, 0.1, and 0.15 mm) were tested at different environmental temperatures ranging from 50 to 70 °C. The study attempted to find the optimal combination of the printing parameters to achieve the best fatigue behaviour of the FDM ABS specimen. The experiential results showed that the specimen with 0° building orientation, 0.8 mm filament width, and 0.15 mm layer thickness vibrated for the longest time before the fracture at each temperature. Both a larger nozzle size and thicker layer height can increase the fatigue life. It was concluded that printing defects significantly decreased the fatigue life of the 3D-printed ABS beam.

Keywords: 3D printing; ABS; fatigue; thermo-mechanical loads; building orientation; nozzle size; layer thickness

Citation: He, F.; Khan, M. Effects of Printing Parameters on the Fatigue Behaviour of 3D-Printed ABS under Dynamic Thermo-Mechanical Loads. *Polymers* **2021**, *13*, 2362. https://doi.org/10.3390/polym13142362

Academic Editors: Emin Bayraktar, S. M. Sapuan and R. A. Ilyas

Received: 11 June 2021
Accepted: 14 July 2021
Published: 19 July 2021

Publisher's Note: MDPI stays neutral with regard to jurisdictional claims in published maps and institutional affiliations.

1. Introduction

Three-dimensional (3D) printing, also known as additive manufacturing (AM), is a rapid prototyping technology that was developed in the 1980s. This technology is based on the computer-aided design (CAD) model and helps to construct physical objects using powdered metal or plastic and other bondable materials [1]. Fused deposition modelling (FDM), as an AM process, was initially developed in the early 1990s. It develops the components using layer by layer deposition with the thermoplastic material filaments extruded through a nozzle [2]. Due to its safe, fast, and effective operation, freedom of customisation, and cost-effectiveness, FDM has become the most widely used AM process in customised and low-volume production industries [3]. Apart from functional prototyping, FDM is also used in aerospace, automotive, medical and biomechanical sectors [4–8].

Several different thermoplastic polymer materials are used in FDM. Acrylonitrile butadiene styrene (ABS) is the most commonly used plastic because of its low expense, high strength, and temperature resistance [9]. Although the differences in the mechanical properties between FDM ABS and conventionally manufactured ABS have been discussed in the literature [10–15], the most common conclusion was that FDM ABS has 11% to 37% reduced modulus and 22% to 57% reduced strength [16,17]. Some studies evaluated the different printing parameter effects on FDM ABS mechanical properties (strength and modulus) [18–23]. Some studies attempted to improve the mechanical properties of FDM polymers by optimising the 3D printing parameters [13,24–26]. However, only a few studies focused on the fatigue behaviour of FDM ABS [10,27–31].

As a result of the propagation of cracks due to a repetitive or cyclic load, fatigue failure is normal in practical structures where the cyclic stresses are typically significantly less than yielding strength. Therefore, it is significant to evaluate the fatigue performance of the material. Safai et al. Shanmugam et al. and He et al. reviewed the fatigue behaviour of FDM polymeric materials [8,32,33]. The results of their overview show that determining the best printing parameter combinations for fatigue strength is a challenge because of the synergism between variables and 3D printed material properties. The experimental studies are essential to understand how these parameters affect the fatigue behaviour. Therefore, some studies carried out the experimental works for different 3D printed polymers with variable parameters. Letcher and Waytashek and Afrose et al. all reported that 45° building orientation generated the maximum fatigue strength for 3D printed polylactic acid (PLA) in tension fatigue tests [34–36]. The results of a tension fatigue test for FDM ABS had the same conclusion [10,28,29]. However, the 0°(X)/90°(Y) raster orientation had the best fatigue life in compact tension test for FDM PLA [37]. X orientation provided the highest fatigue strength for FDM ABS in a tension fatigue test [38]. For other printing parameters, some studies carried out the flexural fatigue tests for FDM PLA bar and the results showed that larger layer thickness and nozzle size both increased the fatigue life [39–41]. However, similar research has not been carried out on FDM ABS. Jap et al. listed the potential printing parameters, depicted in Figure 1 (dominant factors shown in red), which may affect the fatigue life of FDM ABS through many complex mechanisms and proposed that raster orientation is a critical parameter that is related to the tensile strength of FDM ABS [27]. Sharma and Ziemian proposed that the 0° or +45°/−45° raster orientation achieved the most fatigue cycles to failure in a tension–tension fatigue test [10,28,29]. By contrast, Hart and Wetzel confirmed that 90° laminae orientation had the highest fracture toughness [30]. Padzi et al. carried out the fatigue test for ABS dog bone specimens and concluded that the 3D printed part had a lower fatigue life [31].

After careful judgement, surprisingly, although there are several printing parameters that may affect the fatigue strength, only a few studies have tested the relationship between them and fatigue performance [10,27–31]; furthermore, these studies only investigated one or two specific parameters [38]. Safai et al. and Shanmugam et al. presented an overview of the fatigue behavior of 3D printed polymers. However, previous studies only performed the tension–tension or rotating bending fatigue tests for FDM ABS [32,33], and as such, although Puigoriol-Forcada et al. investigated the flexural fatigue behaviors of FDM polycarbon (PC) and proposed that build orientation significantly affected the fatigue life due to the inner anisotropy [42], there have been no studies on the bending/flexure fatigue test for FDM ABS. Additionally, no study has considered the thermo-mechanical loading conditions.

The coupled thermo-mechanical load is more common in an actual working environment, such as the 3D printed flex sensor and capacitive interface device working under temperature load [43]. It even forms part of the development of the next-generation space exploration vehicle. NASA launched a FDM CubeSat Trailblazer in November 2013 to demonstrate the durability of 3D-printed devices with extreme thermal cycling [44]. We evaluated the relationship between the several different critical printing parameters and the bending fatigue performance for the FDM ABS beam under different thermo-mechanical loading conditions. Compared with the prior research, the extra thermo loads and dynamic bending stress simulated the practical working conditions better. Building orientation, nozzle size and layer thickness were selected as the significant variables in the research, because the adjustment of these parameters can affect the printing quality, which related to the air-void defects in the specimen. In the study, the specimens with different printing parameters were tested under bending stress with the thermo loads. The fatigue life was measured to find the optimal printing parameter combination to guide the future FDM process and provide a better fatigue life for the FDM structure.

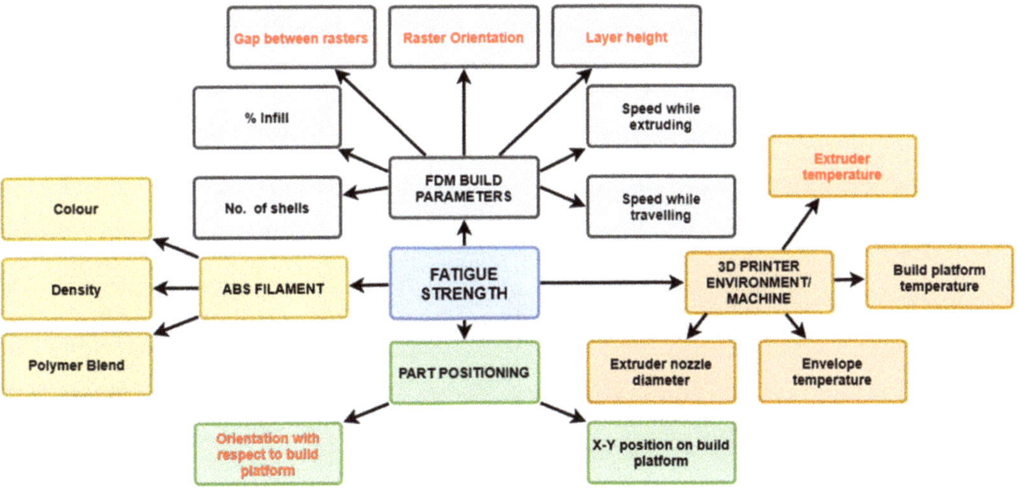

Figure 1. Critical parameters relating to the fatigue performance of the fused deposition modelling (FDM) structure [27].

2. Methodology

2.1. Design of Experiment

The current study focused on three significant parameters (building orientation, nozzle size and layer thickness) and investigated their critical effects on the mechanical behaviour of 3D-printed ABS, as already mentioned in the introduction. These parameters also affect the other aspects of the FDM process, such as the manufacturing time and cost.

1. Building orientation: this can be defined as the path on which the nozzle moves on each layer of the FDM part. It can be set in the 3D printer's software;
2. Nozzle size: this determines the diameter of the filament extruded from the printer's nozzle. Different diameter nozzles can be replaced to print different specimens;
3. Layer thickness: this defines the height of each layer and the number of layers each part has. It can be set in the 3D printer's software.

Three values, which are typically used as the default setting for 3D printing, were evaluated for each parameter to cover the typical range of printing parameters. This facilitated a comprehensive evaluation of the experimental results. In addition, the nozzle size and layer thickness were derived from the profile of the default 3D. These values not only provided excellent printing quality but also covered the typical setting range. The selected values for each parameter are listed in Table 1 and depicted in Figures 2–4.

Table 1. Printing parameters.

Building Orientations	Nozzle Size (mm)	Layer Thickness (mm)
0° (X)	0.4	0.05
±45° (XY)	0.6	0.10
90° (Y)	0.8	0.15

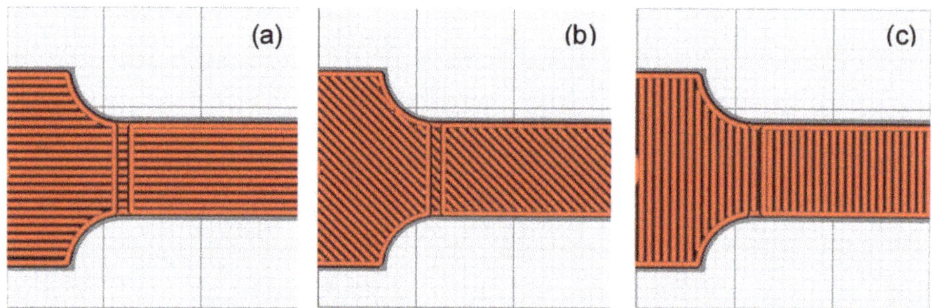

Figure 2. Printing directions: (**a**) X direction; (**b**) XY direction; (**c**) Y direction.

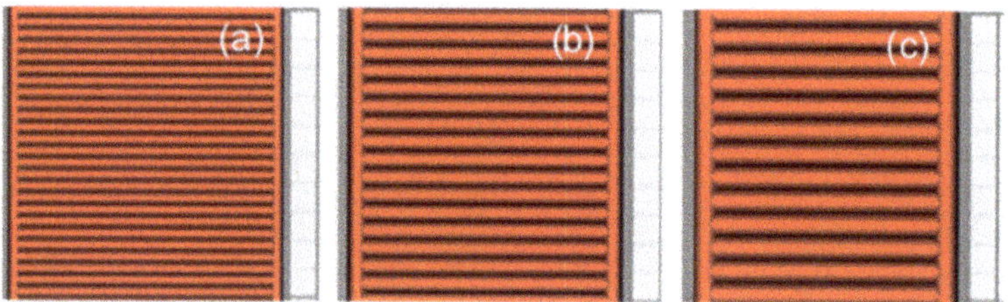

Figure 3. Nozzle size: (**a**) 0.4 mm; (**b**) 0.6 mm; (**c**) 0.8 mm.

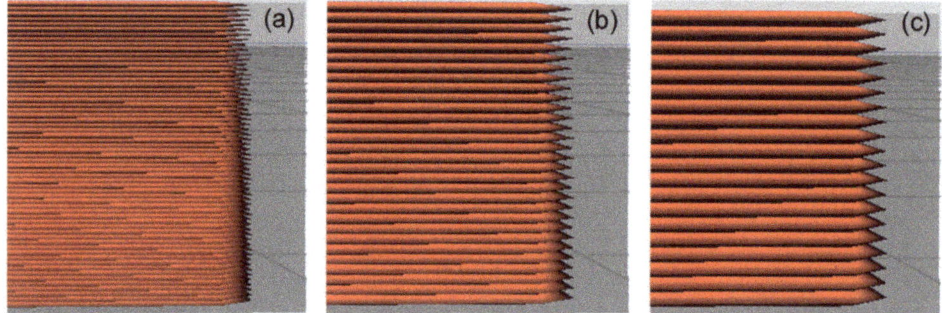

Figure 4. Layer thickness: (**a**) 0.05 mm (**b**) 0.10 mm (**c**) 0.15 mm.

Because the mechanical properties of ABS change a lot when temperature changes, while the practical applications may work under different environmental temperature, a comparison of fatigue life under different temperature is significant in the research. Therefore, three different environmental temperatures, 40, 50 and 60 °C, were selected for this purpose. 60 °C was chosen as the upper limit of temperature because ABS is sensitive to the temperature. The mechanical strength decreases when increasing temperature. The structural strength is not strong enough to support the experiment when environmental temperature is over 60 °C due to softening.

Because each printing parameter had three options as shown in Table 1, there were a total of 27 combinations of the printing parameters. Each combination was tested under three different temperatures, respectively. Consequently, 81 different types of specimen were printed for these parameters and the environmental temperature values. Three similar specimens were manufactured for each combination and tested under the same conditions for confirming the experimental repeatability of the results obtained.

2.2. Materials

In this study, a red ABS filament fabricated by Ultimaker® (Utrecht, Netherlands) was chosen as the raw material because the crack growth path and plastic zones can be clearly observed in this colour. More details of this material are presented in Table 2.

Table 2. Ultimaker® acrylonitrile butadiene styrene (ABS) filament specifications [45].

Filament Specification	Value
Diameter	2.85 ± 0.10 mm
Tensile modulus	1618.5 MPa (ISO 527) [46]
Elongation at yield	3.5% (ISO 527)
Flexural modulus	2070 MPa (ISO 178) [47]
Vicat softening temperature	97 °C
Melting temperature	225–245 °C

2.3. Specimen Preparation

Currently, ASTM D7774 [48] is the standard test method for flexural fatigue properties of plastics. Following the standard, the specimen is cyclically loaded equally in the positive and negative directions to a specific stress or strain level at a uniform frequency until the specimen ruptures or yields. This means the quasi-static loads are applied in the test.

However, the 3D-printed polymers, rather than conventional manufactured polymers, were tested in this research. Extra thermo loads and dynamic mechanical loads were applied during the tests. This means the current standard that focuses on the bending fatigue test of polymers may be not suitable in this research. Therefore, a 3D printing ABS specimen was designed using the CATIA v5 CAD software. The geometry of the specimen is depicted in Figure 5: 150 × 10 × 5 mm and was the same as that used in previous research [49]; therefore, the experimental results of the current study and those of the previous study can be compared. An initial seeded crack with a depth of 0.5 mm was crafted close to the fixed end of the beam to ensure that the maximum stress concentration occurred at the same locations for all the specimens. Therefore, crack propagation in all the experiments was the same and closely resembled fatigue failures of an actual scenario. The test results thus obtained were then compared.

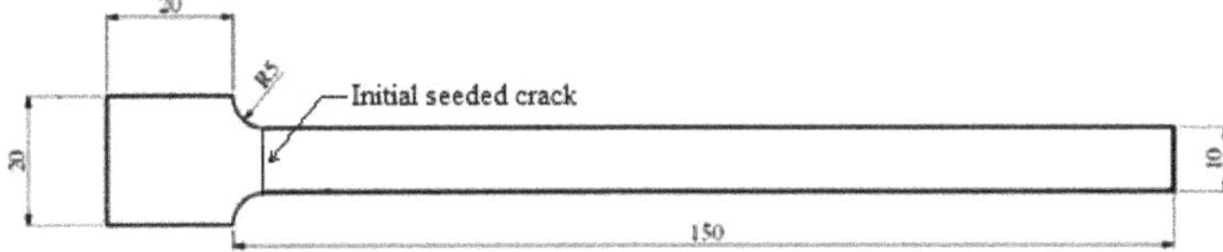

Figure 5. Geometry of specimen.

The specimen's CAD model was converted to an STL file and imported to the Ultimaker CURA 4.6 software. Apart from the selected parameters presented in Table 2, there were a series of parameters in CURA. Because these parameters in CURA were not the focus of the current study, all these settings were held at recommended or default values during the printing process. Further, an infill density of 100% was selected. The appropriate set nozzle temperature was selected as 245 °C based on the recommendation of the printing setup. The bed temperature was set to 90 °C, which was the default value of the printer. Finally, the Ultimaker 2+ printer was used to print all the specimens, as shown in Figure 6.

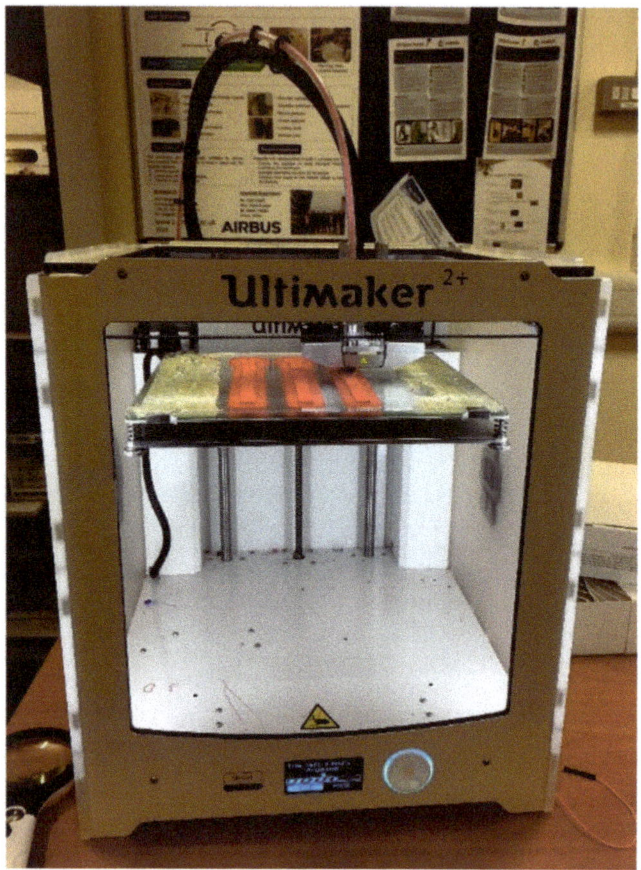

Figure 6. Three-dimensional (3D) printing by Ultimaker 2+ printer.

2.4. Experimental Scheme

The experiments were primarily divided into two parts. First, the number of cycles until the fracture of cantilever beams were evaluated using bending vibration tests. Then, dynamic mechanical analysis tests were performed for the specimens cut from the previous beams.

2.5. Experimental Setup and Procedure

The experimental setup is illustrated in Figure 7. The ABS specimen was fixed on the V55 shaker manufactured by Data Physics (Hailsham, UK). A mica band heater was used to provide the specimen's constant thermal loads throughout the experiment. First, the specimen's natural frequency was measured three times using an accelerometer in an impact test. The first-order natural frequency of the specimen can be calculated using Equation (1):

$$f = \frac{i - j}{t_i - t_j} \tag{1}$$

where t_i and t_j denote the time of the ith and jth peak amplitude, respectively. The vibration under natural frequency provided the initial maximum amplitude for the beam, and this significantly shortened the experiment time. The accelerometer recorded the acceleration and time data, imported them into Signal Express software via a DAQ card. The signal generator produced the sinusoidal output into the power amplifier. The power amplifier then transferred the signal into the shaker, following which the shaker vibrated with a 2 mm amplitude under the measured natural frequency of the beam. This implied that the beam was under resonation initially, following which the pre-seeded crack growth started. The shaker was running continuously until the beam's final fracture occurred.

Figure 7. Experimental setup.

Compared with the traditional fatigue failure approach, which is defined as the number of cycles at which the stiffness of the material reduces by 50% (N_{f50}), the current study directly registered the number of cycles from the beginning of the run to the occurrence of fracture for the specimens because it was difficult to measure the stiffness for the cracked specimen. The results were recorded and then compared with each other. The 3D printing parameter combination corresponding to most cycles was determined as the optimal choice. The beam section after the fracture was captured using a 200× Dino-Lite microscope (AnMo Electronics Corporation, Hsinchu, China).

The DMA test was then performed using Q800 from TA Instruments (New Castle, TX, USA). The specimen was fixed in the chamber of Q800 by the dual cantilever clamp. The temperature ramp rate was 3.00 °C/min. The load oscillation frequency was 1 Hz. This measured the storage modulus of the parts cut from the previous bending fatigue specimens under the temperature range of 30 to 70 °C.

3. Results and Discussion

3.1. DMA Tests

The storage modulus of each specimen cut from the cracked cantilever beams at temperature ranging from 30 to 70 °C was measured by DMA. The statistical significance was determined using MATLAB. A Kruskal–Wallis one-way analysis of variance (Kruskal–Wallis test) with three independent variables (building orientation, nozzle size and layer thickness) was conducted without the consideration of temperature because the data group did not meet the normal distribution requirement. Each printing parameter group contained the sample data, which was the constant in that parameter, but with the different values in other parameters. This tested the hypothesis that the storage modulus of different parameters were equal against the alternative hypothesis that at least one group was different from the others.

The variation of the storage modulus with temperature and statistical data is depicted in Figures 8 and 9. The mean storage moduli for different printing parameters are listed in Table 3. Although calculating the mean storage modulus was not the primary purpose of this study, this value can be used as a reference since it significantly affects the performance of structural fatigue. Higher storage modulus means the higher stiffness. It may potentially lead to longer fatigue life.

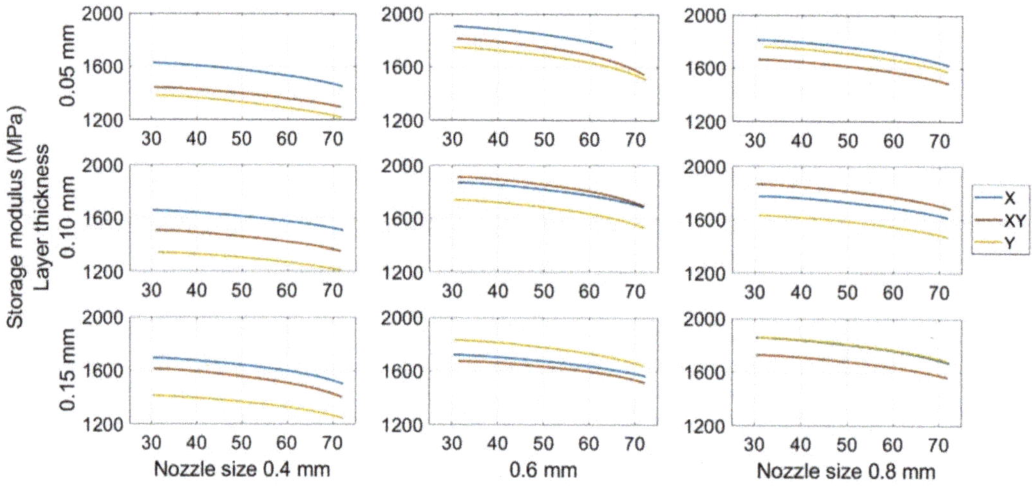

Figure 8. Detailed storage modulus change for different printing parameters from 30 to 70 °C.

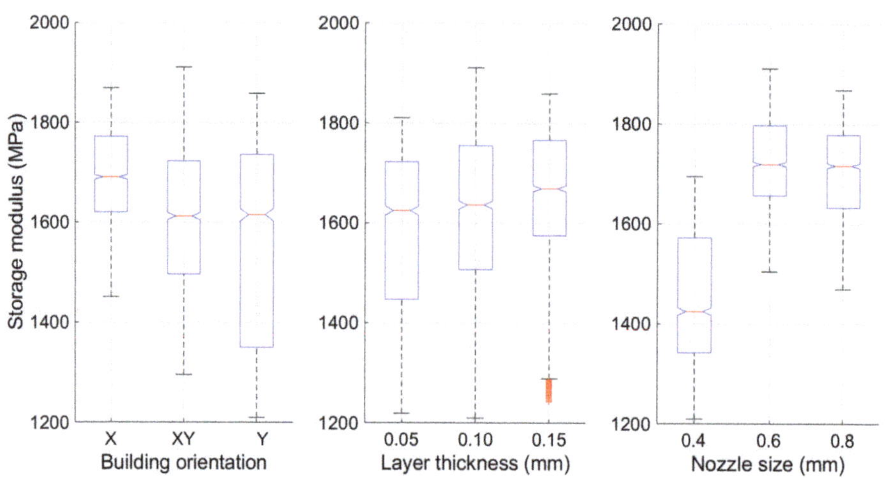

Figure 9. Statistical analysis of the storage modulus for different printing parameters.

Table 3. Mean storage modulus for different printing parameters.

Printing Parameters	Group	Mean Storage Modulus (MPa)	*p*-Value
Building Orientation	X	1692	
	XY	1616	2.34×10^{-115}
	Y	1561	
Nozzle Size (mm)	0.4	1450	
	0.6	1722	0
	0.8	1708	
Layer Thickness (mm)	0.05	1590	
	0.10	1622	2.30×10^{-31}
	0.15	1648	

As can be observed, the storage modulus always decreased with the increase in temperature during the DMA test for all the specimens. This is similar to other materials [49–51]. For the mean storage modulus under the different printing parameter settings, the X orientation provided the highest mean storage modulus, that is, 1692 MPa. The mean storage modulus for the specimen with the Y orientation was only 1561 MPa. The XY building orientation between these two orientations had a mean storage modulus of 1616 MPa. The variation in the storage modulus was found to be similar to that reported in previous studies [36,52–54].

The influence of the layer thickness appeared to be insignificant. Specimens with a layer thickness of 0.15 mm could provide a mean storage modulus of 1648 MPa, which is only approximately 60 MPa more than that provided by a 0.05 mm-thick layer specimen (1590 MPa). Whereas the results showed the difference when comparing with the previous works. Other previous research presented that lower layer thickness seemed to provide the higher strength [14,54]. However, the nozzle size was found to have a significant influence on the storage modulus. The specimens printed with 0.6 mm and 0.8 mm nozzles had a greatly improved storage modulus, that is, over 1700 MPa, as compared with the 1450 MPa mean storage modulus provided by the 0.4 mm nozzle size.

A statistical analysis of the data also indicated a similar conclusion. Compared with the *p*-value for different printing parameters, the *p*-value for the layer thickness of 2.30×10^{-31} was greater than the other two parameters. It confirmed that the storage modulus of FDM ABS is not significantly impacted with the change in layer thickness compared with other two parameters.

3.2. Bending Fatigue Test

Each group of three specimens was tested with a bending fatigue, and the mean cycles to fracture were recorded. The outliers inside each set were filtered and removed. Figure 10 shows the mean value of the number of cycles to the fracture counted for each parameter combination at different environmental temperatures. As can be seen, the specimen with an X raster orientation, nozzle size of 0.8 mm, and layer thickness of 0.15 mm printing had the longest fatigue life until the fracture at different environmental temperatures.

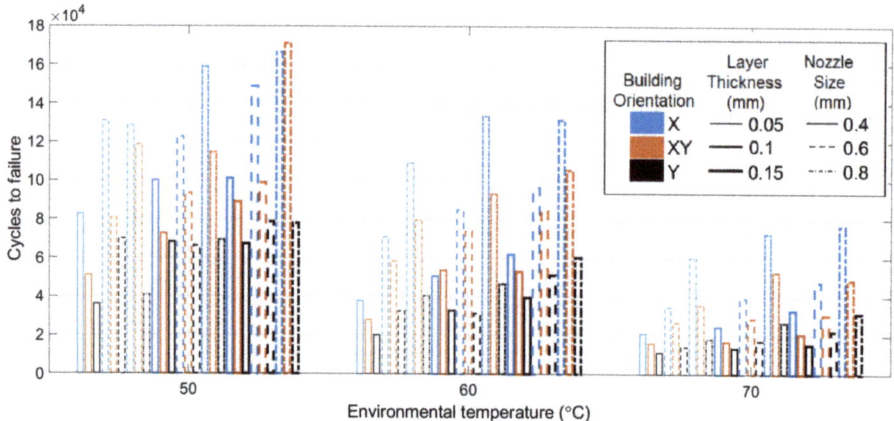

Figure 10. Mean number of cycles of specimens at different temperatures.

3.2.1. Analysis of Variance

The fatigue life results were analysed through the Kruskal–Wallis test at a 95% confidence interval for each parameter. The study calculated the mean number of cycles for each constant parameter as one data group. The Kruskal–Wallis test tested the hypothesis that all group means are equal against the alternative hypothesis that at least one group is different from the others. It was used to determine whether the number of cycles from each group of parameters had a common mean. It can then find out whether different parameters have different influences on the beam's fatigue life.

Table 4 lists the F and p-values for each parameter. The p-value for the building orientation is 5×10^{-5}, for the nozzle size is 9.90×10^{-4}, and for the layer thickness is 0.164, and for the temperature is 6.96×10^{-11}.

Table 4. Statistical parameter values.

Parameter	Chi-Square	p-Value
Building Orientation	17.79	0.0001
Nozzle Size	11.81	0.0027
Layer Thickness	3.56	0.1684
Temperature	41.24	1.11×10^{-9}

Based on the p-values listed in the table above, we can confirm that temperature has the most significant influence on the fatigue strength, statistically. Building orientation and nozzle size also critically affect fatigue life. However, layer thickness did not show a significant influence on the range of values tested because the p-value 0.1684 exceeded 0.05.

3.2.2. Temperature Influence on Fatigue Life

Figure 11 show the variation in the fatigue life at different environmental temperatures. As can be seen, temperature has a significant impact on fatigue life. Notably, all the curves show similar decreasing trends, confirming that the fatigue resistance decreases when the temperature increases from 50 to 70 °C. The prior research focusing on conventional manufactured ABS had the same conclusion [55–57]. Table 5 presents the statistical data regarding the influence of temperature. The mean data was calculated by all specimens with the same temperatures regardless of other parameter settings. The subsequent statistical analysis of other parameters was also based on the same processing. By increasing the environmental temperature from 50 to 70 °C, the mean number of cycles until failure drops from 96,545 to 31,774 drastically.

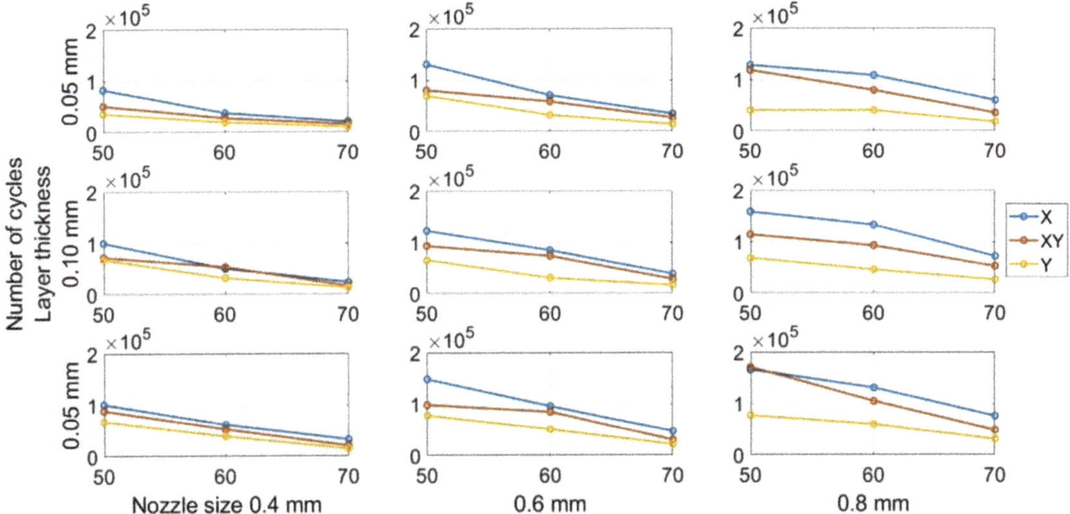

Figure 11. Effect of temperature on fatigue life.

Table 5. Mean number of cycles until fracture at different temperatures.

Environmental Temperature (°C)	Mean Number of Cycles until the Fracture	Standard Deviation
50	96,545	36,969
60	65,312	31,163
70	31,773	17,747

This most significant influence of temperature on fatigue life was due to the deterioration of the mechanical properties with the increase in temperature. The FDM ABS appeared to be glassy at the lower temperatures in terms of the molecular microstructure. In other words, the rotation of the molecular chain ceased to occur and, therefore, the coil could not uncoil and lengthen [58]. The side group and the chain elements could move in a small area. By increasing the environmental temperature, the energy of the moving units and movable space increased. Additionally, the van der Waals force between molecules decreased. The chemical bonds between the molecules were easy to break, and as such, the chain slip occurred easily [59]. Therefore, it can be concluded that microstructural fractures, crack initiation, and propagations are easier in FDM ABS at higher temperatures. Furthermore, such changes in the microstructural properties influence the mechanical behaviour of the specimens. Finally, both the change in the microstructural and mechanical behaviour in FDM ABS contribute significantly to the fatigue life of the beam. The change in the mechanical behaviour is further evidenced from the DMA results presented later herein.

3.2.3. Building Orientation Influence

The building orientation is the most critical printing parameter affecting the fatigue life of the FDM ABS beam. Figure 12 show the variation in the fatigue life for different building orientations. The specimens with the X building orientation have the highest number of cycles in all the subplots. The Y building orientation provides the worst fatigue performance, whereas the performance provided by the XY building orientation lies between them. Table 6 lists the average fatigue life for groups with different building orientations. The X building orientation provides the longest fatigue life with an average of 96,545 cycles until fracture. However, the number of cycles is only 40,699 for beams with the Y building orientation. It is less than half the fatigue life of the X orientation. The experimental results are different with those presented in most previous studies, which performed the tensile fatigue test and reported that X orientation had the best fatigue strength [10,28,29,34–36]. The difference between tension and bending fatigue tests may cause this different result.

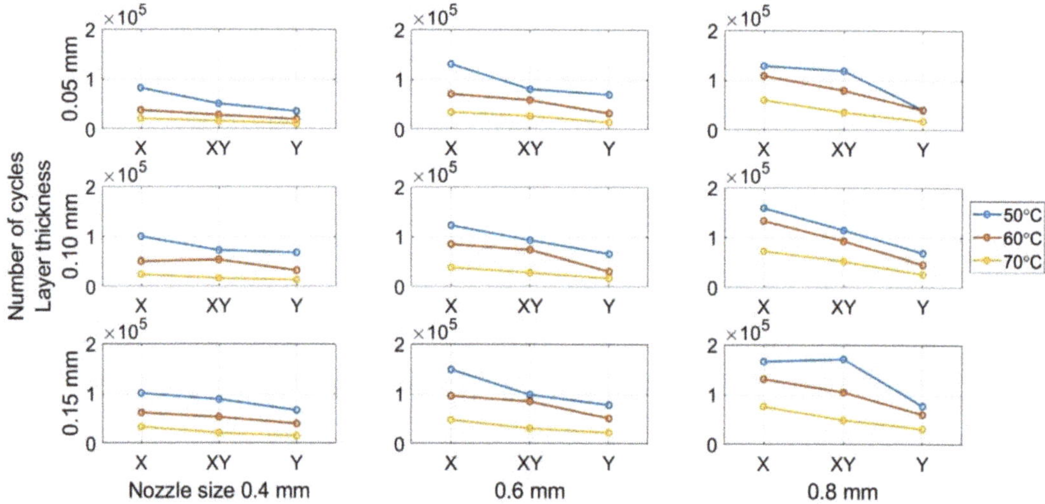

Figure 12. Building orientation effect on fatigue life.

Table 6. Mean number of cycles until fracture for different building orientations.

Building Orientation	Mean Number of Cycles until the Fracture	Standard Deviation
X	86,270	43,307
XY	66,659	37,338
Y	40,699	21,850

The results are reasonable. The initial seeded crack is lateral on the beam, the same as that in the case of the specimen with a Y building orientation. In fracture mechanics, the crack initiates and propagates from micro-cracks on or in the structure [60]. These micro-cracks can be represented by 3D printing defects in the research. The micro-voids occur between the filaments in the structure. The presence of these air voids between the fibres leads to stress concentration when the beam vibrates.

Furthermore, the bonds between the filaments are the weakest in the areas where the voids are present. The occurrence of crazing is earlier in the area with the voids. Notably, the direction of the void area is the same as that of the initial-seeded crack. This provides an excellent crack path that leads to a quick crack propagation. On the contrary, due to the beam's vibration, the bending stress is longitudinal and acts on the Y orientation voids vertically, thereby accelerating the crack growth and decreasing the fatigue life.

3.2.4. Nozzle Size Influence

Figure 13 show the variation tendency of fatigue life for different nozzle sizes. As can be seen, the fatigue strength has an increasing trend when the nozzle size increases for all the specimens. The tests had the similar conclusion with FDM PLA [39–41]. Table 7 present the mean fatigue life for the specimen group with different nozzle sizes. The mean number of cycles increases from 45,195 for the 0.4-mm nozzle size to 84,087 for the 0.8 mm nozzle size. The nozzle size also significantly influences the manufacturing time.

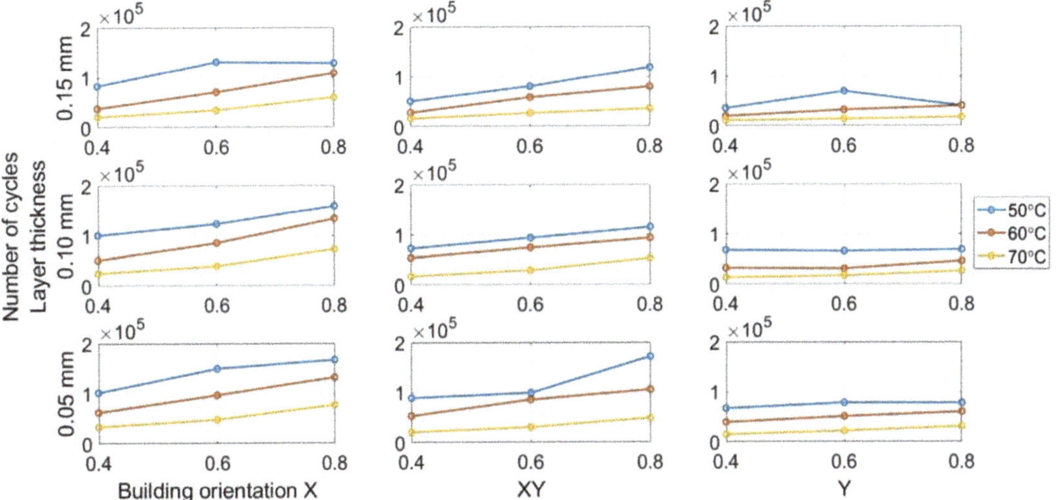

Figure 13. Influence of nozzle size on fatigue life.

Table 7. Mean number of cycles until fracture for different nozzle sizes.

Nozzle Size (mm)	Mean Number of Cycles until the Fracture	Standard Deviation
0.4	45,195	27,329
0.6	64,346	36,201
0.8	84,087	44,604

The effect of nozzle size on the fatigue life was similar to that in the case of a previously reported micro air voids theory. We assumed the spacing between the void defects during the printing process to be almost constant between the filaments, regardless of the thickness of the filament. This implies that the beam printed with a 0.8 mm-nozzle had fewer micro-voids than that printed with a 0.4-mm nozzle. Fewer void defects increase the global density of the specimen, thus reducing the stress concentration. The specimen then has a higher storage modulus, and the crack propagation becomes considerably difficult. Therefore, the overall fatigue life increases with the increase in the size of the nozzle.

3.2.5. Influence of Layer Thickness

We can verify the results obtained from the analysis of variance (ANOVA) test in Section 3.2.1 based on the data presented in Figure 14. The fatigue life only increases slightly when the layer thickness changes from 0.05 mm to 0.15 mm for all the specimens. A similar trend was reported in FDM PLA research [39–41]. Despite the ANOVA results regarding the layer thickness, the influence of layer thickness on the average fatigue life is not entirely clear based on the presented data. The statistical results are presented in Table 8 demonstrate the potential relationship between layer thickness and fracture resistance. The average fatigue life of the beam with a 0.15-mm layer thickness (74,463 cycles) is slightly higher than 0.1-mm (65,196 cycles) and 0.05-mm (53,969) layer thicknesses.

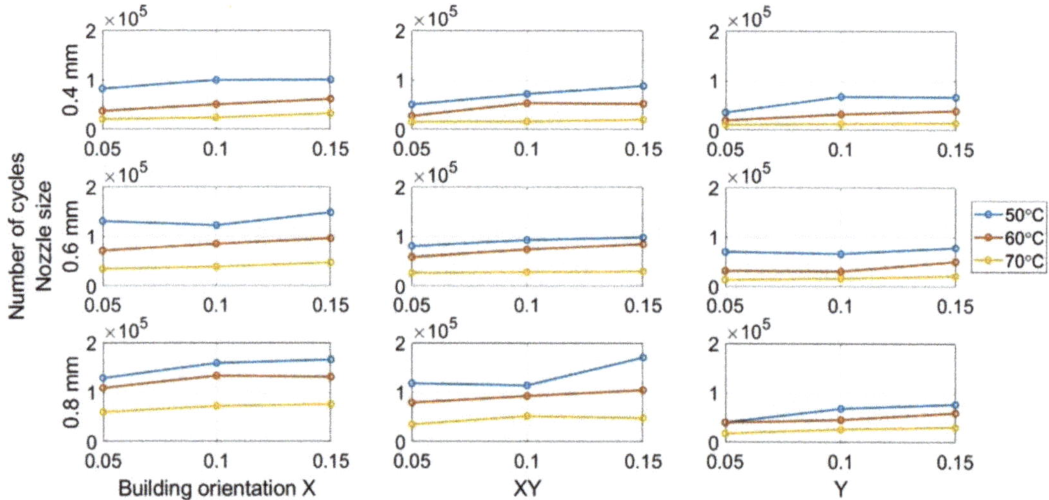

Figure 14. Layer thickness influence on fatigue life.

Table 8. Mean number of cycles until fracture for different layer thickness.

Layer Thickness (mm)	Mean Number of Cycles until the Fracture	Standard Deviation
0.05	53,969	35,857
0.10	65,196	38,314
0.15	74,463	43,126

The possible cause of the effect of layer thickness may be the same as that in the case of nozzle size. Since micro air voids exist between each layer, the size of the voids is also almost constant and does not relate to the layer thickness. As a result, the beam with a thicker layer has fewer microvoids. Therefore, a 0.15-mm thick layer provides the best fatigue performance.

3.3. Technological Recommendations for 3D Printing

The results obtained for fatigue life, as presented above, can help us obtain the optimal 3D printing parameter combination to ensure a better bending fatigue life within a certain parameter range investigated in this research. These parameters are listed in Table 9. A nozzle size of 0.8 mm and a layer thickness of 0.15 mm also had a significant impact on the fatigue life as the use of these parameters led to a significant reduction in the manufacturing time of the parts.

Table 9. Optimal printing parameter combination for a longer fatigue life.

Parameter	Optimal Level for Greater Resistance to Fatigue
Building Orientation	X
Nozzle Size	0.8 mm (The maximum nozzle provided by Ultimaker®)
Layer Thickness	0.15 mm

3.4. Fractography

Different pictures of the specimen fracture section were taken with a Dino-Lite digital microscope (AnMo Electronics Corporation, Hsinchu, China) after the bending fatigue tests. Figure 15 shows which surface is captured by the microscope. All the pictures in the following figures show the different levels of stress-whitening. Furthermore, it can be seen that the mixture of ductile and brittle break occurs during the fracture for all the specimens. The white area represents the ductile fracture, but the red area represents the brittle fracture.

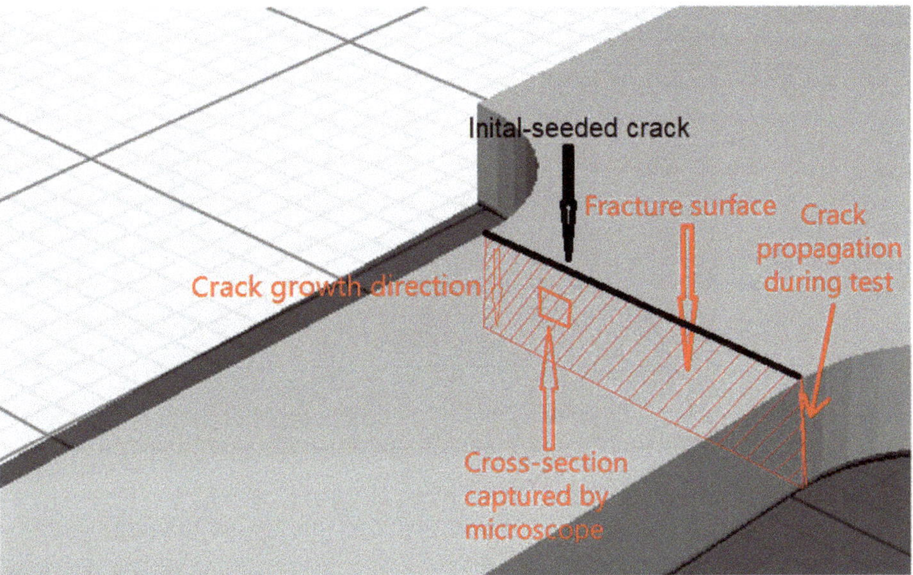

Figure 15. A schematic to show which cross-section was captured by the Dino-Lite digital microscope.

As a universal phenomenon in polymers, stress-whitening is caused by microvoids and crazes [61]. The polymer chains reorganize during tension when the beam is vibrating. The straightening, slipping, and shearing of the fillers and the polymer chains during movement within the plastic's microenvironment may lead to the formation of occlusions or holes in some cases. When these occlusions are combined, microvoids are produced. The transmitted light is scattered when the microvoids cluster to a size greater than or equal to the wavelength of light (380–750 nm). The microvoids change the refractive index of the plastic, causing the object to appear white.

In other words, the defects (voids) of the 3D printing ABS cause local stress concentration and these areas are more likely to induce crazes. Figures 16–18 show the specimens with different nozzle sizes. Similarly, printing defects occur between each filament. However, the distance between each void, representing the nozzle size, varies in the horizontal direction. As can be seen from the aforementioned figures, all the pictures show some bright white lines extending from the printing voids along the direction of crack growth, confirming that the crack propagates along the microvoids and crazes. Then, the void numbers in the cross-section of specimens were calculated approximately. For the 0.15 mm layer thickness specimen, the 0.4 mm nozzle printed a total 500 air voids. This was more than the 333 voids of 0.6 mm and 250 voids of 0.4 mm. This can also be visually observed from the Figures 16–18. Because the number of voids within the same area cross-section of a 0.4-mm nozzle (61) is more than that of a 0.6 (40) and 0.8 mm nozzle (31), consequently, the number of stress whitening areas is also greater in the former. This results in a higher fatigue crack growth rate and short fatigue life.

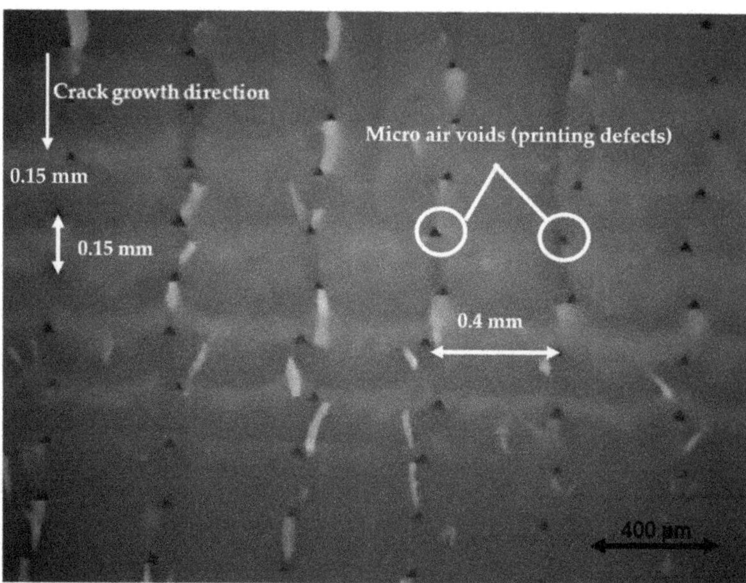

Figure 16. Specimen with a nozzle size of 0.4 mm, X building orientation, and layer thickness of 0.15 mm.

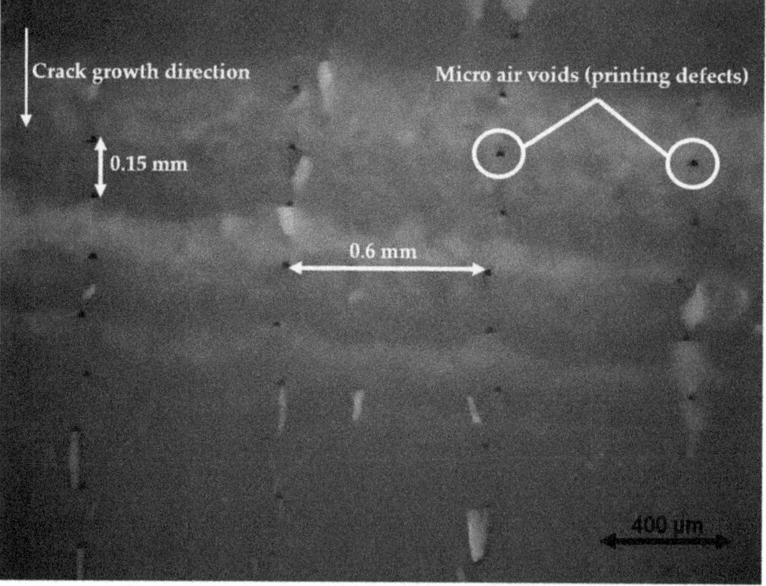

Figure 17. Specimen with a nozzle size of 0.6 mm, X building orientation, and layer thickness of 0.15 mm.

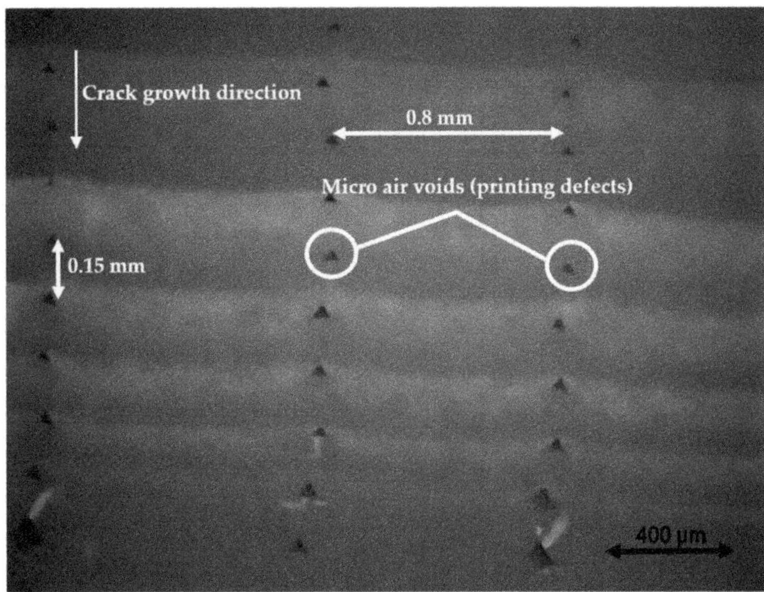

Figure 18. Specimen with a nozzle size of 0.8 mm, X building orientation, and a layer thickness of 0.15 mm.

Furthermore, from the figures mentioned above, it can be observed that all the pictures show that the crack growth rate is time dependent. The white and red areas alternately appear along the crack propagation direction. This means the crack growth dominated by the ductile break and brittle break alternately occur. The layer-by-layer 3D-printed structure may lead to the phenomenon. Furthermore, the brittle break rate in the red area is higher than the ductile break in the white area. It is also reflected on the global colour distribution along the crack propagation direction on the section. Colour near the beam surface is significantly whiter than the bottom area. The ductile break gradually changes to a brittle break during crack propagation. This implies that the initial crack growth rate is less than the latter because the crack tip's stress may increase along with the depth of the crack. The greater the stress, the faster and easier the brittle fracture.

Figure 19 show the cross-sectional area for the XY building orientation. Compared to Figure 16, Figure 19 shows a larger area of voids. Furthermore, the larger stress whitening area connects the voids. These larger area voids decreased the strength of a structure which led to the decreased fatigue life than X building orientation. However, from Figure 20, which presents the fracture section of the Y printing orientation, we can see that there is almost no stress whitening in this section. This is because this section has the weakest bonds between the two filaments. Therefore, the brittle fracture continues along this section. The voids between each filament and layer, represented by dark red, occupy a large area of the fracture surface, thereby speeding up the crack growth rate.

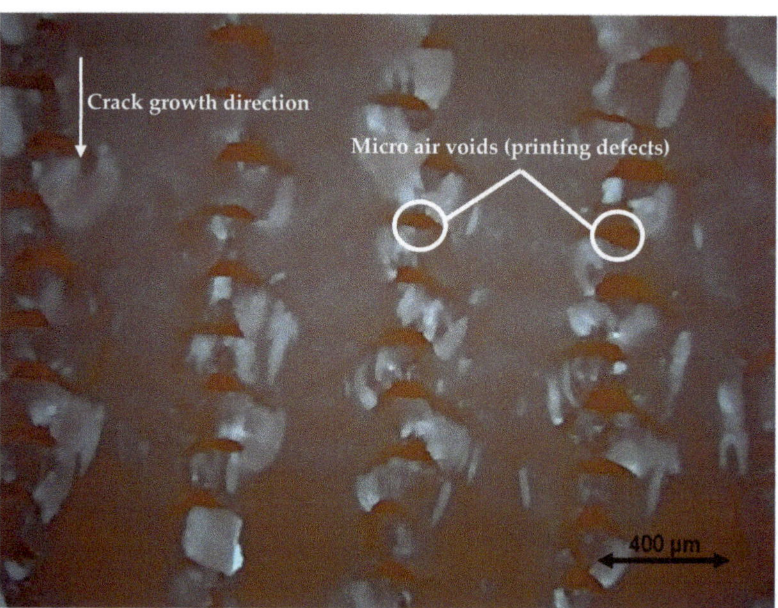

Figure 19. Specimen with nozzle sizes of 0.4 mm, XY building orientation and layer thickness of 0.15 mm.

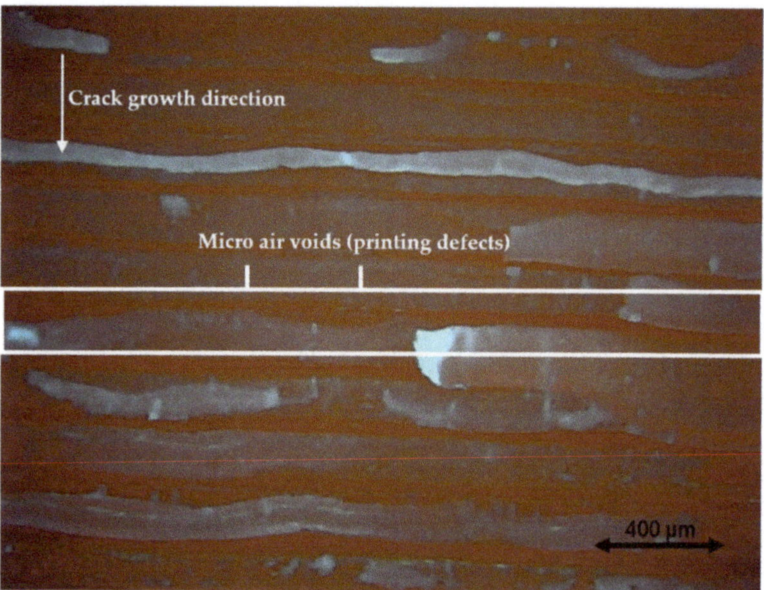

Figure 20. Specimen with a nozzle size of 0.4 mm, Y building orientation, and layer thickness of 0.15 mm.

The inter-filament bonding strength along the crack path in the specimen with a Y building orientation is also extremely weak. As can be seen from Figures 20 and 21, some loose filaments are peeled off the fractured surface.

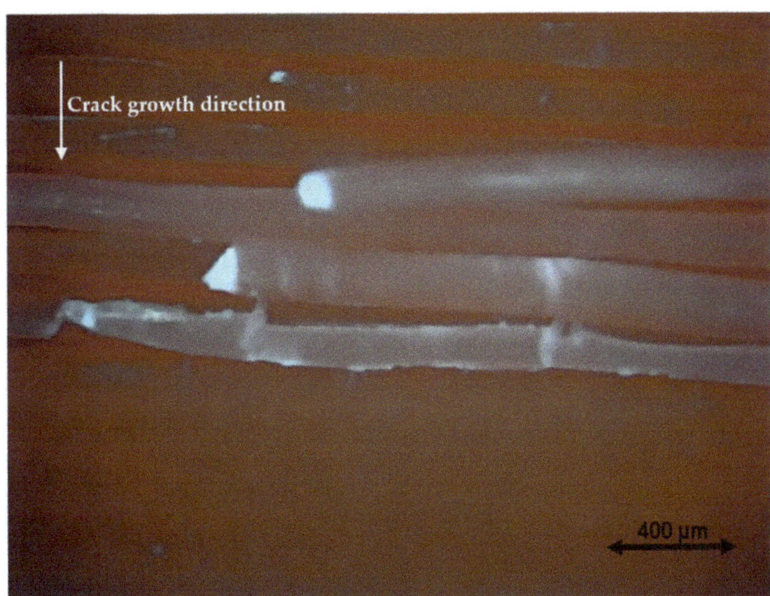

Figure 21. Specimen with a nozzle size of 0.8 mm, Y building orientation, and layer thickness of 0.15 mm.

For the layer thickness effect, when Figures 22 and 23 are compared, we can see that the section with a 0.15 mm layer thickness has fewer defects (dark red area) between the layers in the unit section. This may be the reason why it has a longer fatigue life.

Figure 22. Specimen with a nozzle size of 0.6 mm, Y building orientation, and layer thickness of 0.1 mm.

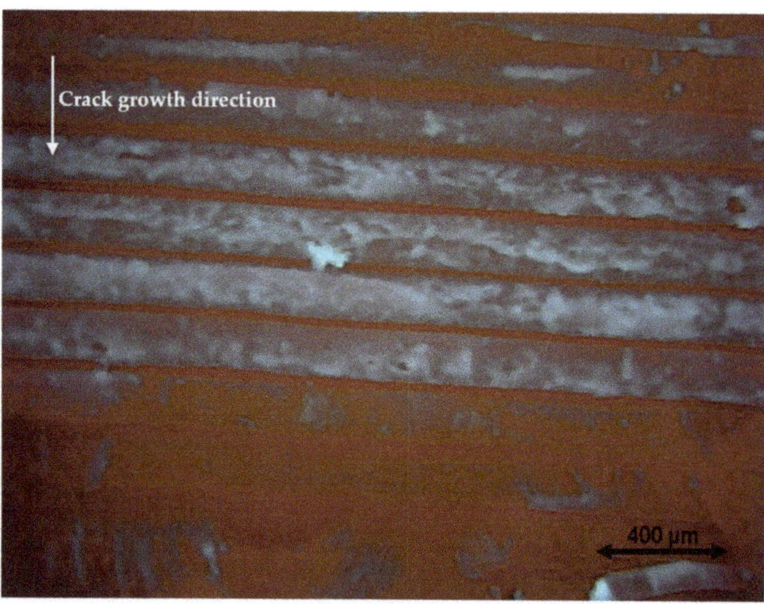

Figure 23. Specimen with a nozzle size of 0.6 mm, Y building orientation, and layer thickness of 0.15 mm.

4. Conclusions

The influence of printing parameters (building orientation, nozzle size, and layer thickness) and environmental temperature on the bending fatigue life of a FDM ABS beam was investigated. The following conclusions were drawn:

1. The environmental temperature has the greatest influence on the fatigue performance, followed by the building orientation and nozzle size, whereas the layer thickness does not show a significant influence with a 0.164 p-value of ANOVA;
2. A combination of the following parameters provides the longest fatigue life among the tested values: X building orientation, 0.8 mm nozzle size, and 0.15 mm-thick layer;
3. Higher temperature reduces the fatigue life possibly due to more active molecular movement. The mean fatigue life of beams under 70 °C is approximate 32,000 cycles. It is only one third of that under 50 °C.
4. Parts with a Y building orientation have the lowest mean fatigue life (around 41,000 number of cycles), compared with 86,000 number of cycles of X building orientation. The reason is that the micro air voids in a Y building orientation beam have the same direction as the initial-seeded transverse crack. This reduces the strength of the potential fracture surface.
5. Both a larger nozzle size and thicker layer height decrease the beams' micro void space and quantity per unit area in the potential crack path and lead to a higher fatigue resistance.

It was found that printing void defects fundamentally affect the fatigue life of the FDM structure. Future work will extend the analytical modelling of the relationship between these printing parameters and the bending fatigue life of the ABS beam.

Author Contributions: Conceptualization, F.H.; methodology, F.H.; software, F.H.; validation, F.H.; formal analysis, F.H.; investigation, F.H.; resources, M.K.; data curation, F.H.; writing—original draft preparation, F.H.; writing—review and editing, F.H. and M.K.; visualization, F.H.; supervision, M.K.; project administration, M.K.; funding acquisition, M.K. All authors have read and agreed to the published version of the manuscript.

Funding: This research received no external funding.

Institutional Review Board Statement: Not applicable.

Informed Consent Statement: Not applicable.

Data Availability Statement: The data presented in this study are available on request from the corresponding author.

Conflicts of Interest: The authors declare no conflict of interest.

References

1. Wang, X.; Jiang, M.; Zhou, Z.; Gou, J.; Hui, D. 3D printing of polymer matrix composites: A review and prospective. *Compos. Part B Eng.* **2017**, *110*, 442–458. [CrossRef]
2. Masood, S.H. *Advances in Fused Deposition Modeling*; Elsevier: Amsterdam, The Netherlands, 2014; Volume 10, ISBN 9780080965338.
3. Vyavahare, S.; Teraiya, S. Fused deposition modelling: A review. *Rapid Prototyp. J.* **2020**, *1*, 176–201. [CrossRef]
4. Ning, F.; Cong, W.; Qiu, J.; Wei, J.; Wang, S. Additive manufacturing of carbon fiber reinforced thermoplastic composites using fused deposition modeling. *Compos. Part B Eng.* **2015**, *80*, 369–378. [CrossRef]
5. Zein, I.; Hutmacher, D.W.; Tan, K.C.; Teoh, S.H. Fused deposition modeling of novel scaffold architectures for tissue engineering applications. *Biomaterials* **2002**, *23*, 1169–1185. [CrossRef]
6. Long, J.; Gholizadeh, H.; Lu, J.; Seyfoddin, A. Review: Application of Fused Deposition Modelling (FDM) Method of 3D Printing in Drug Delivery. *Curr. Pharm. Des.* **2017**, *23*, 433–439. [CrossRef] [PubMed]
7. Yadav, D.K.; Srivastava, R.; Dev, S. Design & fabrication of ABS part by FDM for automobile application. *Mater. Today Proc.* **2020**, *26*, 2089–2093.
8. He, F.; Kumar, V.; Khan, M.A. Evolution and New Horizons in Modelling Crack Mechanics of Polymeric Structures. *Mater. Today Chem.* **2020**, *20*, 100393. [CrossRef]
9. Peterson, A.M. Review of acrylonitrile butadiene styrene in fused filament fabrication: A plastics engineering-focused perspective. *Addit. Manuf.* **2019**, *27*, 363–371. [CrossRef]
10. Ziemian, C.; Sharma, M.; Ziemi, S. Anisotropic Mechanical Properties of ABS Parts Fabricated by Fused Deposition Modelling. In *Mechanical Engineering*; Gokcek, M., Ed.; IntechOpen Limited: London, 2012.
11. Torrado, A.R.; Shemelya, C.M.; English, J.D.; Lin, Y.; Wicker, R.B.; Roberson, D.A. Characterizing the effect of additives to ABS on the mechanical property anisotropy of specimens fabricated by material extrusion 3D printing. *Addit. Manuf.* **2015**, *6*, 16–29. [CrossRef]
12. Ahn, S.; Montero, M.; Odell, D.; Roundy, S.; Wright, P.K.; Montero, M.; Wright, P.K. Anisotropic material properties of fused deposition modeling ABS. *Rapid Prototyp. J.* **2002**, *8*, 248–257. [CrossRef]
13. Lee, B.H.; Abdullah, J.; Khan, Z.A. Optimization of rapid prototyping parameters for production of flexible ABS object. *J. Mater. Process. Technol.* **2005**, *169*, 54–61. [CrossRef]
14. Hibbert, K.; Warner, G.; Brown, C.; Ajide, O.; Owolabi, G.; Azimi, A. The Effects of Build Parameters and Strain Rate on the Mechanical Properties of FDM 3D-Printed Acrylonitrile Butadiene Styrene. *Open J. Org. Polym. Mater.* **2019**, *9*, 1–27. [CrossRef]
15. Tymrak, B.M.; Kreiger, M.; Pearce, J.M. Mechanical properties of components fabricated with open-source 3-D printers under realistic environmental conditions. *Mater. Des.* **2014**, *58*, 242–246. [CrossRef]
16. Rodriguez, J.F.; Thomas, J.P.; Renaud, J. Mechanical behavior of acrylonitrile butadiene styrene (ABS) fused deposition materials. Experimental investigation. *Rapid Prototyp. J.* **2001**, *7*, 148–158. [CrossRef]
17. Rodríguez, J.F.; Thomas, J.P.; Renaud, J.E. Mechanical behavior of acrylonitrile butadiene styrene fused deposition materials modeling. *Rapid Prototyp. J.* **2003**, *9*, 219–230. [CrossRef]
18. Wu, W.; Geng, P.; Li, G.; Zhao, D.; Zhang, H.; Zhao, J. Influence of layer thickness and raster angle on the mechanical properties of 3D-printed PEEK and a comparative mechanical study between PEEK and ABS. *Materials* **2015**, *8*, 5834–5846. [CrossRef]
19. Shubham, P.; Sikidar, A.; Chand, T. The influence of layer thickness on mechanical properties of the 3D printed ABS polymer by fused deposition modeling. *Key Eng. Mater.* **2016**, *706*, 63–67. [CrossRef]
20. Cai, L.; Byrd, P.; Zhang, H.; Schlarman, K.; Zhang, Y.; Golub, M.; Zhang, J. Effect of Printing Orientation on Strength of 3D Printed ABS Plastics. In *TMS 2016 145th Annual Meeting & Exhibition*; The Minerals, Metals & Materials Society (TMS), Ed.; Springer International Publishing: Cham, Switzerland, 2016; pp. 199–204. ISBN 978-3-319-48254-5.
21. Khabia, S.; Jain, K.K. Comparison of mechanical properties of components 3D printed from different brand ABS filament on different FDM printers. *Mater. Today Proc.* **2019**, *26*, 2907–2914. [CrossRef]
22. Croccolo, D.; De Agostinis, M.; Olmi, G. Experimental characterization and analytical modelling of the mechanical behaviour of fused deposition processed parts made of ABS-M30. *Comput. Mater. Sci.* **2013**, *79*, 506–518. [CrossRef]
23. Dawoud, M.; Taha, I.; Ebeid, S.J. Mechanical behaviour of ABS: An experimental study using FDM and injection moulding techniques. *J. Manuf. Process.* **2016**, *21*, 39–45. [CrossRef]
24. Bellehumeur, C.; Li, L.; Sun, Q.; Gu, P. Modeling of bond formation between polymer filaments in the fused deposition modeling process. *J. Manuf. Process.* **2004**, *6*, 170–178. [CrossRef]

25. Abbott, A.C.; Tandon, G.P.; Bradford, R.L.; Koerner, H.; Baur, J.W. Process-structure-property effects on ABS bond strength in fused filament fabrication. *Addit. Manuf.* **2018**, *19*, 29–38. [CrossRef]
26. Chaudhry, M.S.; Czekanski, A. Evaluating FDM process parameter sensitive mechanical performance of elastomers at various strain rates of loading. *Materials* **2020**, *13*, 3202. [CrossRef]
27. Jap, N.S.F.; Pearce, G.M.; Hellier, A.K.; Russell, N.; Parr, W.C.; Walsh, W.R. The effect of raster orientation on the static and fatigue properties of filament deposited ABS polymer. *Int. J. Fatigue* **2019**, *124*, 328–337. [CrossRef]
28. Ziemian, C.W.; Ziemian, R.D.; Haile, K. V Characterization of stiffness degradation caused by fatigue damage of additive manufactured parts. *Mater. Des.* **2016**, *109*, 209–218. [CrossRef]
29. Ziemian, S.; Okwara, M.; Ziemian, C.W. Tensile and fatigue behavior of layered acrylonitrile butadiene styrene. *Rapid Prototyp. J.* **2015**, *21*, 270–278. [CrossRef]
30. Hart, K.R.; Wetzel, E.D. Fracture behavior of additively manufactured acrylonitrile butadiene styrene (ABS) materials. *Eng. Fract. Mech.* **2017**, *177*, 1–13. [CrossRef]
31. Padzi, M.M.; Bazin, M.M.; Muhamad, W.M.W. Fatigue Characteristics of 3D Printed Acrylonitrile Butadiene Styrene (ABS). *IOP Conf. Ser. Mater. Sci. Eng.* **2017**, *269*, 012060. [CrossRef]
32. Safai, L.; Cuellar, J.S.; Smit, G.; Zadpoor, A.A. A review of the fatigue behavior of 3D printed polymers. *Addit. Manuf.* **2019**, *28*, 87–97. [CrossRef]
33. Shanmugam, V.; Das, O.; Babu, K.; Marimuthu, U.; Veerasimman, A.; Johnson, D.J.; Neisiany, R.E.; Hedenqvist, M.S.; Ramakrishna, S.; Berto, F. Fatigue behaviour of FDM-3D printed polymers, polymeric composites and architected cellular materials. *Int. J. Fatigue* **2021**, *143*, 106007. [CrossRef]
34. Letcher, T.; Waytashek, M. Material property testing of 3D-printed specimen in pla on an entry-level 3D printer. In Proceedings of the ASME International Mechanical Engineering Congress and Exposition, Proceedings (IMECE), Montreal, QC, Canada, 14–20 November 2014; Volume 2A.
35. Afrose, M.F.; Masood, S.H.; Iovenitti, P.; Nikzad, M.; Sbarski, I. Effects of part build orientations on fatigue behaviour of FDM-processed PLA material. *Prog. Addit. Manuf.* **2016**, *1*, 21–28. [CrossRef]
36. Afrose, M.F.; Masood, S.H.; Nikzad, M.; Iovenitti, P. Effects of Build Orientations on Tensile Properties of PLA Material Processed by FDM. *Adv. Mater. Res.* **2014**, *1044*, 31–34. [CrossRef]
37. Arbeiter, F.; Spoerk, M.; Wiener, J.; Gosch, A.; Pinter, G. Fracture mechanical characterization and lifetime estimation of near-homogeneous components produced by fused filament fabrication. *Polym. Test.* **2018**, *66*, 105–113. [CrossRef]
38. Lee, J.; Huang, A. Fatigue analysis of FDM materials. *Rapid Prototyp. J.* **2013**, *19*, 291–299. [CrossRef]
39. Jerez-Mesa, R.; Travieso-Rodriguez, J.A.; Llumà-Fuentes, J.; Gomez-Gras, G.; Puig, D. Fatigue lifespan study of PLA parts obtained by additive manufacturing. *Procedia Manuf.* **2017**, *13*, 872–879. [CrossRef]
40. Travieso-Rodriguez, J.A.; Jerez-Mesa, R.; Llumà, J.; Traver-Ramos, O.; Gomez-Gras, G.; Rovira, J.J.R. Mechanical properties of 3D-printing polylactic acid parts subjected to bending stress and fatigue testing. *Materials* **2019**, *12*, 3859. [CrossRef]
41. Gomez-Gras, G.; Jerez-Mesa, R.; Travieso-Rodriguez, J.A.; Lluma-Fuentes, J. Fatigue performance of fused filament fabrication PLA specimens. *Mater. Des.* **2018**, *140*, 278–285. [CrossRef]
42. Puigoriol-Forcada, J.M.; Alsina, A.; Salazar-Martín, A.G.; Gomez-Gras, G.; Pérez, M.A. Flexural fatigue properties of polycarbonate fused-deposition modelling specimens. *Mater. Des.* **2018**, *155*, 414–421. [CrossRef]
43. Leigh, S.J.; Bradley, R.J.; Purssell, C.P.; Billson, D.R.; Hutchins, D.A. A Simple, Low-Cost Conductive Composite Material for 3D Printing of Electronic Sensors. *PLoS ONE* **2012**, *7*, e49365. [CrossRef]
44. Espalin, D.; Muse, D.W.; MacDonald, E.; Wicker, R.B. 3D Printing multifunctionality: Structures with electronics. *Int. J. Adv. Manuf. Technol.* **2014**, *72*, 963–978. [CrossRef]
45. Ultimaker Mechanical properties Injection molding 3D printing Technical Data Sheet ABS. Available online: https://support.ultimaker.com/hc/en-us/article_attachments/360010199279/TDS_ABS_v3.011_-en.pdf (accessed on 18 July 2021).
46. ISO 527-1, Plastics– Determination of tensile properties. In *Part 1: General principles*; ISO Copyright Office: Geneva, Switzerland, 2019; p. 26.
47. *ISO 178:2019, Plastics — Determination of flexural properties*; ISO Copyright Office: Geneva, Switzerland, 2019; p. 25.
48. *ASTM D7774-17 Standard Test Method for Flexural Fatigue Properties of Plastics*; ASTM International: West Conshohocken, PA, USA, 2017. [CrossRef]
49. Baqasah, H.; He, F.; Zai, B.A.; Asif, M.; Khan, K.A.; Thakur, V.K.; Khan, M.A. In-situ dynamic response measurement for damage quantification of 3D printed ABS cantilever beam under thermomechanical load. *Polymers* **2019**, *11*, 2079. [CrossRef]
50. Khan, M.A.; Khan, S.Z.; Sohail, W.; Khan, H.; Sohaib, M.; Nisar, S. Mechanical fatigue in aluminium at elevated temperature and remaining life prediction based on natural frequency evolution. *Fatigue Fract. Eng. Mater. Struct.* **2015**, *38*, 897–903. [CrossRef]
51. Zai, B.A.; Khan, M.A.; Khan, K.A.; Mansoor, A. A novel approach for damage quantification using the dynamic response of a metallic beam under thermo-mechanical loads. *J. Sound Vib.* **2020**, *469*, 115134. [CrossRef]
52. Schirmeister, C.G.; Hees, T.; Licht, E.H.; Mülhaupt, R. 3D printing of high density polyethylene by fused filament fabrication. *Addit. Manuf.* **2019**, *28*, 152–159.
53. Riddick, J.C.; Haile, M.A.; Von Wahlde, R.; Cole, D.P.; Bamiduro, O.; Johnson, T.E. Fractographic analysis of tensile failure of acrylonitrile-butadiene-styrene fabricated by fused deposition modeling. *Addit. Manuf.* **2016**, *11*, 49–59. [CrossRef]

54. Rankouhi, B.; Javadpour, S.; Delfanian, F.; Letcher, T. Failure Analysis and Mechanical Characterization of 3D Printed ABS with Respect to Layer Thickness and Orientation. *J. Fail. Anal. Prev.* **2016**, *16*, 467–481. [CrossRef]

55. Mura, A.; Ricci, A.; Canavese, G. Investigation of fatigue behavior of ABS and PC-ABS polymers at different temperatures. *Materials* **2018**, *11*, 1818. [CrossRef] [PubMed]

56. Kim, H.S.; Wang, X.M.; Abdullah, N.A.H.N. Effect of temperature on fatigue crack growth in the polymer abs. *Fatigue Fract. Eng. Mater. Struct.* **1994**, *17*, 361–367. [CrossRef]

57. Kim, H.O.S.; Wang, X.M. Temperature and Frequency Effects on Fatigue Crack Growth in Acrylonitrile-Butadiene-Styrene (ABS). *J. Appl. Polym. Sci.* **1995**, *57*, 811–817. [CrossRef]

58. Introduction to Polymers. Available online: https://www.open.edu/openlearn/science-maths-technology/science/chemistry/introduction-polymers/content-section-2.5.3 (accessed on 18 July 2021).

59. Tuttle, M.E. *A Brief Introduction to Polymeric Materials*; University of Washington: Seattle, WA, USA, 1999.

60. Wang, C.H. *Introduction To Fracture Mechanics*; DSTO Aeronautical and Maritime Research Laboratory: Kensington, NSW, Australia, 1996.

61. Bucknall, C.B.; Smith, R.R. Stress-whitening in High-impact Polystyrenes. *Polymer* **1965**, *6*, 437–446. [CrossRef]

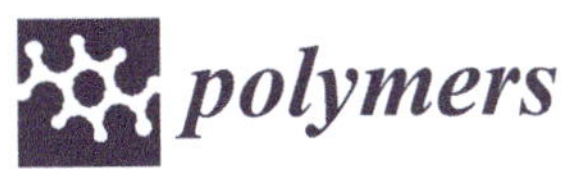 *polymers*

Article

Optimization of FFF Process Parameters by Naked Mole-Rat Algorithms with Enhanced Exploration and Exploitation Capabilities

Jasgurpreet Singh Chohan [1], Nitin Mittal [2], Raman Kumar [1], Sandeep Singh [3], Shubham Sharma [4,*], Shashi Prakash Dwivedi [5], Ambuj Saxena [5], Somnath Chattopadhyaya [6], Rushdan A. Ilyas [7,8], Chi Hieu Le [9] and Szymon Wojciechowski [10,*]

1   Department of Mechanical Engineering, Chandigarh University, Mohali 140413, India; jasgurpreet.me@cumail.in (J.S.C.); ramankakkar@gmail.com (R.K.)
2   Department of Electronics and Communication Engineering, Chandigarh University, Mohali 140413, India; mittal.nitin84@gmail.com
3   Department of Civil Engineering, Chandigarh University, Mohali 140413, India; drsandeep1786@gmail.com
4   Department of Mechanical Engineering, IK Gujral Punjab Technical University, Main Campus, Kapurthala 144603, India
5   G.L. Bajaj Institute of Technology & Management, Greater Noida 201310, India; spdglb@gmail.com (S.P.D.); ambuj.saxena1@gmail.com (A.S.)
6   Department of Mechanical Engineering, Indian Institute of Technology (ISM), Dhanbad 826004, India; somnathchattopadhyaya@iitism.ac.in
7   Faculty of Engineering, School of Chemical and Energy Engineering, Universiti Teknologi Malaysia, Johor Bahru 81310, Malaysia; ahmadilyas@utm.my
8   Centre for Advanced Composite Materials, Universiti Teknologi Malaysia, Johor Bahru 81310, Malaysia
9   Faculty of Engineering and Science, University of Greenwich, Kent ME4 4TB, UK; c.h.le@gre.ac.uk
10   Faculty of Mechanical Engineering and Management, Poznan University of Technology, 60-965 Poznan, Poland
*   Correspondence: shubham543sharma@gmail.com (S.S.); sjwojciechowski@o2.pl (S.W.)

Citation: Chohan, J.S.; Mittal, N.; Kumar, R.; Singh, S.; Sharma, S.; Dwivedi, S.P.; Saxena, A.; Chattopadhyaya, S.; Ilyas, R.A.; Le, C.H.; et al. Optimization of FFF Process Parameters by Naked Mole-Rat Algorithms with Enhanced Exploration and Exploitation Capabilities. *Polymers* **2021**, *13*, 1702. https://doi.org/10.3390/polym 13111702

Academic Editor: Carola Esposito Corcione

Received: 2 May 2021
Accepted: 21 May 2021
Published: 23 May 2021

Publisher's Note: MDPI stays neutral with regard to jurisdictional claims in published maps and institutional affiliations.

Abstract: Fused filament fabrication (FFF) has numerous process parameters that influence the mechanical strength of parts. Hence, many optimization studies are performed using conventional tools and algorithms. Although studies have also been performed using advanced algorithms, limited research has been reported in which variants of the naked mole-rat algorithm (NMRA) are implemented for solving the optimization issues of manufacturing processes. This study was performed to scrutinize optimum parameters and their levels to attain maximum impact strength, flexural strength and tensile strength based on five different FFF process parameters. The algorithm yielded better results than other studies and successfully achieved a maximum response, which may be helpful to enhance the mechanical strength of FFF parts. The study opens a plethora of research prospects for implementing NMRA in manufacturing. Moreover, the findings may help identify critical parametric levels for the fabrication of customized products at the commercial level and help to attain the objectives of Industry 4.0.

Keywords: fused-filament fabrication; mechanical strength; naked mole-rat algorithm; optimization; process parameters

1. Introduction

In the last few decades, the manufacturing sector has witnessed a paradigm shift from conventional to advanced manufacturing techniques in both developed and developing nations. Another breakthrough was achieved by researchers after developing additive manufacturing (AM) technologies, which is gradually taking the present industry into the era of Industry 4.0 [1]. The ever-increasing demand for good quality and customized products at lower cost has brought additive manufacturing technologies into the limelight.

AM primarily solves the problem of long delay periods between customer demands and product development. During new product development, conventionally, it must pass through many stages, which are eliminated in AM, even for complex parts [2]. This technology used digital fabrication methods where the computer-generated design of the required product is created through a layer-by-layer manufacturing technique. There is no requirement for jigs, fixtures, dies and other tools for processing. Researchers have even successfully developed multiple valve attachments for ventilators during emergency conditions, such as pandemic occurred in 2020 [3]. Thus, the rapid delivery of high-quality customized products while maintaining minimum cost is the reason for the supremacy of these techniques.

Fused-filament fabrication (FFF) is the most popular AM technology, which works on the principle of layer-by-layer manufacturing as designated by ASTM [4]. This technology can transfer conceptual digital drawing into an actual product within a few hours using a wide range of polymers and polymer composite filaments as raw material. The fiber and particle-reinforced composites find vast applications in the areas of medicine, aerospace, automobile and electronics [5,6]. Moreover, carbon-reinforced composites fabricated using FFF are more cost-efficient than powdered alloys used in metal printers [7]. Due to its material flexibility, easy transportation, minimal environmental degradation, lower installation and material cost [8], researchers have used it extensively while developing smart materials, metamaterials, implants, and other biomaterials for medical applications. Medical 3D printing is gaining popularity as customized anatomical models are prepared with higher accuracy and minimum lead time [9]. The inclusion of materials with sensing capabilities in 3D printing has realized the dream of producing smart materials [10]. FFF uses two-dimensional motion (X- and Y-direction) of heated extrusion nozzle, moved by stepper motors and controlled by a microprocessor. The feedstock filament is passed by rotating rollers into the extrusion nozzle, which heats it slightly lower than the melting point, as shown in Figure 1. This semi-molten polymer bead is deposited on the build platform; after one layer is deposited, the build platform moves downward (Z-direction), and the next layer of materials is deposited. The motion of the nozzle and build platform are synchronized by the microprocessor according to the product design [11]. The part is immediately cooled and cleaned after completion, after which it can be used for the desired application.

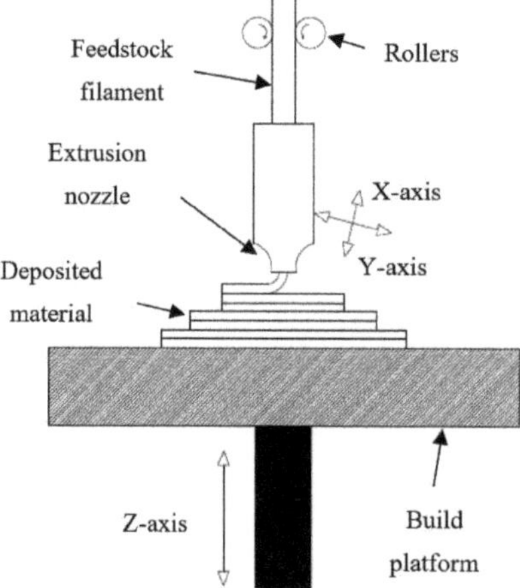

Figure 1. Schematic of the Fused Filament Fabrication (FFF) process.

Despite advantages, FFF has certain process limitations like poor surface finish and mechanical strength than injection molding [12]. The surface finish can be enhanced by postprocessing methods, including mechanical finishing, chemical finishing and vapor smoothing. However, there is no method available, which can improve the mechanical strength of FFF parts. This aspect limits the usability of these components for certain applications where flexural strength, compressive strength and tensile strength are mandatory [13]. On the contrary to postprocessing, the FFF part shows significant variation in mechanical stability while manufactured at different conditions. There are numerous studies, which have found that process parameters like layer thickness, raster angle, build temperature, orientation angle, infill density, infill pattern and layer gap have a significant impact on the mechanical stability of FFF parts [14,15]. As a general notion, the filament layers deposited in the direction of tensile force and compressive force results in higher mechanical strength [11]. Honeycomb structure with 100% infill density has yielded maximum tensile strength, as reported by Fernandez-Vicente et al. [16]. However, complex issues arise when estimating strength is done for intricate designs manufactured using different build strategies. In addition to conventional optimization techniques, researchers have utilized advanced algorithms to identify optimum parameter settings to attain higher mechanical stability. Section 2 presents different optimization algorithms used for mechanical strength enhancement of FFF parts. Afterward, in Sections 3 and 4, the novel naked mole-rat algorithm (NMRA) and its variants are discussed. Section 5 elaborates the case study performed to test the performance of advanced algorithms followed by simulation results and discussion in Section 6.

2. Literature Review

Most of the studies, which reported using advanced algorithms for the FFF process were performed to enhance surface finish and mechanical strength of parts by optimizing various process parameters. Initially, conventional techniques were used for optimizing tensile strength based on orientation angle. At the same time, results were compared with finite element analysis (FEA) data. The simulated results were in strong correlation with experimental data, thus validating the FEA model. The maximum tensile strength has occurred at a 45° orientation angle [17]. Panda et al. [18] studied the impact of several process parameters of FFF-like layer thickness, raster width, raster angle, orientation, and air gap on tensile strength, flexural strength and impact strength of standard test samples. Experiments were carried out, and bacterial foraging technique was implemented after ANOVA tests. This advanced optimization technique was robust enough to predict the response for parameters outside the range of mathematical models formulated by experimental data.

Rayegani and Onwubolu [19] proposed a relation between input parameters of FFF with tensile strength using the hybrid group method of data handling. The raster angle, air gap, orientation angle, and raster width varied during experimentation. Results were compared with modeled data, which showed a strong correlation between experimental and predicted tensile strength. Goudswaard et al. [20] tested and compared the efficacy of evolutionary algorithm, particle swarm optimization and Simulated Annealing during optimizing tensile strength of FFF parts. It was clear that particle swarm optimization outclassed both evolutionary and simulated annealing algorithms in quality and consistency. Liu et al. [13] solved the problems of the anisotropic behavior of FFF materials using hybrid deposition path planning and topology optimization technique. The anisotropy is induced by tool paths generated during the slicing of part, which results in the variation of mechanical properties in different loading directions. The hybrid tool path was developed using the techniques above, which performed better than regular infill strategies, such as crisscross and contour offset. The simulated results were compared with actual parts fabricated using a new algorithm, which validated its efficacy.

Genetic programming technique [1] was used for optimizing fatigue behavior of FFF parts considering six input parameters with three levels of each. Contour number was found to be a significant parameter than the other five. At the same time, genetic programming outperformed response surface methodology during the prediction of fatigue.

The artificial neural networks technique was successfully implemented to enhance the creep behaviors of FFF parts after investigating the impact of five input parameters [21]. The best creep compliance occurred at 0.127 mm layer thickness, zero raster angle and air gap, 17.188° deposition angle, 0.4572 mm width with 10 contours. ANN emerged as the most popular optimization tool for manufacturing processes. It can be applied to nonlinear and multitudinous cases, which was validated by this study. Raju et al. [22] utilized hybrid particle swarm and bacterial foraging techniques for four FFF process parameter optimization while studying their impact on hardness, mechanical strength and surface finish of test samples. Compared to conventional optimization tools, the hybrid tool yielded a 7.44% higher response in all the output parameters. Moreover, this tool also performed better than the other algorithms like conventional particle swarm and bacterial foraging. Rao and Rai [23] solved the single and multiobjective optimization issues of FFF using simple and non-dominated TLBO techniques, respectively. Five case studies were discussed with different objective functions and parameter bounds. The response parameters, such as compressive strength, sliding wear, flexural strength, impact strength and tensile strength, were investigated using advanced optimization tools. All the mathematical models were created by previous researchers during experimentations, which were employed as input to study and compare the performance of the optimization tools, which proves highly efficient than previous results.

Malviya and Desai [24] coupled two techniques, i.e., artificial neural networks and Bayesian algorithm, to optimize orientation angle. A significant impact on tensile strength parameters was experienced. The proposed methodology has successfully achieved desired goals despite the anisotropic behavior of FFF parts. Yadav et al. [25] discussed the impact of infill density, material density and nozzle temperature on the tensile strength of FFF parts manufactured by different materials. They used a hybrid genetic algorithm-artificial neural network technique for the optimization. The experimental data showed strong similarity with predicted results with an error of less than 3%, thus validating the efficiency of the hybrid algorithm. Natarajan et al. [26] used a Non-dominated sorting modified TLBO for conventional machining processes. The impact of four input parameters was studied during the machining of polytetrafluoroethylene with a cemented carbide tool. This algorithm performed better than the other six algorithms by yielding more uniform results and non-dominated solutions. While comparing predicted and experimental results, only 3.7% error was observed while validating the efficacy of this tool. Some researchers also used different algorithms to solve a single objective function to compare and identify the most efficient algorithm. Saad et al. [27] tested four different algorithms for minimizing the surface roughness of FFF parts. The symbiotic organism search algorithms were most effective as minimum surface roughness was predicted at optimum parameter settings, further validated by experimental results.

Another research reported using the firefly algorithm to optimize the wear rate of copper-plated FFF parts. The mathematical model was designed based on the raster angle, air gap, voltage and time to predict the wear rate [28]. A Multicriteria genetic algorithm was utilized by Pandey et al. [2] to predict optimum orientation angle against two objective functions, i.e., build time and surface roughness. The algorithm was robust enough to predict the surface roughness and build time for parts of any geometry.

The research, development and testing of the efficacy of new algorithms is a never-ending process as advanced versions of algorithms are rapidly developed by researchers. This study focuses on optimizing process parameters of FFF to attain maximum mechanical strength using NMRA [29] and its variants, which are discussed in the next section.

3. Naked Mole-Rat Algorithm Variants

Researchers are motivated by the behavioral characteristics of naked mole rats (NMRs) to design an optimization technique termed NMRA [29]. The algorithm reproduces the natural mating patterns of NMRs for designing algorithms. NMRs are classified into two types, i.e., breeders and worker NMRs. The breeders are the most efficient NMRs in the group, and these are only intended for mating with the queen. Workers are doing the additional tasks, and the best performing workers will move on to the breeding group. The best executor of the breeder's pool will be the queen's mating partner.

There are three steps to the algorithm. The NMR group is initially created, and the NMR group is then divided into employees and breeders. The group of breeders is selected based on the probability of breeding. The NMRA steps are as follows:

3.1. Initialization

Initially, n NMRs are generated at random. The D-dimensional vector space is represented by each NMR in the range of $[1, 2, \dots \dots, n]$. Each NMR shall be initialized as

$$NMR_{i,j} = NMR_{min,j} + U(0,1) \times \left(NMR_{min,j} - NMR_{max,j} \right) \tag{1}$$

where $i \in [1, 2, \dots \dots, n]$, $j \in [1, 2, \dots \dots, D]$, and $U(0,1)$ is a uniform random number. After initialization, the objective function and fitness values are determined, and breeders (B) and workers (W) based on their fitness function and the best solution d is selected.

3.2. Worker Phase

To switch to the breeder group and to get a chance to mate with the queen, each worker tries to improve his fitness. The new fitness value of NMR is calculated, and the updated solution is selected and recorded if the fitness is better than the previous one. The previous solution will otherwise be chosen.

The equation used to generate the updated solution is as follows:

$$w_i^{t+1} = w_i^t + \lambda \left(w_j^t - w_k^t \right) \tag{2}$$

where ith worker is represented by w_i^t in the tth iteration. The random solution (jth and kth worker) chosen from the worker's group is w_j^t and w_k^t. w_i^{t+1} represents a new fitness solution. A uniform distribution ranges from $[0, 1]$ provides the value of λ.

3.3. Breeder Phase

The breeder needs to update the NMRs to be chosen as their partner or to remain a breeder. The breeder's NMRs are updated with the best d overall reproductive probability (b_p). This b_p is a random $[0, 1]$ number. If breeders do not update fitness, they may be removed from the worker's pool:

$$b_i^{t+1} = (1 - \lambda)b_i^t + \lambda\left(d - b_i^t\right) \tag{3}$$

where in tth iteration b_i^t represents ith breeder, λ controls the mating frequency and assists in the next iteration in identifying a new breeder b_i^{t+1}. At first, b_p is set to 0.5 for breeders.

The entire search process is repeated iteratively until the termination criteria is satisfied. Consequently, the best breeder chosen from the whole population is the possible solution to the problem under study.

Because of its linear nature, NMRA has recently gained interest among researchers. The revised NMRA improves performance by improving both its basic exploitation and exploration capabilities. The elite opposition-based learning (EOBL) strategy [30] improves the exploration of basic NMRA. Exploitation is improved by local neighborhood search (LNS) with information on the best solution to date in the small neighborhood of the solution [25]. The following are the main changes:

3.3.1. NMRA Version 1.0

NMRV 1.0 has improved NMRA's exploration capabilities through implementing the EOBL Strategy [30]. The opposition-based NMRA, also called NMRV 1.0, Algorithm 2 detailed the pseudocode of NMRV 1.0.

Algorithm 1 Pseudocode of NMRA

Input: **Define objective function** f(NMR), **NMR = (NMR1, NMR2, . . ., NMR**$_D$**)**
Output: **Identify current best solution** d;
Initialization: **Initialize NMRs: n, breeders** B**:** $n/5$**, workers** W**:** $B - n$
                              Describe breeding probability: bp
                    while iterations $< it_{max}$
for i = 1: w
$$w_i^{t+1} = w_i^t + \lambda\left(w_j^t - w_k^t\right)$$
evaluate w_i^{t+1}
                              end for
for i = 1: B
if rand > bp
$$b_i^{t+1} = (1 - \lambda)b_i^t + \lambda\left(d - b_i^t\right)$$
end if
evaluate b_i^{t+1}
end for
combine the updated breeder and worker population
estimate the NMRs
update the overall best d
update iteration count
end while
update final best d
end

Algorithm 2 Pseudocode of NMRV 1.0

Input: **Define objective function** f(NMR), **NMR = (NMR1, NMR2, . . ., NMR**$_D$**)**
Output: **Identify current best solution** d;
Initialization: **Initialize NMRs: n, breeders** B**:** $n/5$**, workers** W**:** $B - n$
                              Describe breeding probability: bp
                    Update the current population with EOBL;
while iterations $< it_{max}$
for i = 1: w
$$w_i^{t+1} = w_i^t + \lambda\left(w_j^t - w_k^t\right)$$
evaluate w_i^{t+1}
                              end for
for i = 1: B
if rand > bp
$$b_i^{t+1} = (1 - \lambda)b_i^t + \lambda\left(d - b_i^t\right)$$
end if
evaluate b_i^{t+1}
end for
combine the updated breeder and worker population
estimate the NMRs
update the overall best d
update iteration count
end while
update final best d
end

3.3.2. NMRA Version 2.0

The local neighborhood search (LNS) model has been employed [30] to enhance the exploitation capacity of NMRA. In the basic NMRA, workers change their position during the employee phase according to local information and their past experience. The new best solution is restructured with LNS in the upgraded NMRA version 2.0, and the worker phase is updated as a solution:

$$L_i^t = w_i^t + u * \left(w_{n\_opt} - w_i^t\right) + v * \left(w_p^t - w_q^t\right) \tag{4}$$

where w_{n_opt} is the best solution in the neighborhood of w_i^t and $u, v \in \text{rand()}$ are the scaling factors, and $p, q \in [i - r, i + r]$ ($p \neq q \neq i$) neighborhood. The updated solution using LNS in the worker phase is given by:

$$w_i^{t+1} = L_i^t + \lambda \left(w_j^t - w_k^t\right) \tag{5}$$

where $\lambda \in \text{rand()}$ is a scaling factor, L_i^t is the LNS updated best solution. The pseudocode of NMRV 2.0, i.e., LNS-based NMRA, is shown in Algorithm 3.

Algorithm 3 Pseudocode of NMRV 2.0

Input: **Define objective function** f**(NMR), NMR = (NMR1, NMR2, . . ., NMR$_D$)**
Output: **Identify current best solution d;**
Initialization: **Initialize NMRs: n, breeders B: $n/5$, workers W: $B - n$**
  Describe breeding probability: *bp*
  while iterations $< it_{max}$
for i = 1: w
$L_i^t = w_i^t + u * \left(w_{n\_opt} - w_i^t\right) + v * \left(w_p^t - w_q^t\right)$
$w_i^{t+1} = L_i^t + \lambda \left(w_j^t - w_k^t\right)$
evaluate w_i^{t+1}
  end for
for i = 1: B
if rand $>$ bp
$b_i^{t+1} = (1 - \lambda)b_i^t + \lambda\left(d - b_i^t\right)$
end if
evaluate b_i^{t+1}
end for
combine the updated breeder and worker population
estimate the NMRs
update the overall best d
update iteration count
end while
update final best d
end

3.3.3. NMRA Version 3.0

NMRA version 3.0 has two modifications in basic NMRA to balance the exploitation and exploration capacity [30]. First, the exploration tendency is improved by the EOBL strategy. Second, the exploitation capability is improved by LNS. The pseudocode of NMRV 3.0 is given in Algorithm 4.

Algorithm 4 Pseudocode of NMRV 3.0

Input: **Define objective function** $f(NMR)$, $NMR = (NMR1, NMR2, \ldots, NMR_D)$
Output: **Identify current best solution** d;
Initialization: **Initialize NMRs:** n, **breeders** $B: n/5$, **workers** $W: B - n$
         Describe breeding probability: bp
      Update the current population with EOBL;
while iterations $< it_{max}$
for $i = 1$: w
 $L_i^t = w_i^t + u * \left(w_{n\_opt} - w_i^t\right) + v * \left(w_p^t - w_q^t\right)$
          $w_i^{t+1} = L_i^t + \lambda \left(w_j^t - w_k^t\right)$
evaluate] w_i^{t+1}
       end for
for $i = 1$: B
if rand $> bp$
 $b_i^{t+1} = (1 - \lambda)b_i^t + \lambda\left(d - b_i^t\right)$
end if
evaluate b_i^{t+1}
end for
combine the updated breeder and worker population
estimate the NMRs
update the overall best d
update iteration count
end while
update final best d
end

---

4. Case Study

The simulation results were compared with the optimization study carried out by Panda et al. [18], where the influence of five FFF parameters was optimized to improve the mechanical properties. The process parameters utilized for this study are:

$$x_1 \; Layer \; thickness \; (in \; mm)$$
$$x_2 \; Building \; orientation \; (in \; degree)$$
$$x_3 \; Raster \; angle \; (in \; degree)$$
$$x_4 \; Raster \; width \; (in \; mm)$$
$$x_5 \; Air \; gap \; (in \; mm)$$

The primary motivation for selecting the above-mentioned input parameters is the literature review, which has found a good correlation between abovesaid parameters with mechanical properties of FFF components [2,19]. The output variables used for this case study are impact strength (IS) in MJ/m^2, flexural strength (FS) in MPa and Tensile Strength (TS) in MPa. The main reason for selecting these output parameters is that these components undergo different types of loading conditions for medical, aerospace and automobile applications. Therefore, it is mandatory to evaluate all these mechanical properties utilizing an advanced optimization algorithm. These response parameters are considered as an objective function for the present case study.

The objective function is shown in Equations (6)–(8):

$$TS = 13.5625 + 0.7156x_1 - 1.3123x_2 + 0.9760x_3 + 0.5183x_5 + 1.1671x_1^2 - 1.3014x_2^2 - 0.4363x_1x_3 + 0.4364x_1x_4$$
$$-0.4364x_1x_5 + 0.4364x_2x_3 + 0.4898x_2x_5 - 0.5389x_3x_4 + 0.5389x_3x_5 - 0.5389x_4x_5 \tag{6}$$

$$FS = 29.9178 + 0.8719x_1 - 4.8741x_2 + 2.4251x_3 - 0.9096x_4 + 1.6626x_5 - 1.7199x_1x_3 + 1.7412x_1x_4 - 1.1275x_1x_5$$
$$+1.0621x_2x_5 + 1.0621x_3x_5 + 1.0408x_4x_5 \tag{7}$$

$$IS = 0.401992 + 0.034198x_1 + 0.008356x_2 + 0.013673x_3 + 0.021383x_1^2 + 0.008077x_2x_4 \tag{8}$$

It is required to maximize each type of strength; thus, maximization of objective functions is performed.

The lower and upper bounds of process parameters are identified based on constraints as most commercially available FFF machines can manufacture components within the following parameter bounds. The parameter bounds for objective function are expressed in Table 1 as follows:

Table 1. Parameter bounds for the objective function.

Parameter	Lower Bound	Upper Bound
x_1 *Layer thickness (in mm)*	0.127	0.254
x_2 *Building orientation (in degree)*	0	30
x_3 *Raster angle (in degree)*	0	60
x_4 *Raster width (in mm)*	0.4064	0.5064
x_5 *Air gap (in mm)*	0	0.008

5. Simulation Results and Discussion

In the FFF process, to resolve this optimization problem, the NMRA variants are utilized. Since heuristic algorithms are stochastic optimization methods, these must be executed at least 10 times to generate meaningful statistics. For this purpose, the simulations are performed 30 times. Population sizes vary from 20 to 40, and the number of iterations varies from 100 to 500. The NMRA variants have been selected for efficiency check by the FFA [31], FPA [32], NBA [33], SCA [34] and SSA [35] algorithms. Table 2 details the parameter settings used to compare the results.

Table 2. Parameter values.

Algorithm	Parameters
FFA	$NP = 20 - 40$, $dim = 5$, $it_{max} = 100 - 500$; $\beta_0 = 1$; $\beta_{min} = 0.2$; $\alpha = 0.5$; $\gamma = 1$
FPA	$NP = 20 - 40$, $dim = 5$, $it_{max} = 100 - 500$; $p = 0.7$
NBA	$NP = 20 - 40$, $dim = 5$, $it_{max} = 100 - 500$; $A=0.5$; $r = 0.5$; $\alpha = \gamma = 0.9$; $f_{min} = 0$; $f_{max} = 1.5$
SCA	$NP = 20 - 40$, $dim = 5$, $it_{max} = 100 - 500$; $a = [2 - 0]$
SSA	$NP = 20 - 40$, $dim = 5$, $it_{max} = 100 - 500$; $c_1 = [2 - 0]$
NMRA	$NP = 20 - 40$, $dim = 5$, $it_{max} = 100 - 500$; $b_p = 0.5$
NMRV 1.0	$NP = 20 - 40$, $dim = 5$, $it_{max} = 100 - 500$; $b_p = 0.5$
NMRV 2.0	$NP = 20 - 40$, $dim = 5$, $it_{max} = 100 - 500$; $b_p = 0.5$
NMRV 3.0	$NP = 20 - 40$, $dim = 5$, $it_{max} = 100 - 500$; $b_p = 0.5$

Here, NP is the population size, dim is the dimension of the problem, it_{max} is the number of iterations

The optimum results attained by simulated algorithms with population size 40 and maximum amount of iterations, i.e., 500, are shown in Tables 3–5 for *TS*, *FS* and *IS*, respectively. It is clear from the results that the NMRV 2.0 and 3.0 algorithm's fitness values are slightly better than others for *TS* and *FS* fitness functions, respectively. Outcomes of all competitive algorithm's fitness values are nearly similar to competitive algorithms for *IS* fitness function. The convergence rate for *TS*, *FS* and *IS* are drawn in Figures 2–4, respectively, which is also better than others. The performance of simulated algorithms for FFF for *TS*, *FS* and *IS* are specified in Tables 3–5, respectively. This demonstrates that NMRV 3.0 algorithm's standard deviation is better than others that demonstrates improved exploitation and exploration capabilities of NMRV 3.0 for process parameters optimization of FFF as confirmed by box-plots shown in Figures 5–7.

Table 3. Statistics of simulated algorithms with population size 40 over 30 independent runs for 500 iterations for Tensile Strength (*TS*) estimation.

Algorithm	Worst	Best	Average	Median	Std. Dev.
FFA	174.8324	174.9215	174.9123	174.9215	0.021354
FPA	174.9215	174.9215	174.9215	174.9215	1.20×10^{-10}
NBA	174.9215	174.9215	174.9215	174.9215	1.17×10^{-11}
SCA	174.61	174.9215	174.7647	174.757	0.143845
SSA	174.71	174.9215	174.897	174.9215	0.057414
NMRA	174.9215	174.9215	174.9215	174.9215	8.26×10^{-14}
NMRV 1.0	174.9215	174.9215	174.9215	174.9215	8.67×10^{-14}
NMRV 2.0	174.9215	174.9215	174.9215	174.9215	7.97×10^{-14}
NMRV 3.0	174.9215	174.9215	174.9215	174.9215	8.24×10^{-14}

Table 4. Statistics of simulated algorithms with population size 40 over 30 independent runs for 500 iterations for Flexural Strength (*FS*) estimation.

Algorithm	Worst	Best	Average	Median	Std. Dev.
FFA	162.6744	162.6744	162.6744	162.6744	5.78×10^{-14}
FPA	162.6744	162.6744	162.6744	162.6744	5.78×10^{-14}
NBA	162.1491	162.6744	162.6569	162.6744	0.095915
SCA	162.6744	162.6744	162.6744	162.6744	5.78×10^{-14}
SSA	162.6744	162.6744	162.6744	162.6744	5.78×10^{-14}
NMRA	162.6744	162.6744	162.6744	162.6744	5.78×10^{-14}
NMRV 1.0	162.6744	162.6744	162.6744	162.6744	5.78×10^{-14}
NMRV 2.0	162.6744	162.6744	162.6744	162.6744	5.78×10^{-14}
NMRV 3.0	162.6744	162.6744	162.6744	162.6744	5.71×10^{-14}

Table 5. Statistics of simulated algorithms with population size 40 over 30 independent runs for 500 iterations for Impact Strength (*IS*) estimation.

Algorithm	Worst	Best	Average	Median	Std. Dev.
FFA	1.605824	1.605824	1.605824	1.605824	2.26×10^{-16}
FPA	1.605824	1.605824	1.605824	1.605824	2.26×10^{-16}
NBA	1.605824	1.605824	1.605824	1.605824	2.26×10^{-16}
SCA	1.605824	1.605824	1.605824	1.605824	2.26×10^{-16}
SSA	1.605824	1.605824	1.605824	1.605824	2.26×10^{-16}
NMRA	1.605824	1.605824	1.605824	1.605824	2.26×10^{-16}
NMRV 1.0	1.605824	1.605824	1.605824	1.605824	2.26×10^{-16}
NMRV 2.0	1.605824	1.605824	1.605824	1.605824	2.26×10^{-16}
NMRV 3.0	1.605824	1.605824	1.605824	1.605824	2.26×10^{-16}

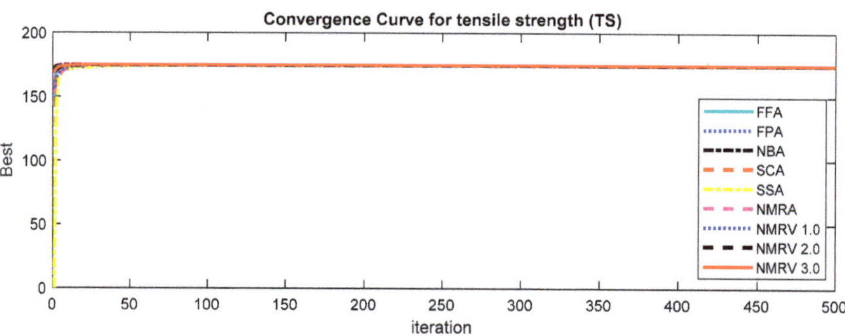

Figure 2. Convergence graph of simulated algorithms with population size 40 over 30 runs for 500 iterations for Tensile Strength (*TS*) estimation.

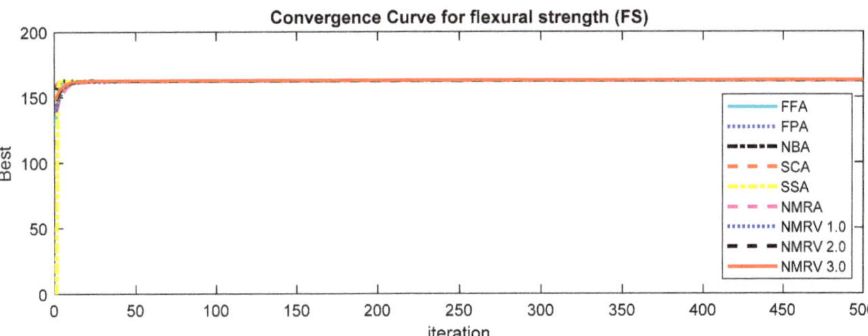

Figure 3. Convergence graph of simulated algorithms with population size 40 over 30 runs for 500 iterations for Flexural Strength (*FS)* estimation.

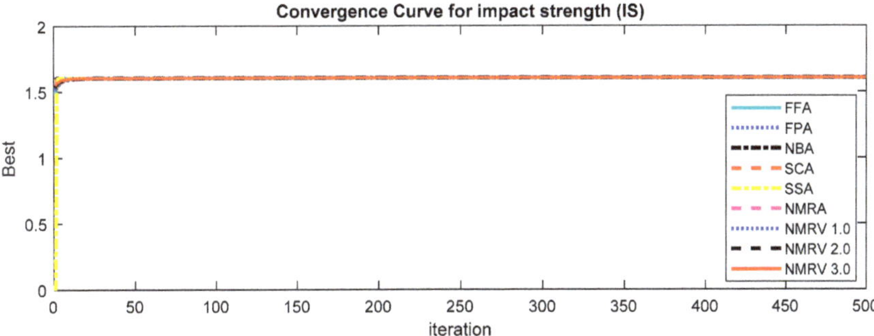

Figure 4. Convergence graph of simulated algorithms with population size 40 over 30 runs for 500 iterations for Impact Strength (*IS)* estimation.

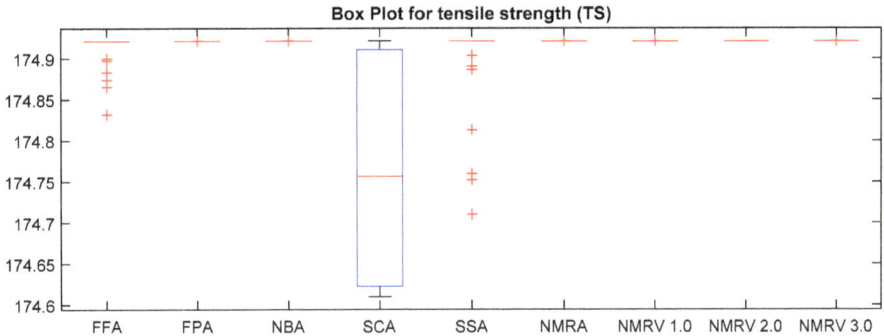

Figure 5. Box plot of simulated algorithms with population size 40 over 30 runs for 500 iterations for *TS* estimation.

The performance evaluation of simulated algorithms with population size 40 and maximum amount of iterations, i.e., 100, are specified in Tables 6–8 for *TS, FS* and *IS*, respectively. It is clear from the statistics that NMRV 3.0 algorithm's fitness values are slightly better than others for *TS, FS* and *IS* fitness functions.

To further check the effectiveness of proposed variants, the simulations were carried out for different numbers of iterations and population sizes [36]. The performance evaluation of simulated algorithms with population size 20 and maximum amount of iterations, i.e., 100, are given in Tables 9–11. It is clear from the results that NMRV 3.0 algorithm's fitness values are slightly better than others for *TS, FS* and *IS* fitness functions.

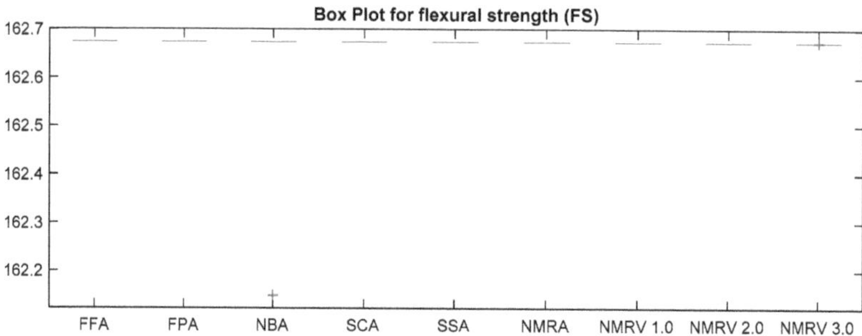

Figure 6. Box plot of simulated algorithms with population size 40 over 30 independent runs for 500 iterations for *FS* estimation.

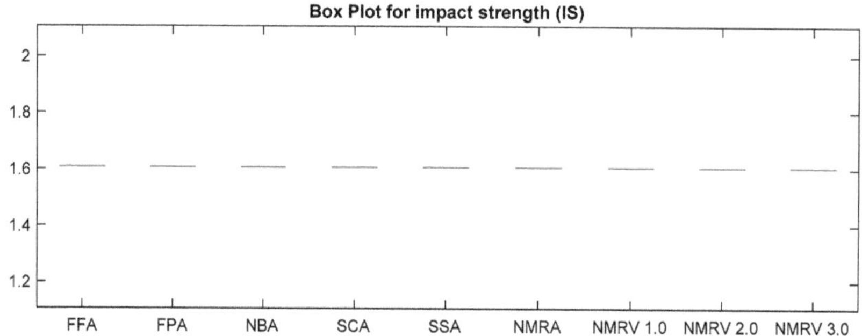

Figure 7. Box plot of simulated algorithms with population size 40 over 30 independent runs for 500 iterations for *IS* estimation.

Table 6. Statistics of simulated algorithms with population size 40 over 30 runs for 100 iterations for *TS* estimation.

Algorithm	Worst	Best	Average	Median	Std. Dev.
FFA	174.1897	174.9215	174.731	174.7742	0.194092
FPA	174.6641	174.9215	174.8728	174.9107	0.0739
NBA	174.6234	174.9215	174.8686	174.9215	0.112
SCA	174.4651	174.9213	174.6961	174.6229	0.140252
SSA	174.6239	174.9215	174.8157	174.877	0.117193
NMRA	174.9215	174.9215	174.9215	174.9215	4.31×10^{-6}
NMRV 1.0	174.9215	174.9215	174.9215	174.9215	3.02×10^{-7}
NMRV 2.0	174.9215	174.9215	174.9215	174.9215	5.48×10^{-7}
NMRV 3.0	174.9215	174.9215	174.9215	174.9215	4.21×10^{-8}

Table 7. Performance of simulated algorithms with population size 40 over 30 runs for 100 iterations for *FS* estimation.

Algorithm	Worst	Best	Average	Median	Std. Dev.
FFA	162.5976	162.6744	162.6555	162.6574	0.0187
FPA	162.615	162.6744	162.6697	162.6744	0.0152
NBA	162.1491	162.6744	162.6044	162.6744	0.181637
SCA	162.6744	162.6744	162.6744	162.6744	5.78×10^{-7}
SSA	162.6744	162.6744	162.6744	162.6744	5.78×10^{-7}
NMRA	162.6744	162.6744	162.6744	162.6744	6.65×10^{-6}
NMRV 1.0	162.6744	162.6744	162.6744	162.6744	6.25×10^{-7}
NMRV 2.0	162.6744	162.6744	162.6744	162.6744	2.25×10^{-6}
NMRV 3.0	162.6744	162.6744	162.6744	162.6744	4.11×10^{-7}

Table 8. Performance of simulated algorithms with population size 40 over 30 runs for 100 iterations for *IS* estimation.

Algorithm	Worst	Best	Average	Median	Std. Dev.
FFA	1.604239	1.605824	1.605749	1.605824	3.02×10^{-4}
FPA	1.605824	1.605824	1.605824	1.605824	2.26×10^{-10}
NBA	1.605824	1.605824	1.605824	1.605824	2.26×10^{-10}
SCA	1.605824	1.605824	1.605824	1.605824	2.26×10^{-10}
SSA	1.605824	1.605824	1.605824	1.605824	2.26×10^{-10}
NMRA	1.605824	1.605824	1.605824	1.605824	2.52×10^{-10}
NMRV 1.0	1.605824	1.605824	1.605824	1.605824	1.14×10^{-10}
NMRV 2.0	1.605824	1.605824	1.605824	1.605824	9.56×10^{-10}
NMRV 3.0	1.605824	1.605824	1.605824	1.605824	8.12×10^{-11}

Table 9. Performance of simulated algorithms with population size 20 over 30 independent runs for 100 iterations for *TS* estimation.

Algorithm	Worst	Best	Average	Median	Std. Dev.
FFA	174.1762	174.9215	174.724	174.7416	0.189432
FPA	174.6634	174.9215	174.8716	174.9093	0.0727
NBA	174.6231	174.9209	174.8679	174.9128	0.159
SCA	174.4648	174.9213	174.6948	174.6218	0.150682
SSA	174.6236	174.9214	174.8149	174.8637	0.127843
NMRA	174.9214	174.9215	174.9215	174.9215	4.38×10^{-5}
NMRV 1.0	174.9215	174.9215	174.9215	174.9215	3.13×10^{-7}
NMRV 2.0	174.9215	174.9215	174.9215	174.9215	5.53×10^{-7}
NMRV 3.0	174.9215	174.9215	174.9215	174.9215	4.52×10^{-8}

Table 10. Performance of simulated algorithms with population size 20 over 30 independent runs for 100 iterations for *FS* estimation.

Algorithm	Worst	Best	Average	Median	Std. Dev.
FFA	162.218	162.6709	162.5367	162.5618	0.119
FPA	155.0595	162.6744	162.3814	162.3684	1.39
NBA	162.1491	162.6744	162.6219	162.6172	0.160298
SCA	162.1491	162.6744	162.6044	162.6059	0.182
SSA	162.6744	162.6744	162.6744	162.6744	5.78×10^{-4}
NMRA	162.6744	162.6744	162.6744	162.6744	2.38×10^{-5}
NMRV 1.0	162.6744	162.6744	162.6744	162.6744	1.02×10^{-5}
NMRV 2.0	162.6743	162.6744	162.6743	162.6743	9.03×10^{-5}
NMRV 3.0	162.6744	162.6744	162.6744	162.6744	1.72×10^{-6}

Table 11. Performance of simulated algorithms with population size 20 over 30 independent runs for 100 iterations for *IS* estimation.

Algorithm	Worst	Best	Average	Median	Std. Dev.
FFA	1.602716	1.605824	1.605407	1.605812	7.42×10^{-4}
FPA	1.599358	1.605824	1.605184	1.605806	1.74×10^{-3}
NBA	1.605824	1.605824	1.605824	1.605824	2.26×10^{-9}
SCA	1.605824	1.605824	1.605824	1.605824	2.26×10^{-9}
SSA	1.605824	1.605824	1.605824	1.605824	2.26×10^{-9}
NMRA	1.605823	1.605824	1.605824	1.605824	4.68×10^{-8}
NMRV 1.0	1.605824	1.605824	1.605824	1.605824	2.22×10^{-9}
NMRV 2.0	1.605824	1.605824	1.605824	1.605824	3.08×10^{-9}
NMRV 3.0	1.605824	1.605824	1.605824	1.605824	1.71×10^{-9}

6. Confirmatory Experiments

To validate the findings of NMRA, the confirmatory experiments were planned to fabricate sample parts at optimized parameters followed by mechanical testing. Five samples were manufactured using the FFF process at predefined conditions as suggested by NMRA. The samples are prepared for confirmation of *TS* and *FS* of parts fabricated at optimum parameter settings. The samples for *TS* and *FS* are prepared as per ASTM D638 and ASTM D790 standards. The dimensions of *TS* and *FS* samples are shown in Figure 8.

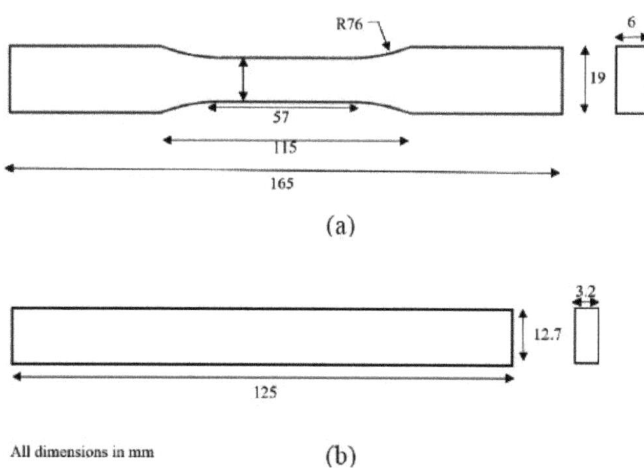

Figure 8. Dimensions of test parts for (**a**) *TS* (**b**) *FS*.

Acrylonitrile butadiene styrene (ABS) has been extensively used for automobile and electronic components and hence it is most popular raw material used in FFF printers. The filament material selected for the fabrication of test samples in the present study is ABS P400 with a 1.75 mm diameter supplied by Robokits, India. Table 12 displays the FFF process parameters selected during the fabrication of test samples.

Table 12. Process parameters selected for the fabrication of test parts.

Parameter	Details
Layer height	0.12 mm
Building orientation	0°
Raster angle	60°
Raster width	0.4064 mm
Air gap	0.008 mm
Infill pattern	Cubic
Infill density	50%
Printing speed	50 mm/s

The FFF machine used for the fabrication of test parts was Model I3 supplied by Prusa Research, Prague, Czech Republic and Universal testing machine (UTM) was supplied by Shanta Engineering Pvt. Ltd., Pune, India, as shown in Figure 9. The UTM was operated at 50 mm/min strain rate with gradual bending load subjected at three points during flexural tests (see Figure 9d). During tensile testing, the samples were held in two jaws with 52 mm grip separation. The tensile force was exerted gradually at 50 mm/min until the samples were fractured. Afterward, the fracture points of both tensile and flexural test samples were studied using scanning electron microscope (SEM) images which were generated by Model IT500HR supplied by Jeol Ltd. Tokyo, Japan.

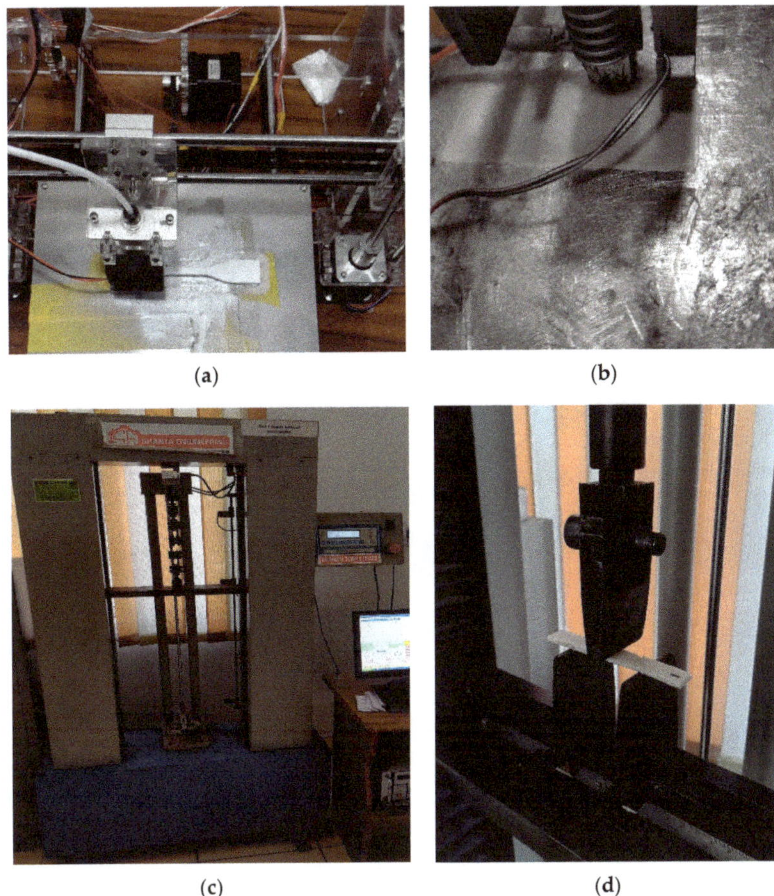

Figure 9. (**a**) FFF printer (**b**) *FS* sample during the fabrication (**c**) Universal Testing Machine (UTM) used for testing (**d**) *FS* sample during testing.

The mechanical strength of parts was measured and compared with simulated results, as shown in Figure 10. The experimental results were compared with predictions made by the NMRA algorithm at suggested parametric settings. Notably, experimental results were in significant agreement with the modeled data when experiments were performed at optimum parameter settings. The output of NMRA yielded significant and promising results, which may be beneficial for deciding 3D printing and part manufacturing conditions at the commercial level. During mass production and part production, the optimized parametric settings would be utilized to attain desired mechanical properties in products.

In addition, the SEM images retrieved after mechanical testing reveal the conditions at the location of breakage. Figure 11a shows the SEM micrographs of tensile test parts, indicating higher flexibility in the layers at the point of failure. A higher elongation was experienced for these parts as an optimized combination of parameters was used. In Figure 9b, the flexural test sample was photographed. It can be confirmed that failure occurs after extreme elongation due to flexibility induced in parts. The mechanical strength and flexibility attained at this location reveal the strategic selection of parametric levels.

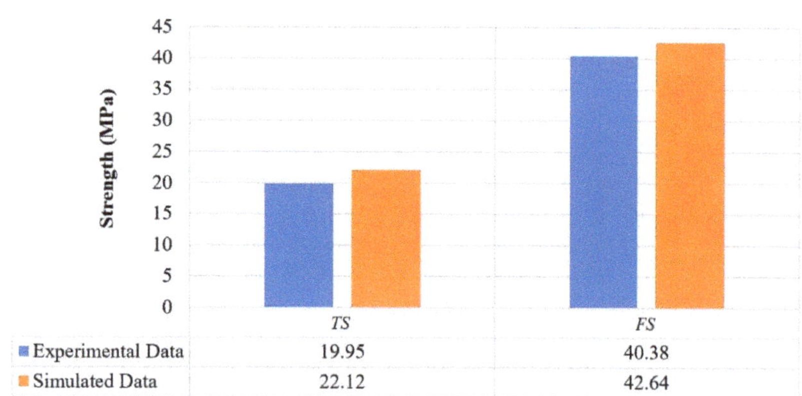

Figure 10. Comparison of experiments and simulated results of mechanical strength.

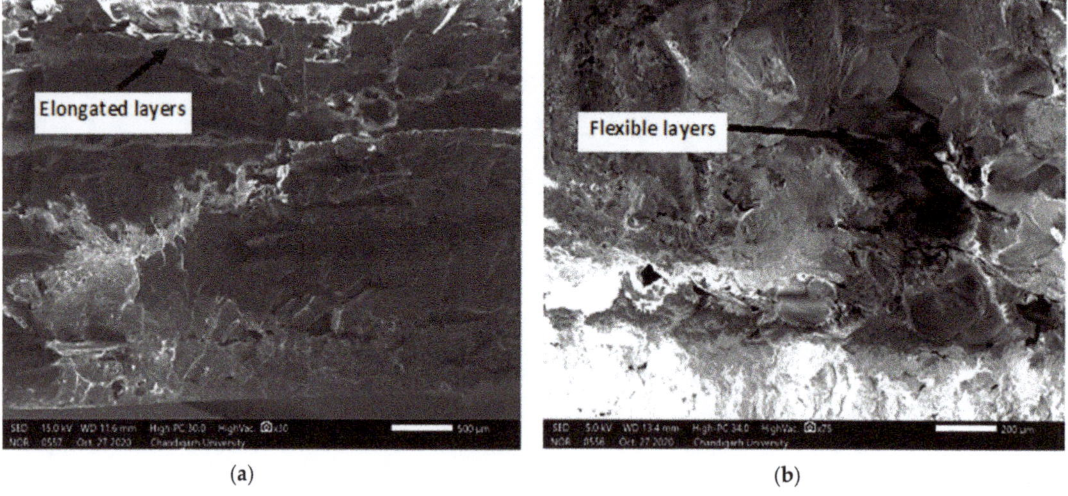

(a)

(b)

Figure 11. SEM images of point of failure of parts during (**a**) tensile test (**b**) flexural test.

7. Conclusions and Future Scope

The mechanical strength of FFF parts is significantly lower than conventionally prepare thermoplastics components, which hinders their applicability for certain applications. Thus, we must identify optimum process parameters to enhance *TS, FS* and *IS* of FFF parts. The performance of advanced naked mole-rat algorithm variants was tested to solve FFF issues of poor mechanical strength. The results indicated a significant enhancement of tensile, flexural and impact strength than previous studies using NMRV 3.0. The study could be further extended to identify optimum parameter settings to achieve maximum dimensional accuracy and surface finish of FFF components. Moreover, the efficacy of the enhanced versions of NMRA algorithms could be further tested for optimizing parameters of the other additive manufacturing techniques, such as stereolithography, electron beam melting and selective laser melting.

Author Contributions: Conceptualization, J.S.C.; N.M.; R.K.; S.S. (Sandeep Singh); S.S. (Shubham Sharma); methodology, J.S.C.; N.M.; R.K.; S.S. (Sandeep Singh); S.S. (Shubham Sharma); software, J.S.C.; N.M.; R.K.; S.S. (Sandeep Singh); S.S. (Shubham Sharma); S.P.D.; A.S.; S.C.; R.A.I.; validation, J.S.C.; N.M.; S.S. (Shubham Sharma); S.C.; R.A.I.; C.H.L.; S.W.; formal analysis, J.S.C.; N.M.; S.S. (Shubham Sharma); S.P.D.; A.S.; S.C.; R.A.I.; C.H.L.; S.W.; investigation, J.S.C.; N.M.; R.K.; S.S. (Sandeep Singh); S.S. (Shubham Sharma); resources, J.S.C.; N.M.; S.S. (Shubham Sharma); R.A.I.; C.H.L. and S.W.; writing—original draft preparation, J.S.C.; N.M.; R.K.; S.S. (Sandeep Singh); S.S. (Shubham Sharma); writing—review and editing, J.S.C.; N.M.; R.K.; S.S. (Sandeep Singh); S.S. (Shubham Sharma); S.P.D.; A.S.; S.C.; R.A.I.; C.H.L. and S.W.; supervision, J.S.C.; N.M.; R.K.; S.S. (Sandeep Singh); S.S. Shubham Sharma); funding acquisition, S.S. (Shubham Sharma); S.C.; R.A.I.; C.H.L.; S.W. All authors have read and agreed to the published version of the manuscript.

Funding: This research received no external funding.

Institutional Review Board Statement: Not applicable.

Informed Consent Statement: Not applicable.

Data Availability Statement: The data presented in this study are available on request from the corresponding author.

Conflicts of Interest: The authors declare no conflict of interest.

References

1. Mishra, S.B.; Mahapatra, S.S. An experimental investigation on strain-controlled fatigue behaviour of FDM build parts. *Int. J. Product. Qual. Manag.* **2018**, *24*, 323–345. [CrossRef]
2. Pandey, P.M.; Thrimurthulu, K. Optimal part deposition orientation in FDM by using a multicriteria genetic algorithm. *Int. J. Prod. Res.* **2004**, *42*, 4069–4089. [CrossRef]
3. Pearce, J.M. A review of open-source ventilators for COVID-19 and future pandemics. *F1000Research* **2020**, *9*, 218. [CrossRef] [PubMed]
4. ASTM Committee F42 on Additive Manufacturing Technologies. Subcommittee F42. 91 on Terminology. In *Standard Terminology for Additive Manufacturing Technologies*; ASTM International: West Conshohocken, PA, USA, 2012.
5. Blanco, I. The Use of Composite Materials in 3D Printing. *J. Compos. Sci.* **2020**, *4*, 42. [CrossRef]
6. Wang, X.; Jiang, M.; Zhou, Z.; Gou, J.; Hui, D. 3D printing of polymer matrix composites: A review and prospective. *Compos. Part B Eng.* **2017**, *110*, 442–458. [CrossRef]
7. Chapiro, M. Current achievements and future outlook for composites in 3D printing. *Reinf. Plast.* **2016**, *60*, 372–375. [CrossRef]
8. Kim, G.D.; Oh, Y.T. A benchmark study on rapid prototyping processes and machines: Quantitative comparisons of mechanical properties, accuracy, roughness, speed, and material cost. *Proc. Inst. Mech. Eng. Part B J. Eng. Manuf.* **2008**, *222*, 201–215. [CrossRef]
9. Aimar, A.; Palermo, A.; Innocenti, B. The Role of 3D Printing in Medical Applications: A State of the Art. *J. Healthc. Eng.* **2019**, *2019*, 5340616. [CrossRef]
10. Khoo, Z.X.; Teoh, J.E.M.; Liu, Y.; Chua, C.K.; Yang, S.; An, J.; Leong, K.F.; Yeong, W.Y. 3D printing of smart materials: A review on recent progresses in 4D printing. *Virtual Phys. Prototyp.* **2015**, *10*, 103–122. [CrossRef]
11. Ahn, S.; Montero, M.; Odell, D.; Roundy, S.; Wright, P.K. Anisotropic material properties of fused deposition modeling ABS. *Rapid Prototyp. J.* **2002**, *8*, 248–257. [CrossRef]
12. Rane, R.; Kulkarni, A.; Prajapati, H.; Taylor, R.; Jain, A.; Chen, V. Post-Process Effects of Isothermal Annealing and Initially Applied Static Uniaxial Loading on the Ultimate Tensile Strength of Fused Filament Fabrication Parts. *Materials* **2020**, *13*, 352. [CrossRef] [PubMed]
13. Liu, J.; Ma, Y.; Qureshi, A.J.; Ahmad, R. Light-weight shape and topology optimization with hybrid deposition path planning for FDM parts. *Int. J. Adv. Manuf. Technol.* **2018**, *97*, 1123–1135. [CrossRef]
14. Gordelier, T.J.; Thies, P.R.; Turner, L.; Johanning, L. Optimising the FDM additive manufacturing process to achieve maximum tensile strength: A state-of-the-art review. *Rapid Prototyp. J.* **2019**, *25*, 953–971. [CrossRef]
15. Dey, A.; Yodo, N. A Systematic Survey of FDM Process Parameter Optimization and Their Influence on Part Characteristics. *J. Manuf. Mater. Process.* **2019**, *3*, 64. [CrossRef]
16. Fernandez-Vicente, M.; Calle, W.; Ferrandiz, S.; Conejero, A. Effect of Infill Parameters on Tensile Mechanical Behavior in Desktop 3D Printing. *3D Print. Addit. Manuf.* **2016**, *3*, 183–192. [CrossRef]
17. Hambali, R.H.; Smith, P.; Rennie, A. Determination of the effect of part orientation to the strength value on additive manufacturing FDM for end-use parts by physical testing and validation via three-dimensional finite element analysis. *Int. J. Mater. Eng. Innov.* **2012**, *3*, 269. [CrossRef]
18. Panda, S.K.; Padhee, S.; Sood, A.K.; Mahapatra, S.S. Optimization of Fused Deposition Modelling (FDM) Process Parameters Using Bacterial Foraging Technique. *Intell. Inf. Manag.* **2009**, *1*, 89–97. [CrossRef]

19. Rayegani, F.; Onwubolu, G.C. Fused deposition modelling (FDM) process parameter prediction and optimization using group method for data handling (GMDH) and differential evolution (DE). *Int. J. Adv. Manuf. Technol.* **2014**, *73*, 509–519. [CrossRef]
20. Goudswaard, M.; Nassehi, A.; Hicks, B. Towards the democratisation of design: The implementation of metaheuristic search strategies to enable the auto-assignment of manufacturing parameters for FDM. *Procedia Manuf.* **2019**, *38*, 383–390. [CrossRef]
21. Mohamed, O.A.; Masood, S.H.; Bhowmik, J.L. Influence of processing parameters on creep and recovery behavior of FDM manufactured part using definitive screening design and ANN. *Rapid Prototyp. J.* **2017**, *23*, 998–1010. [CrossRef]
22. Raju, M.; Gupta, M.K.; Bhanot, N.; Sharma, V.S. A hybrid PSO–BFO evolutionary algorithm for optimization of fused deposition modelling process parameters. *J. Intell. Manuf.* **2018**, *30*, 2743–2758. [CrossRef]
23. Rao, R.V.; Rai, D.P. Optimization of fused deposition modeling process using teaching-learning-based optimization algorithm. *Eng. Sci. Technol. Int. J.* **2016**, *19*, 587–603. [CrossRef]
24. Malviya, M.; Desai, K. Build Orientation Optimization for Strength Enhancement of FDM Parts Using Machine Learning based Algorithm. *Comput. Des. Appl.* **2019**, *17*, 783–796. [CrossRef]
25. Yadav, D.; Chhabra, D.; Garg, R.K.; Ahlawat, A.; Phogat, A. Optimization of FDM 3D printing process parameters for multi-material using artificial neural network. *Mater. Today Proc.* **2020**, *21*, 1583–1591. [CrossRef]
26. Natarajan, E.; Kaviarasan, V.; Lim, W.H.; Tiang, S.S.; Parasuraman, S.; Elango, S. Non-dominated sorting modified teaching–learning-based optimization for multi-objective machining of polytetrafluoroethylene (PTFE). *J. Intell. Manuf.* **2019**, *31*, 911–935. [CrossRef]
27. Saad, M.S.; Nor, A.M.; Baharudin, M.E.; Zakaria, M.Z.; Aiman, A. Optimization of surface roughness in FDM 3D printer using response surface methodology, particle swarm optimization, and symbiotic organism search algorithms. *Int. J. Adv. Manuf. Technol.* **2019**, *105*, 5121–5137. [CrossRef]
28. Mishra, S.B.; Khan, M.S.; Acharya, E.; Nanda, D. Parametric Appraisal of Wear Behavior of Coated FDM Build parts using Firefly Algorithm. *Mater. Today Proc.* **2018**, *5*, 17968–17973. [CrossRef]
29. Salgotra, R.; Singh, U. The naked mole-rat algorithm. *Neural Comput. Appl.* **2019**, *31*, 8837–8857. [CrossRef]
30. Singh, P.; Mittal, N.; Singh, U.; Salgotra, R. Naked Mole-Rat Algorithm with Improved Exploration and Exploitation Capabilities to Determine 2D and 3D Coordinates of Sensor Nodes in WSNs. *Arab. J. Sci. Eng.* **2021**, *46*, 1155–1178. [CrossRef]
31. Yang, X.S. *Nature-Inspired Metaheuristic Algorithms*; Luniver Press: Frome, UK, 2008.
32. Yang, X.-S. Flower Pollination Algorithm for Global Optimization. In Proceedings of the Transactions on Petri Nets and Other Models of Concurrency XV, Hamburg, Germany, 27–29 June 2012; Springer: Berlin/Heidelberg, Germany, 2012; pp. 240–249.
33. Meng, X.-B.; Gao, X.; Liu, Y.; Zhang, H. A novel bat algorithm with habitat selection and Doppler effect in echoes for optimization. *Expert Syst. Appl.* **2015**, *42*, 6350–6364. [CrossRef]
34. Mirjalili, S. SCA: A Sine Cosine Algorithm for solving optimization problems. *Knowl. Based Syst.* **2016**, *96*, 120–133. [CrossRef]
35. Mirjalili, S.; Gandomi, A.H.; Mirjalili, S.Z.; Saremi, S.; Faris, H.; Mirjalili, S.M. Salp Swarm Algorithm: A bio-inspired optimizer for engineering design problems. *Adv. Eng. Softw.* **2017**, *114*, 163–191. [CrossRef]
36. Chohan, J.S.; Mittal, N.; Kumar, R.; Singh, S.; Sharma, S.; Singh, J.; Rao, K.V.; Mia, M.; Pimenov, D.Y.; Dwivedi, S.P. Mechanical Strength Enhancement of 3D Printed Acrylonitrile Butadiene Styrene Polymer Components Using Neural Network Optimization Algorithm. *Polymers* **2020**, *12*, 2250. [CrossRef]

Review

Critical Review of Biodegradable and Bioactive Polymer Composites for Bone Tissue Engineering and Drug Delivery Applications

Shubham Sharma [1,2,*], P. Sudhakara [1,*], Jujhar Singh [3], R. A. Ilyas [4,5], M. R. M. Asyraf [6,*] and M. R. Razman [7,*]

1. Regional Centre for Extension and Development, CSIR-Central Leather Research Institute, Leather Complex, Kapurthala Road, Jalandhar 144021, India
2. PhD Research Scholar, IK Gujral Punjab Technical University, Jalandhar-Kapurthala, Highway, VPO, Ibban 144603, India
3. IK Gujral Punjab Technical University, Jalandhar-Kapurthala, Highway, VPO, Ibban 144603, India; jujharsingh2085@gmail.com
4. School of Chemical and Energy Engineering, Faculty of Engineering, Universiti Teknologi Malaysia, Johor Bahru 81310, Malaysia; ahmadilyas@utm.my
5. Centre for Advanced Composite Materials, Universiti Teknologi Malaysia, Johor Bahru 81310, Malaysia
6. Department of Aerospace Engineering, Faculty of Engineering, Universiti Putra Malaysia (UPM), Serdang 43400, Malaysia
7. Research Centre for Sustainability Science and Governance (SGK), Institute for Environment and Development (LESTARI), Universiti Kebangsaan Malaysia (UKM), Bangi 43600, Malaysia
* Correspondence: shubham543sharma@gmail.com or shubhamsharmacsirclri@gmail.com (S.S.); sudhakarp@clri.res.in (P.S.); asyrafriz96@gmail.com (M.R.M.A.); mrizal@ukm.edu.my (M.R.R.); Tel.: +91-7009239473 (S.S.)

Citation: Sharma, S.; Sudhakara, P.; Singh, J.; Ilyas, R.A.; Asyraf, M.R.M.; Razman, M.R. Critical Review of Biodegradable and Bioactive Polymer Composites for Bone Tissue Engineering and Drug Delivery Applications. *Polymers* **2021**, *13*, 2623. https://doi.org/10.3390/polym13162623

Academic Editors: Emin Bayraktar and S. M. Sapuan

Received: 8 July 2021
Accepted: 31 July 2021
Published: 6 August 2021

Publisher's Note: MDPI stays neutral with regard to jurisdictional claims in published maps and institutional affiliations.

Abstract: In the determination of the bioavailability of drugs administered orally, the drugs' solubility and permeability play a crucial role. For absorption of drug molecules and production of a pharmacological response, solubility is an important parameter that defines the concentration of the drug in systemic circulation. It is a challenging task to improve the oral bioavailability of drugs that have poor water solubility. Most drug molecules are either poorly soluble or insoluble in aqueous environments. Polymer nanocomposites are combinations of two or more different materials that possess unique characteristics and are fused together with sufficient energy in such a manner that the resultant material will have the best properties of both materials. These polymeric materials (biodegradable and other naturally bioactive polymers) are comprised of nanosized particles in a composition of other materials. A systematic search was carried out on Web of Science and SCOPUS using different keywords, and 485 records were found. After the screening and eligibility process, 88 journal articles were found to be eligible, and hence selected to be reviewed and analyzed. Biocompatible and biodegradable materials have emerged in the manufacture of therapeutic and pharmacologic devices, such as impermanent implantation and 3D scaffolds for tissue regeneration and biomedical applications. Substantial effort has been made in the usage of bio-based polymers for potential pharmacologic and biomedical purposes, including targeted deliveries and drug carriers for regulated drug release. These implementations necessitate unique physicochemical and pharmacokinetic, microbiological, metabolic, and degradation characteristics of the materials in order to provide prolific therapeutic treatments. As a result, a broadly diverse spectrum of natural or artificially synthesized polymers capable of enzymatic hydrolysis, hydrolyzing, or enzyme decomposition are being explored for biomedical purposes. This summary examines the contemporary status of biodegradable naturally and synthetically derived polymers for biomedical fields, such as tissue engineering, regenerative medicine, bioengineering, targeted drug discovery and delivery, implantation, and wound repair and healing. This review presents an insight into a number of the commonly used tissue engineering applications, including drug delivery carrier systems, demonstrated in the recent findings. Due to the inherent remarkable properties of biodegradable and bioactive polymers, such as their antimicrobial, antitumor, anti-inflammatory, and anticancer activities, certain materials have gained significant interest in recent years. These systems are also actively being researched to improve therapeutic activity and mitigate adverse consequences. In this article, we also present the main drug delivery systems reported in the literature and the main methods available to impregnate the polymeric scaffolds with drugs, their properties, and their respective benefits for tissue engineering.

Keywords: drug delivery; biodegradable polymers; polymeric scaffolds; natural bioactive polymers; antimicrobial properties; anticancer activity; tissue engineering

1. Introduction

Bone tissue reconstruction represents one of the biggest challenges for medicine due to the existence of serious global health problems, such as diseases, defects, trauma, the rise of obesity, and sedentary lifestyles [1–4]. Bone tissue engineering is a recent field of research associated with regenerative medicine, and applies the principles of engineering and the life sciences toward the development of biological substitutes that restore, maintain, or improve tissue function [5–7]. Until recently, bone tissue reconstruction was represented by bone grafts, which present several limitations, such as disease transfer and cost. At present, a new generation of development is required in medicine that comprises not only physical support for bone formation, but also the presence of biochemical agents to promote the formation of the bone. One of the biggest advantages of this system is the fact that it enables controlled delivery of the drugs to the affected tissue [1,8,9].

To date, numerous porous nanocomposite scaffold materials have been investigated. However, these materials still present challenges due to their capability for regeneration and remodeling, and for mimicking the complicated physiochemical attributes of bone. In addition, the functionality of the scaffolds has been studied by loading biomolecules (drugs, growth factors (GFs)) onto the scaffolds to treat bone disorders or to act on the surrounding tissues [10–12].

Three-dimensional bone bioactive nanocomposite scaffolds can be fabricated from a wide variety of bulk biomaterials, such as bioceramic tricalciumphosphate (TCP), hydroxyapatite (HA), and bioglass (BG); or biodegradable polymer—collagen, chitosan, alginate, fibrin, polyesters, and polyethylene glycol (PEG) [13–18]. It was demonstrated that their composites represent a suitable alternative because they combine the advantages of both bioactive ceramics and biodegradable polymers for bone tissue engineering. The reason for this is simple: ceramics present weak mechanical properties due to brittleness (hard material with small elongation to failure) and the polymers present a deficiency in their compressive modulus compared with native bone tissue (polymers are typically too soft) [8]. Thus, these systems can reduce the disadvantages and offer new advantages in the case of bone tissue reconstruction. The Word-Cloud info-graphic plugin's aim should be to provide succinct visually graphics representations of these kind of contextual features for better accessibility throughout intrusion-network mapping analysis of the existing-review as exhibits in the Figure 1. A word-cloud visualization of the keywords examined in this article is shown in Figure 1.

In tissue regeneration, the use of individual component scaffolds is widespread. Nevertheless, in certain circumstances, a single polymer is not able to fulfil all of the necessary criteria in several tissue regeneration applications. The bone matrix is a collagenous and apatite-based organic or inorganic composite. In addition to bone-tissue engineering, biocompatible composite scaffolds with an apatite element have been formed [19,20]. The most prevalently utilized material is perhaps hydroxy-apatite (HAP), which vaguely resembles the natural ingredients of bone. In addition, calcium phosphate (CaP) variants and bioglass have also been employed due to their excellent biocompatibility [21,22]. For instance, a PLGA/HAP-based nanofibrous composite scaffold was previously developed by polymer coating hydroxyapatite on to the PLGA scaffolds using a variety of methods [23,24]. Hydroxyapatite in composite scaffolds dramatically enhances proteinous adsorption capabilities, represses apoptosis cell death, and tends to create a more desirable microenvironment for bone tissue regeneration [25]. Nanohydroxyapatite polymeric composite scaffolds have been produced to imitate the nanosized characteristics of an organic mineral, in addition to emulating the inorganic–organic essence of natural bone [26–28]. Numerous inorganic elements have also been used to develop biotic and abiotic composite scaffolds. Lei et al. [29], for instance, utilized the temperature-dependent mediated phase-separation (TMPS) method to produce nanofibrous gelatinous silica hybrid scaffolds.

Figure 1. Infographic word-cloud visualization of the semantic clustering network of the keywords for physicomechanical, thermostability, and in vitro drug release studies of bioactive/biodegradable polymeric materials with remarkable biocompatibility in biomedical research.

Due to their ability to mimic both the nanometer-scale architectural design and chemical content of innate organic bone extracellular matrices, mineralized nanofibrous scaffolds have been recommended due to their potential for scaffold material restoration. Macroporous nanofiber scaffolds for bone-like apatite deposition have been developed by incubating them in simulated body fluid (SBF) with an ionic electrolyte concentration and a pH closely comparable to that of human blood plasma [30,31]. After an acceptable period of incubation in SBF, it was found that CaP could be deposited immediately upon the substratum, and a homogeneous, fibrous, compact and dense layer of nano-apatite was developed to portray an intrinsic interlayer of a porous wall surface without clogging the macropores [32–36].

An electro-deposition methodology was previously proposed to reduce the mineralization period to below 1 hour [32]. In terms of the dissolution rate, the chemical composition, and the microstructure of CaP developed on electrospinning-based PLLA-fiber skinny-matrices, a previous study [33] recently contrasted a novel electro-deposition technique with a well-known SBF incubation approach. Based on the mineralization of a fibrillar matrix, electro-deposition appears to have been two–three orders of magnitude quicker than the SBF technique. The aim of the study was to reduce the mineralization period from fourteen days to one hour while achieving the same degree of mineralization (Figure 2). Steadily increasing the fiber diameter led to faster mineral deposition in the electro-deposition technique compared to the relatively slower mineralization of the SBF incubation approach. The chemical structure and morphological characteristics of CaP can be obtained by altering the electric-deposition prospective and electrolyte temperatures to tune the blend of brushite and hydroxyapatite. The SBF technique can only generate a minimal HAP. Mineralized electrospinning-based PLLA fibrous scaffolds acquired by either technique strengthen the proliferative and osteogenesis differentiation of pre-osteoblastic MC3T3-E1 cells to a level comparable with that of the neat PLLA matrix.

Figure 2. Microstructure of mineralized PLLA matrices: (**a**) electro-deposition at 3 V, 60 °C for 1 h; (**b**) high-magnification picture of (**a**,**c**) mineralized in 1.5 SBF for 12 days; (**d**) magnified image of (**c**,**e**) mineralized in 1.5 SBF for one month; and (**f**) magnified picture of (**e**). Reproduced with permission from [33].

Mamidi and Delgadillo [34] adopted an ionic gelation strategy to develop chitosan (CS) nanocomposite hydrogel nanoparticles (CNPs). The authors observed disassembling of the CNPs' structure at 55 °C. In addition, the CNPs showed good cell viability against human fibroblast cells, as exhibited in Figure 3.

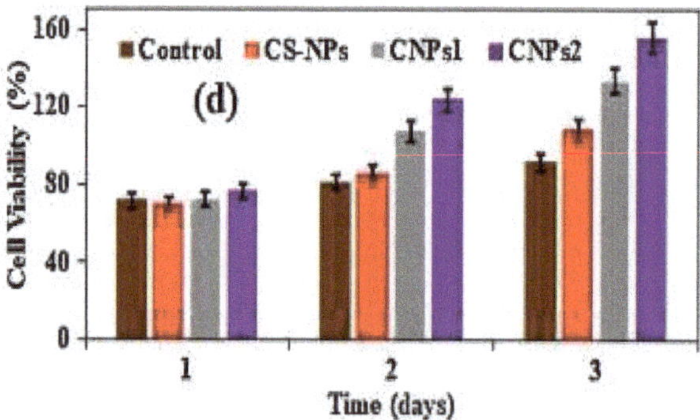

Figure 3. Cell viability of chitosan (CS) nanocomposite hydrogel nanoparticles (CNPs). Reproduced with permission from [34].

Therefore, CNPs provide a better pH and temperature-triggered drug delivery platform for the GI tract and colon-targeted drug delivery, in addition to the highest drug release under specific pH and temperature values, as displayed in Figure 4.

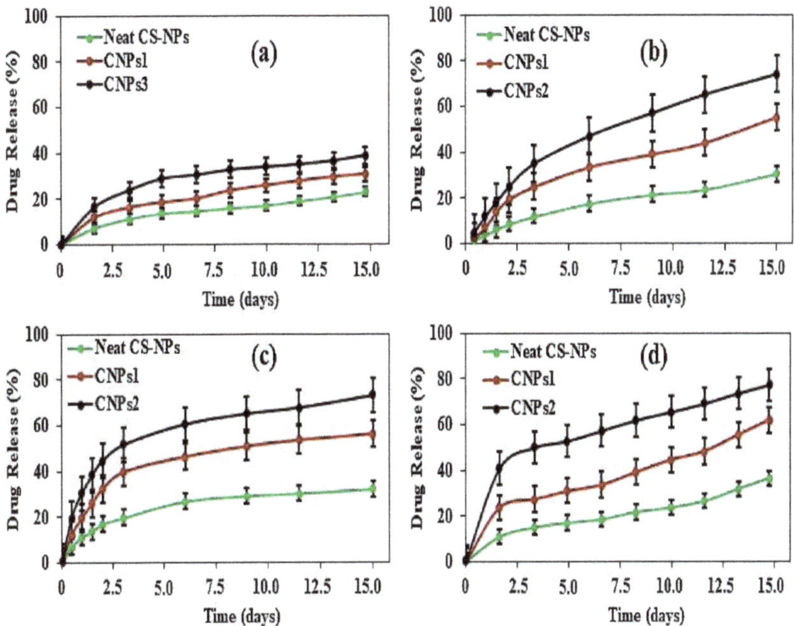

Figure 4. Diclofenac-embedded pure CS-NPs, CNPs1, and CNPs2 in vitro release of drugs studied at 37 °C in varying acid levels and environmental conditions: (**a**) pH 2, (**b**) pH 6, (**c**) pH 7.4, and (**d**) pH 9. The results are presented as an average deviation of ±3. Reproduced with permission from [34].

In another study, Mamidi et al. [35] developed PAPMA-CNOs and AN-PEEK biopolymer matrix composites. The authors observed that interactions between CNO and DOX played a vital role in controlling the drug release from the thin film. This film possessed a tensile strength of 891 MPa, an elastic modulus of 43.2 GPa, and a toughness value of 164.5 J/g. This thin film is used in numerous medical applications. To examine the fracture deformation phenomenon during the tensile test, the researchers analyzed the delamination zone under tensile loading. The sheets had a relatively uniform homogeneous and compact morphology, as illustrated in Figure 5.

Polymeric materials, among other materials, have been formed as tissue engineering scaffolds. These materials have relatively higher processing and handling versatility, processability, adaptability, degradability, and biocompatibility, which can be augmented via structural design analysis [36]. Polymeric materials (such as natural polymers, natural polymeric-derived composites, and synthetic polymeric materials, in addition to synthetic polymeric materials made of natural monomeric units and amended to natural moieties) are therefore the primary scaffold materials used in tissue engineering [37]. This descriptive analysis summary is not intended to be a rigorous exhaustive analysis of all of the polymeric materials used in biomedical applications, bioengineering, regenerative medicine, bone regeneration, or tissue regeneration.

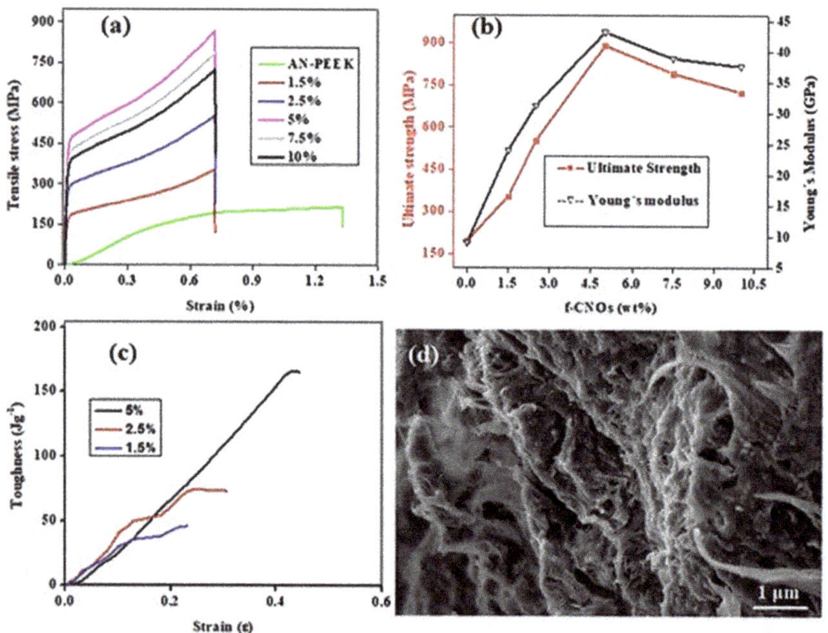

Figure 5. Physicomechanical analysis of biopolymer composite films employing quantitative techniques: (**a**) load displacement; (**b**) ultimate strength elastic moduli; and (**c**) impact–strength–strain graphs from uni-axial tensile nanocomposite thin films. (**d**) SEM micrographs of the fractography of biopolymeric nanocomposites comprising 5% of f-CNOs, produced by tear testing on perforated grooved specimens. Reproduced with permission from [35].

1.1. Advances in Biodegradable Polymers for Biomedical Applications

The biodegradable polymers can be classified as natural polymers or synthetic polymers based on their origin. Natural polymers are further classified as polysaccharides, polypeptides, and polyesters, depending on their repeating units. The polysaccharide contains d-glucopyranoside repeating units, e.g., starch, cellulose, chitin, and chitosan. Naturally occurring polypeptides contain amino acid repeating units. Among the naturally occurring polypeptides, gelatin is extensively used in pharmaceutical and biomedical applications. Naturally occurring polyesters, for example, polyhydroxybutyrate (PHB) and polyhydroxyvalerate (PHV), are potential candidates for biomedical applications [38]. Natural materials are generally biocompatible, and exhibit mechanical properties comparable to those of native tissues. However, these materials also suffer from disadvantages, such as limited control over physicochemical properties and difficulties in modifying degradation rates. Purification and sterilization of these biomaterials after isolation from different sources is relatively cumbersome [39].

In oral drug delivery, cellulose derivatives, for example, cellulose ethers such as ethyl cellulose, methyl cellulose, hydroxypropyl methylcellulose (HPMC), and hydroxypropyl cellulose, have been used in the form of coatings. Similarly, synthetic polymers such as poly(acrylates), poly(methacrylates), poly(methyl methacrylates), poly(hydroxyethyl methacrylates), and copolymers thereof have been extensively used [40].

Stimuli-sensitive polymers are another class of polymers that interact and respond to the environmental conditions, such as temperature, light, salt concentration, and pH [41]. pH-sensitive polymers are used to develop smart delivery systems due to the variation in the physiological pH in different parts of the body. However, although these polymers are biocompatible, they are not biodegradable. The use of polymeric carriers in injectable drug delivery systems requires biodegradable polymers that degrade into nontoxic and safe products.

PLGA and PLLA are used to encapsulate drugs in the form of microparticles or nanoparticles to increase the circulation time and the bioavailability of the drug [42]. New drug delivery systems have been developed for chronic diseases and/or conditions that require sustained drug delivery.

The sustained release of drugs was achieved initially by drug diffusion from polymeric microspheres followed by polymer degradation [43].

The development of injectable in situ semisolid drug depots has been explored as an alternative delivery system. Biodegradable polymers can be used in the form of an injectable matrix or a depot for drug delivery, and as injectable scaffolds in tissue engineering [44]. Atrix laboratories developed ATRIGEL® technology, in which sustained release of Leuprorelin acetate was achieved through a PLGA depot formed in situ.

The literature indicates that polymer-based scaffolds have been developed that can be used in tissue engineering as supports for cell attachment and proliferation [45–50]. These biodegradable scaffolds can be simultaneously used as cell support and for controlled delivery of biologically active proteins, such as growth factors and cytokines [51]. In addition to biodegradability, the polymer should satisfy the following requirements for processability:

i.   It must be liquid so that it can appropriately fill the cavities and replicate the patterns present on mold with high fidelity.
ii.   It must contain functional groups to enable cross-linking during processing.

PEG and PLA block macromers containing terminal (meth) acrylate groups were photopolymerized to yield highly cross-linked biodegradable materials [52]. In another effort to incorporate vinyl functionality in biodegradable aliphatic polyesters, di- and triblock copolymers of ε-caprolactone (CL), glycolide (GA), and lactide (LA) with ethylene oxide (EO) bearing terminal vinyl groups were developed. These polymers were further processed by UV-micro-embossing to fabricate biodegradable scaffolds. However, these polymers must be processed at 65 °C because they are not liquids at room temperature [53,54].

Photo-patternable biodegradable 2-hydroxylethyl methacrylate (HEMA) conjugated poly(ε-caprolactone-Co-RS-β-malic-acid) (PCLMAc) copolymers were synthesized to fabricate biodegradable scaffolds. However quantitative conjugation of carboxylic acid groups to HEMA ester is difficult and its synthesis involves multiple steps. Furthermore, the resulting polymers are not liquids at room temperature and need to be processed at 60 °C [55].

Liquid photo-patternable polyurethane diacrylates were developed to fabricate biocompatible scaffolds by UV-micro-embossing. The fabricated scaffolds exhibited cytotoxicity due to the presence of unreacted monomers and the photo initiator used during polymer synthesis. The polymers were rendered biocompatible after the residues were leached out by repeated extractions with methanol, which is not a highly desirable solvent [56]. Other examples of photopolymerizable and degradable polymers developed to date include poly(propylene fumarate) (PPF), photo-cross-linkable poly(anhydride), polyethylene glycol, and cross-linkable poly(saccharide). However, synthesis of these polymers involves multiple reactions and purification steps [57].

Thus, biodegradable polymers are extensively used in numerous biomedical applications, including as drug carriers, tissue regeneration, regenerative medication, gene therapy, temporary implantable devices, and coatings on implants. The basic criteria for choosing a polymer to be used as a degradable biomaterial are: (i) the mechanical properties and degradation rate should match the needs of an application so that adequate strength remains until the surrounding tissue has been cured; (ii) biocompatibility; (iii) non-toxic degradation products; (iv) shelf life or stability; (v) processability and cost. For drug delivery applications, the time of release governs the type of polymer, and the shape and size of the device. For example, lactide and glycolide polymers are clinically approved polymers that can be used in any application [58,59].

Biodegradable plastics are also widely used in agricultural areas. The main motive for using biopolymers in agricultural sectors is the rising utilization of polymers in agriculture, which has enabled farmers to enhance their crop production. Some of the plastics employed in agriculture are recyclable, such as silage films, greenhouse sheet layers, fertilizer satchels, tubing pipelines, and other polymer materials, whereas others are difficult to recycle, such as thin mulching films, thin low tunnel films, and direct covering films. These covering layer sheets are very slender and flimsy, and are often heavily degraded with dust, dirt, and unwanted substances. Thus, an attractive alternative for nonrecyclable plastic waste is biodegradation. Consequently, the utilization of biodegradable plastics is increasing in agricultural applications. These applications primarily include mulching films, plant pots, and compost bags [60,61].

Although biodegradable polymers can be used in different applications, such as the packaging, medical, and agricultural fields, the commercialization of biodegradable polymers is often hampered due to competition with commodity plastics, which are cheap and familiar to the customer. In addition, the infrastructure needs to be developed for the disposal of biodegradable polymers in bioactive environments, which requires capital investment [62]. Moreover, the biodegradable polymers currently available possess inferior physical properties, such as poor strength and dimensional stability, and their processing is technically difficult [63].

Synthetic biodegradable polymerics are regarded as being biodegradable, biocompatible, and highly safe. As a result, they are widely used in biomedical applications, and particularly in the areas of controlled drug carrier systems and tissue regeneration. Due to the degradable nature of polymeric implants, there is no requirement of surgical intervention to remove the implant at the end of its functional life [64]. In tissue engineering applications, synthetic polymers are mainly used in scaffolds, which provide suitable mechanical support and show favorable surface properties, such as adhesion, proliferation, and differentiation of cells [65].

Polylactic acid is produced either through the fermentation process of carbohydrate crops (such as corn, sugar beets, tapioca roots, wheat, barley, and sugarcane) or chemical synthesis [66]. The fermentation process is preferred over the synthetic route because the latter is unable to produce the desirable L-isomer, in addition to its high manufacturing cost. In contrast, the fermentation process produces the L-isomer with a high purity (99.5%). In general, PLA is produced from the pure L-isomer.

PLA is principally produced via different processes: condensation polymerization of lactic acid (LA), condensation reaction in an azeotropic solution, and ring opening polymerization of an intermediate called lactide. The first method (polycondensation) involves the esterification of monomers in the presence of suitable solvents, and water (byproduct) is removed azeotropically under reduced pressure (vacuum) and high temperature. Tin (II) chloride is the most commonly used catalyst in this method and can be recovered at the end of the reaction. This method is the least expensive route but cannot produce solvent-free high molecular weight PLA having superior mechanical properties. The second method involves the condensation reaction of lactic acid in an azeotropic solution. The third method appears to be the most commonly employed procedure for producing higher molecular weight PLA. This approach involves three steps: (i) condensation of lactic acid monomeric subunits; (ii) depolymerization of the PLA to the lactic acid; and (iii) the cyclization ring opening polymerization of the lactide unit in the presence of metal catalysts, resulting in PLA with a high molecular weight [67].

Furthermore, PLA exists in three stereoforms: PLLA, PDLA, and PDLLA. Of these, PLLA and PDLA are semicrystalline polymers that show a high tensile strength and low elongation, whereas PDLLA is a more amorphous polymer and shows a random distribution of both of the isomers [68]. The ROP of the L-lactide unit can be performed via melting or a suspension solution using stannous octoate (SnOct2) as the initiator, which avoids racemization at high temperature and transesterification.

Polylactic acid has a vast range of applications. However, it cannot be used in flexible films due to poor ductility, and poor thermal and barrier properties [69]. PLA is also extensively used in medical applications due to its unique characteristics, such as biodegradability, biocompatibility, ecofriendliness, and thermoplastic processability. Moreover, PLA works very well and offers outstanding properties at a low price. It is used for preparing various devices, such as degradable sutures, nanoparticles, and drug-releasing microparticles. The use of biodegradable polymers rather than nondegradable polymers in medical applications has the advantage that it eliminates the need to remove implants; the biodegradable polymers remain temporarily in the body and disappear on degradation [70]. The physical properties of PLA (such as transparency or the mechanical properties) are comparable to those of polystyrene and poly(ethylene-terephthalate), but due to its high cost (compared to PP, PE, PS, etc.), brittleness, low viscosity, medium gas barrier properties, and high moisture sensitivity, its use is restricted to specific applications. Thus, efforts are being made to improve the properties of PLA by blending. Investigations have been conducted on blends of PLA with other polymers, such as poly(ε-caprolactone), poly(hydroxyl butyrate), polyethylene glycol, and poly(hexamethylene succinate). However, the produced blends were immiscible and resulted in poor mechanical properties [71–73].

Polycaprolactone (PCL) is a synthetic linear polyester derived from crude oil. It is resistant to water, oil, solvents, and chlorine. The average molecular weight of PCL ranges from 3000 to 90,000 g/mol. With an increase in molecular weight, the crystallinity of PCL tends to decrease. It is fully biodegradable under composting conditions and mainly used in the biomedical field. It acts as a stiffening material for shoes and orthopedic splints, and in completely biodegradable compostable bags, fibers, and sutures. It is also used in thermoplastic polyurethanes, adhesives, resins, etc. [74,75]. PCL is also used in tissue engineering [76].

PCL is readily biodegradable in diverse environments, such as marine water, soil, sewage sludge, and compost ecosystems; hence, it is widely used in drug delivery systems. The biodegradation of PCL occurs through either enzymes, simple hydrolysis, or both.

There are several parameters that influence the biodegradation of PCL, such as the molecular weight, crystallinity, thickness of the films, and degradation parameters. The microorganisms secrete extracellular depolymerases that degrade the polymer.

The enzymatic degradation of polycaprolactone has been studied mainly in the presence of lipase enzymes, which help in accelerating the biodegradation of PCL [77], for example, Rhizopus delemer lipase [78], Rhizopus arrhizus lipase [79], and Pseudomonas lipase [80,81].

Studies have been performed on the biodegradation of PCL [82]. Chen et al. (2000) observed that the enzyme lipase can accelerate the degradation of polycaprolactone microparticles, and the degradation rate of PCL is not significantly influenced by its surface area [83]. Murphy et al. (1996) revealed that the depolymerized enzyme produced by *Fusarium moniliforme* is cutinase [84]. Oda et al. (1995) isolated five fungal strains having the capability to degrade two polymers: PHB and PCL [85]. One of the fungal strains was identified as Paecilomyces lilacinus. The degradation of polycaprolactone was also studied using the bacteria Alcaligenes faecalis [82,85,86]. Abdel-Motaal et al. (2014) found that Pseudomonas japonica-Y7-09 (yeast) produced the extracellular enzyme cutinase, which degraded PCL by 93.33% in 15 days [87]. The mechanism of biodegradation of PCL has also been studied in detail. It is believed that PCL depolymerases preferentially attack the amorphous areas of polymers and degradation occurs due to endo- and exo-cleavage [82].

The various physicochemical and physicomechanical characteristics of polycaprolactone appear to be modified by either copolymerization or by efficiently blending with other polymers. The copolymerization helps to alter the chemical property of PCL, which further affects numerous properties, such as crystallinity and solubility, resulting in a modified polymer that has the desired attributes for drug delivery. By comparison, blending helps to change the physical properties and biodegradation, in addition to the mechanical properties, resulting in polymers that are preferred for tissue engineering. PCL has been found to be compatible with natural polymers (starch, hydroxyl apatite, and chitosan), polyethylene oxide (PEO), and polylactic acid and polylactic co-glycolic acid (PLGA). These modifications are useful in formulations for drug delivery [88].

Polyglycolic acid is produced by the polycondensation reaction between glycol and aliphatic dicarboxylic acids. The constituents are derived from renewable resources, such as glycol obtained from glycerol, and organic acids are obtained via fermentation. PGA is a soft and biodegradable material, and possesses good sensitivity and a high melting point (approximately 200 °C). It has excellent material properties similar to those of aromatic PET. PGA is commercially produced by Dupont, either in the form of an aliphatic-aromatic copolymer (Biomax®) or as aramid fibers (Kevlar®) [89].

PGA and its copolymers are both widely utilized in medical applications for degradable and absorbable sutures. They can easily degrade in aqueous surroundings, such as body fluids, via hydrolysis of the ester backbone. Furthermore, the degraded products of PGA are metabolized to CO_2 and water [90,91].

Microbial polymers/polyhydroxyalkanoates (PHAs) are biodegradable biopolyesters that are completely synthesized by microorganisms, such as bacteria and fungi, in addition to some plants. There are many bacteria that can synthesize PHA, such as those found in activated sludge, oceans, or extreme environments. More than 30% of soil-inhabiting bacteria are capable of synthesizing PHA [83,92]. Some examples of PHA-producing bacteria are Alcaligenes latus, Pseudomonas oleovorans, and Azotobacter vinelandii [93]. Microorganisms such as Ralstonia eutropha and recombinant Escherichia coli are capable of accumulating PHA in quantities of as much as 90% (w/w) of their dry cell mass in a nutrient-limited media, i.e., media that is deprived of essential nutrients such as nitrogen, phosphorus, or oxygen, but in which an excess of carbon is present. The most general limitation is observed with nitrogen (*Azotobacter* spp.), but the most efficient limitation is that of oxygen. Due to the insolubility of PHA in water, it accumulates as carbon or an energy source within the intracellular granules [83,92,94].

Poly3-hydroxy-butyrate (PHB) is the most common and well-studied polymer of the polyhydroxyalkanoate family. It has been reported that this bacterium can accumulate PHB intracellularly. It is a homopolymer made up of 3-hydroxybutyric (3HB) acid molecules. The molecules are joined by ester bonds formed between the 3-hydroxyl group of one monomer and the carboxylic group of another. Numerous other bacteria have been identified as accumulating PHB in their cells, both aerobically and anaerobically. However, PHB possesses poor physical properties, and is too stiff and brittle to be used in most commodity products. It was subsequently found that PHA in activated sludge contains monomers other than 3-hydroxybutyric acid (3HB), such as 3-hydroxyvalerate (3HV). The incorporation of a few percent of 3HV units in the polymer helps to improve the flexibility and also reduces the brittleness. Numerous companies, such as ICI and, subsequently, Zeneca and Monsanto, started production of PHBV at an industrial scale [95–98]. Their production capacity has increased to 900 million tons per year. The commercial PHA produced by Tianan contains about 5% ester of valeric acid, although some experimental grades contain up to 15% valerate. Valerate improves the flexibility of the polymer.

PHAs are thermoplastic and/or elastomeric, biocompatible, non-toxic, enantiomerically pure, optically active (i.e., possess only the R-configuration), piezoelectric (i.e., assist in wound repair and healing), and also induce bone regeneration and formation. They show better resistance to UV degradation than polypropylene (PP) but are less solvent resistant. The most important characteristic is that they are completely biodegradable. Due to these properties, PHAs are widely used in biomedical applications, such as orthopedy (screws, bone graft substitutes, and scaffolds for cartilage engineering), cardiovascular system devices, wound management (sutures, dressings, and dusting powders), urological stents, and controlled drug delivery (tablets, micro-carriers, and implants). Like PVC and PET, PHAs also exhibit good barrier properties, so they are also used in packaging applications, such as shampoo bottles, cosmetic containers, milk cartons and films, cover for cardboard and paper, pens, combs, bullets, and moisture barriers in nappies and sanitary towels. PHA may help to address the problems of environmental pollution caused by nondegradable synthetic polymers [99–102].

There are some drawbacks of using these polymers: (i) the cost of producing PHAs is very high compared to conventional petroleum-based plastics; (ii) the processing of PHAs is more difficult than conventional petroleum-based plastics due to their slow crystallization process; (iii) their mechanical and thermal characteristics are not consistent compared to those of petrochemical plastics; (iv) they are further required to be developed for a wider range of applications and large-scale production; (v) the quality and uniformity of PHA must be optimized.

The two polyesters, namely, PLA and PHA, have their own advantages and disadvantages. Typically, polylactic acid (PLA) is cheaper than PHA. Therefore, the application research of PLA is more advanced than that of PHA [83,92].

Roy et al. (2008) examined the biodegradation of PE containing a pro-oxidant (cobalt stearate) using a consortium of three bacteria, namely, Bacillus cereus, Bacillus pumilus, and Bacillus halodenitrificans [103]. The films were UV irradiated (λmax at 313 nm) and subsequently incubated with the bacteria. The degradation was monitored based on the FTIR, mechanical properties, GC-MS, DSC, TGA, SEM, melt flow index, weight loss, and cfu count. It was observed that there was decrease in the carbonyl index (by FTIR analysis); the formation of low molecular weight compounds (by GC-MS studies); an increase in initial decomposition temperature (TGA); the formation of biofilm on the polymer surface (by SEM analysis); a weight loss of polymer of 8.4%; and an increase in the bacterial count (by cfu count).

Reddy et al. (2008) developed the blends of PLA/PP to create fibers and characterized them via their mechanical properties, and SEM, XRD, and DSC techniques [104]. The blends showed partial compatibility between PLA and PP, and their mechanical properties were inferior to those of the pure polymers. However, the blends showed better resistance to hydrolysis and biodegradation, in addition to better dyeability, than pure PLA. Nishida et al. (2009) prepared blends of PLLA and PP with and without the catalyst MgO, and characterized them using SEC, NMR, FTIR, SEM, and TGA techniques [105]. TGA analysis showed that the addition of MgO to the blend selectively accelerated the depolymerization of the PLLA component in the blend, leading to the generation of L, L-lactide as a main volatile product. Hamad et al. (2011b) blended PLA with PP in different ratios and studied their rheological and mechanical properties [106]. The rheological results of blends revealed that the true viscosity was between that of the pure polymers, whereas the flow activation energy was less than that of the pure polymers. The mechanical tests showed that there was incompatibility between the two polymers. Choudhary et al. (2011) blended PLA with PP in various ratios with and without compatibilizers, i.e., maleic anhydride grafted PP (MAPP) and glycidyl methacrylate [107]. The blends were characterized by mechanical tests, and DSC, TGA, FTIR and SEM techniques. The results revealed that a blend of PLA/PP in the ratio of 90:10 had optimum mechanical properties, which led to improved melt processability of PLA. The interaction between these two polymeric materials can be improved by the addition of a suitable compatibilizer, such as MAPP. MAPP is an effective compatibilizer that mediates the polarity at the interface of two polymers.

1.2. Developments in Bioactive/Biodegradable Polymers with Interfacial Activity-Assisted Surface Functionalization for Drug Delivery

Polymers are macromolecules formed from the combined repetitive monomer units. Many of these polymers excel in transporting the drug to the diseased site and releasing it in a controlled manner [108]. Their structures help protect the drug and thereby increase its bioavailability. Polymers are broadly classified as biodegradable and nonbiodegradable. The latter have some advantages and are also employed in drug delivery [109]. Occasionally, after the drug is released, nonbiodegradable polymers need to be retrieved using invasive methods. In contrast, biodegradable polymers do not require retrieval because they degrade/erode into smaller molecules that are eliminated through different metabolic pathways.

Biopolymers are mainly classified as natural (chitosan and cellulose) and synthetic (PLGA, polyanhydrides), as illustrated in Table 1. Synthetic biopolymers have greater significance because they can be altered to suit specific requirements. The drug release rates can be controlled, and drugs of different physiochemical properties can be accommodated [110] without altering their therapeutic efficacy, thereby making them one of the most researched polymers in the field of drug delivery. Many biopolymers have transitioned from academic curiosities to real-world applications, and many water-soluble polymers (Table 2) play an essential role as drug delivery agents [111].

Table 1. Biodegradable polymers, with their structure and possible applications [110,112].

Biodegradable Polymers	Structure	Uses
Polylactic acid, poly(L-lactide), and poly(DL-lactide) family (PLA)	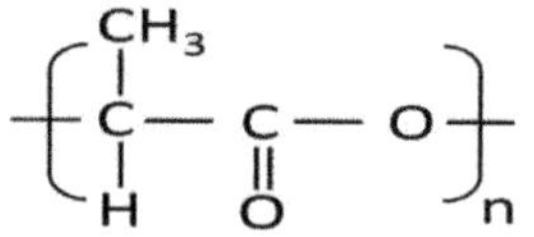	Sutures, prosthetics, drug delivery.
Polyglycolic acid (PGA)	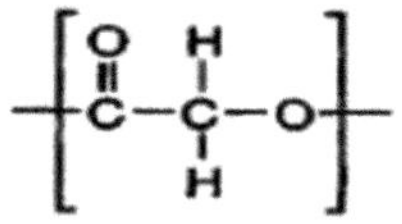	Drug carriers, sutures.
PolyL-lactide-*co*-glycolide family	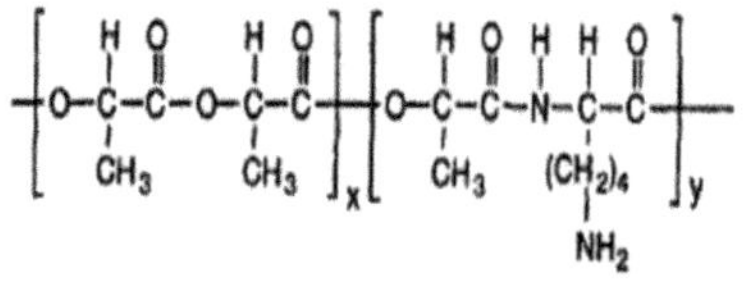	Drug delivery vehicle, sutures.
Poly(ortho esters)	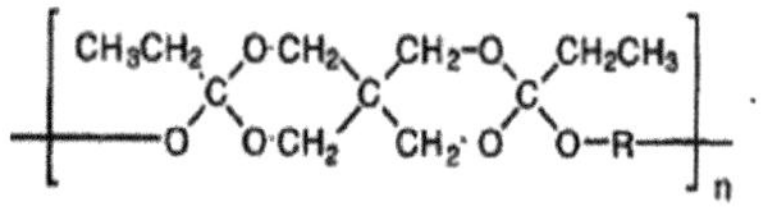	Ointments, drug delivery devices.
Polyanhydrides (e.g., polysebacic anhydride) (highly hydrophilic)	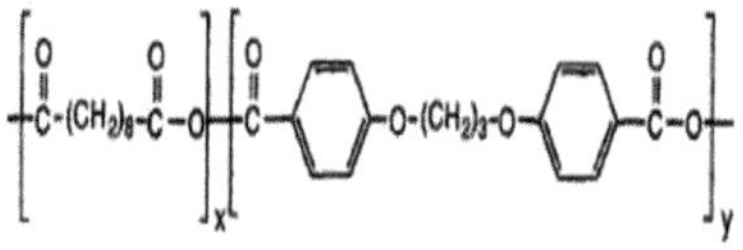	Drug delivery devices.
Polycarbonates	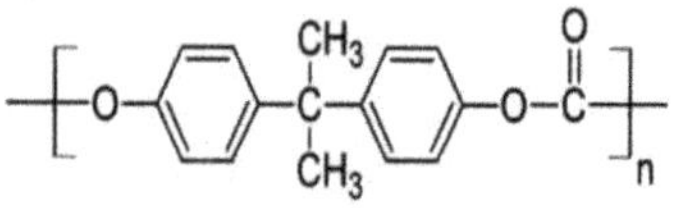	Drug delivery with some modifications, bone repair.
Phosphorus-containing polymers	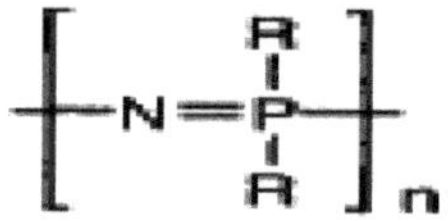	Implantable biomaterials.
Polydioxonone	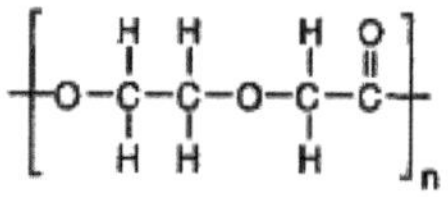	Plastic surgery, drug delivery, and tissue engineering.

Table 2. Water-soluble polymers [110].

Polymer	Structure	Uses
Saccharides (cellulose, dextran, chitin)		Cosmetics and health care products.
Acrylates and acrylamides (HEMA) PHEMA		Contact lenses, catheters.
Polyethylene glycol (PEG)		Drug delivery and as a stealth polymer.

Nonbiodegradable polymers, by comparison, have been employed extensively in biomedical devices, such as catheters, heart valves, prostheses, and dialysis membranes, for a substantial period. All of these polymers are hydrophobic, and their usage in the field of drug delivery has been limited due to their degradability issues. Hence, they mainly act as reservoirs for the drugs, and need to be retrieved surgically or after depletion of the drug.

Synthetic degradable polymers, such as polyesters, mainly display bulk erosion [113]. The polymer matrix becomes porous as time progresses, leading to the release of the drug into its environment due to the matrix's sudden collapse. Polymers, such as polyanhydrides, predominantly experience surface erosion as they undergo hydrolytic bond cleavage to form products that dissolve slowly in water [113,114]. Most of the biodegradable polymers display a combination of bulk and surface erosion, leading to varying drug release profiles. Polymers that predominantly degrade via bulk erosion initially show a first-order release, followed by a slow, constant release phase. Slower degradation of polymers such as PLGA is possible by employing higher molecular weight, leading to a slower, controlled drug release. Some of the well-known water-soluble polymers, such as polyethylene glycol (PEG), are either employed as standalone drug carriers [115] or conjugated to other polymers to provide properties of stealth [116]. They also help increase the stability of the drug.

Among all of these available prospects, PLGA, PEG, and PLA are a few of the FDA-approved polymers [117] and, thus, are most widely incorporated in the field of drug delivery.

For polymers in drug delivery, PLGA is the gold standard as a drug delivery vehicle. It is one of the most extensively employed biodegradable polymers for delivering drugs, proteins, DNA, and other bioactive agents [118]. PLGA is a copolymer produced by the combination of PLA and PGA, and is commercially available as 50:50, 65:35, and 75:25 PLGA, among other variants. The value 75 (in 75:25) denotes the percentage quantity of PLA, and the remainder (25) is that of PGA. Varying the concentration of PLA and PGA helps obtain the desired properties in the copolymer. The higher the concentration of PLA, the slower the degradation rate, with some exceptions [119]. The properties of the polymer can be tuned for the drug by varying the molecular weight, the ratio of PLA and PGA, the amount of the polymer, and the drug, to obtain the desired protection and release profile.

PLGA microparticles loaded with triptorelin were synthesized by Mahboubian et al. using the double emulsion solvent evaporation technique. The effect of various parameters, such as the emulsifying agent, volume of the water phase, and addition of NaCl, was studied [118,119].

Xie and colleagues formulated PLGA NPs to deliver Paclitaxel (PTX) [120]. To evaluate the formulation's effectiveness in passing the blood–brain barrier (BBB) and attaining the desired tissue, they used MDCK/C6 cell lines and varied the additives and coatings. The uptake results indicated that PLGA NPs with additives showed a higher uptake than those with surface coatings. Cell viability was also low when the cells were treated with PLGA NPs compared to the control (no treatment) [120]. Similarly, Averineni et al. encapsulated the drug in PLGA NPs (50:50) and assessed its antitumor activity on BT-549 cell lines. The optimized formulations released the drug for 15 days and had an inhibitory effect for around 7 days with a lower clearance rate [121].

Sengel-Turk et al. synthesized and characterized Meloxicam-loaded PLGA NPs and later studied their efficacy on HT-29 cells. NPs increased the stability of meloxicam and also helped in a sustained release. The formulations showed a higher uptake and had a higher cytotoxic effect on the cells [122]. Schleich and group incorporated dual agents (Paclitaxel and superparamagnetic iron oxide) in PLGA NPs. High cytotoxicity was observed when PTX-encapsulated PLGA NPs were used, but no such toxicity was noticed in iron oxide-loaded NPs. Both NPs showed a high uptake in CT 26 cell lines. In vivo studies showed delayed regrowth of CT26 tumors [123].

Cooper et al. synthesized and optimized Diclofenac-loaded PLGA NPs by employing varying stabilizers and other parameters. The formulated particles showed a low size (<200 nm), high drug entrapment (80%), and high stability [124].

Rafiei et al. developed docetaxel-encapsulated PLGA NPs for intravenous applications. Surface-modified PLGA NPs (PLGA-PEG) were synthesized, and animal model studies showed that both NPs increased the circulation time and concentration of the drug in the blood [125].

Afrooz et al. co-encapsulated PTX and verapamil (VIR) in PLGA NPs and character-ized the formulations for drug loading, zeta, and size. The co-encapsulated NPs showed higher cytotoxicity on MCF-7 cell lines after three days than the free drug [126]. Similarly, Ahmadi et al. co-encapsulated Doxorubicin (DOX) and VIR, which resulted in a similar effect on the MCF-7 cell line [127]. Vakilinezhad et al. co-encapsulated methotrexate and curcumin in PLGA NPs and optimized the formulations. The NPs showed higher cyto-toxicity on SK-Br-3 cell lines when compared to the free drug, and in vivo studies also indicated the inhibition of breast cancer [128].

Surface functionalization in polymer-based NPs is the process of altering the ex-isting polymer surface's physicochemical properties by introducing different materials (hydrophilic polymers) or molecules to enhance its efficacy or to help provide functional groups for further ligand attachments. Surface functionalization methods mainly depend on the surface chemistry and type of ligand, and the NP preparation technique. These are broadly classified under three essential categories, with the interaction between the NP surface and ligand forming the categorization basis.

In the chemical conjugation method (coupling), the ligand and polymer undergo chemical modifications to attach active groups [129,130] with the help of coupling reagents such as EDC, DCC, and NHS.

The noncovalent methods employ the affinity between the ligand and the NP surface to achieve the desired results. Proteins such as streptavidin, having a high affinity to biotin, are incorporated in the NP to noncovalently attach the ligand to the NP [131,132]. Electrostatic interaction or the physical adsorption method are achieved by selecting the appropriate ligand and NP polymer to promote hydrophobic interactions or hydrogen bonding. The NP can also be coated with surface polymers to mediate the interaction between the ligand and the NP [133–135].

Various polymers, such as PEG, PVA, dextran, and chitosan, have been utilized for surface functionalization. PEG is a hydrophilic polymer that has extensively been used due to its ability to protect NPs (via stealth) from rapid renal clearance. Simultaneously, PEG's presence hinders the interaction that is necessary for uptake in the cells [136]. For this reason, noncovalently bonded PEG (cleavable) on the surface of polymer nanocarriers is preferred, because it not only increases circulation time, but also does not hinder drug release and uptake.

The process of attaching PEG, covalently or otherwise, onto polymer molecules, drugs, and macrostructures is known as PEGylation. PEGylation is a standard method employed to increase the stability of different polymer carriers, drugs, and proteins. The addition of PEG has been shown to improve the efficacy of the carriers and drugs.

Ruan et al. employed triblock copolymers of PLA and PEG to load the hydrophobic drug PTX in microspheres of PLA-PEG-PLA. Their work showed that the release of the drug was faster in PEGylated NPs than regular PLGA NPs. Overall, about 50% of drug release was possible in a sustained manner for a period of one month [137]. Similarly, Danafar et al. encapsulated both hydrophobic and hydrophilic drugs into NPs of PLA-PEG-PLA and obtained high encapsulation efficiencies. The drug release profile was biphasic for the hydrophobic drug and triphasic for the hydrophilic drug [138]. Similarly, Dong et al. formulated, characterized, and optimized PTX-loaded MPEG-PLA NPs that took on the core-shell structure and released the drug in a biphasic manner [139].

Danhier et al. studied the effect of PTX encapsulated in PEGlylated PLGA NPs on HeLa cell lines compared to other commercial formulations, such as Taxol and Cremophor EL. The results showed that treatment with PTX-loaded NPs leads to lower viability when compared to the other formulations. In vivo studies also indicated more significant growth inhibition of tumors with the PTX-loaded NPs [140].

Rafiei et al. synthesized, characterized, and optimized Docetaxel-loaded NPs of PLGA and PEGylated PLGA, and found that the drug's release was higher in PEGylated NPs. The higher encapsulation also enhanced the blood concentration of the drug during in vivo studies [141].

Sims et al. used PEGylated PLGA NPs, in addition to other surface-functionalized PLGA NPs, and studied the internalization efficiencies on HeLa cell lines. They found that coating with PEG reduced cellular internalization but increased tissue penetration [142]. In vivo studies of PEGylated PLGA NPs encapsulating curcumin synthesized by Khalil et al. showed that curcumin release was slower in non-PEGylated NPs. Nonetheless, the bioavailability of the drug was significantly higher when delivered through PEGylated NPs. Both NPs were able to increase the mean half-life of the drug [143].

Interfacial activity assists in the surface functionalization of polymer-based NPs, and molecules such as PEG increase the circulation time of the NPs. However, this process is accompanied by a number of drawbacks [144]. Receptor-mediated targeted drug delivery helps overcome most of these issues and is also beneficial, because normal cells/tissues are left unaffected. Attaching ligands onto the NPs requires functional groups that are lacking in polymers such as PLGA. However, PEG can be modified to possess homo- or hetero-bifunctional groups [145]. This process provides the opportunity to attach the desired/possible ligands onto the PEGylated NPs, drugs, or other macromolecules.

Attaching ligands to polymer-based carriers is often performed through chemical conjugation, which requires in-depth knowledge and proper control of all of the variables (ligand, polymer, drug, process variables), which is an expensive and time-consuming [145–147] process. The interfacial activity-assisted surface functionalization technique utilizes the interfacial activity of amphiphilic polymers to incorporate the ligand into the polymer carrier, such as PLGA. The process of functionalization is based on the principle of self-assembly. The hydrophobic part (PLA) of a block copolymer (PLA-PEG) is inserted into the PLGA NP. At the same time, the hydrophilic PEG chain with the ligand remains on the outer surface of the PLGA NP during the solvent evaporation step of NP synthesis. Patil et al. simultaneously functionalized PTX-loaded PLGA NPs with folic acid (FA) and biotin through the PEG-PLA block copolymers. The drug-loaded NPs showed better efficacy on numerous cancer cell lines, and studies also indicated an enhanced accumulation of these NPs during in vivo studies [147].

Similarly, Toti et al. utilized the maleimide end group in their block copolymer of PLA-PEG to conjugate the cRGD peptide, which was then attached to the coumarin 6-loaded PLGA NPs. A significant and high cellular uptake was observed on numerous cell lines. Their study showed a two-fold increase in conjugate-NP accumulation during in vivo studies compared to NPs without the ligand [148].

Roger et al. also employed the same method to conjugate the FA moiety to PTX- loaded PLGA NPs. Cell line studies on Caco-2 showed a five-fold increase in the apparent permeability of PTX when encapsulated in the PLGA polymer carrier. The FA-functionalized NPs underwent an eight-fold rise in transport compared to the free drug, thereby increasing the drug's oral bioavailability [149].

The IAASF technique was also used by Dhoke et al. to conjugate Lactosaminated-Human Serum Albumin peptide on PLGA NPs, loaded with the drug lamivudine. In vivo studies also exhibited the benefits of ligand-based targeting [150].

The targeting of diseased sites/tissues through the ligand–receptor route has gained significant attention in recent decades. Drug delivery by this method not only protects healthy cells, but also helps in increasing the therapeutic effect. The effective use of polymers to carry the drug adds to the efficacy, because sustained drug delivery is possible. Studies have shown that cancer cells overexpress specific receptors and thus can be utilized to deliver the drug. Similarly, activated macrophages also overexpress numerous receptors, including folate (FRβ) [151].

Folate receptors (FRα and FRβ) are among the many receptors known to be overexpressed on cancer cells. Folic acid, a synthetic version of vitamin B9, is known to show affinity towards these receptors. Many researchers have utilized this ligand–receptor route to target and deliver the associated drug [152]. Vortherms et al. conjugated folic acid to PEG, which was later coupled to 3′-azido-3′-deoxythymidine (AZT). In vitro cytotoxicity studies on the A2780/AD cell line that overexpressed folate receptors showed a 20-fold increase in potency compared to free AZT [153]. Xiong et al. synthesized FA-PEG-PLA block copolymers for targeted delivery of PTX on KB cell lines. Cytotoxicity studies indicated that the folic acid conjugated micelle toxicity on the cells continued to increase with the folate content. Cells treated with the free drug displayed lower toxicity than PTX-loaded FA-PEG-PLA micelles [154].

Similarly, Hami et al. synthesized and characterized folate-functionalized PLA- PEG block copolymer micelles and later conjugated DOX to these micelles. Cytotoxicity studies on SKOV3 human ovarian cancer cell lines showed significantly higher toxicity compared to non-targeting micelles [129]. Goren et al., in early research, demonstrated that folic acid conjugated liposomes loaded with DOX showed 10-fold higher toxicity on M109R cells compared to unconjugated liposomes. Inhibitory effects in vivo also indicated a significantly higher effect compared to free DOX [155].

Chandrasekar et al. synthesized folate-dendrimer conjugates to deliver indomethacin to inflammatory regions. The study showed that drug encapsulation increased with an increase in folate content. The drug's half-life increased, and drug exposure to the site was also significantly higher for folic acid-conjugated PAMAM dendrimers compared to the unconjugated formulation [156]. Zhang et al. and Pan et al. conjugated FA to D-α-Tocopherol polyethylene glycol succinate (TPGS). They compared it with other polymer carriers loaded with drugs, such as PTX and DOX. The cytotoxic effects of FA-conjugated carriers on C6 cell lines showed significantly higher inhibition and increased uptake in MCF-7 cells, thereby highlighting the benefits of receptor-mediated targeted drug delivery [157,158].

GCs are steroid hormones released in the body when stimulated by stress. They primarily function as an anti-inflammatory agent and also as an immunosuppressant. They cause severe side effects when deployed in large doses. Some of the most prominent synthetic GCs are DEX, cortisol, and prednisolone. GCs are predominantly used to treat inflammatory conditions, such as asthma and rheumatoid arthritis. They have also been used as a supplementary drug to reduce edema in cancer tumors. Recent studies have indicated the cytotoxic effects on glioma and other cancers [159].

Morita et al. investigated the effect of DEX on C6 cells and found that serum deprivation led to cell death, but DEX's presence further enhanced necrotic death in the glioma cells [160].

Shapiro et al. and Grasso et al. studied the effect of various GCs, such as cortisol and DEX. They concluded that GCs primarily inhibited glioma cell growth [161–163] but sometimes displayed transient inhibition [164]. Gurcay et al. presented the initial studies on prednisolone's effect on primary brain tumor and identified growth inhibitory effects [164].

Kaup et al. studied the inhibitory effects of DEX on different glioma cell lines, such as A172, T98G, and 86HG39, to elucidate the effects of GCs on tumors. The cell lines were subjected to acute (continuous), pre, and combination (pre and acute) treatments. A time-delayed inhibitory effect was noted in all cell lines when subjected to DEX pretreatment. A172 and T98G cell lines displayed significant inhibitory effects when subjected to combination treatment. Acute treatment of DEX had potent inhibitory effects on A172, whereas the effect was negligible on T98G and 86HG39 cell lines [165].

Fan et al. reported a concentration-dependent inhibitory effect of DEX on murine and rodent glioma tumor growth. It was also reported that primary astrocytes (human) and primary neurons (rodent) were not affected by DEX [166].

1.3. Physicomechanical, Thermostability, and Morphological Characteristics of Biodegradable Polymeric Materials

Polyolefins such as polypropylene (PP) and PE are resistant to biodegradation. It has been observed by numerous authors [167–169] that these polymers are not significantly affected by soil burial, whereas Darby [170] and Griffin [171] found that PE is successively degraded in compost. Moreover, they established a relationship between the decrease in the tensile strength and the extent of biodegradability. However, molecular weight was the critical factor in this assessment. Studies with polyethylene samples indicated that bacterial growth decreased with the increase in molecular weight [172]. Potts et al. [173–176] conducted studies on the biodegradation of synthetic polymers, such as polyester, PE, and PS. They found that, among these high molecular weight synthetic polymers, only the aliphatic polyesters and oligomers of PE showed biodegradability, whereas PS did not. Weiland [177] and Khabhaz et al. [178] reported degradation of thermally oxidized PE. It is well known that the low molecular weight polymers and straight chain polymers are more affected by microbial degradation [177,179,180]. Further branching and crystallinity [181,182] have also been observed to reduce the rate of biodegradation.

For biodegradation to take place, some simple organic substances should be added to polymers for rapid decomposition. Photolabile chemical groups (e.g., carbonyl moieties and benzophenone) have been introduced into polymers to accelerate their UV light catalyzed depolymerization in the environment via the free radical process [183,184]. Products containing photosensitizers are affected by light and become biodegradable thereafter. For certain applications and disposal routes, this may be a viable option. However, in many cases the materials will not be exposed to sunlight when discarded or buried, and coatings may obscure the direct exposure to the light that is necessary to initiate the degradation process. Hence the aforesaid approach remains questionable.

In addition to the already-cited polymers, acrylonitrile-butadiene-styrene tercopolymer (ABS), aromatic polyesters (including PET, polyether-urethanes, and most acrylates except poly(alkyl alpha-cyanoacrylate)) are also considered to be resistant to biodegradation.

Although polyethylene (PE) is considered to be resistant to biodegradation, Lee et al. [185] reported biodegradation of PE by phanerochaete and streptomyces species in blends containing 6% starch and oxidants. They reported a decreased molecular weight and changes in mechanical properties due to biodegradation. However, only low molecular weight fragments appeared to be responsible for the observed biodegradation. A degradable PE–starch complex has also been prepared using a transition metal and an antioxidant [186,187].

Blown films containing up to 40% starch have been developed by extrusion and the biodegradability was assessed by measuring the changes in weight loss and chemical composition by FTIR [188–196]. However, Psomiadou et al. [197] reported that more than 30% starch had deleterious implications for the mechanical characteristics of LDPE/starch films used for food packaging [198].

A number of degradable blends of starch and PE have been previously synthesized [199–201], and their environmental weathering [202] and biodegradation has been studied in a compost environment [203], soil [204], and by microbial culture [205,206]. Thermoplastic PE/starch compositions were reported to have the required strength and biodegradability [207,208], and those developed by Chiquet [209] were found to decompose during heating, exposure to UV light or sunlight, and composting.

Chandra and Rustgi [210] investigated the biodegradation of maleated LLDPE and starch blends. They observed that the tensile strength and moduli escalated and percent elongation-at-break reduced as the starch concentration in the mixtures increased. Poly(ethylene-co-acrylic acid) has been frequently used as a compatibilizer in starch composites with LDPE [211,212], HDPE [213], and blown films [214–216] of PE, which have been applied in agricultural mulch and in packaging. Li et al. [217] conducted studies on a starch graft copolymer as a compatibilizer for LLDPE/starch. Many commercial products containing LDPE and starch have appeared in recent years [169,218–220].

A number of reports are available in the literature in which starch modified by various methods was used for the development of biodegradable products. Japanese patents have been filed [221–225] on the production of biodegradable polymeric thin films with superior physicomechanical characteristics utilizing starch derivative compounds. Favis et al. [226] obtained a patent for a composition containing LDPE, alkyl ethers of starch, and a vinyl or acryl polymer as a compatibilizing agent that showed good biodegradability, as exhibited in Figure 6.

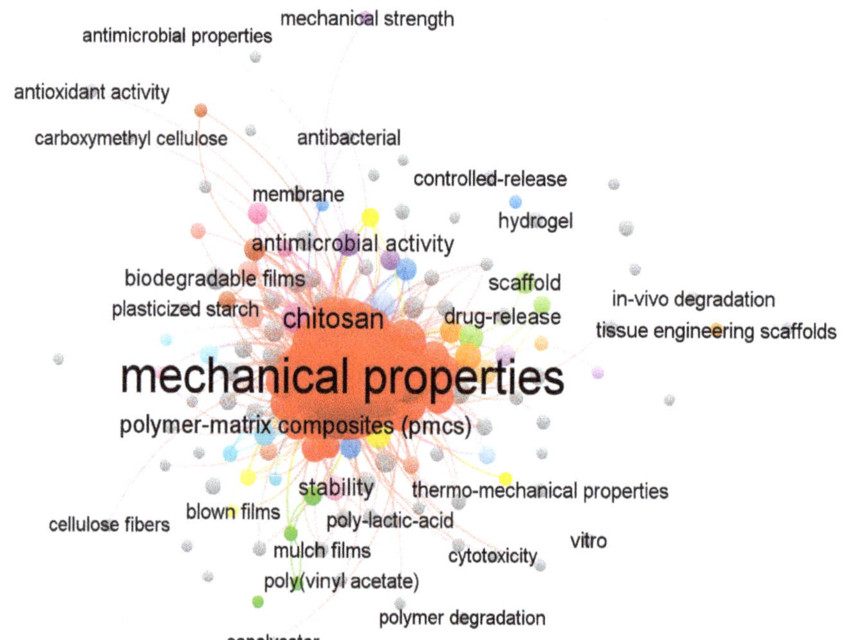

Figure 6. Systematic mapping summary of scientific advances in physicomechanics and thermostability in in vitro drug release studies of biopolymeric materials/biocomposites for biomedical applications.

Shah et al. [227] investigated the initial degrading process of starch-filled LDPE strips compounded with commercially available, well-dried, modified, granular starch (CATO-32). Numerous ecological situations have been reported to have combinatorial synergism impacts on the decomposition rate and processes. Various new technologies have been developed for the manufacture of biodegradable starch PE blends. Muller et al. [228] carried out oil-philicity treatment of starch with a coupling agent, which promoted uniform dispersion of starch in PE and improved the mechanical properties of the blends. Pierre et al. [229] devised a method for combining the blends containing thermoplastic starch (gelatinized starch plasticized with glycerol) in a continuous manner in a co-rotating dual-screw-based extrusion process.

The biodegradation of PE films containing 40% gelatinized corn starch and 15% EAA were investigated in a diverse spectrum of aquatic conditions [230]. The loss-of-starch on degradation was preceded or followed by reduction in the tensile strength, and enabling of the films to disintegrate to test their susceptibility to mechanical strain. By comparison, Shogren et al. [231] reported that films with the same composition exhibited a heterogeneous microstructure with a non-uniform distribution of starch. The compatibility, physicomechanical characteristics, and morphology of LDPE-starch/modified-starch blends used in agricultural mulch and packaging films were studied by Tian et al. [232] for the films. Similar starch-PE agricultural mulch was prepared using a graft copolymer of starch and MMA [233]. The manufactured films showed better biodegradability but a higher production cost than conventional PE [234].

A detailed study on the characteristics of various fatty acid esters of starches and their mixtures with LDPE was carried out by Bikiaris et al. [235] and Thiebaud et al. [236]. The latter observed that the thermal stability and elongation increased, but the tensile strength and water absorption decreased on esterification. Aburto et al. [237] examined the characteristics of octanoated starch and its mixtures with PE. Its blends with LDPE showed better mechanical properties and thermostability, and lower water absorption in comparison with the mixtures of LDPE with plasticized starches. Starch modified by various fatty adds, such as lauric acid, palmetic acid, and stearic acid, has been used to develop bio- and photodegradable PE films [238].

Evangelista et al. [187] investigated the impact of compound blending and starch alteration on the characteristics of starch-filled LDPE. They observed that the cast films of LLDPE comprising starch-octenyl-succinate revealed high tensile strength and percent elongation, but a lower rate of biodegradation than those containing native corn starch. Kshirsagar et al. [239] synthesized starch acetate with varying acetyl content. They investigated the rheology and permeability characteristics of blends of starch and starch acetate with LDPE.

The other synthetic polymers used for blending with starch are polypropylene (PP), polyvinyl alcohol (PVA), polyvinyl acetate (PVAc), some modified polymers, and copolymers [240–244].

The effect of operating factors on physical characteristics of blends of starches with ethylene-vinyl-acetate (EVA) and polyethylene modified with maleic-anhydride (EMA) was observed by Ramkumar et al. [245]. Rheological and morphological analysis of corn-starch and SMA/EPMA mixtures containing 60–70% by weight of the starches was undertaken by Sethamraju et al. [246].

Otey et al. [247] formulated starch-based PE films comprising up to 40% starch using urea and ammonia. Fanta et al. [248] also discussed the effect of urea in the presence of water or aqueous ammonia on the composites. Reis et al. [249] reported the in vitro decomposition of starches/ethylene-vinyl-alcohol polymeric-blends and observed a distinct rheological behavior and mechanical properties. The same authors also carried out a detailed investigation on starch–EM composite films.

Polymer compositions that yield films with good strength [250] and migration-resistant plasticizer [251] are obtained from starch/poly(ethylene-co-vinyl alcohol) [252–254]. Starch-

PVA blends [255–259] are useful as moisture-resistant biodegradable polymers for agricultural mulch, shock absorbing foams, and articles with good dimensional stability.

Starch-based biodegradable packing materials have been developed by extrusion of starch with polyethylene glycol [260] or PVA [261]. Such films have been prepared by mixing starch with poly(vinyl alcohol) [262], vinyl acetate and its copolymers [263–276], and EM. The films also exhibited excellent mechanical strength and flexibility.

Polymer compositions containing starch have also been formulated using bacterially produced poly-3-hydroxybutyrate (PHB) and poly-3-hydroxyvalarate (PHV). Roller and Owen carried out studies on the structural and physicomechanical characteristics of melt-pressed sheets of PHB and PHV filled with various amounts of particulated maize starch granules.

Processing and mechanical properties, and biodegradation, in municipal activated sludge of starch-poly(p-hydroxyl butyrate–covalerate) composites have been extensively studied. The change in composition after biodegradation was quantified by FTIR, and increased starch content was observed to result in more extensive degradation. Lactic acid copolymer-starch compositions useful for films, filaments, and packing materials were found to be biodegradable. It was observed that the film was broken into pieces after two weeks and disappeared after two months.

PP-based biodegradable plastics and blends have been prepared by mixing starch with maleated PP. Some ternary blends of starch and PP with polylactide and ethylene-vinyl acetate copolymer have been reported for disposable diapers, and with PE for the marine environment [276].

Bennett et al. [277] developed a rigid urethane foam formulation containing 10 to 40% starches. These investigations revealed that foams prepared from starch-derived products show better flame resistance but are readily attacked by soil microorganisms.

The incorporation of surface-altered starches to plastics has been shown to be a commercially feasible means of producing conventional biodegradable polymeric materials [278]. Starch in various forms, such as starch xanthate, gelatinized starch, and dried starch, has been incorporated in large amounts as a filler in disposable PVC-based plastics [279,280].

Several modified forms of starch, such as silyl isocyanate modified starch dialdehyde [281], acid hydrolyzed starch [282], and starch derivatives [283,284], have been used to develop thermoplastic mixtures by blending with other plastics. Jeremic et al. [285] developed blends of thermoplastic starch and thermoplastic polymers, such as EM and EVAc, in a twin-screw extruder [286].

Biodegradable plastics with high tensile strength have been obtained from PE, PP, PS, and PVC with 6 to 50% of octenyl succinate starch metal ion complexes [287]. Similarly, biodegradable starches containing polymer compositions with good mechanical properties were developed by Osada et al. [288].

Huang et al. [289] carried out an in-depth study of the development of the technology and product application of the biodegradable plastics based on graft copolymers of starch. Imam et al. [290] studied the morphological and thermo-behavior of poly3-hydroxybutyrate hydroxy-valerate/starch valerate mixtures, and reported that no phase separation was observed.

Several patents based on the use of modified starch for blending with synthetic polymers are available. Biodegradable foams useful as packing material were patented by Jaffs [291]. Bastioli et al. [292–294] developed a composition for biodegradable plastic moldings based on destructured-starch and blends of starch with plastics for strong films, sheets, and fibers. The preparation and use of a biodegradable starch derivative and polymer mixture are discussed in a European patent [295]. Berruezo [296] developed photo/biodegradable high impact polystyrene (HIPS) sheets by mixing modified starch, HIPS, and photo-degrading agents.

1.4. Comparative Analysis of Physicomechanical, Thermostability, Rheological, and Morphological Characteristics of Biopolymeric Materials with Other Materials

In another study, Mamidi et al. [297] presented a new class of PCL/f-CNOs composite. Newly developed composites were characterized using SEM and FTIR techniques. The DOX releasing ability was analyzed under various pH conditions. The authors reported that force spinning provides regular and bead-free nanofibers in the range of 210–596 nm. Results indicated a good control over the DOX release in forcefully spun PCL/f-CNOs fibers, in addition to better tensile strength (3.16 MPa). These results reflect the stability, viability, and cell adhesion characteristics of the reinforcement in the matrix material, as depicted in Figure 7. The newly fabricated composite is used in numerous applications in the biomedical field.

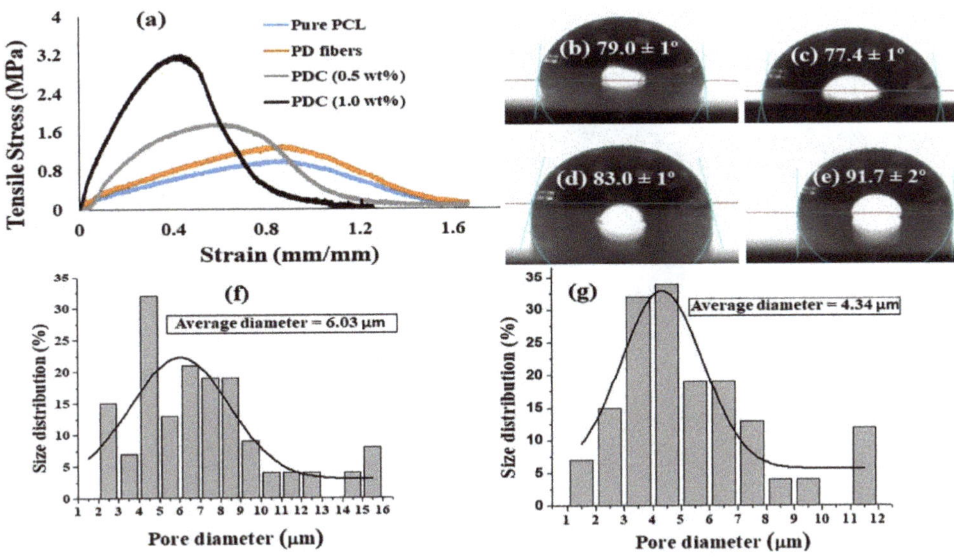

Figure 7. (a) Tensile graphs; (b–e) contact angles of neat PCL, PD, PDC (0.5 wt.%), and PDC (1.0 wt.%); and (f,g) porosity/permeability of PDC (0.5 wt.%) and PDC (1.0 wt.%) composite filaments. Reproduced with permission from [297].

Mamidi et al. [298] prepared PHPMA-CNOs = f-CNOs reinforced BSA nanocomposite fibers using the force-spinning technique with DOX as a drug. The DOX-releasing ability was measured in terms of the concentration of DOX, the incubation temperature, and the pH value. The authors observed 72–95% drug release at the temperature limits of 37–43 °C in fifteen days of study. The mechanical strength of the BSA increased to 18.2 MPa with the addition of f-CNOs, in addition to the water absorption angle and thermal characteristics revealed in Figure 8.

Mamidi et al. [299] prepared PHPMA-based composites that were reinforced with SWCNT. The composites were primed using force spinning followed by the thermal pressure approach. The newly developed composites possessed a tensile strength of 13.7 GPa, an elastic modulus of 243.3 GPa, and a toughness of 1421 J/g. The authors reported that the newly developed composites were the strongest and stiffest composites among all of the nanocomposites available in the literature.

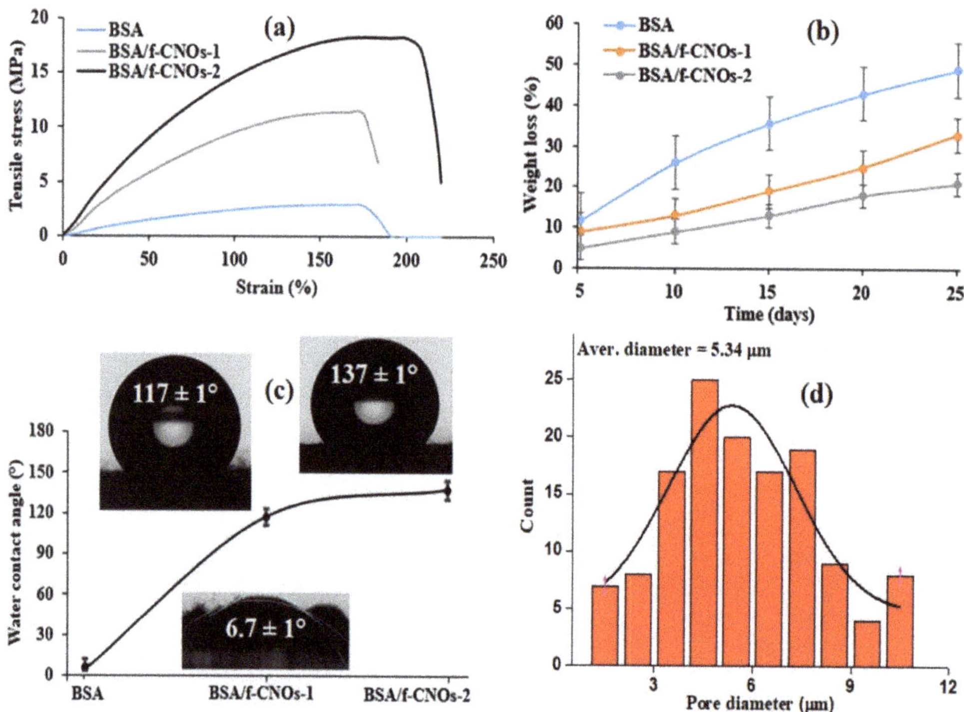

Figure 8. (**a**) Tensile strength; (**b**) degradation; (**c**) water contact angle for virgin BSA, BSA/f-CNOs-1, and BSA/f-CNOs-2 filaments; (**d**) porosity/permeability of biomimetic or biocompatible BSA/f-CNOs-2 nanocomposite fibrils. Reproduced with permission from [298].

Mamidi et al. [300] adopted the force-spinning approach to develop 3D scaffolds of gelatin and zein protein. The authors reported that one unit of gelatin and four units of zein provided better tensile strength and hydrophobic behavior at a water angle of 115°.

Mamidi et al. [301] developed PAPMA-CNOs and (GelMA)/f-CNOs/CD supramolecular hydrogel interfaces using the photo cross-linking approach. The morphology and properties were evaluated. The authors observed that the maximum drug release occurred under acidic conditions over 18 days. In addition, (GelMA)/f-CNOs/CD composites exhibited better properties than the alternative composite. The physical and chemical behaviors, microstructure, biodegradation, and swelling characteristics of hydrogels have been studied. Over an 18 day period, the composite hydrogels exhibited improved controlled release of the drug under acidic environments (pH 4.5 = 99% and pH 6 = 82%). GelMA/f-CNOs/CD supramolecular hydrogels possessed significantly enhanced tensile strength (ultimate strength = 356.1 ± 3.4 MPa), impact strength (K = 51.5 ± 0.24 Jg^{-1}), and elastic moduli (E = 41.8 ± 1.4 GPa) with the addition of f-CNOs. The framework of GelMA/f-CNOs/CD hydrogel augmentation reveals an excellent distribution and extent of polymeric envelopment of f-CNOs throughout GelMA matrices, as displayed in Figure 9. Additionally, the produced hydrogels were found to have significant cell viability when evaluated against human fibroblast cell cultures. Nonetheless, the primed supramolecular hydrogels will form the basis of future targeted drug carrier systems utilizing regulated delivery methods.

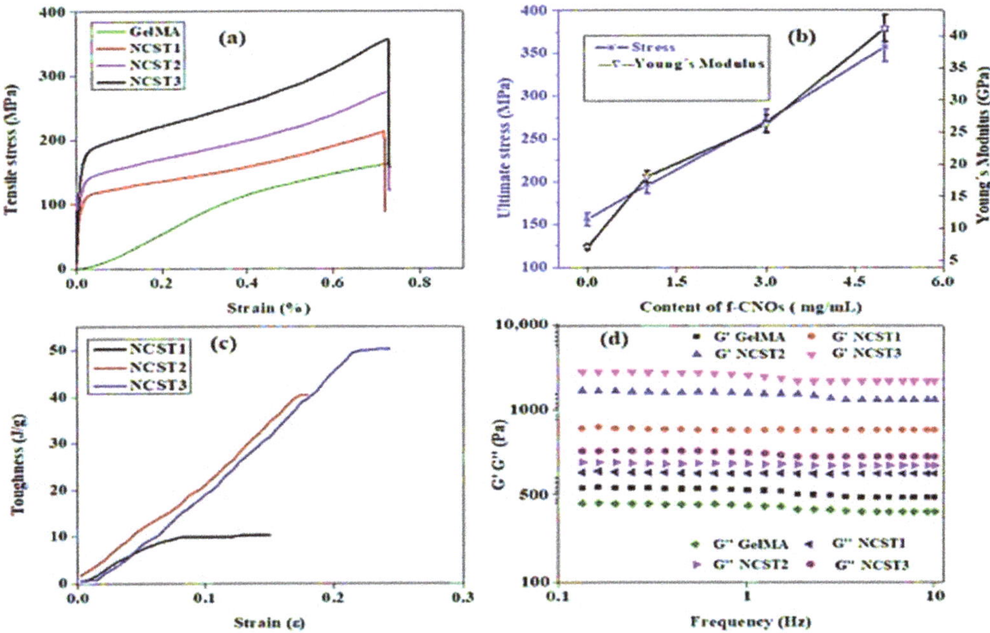

Figure 9. Graphs demonstrating hydrogel physicomechanical characteristics: (**a**) tensile stress; (**b**) tensile strain versus elastic moduli; (**c**) impact strength; and (**d**) storage and loss modulus (G′ G″). Reproduced with permission from [301].

Cole et al. [302] evaluated the mechanical properties of the exopolysaccharide biopolymers by developing methods to obtain biopolymer bonding among the grains of natural occurring materials. The authors examined two treatments in their experiments. The first treatment refers to one week of incubation at 28 °C in yeast extract mannitol media. The purpose of this treatment was to determine cohesive and adhesive properties of the polymer. The second treatment related to the precipitation and resolubilization of the Rhizobium EPS from the media supernatant. The authors observed an increased bond stiffness with an increase in curing time. The modulus was varied within the limits of 0.2 to 3.2 MPa, and cohesive strength was varied from 16 to 62 MPa. In addition, the cohesive strength of the precipitated exopolysaccharide biopolymer was more than that of natural exopolysaccharide.

Nam et al. [303] developed PBS biodegradable composites that were embedded with coir fibers. The influence of alkali treatment on the properties and morphology was analyzed. The highest shear strength was observed in a sample that was soaked in 5% NAOH solution for 72 h at room temperature. In addition, the mechanical properties were improved significantly with the alkali treatment. The composite containing 25% coir fibers resulted in increases of around 55, 142, 46, and 97% in tensile strength, tensile modulus, flexural strength, and flexure modulus, respectively, as shown in Figure 10. The morphology results indicated better interfacial bonding in alkali-treated composites than untreated composites.

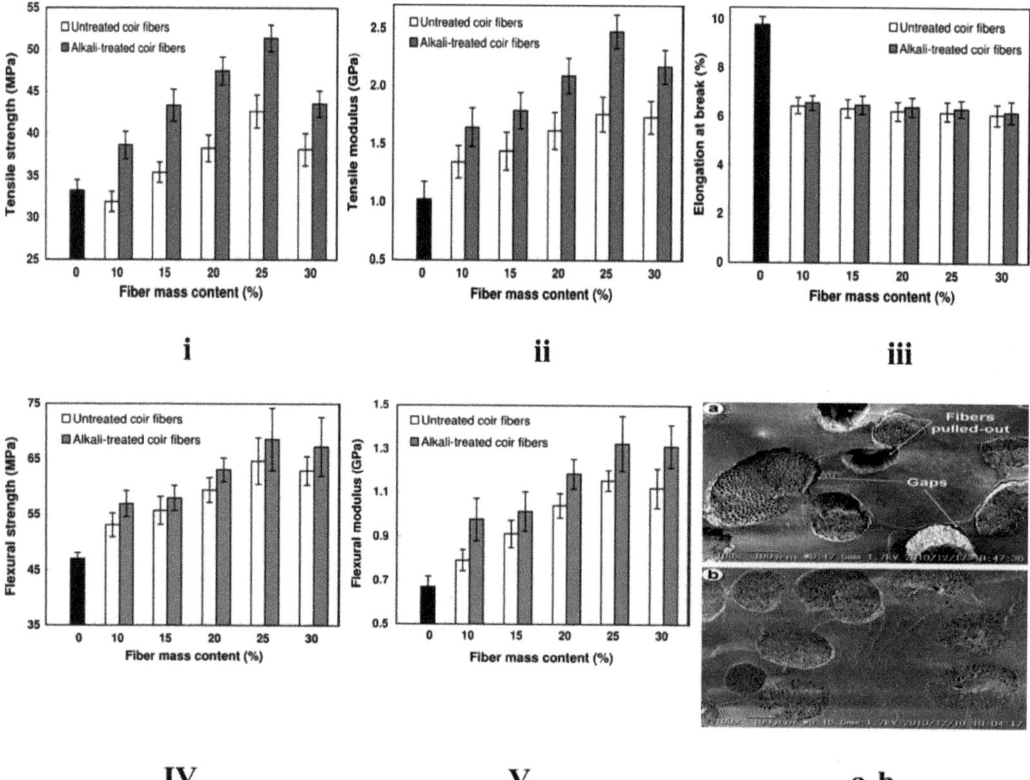

Figure 10. (i) Tensile strength; (ii) tensile moduli; (iii) fracture strain; (IV) moduli of rupture/bending strength; (V) bending modulus of unprocessed and 5N-72-treated coir-fiber/PBS biodegradable composites; and tensile fracturing surfaces of PBS biodegradable composite strengthened with 20% weight concentration of: (a) unprocessed coir fibers, (b) 5N-72 processed coir-fibers. Reproduced with permission from [303].

Treated natural fiber materials have the capability to replace synthetic fiber materials because they are more easily available, less expensive, ecofriendly, renewable, and light weight. Among the available natural fibers, kenaf possesses excellent properties, and jute fiber exhibits high strength and compatibility with biopolymers [303].

Gallo et al. [304] fabricated kenaf fiber-reinforced biopolymer composites using compression molding. The influence of core thickness of the fibers on the properties of polyhydroxyalkanoates was evaluated. Kenaf-reinforced material prevents the combustion of the material and kenaf performs as a carbonizing compound, which provides an insulating layer on the material's surface. Results revealed that monofiber-reinforced composites possess better mechanical properties than double-layered composites. The regular dispersion of the kenaf material in the matrix ensures improved properties of the composite, as shown in Figure 11. The dense microstructure obtained from fire residues indicates that the reinforcing material acts as a carbonizing agent.

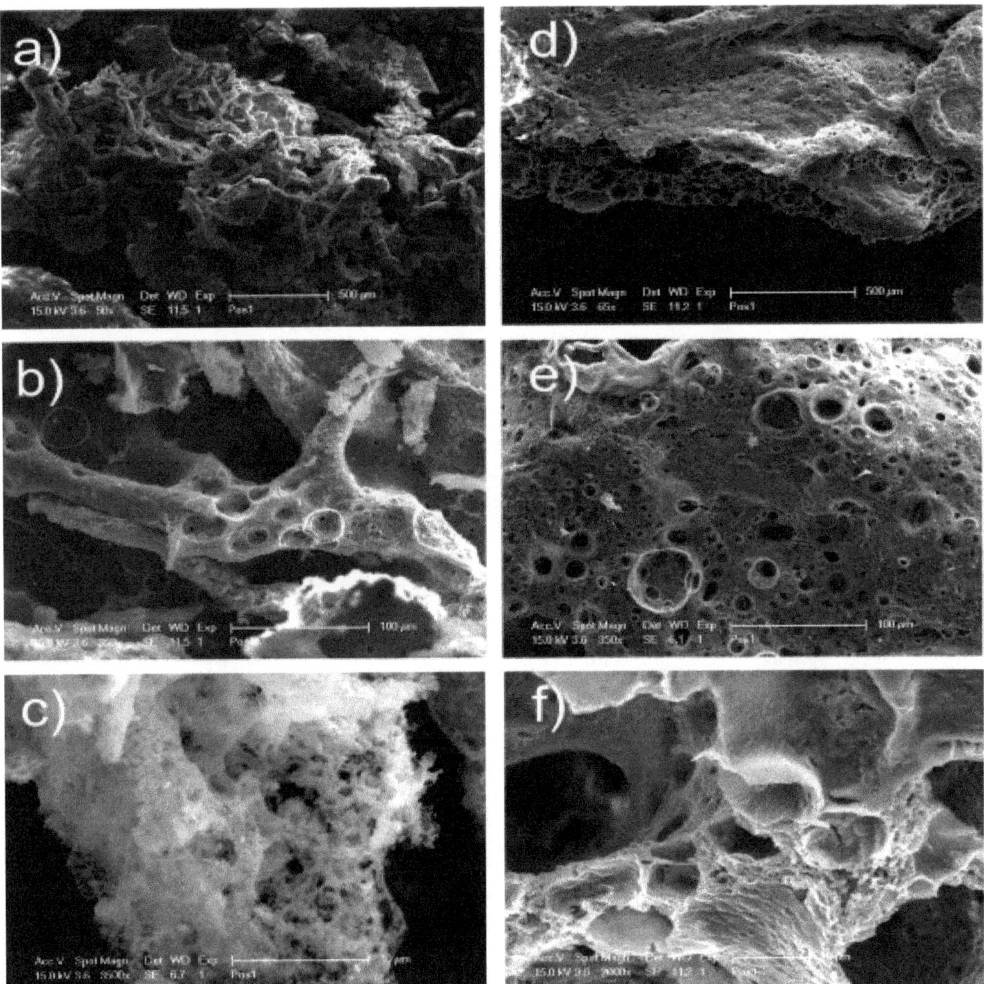

Figure 11. Microstructure of layered 2 (**a–c**) and Lam 1:6 (**d–f**) under numerous varied magnification scales. Reproduced with permission from [304].

Motru et al. [305] developed flax-reinforced PLA polymer matrix composites. During fabrication, flax fibers were varied by weight, in the values of 7.9, 13.6, and 17.6%. The authors observed that mechanical properties were increased with an increase in the fiber content. In contrast, for treated PMCs, the opposite results were observed. The flexural strength remained the same for all of the composites, whereas the maximum UTS was observed in a composite containing 13.6% flax fibers. The impact energy of the newly developed composites was varied within the limits of 25–30 joules, as exhibited in Figure 12. The authors also observed a regular dispersion of the flax fibers within the PLA matrix in all composites.

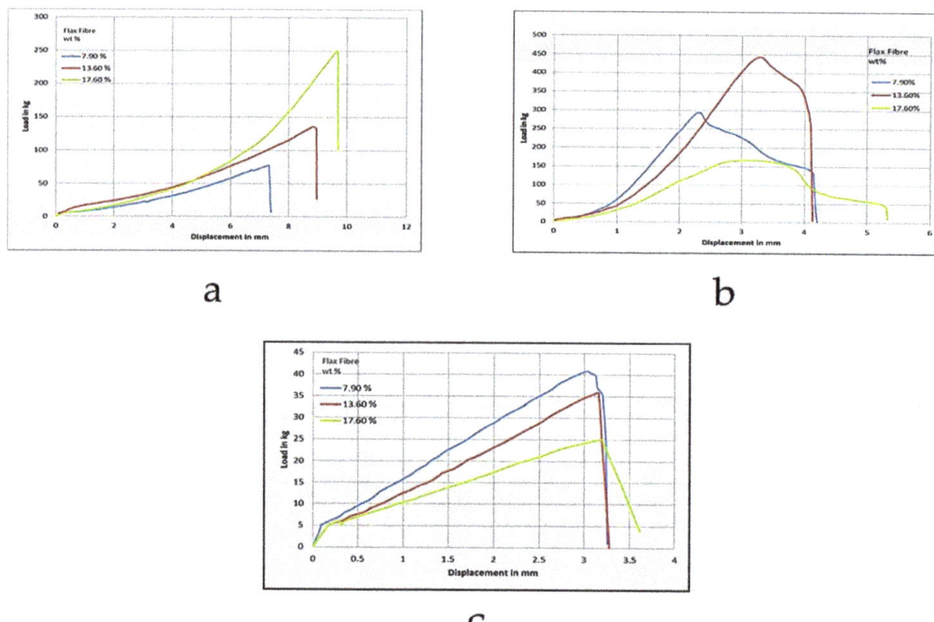

Figure 12. Loading versus displacement graphs from: (**a**) tensile test; (**b**) compression test; and (**c**) bending test. Reproduced with permission from [305].

Russo et al. [306] fabricated kenaf-reinforced polymer matrix composites with PHBV, LDPE, and PBAT as matrices using melt compounding. Kenaf fibers were treated chemically before being added to the matrix to enhance interfacial bonding among the constituents. Cast composites were alkalized and silanized to improve their performance. Thermal properties of the PHBV were not altered due to the addition of kenaf; however, a negative effect was observed in the case of PBAT. In addition, PHBV-based composites possessed better flexural, modulus, and impact strength, irrespective of the type of treatment performed. However, the existence of stiff cellulose fibers reduced these properties in PBAT- and LDPE-based composites.

Alvarez et al. [307] prepared sisal fiber-reinforced polymer matrix composites. The composition of the reinforced fibers was varied from 5 to 15% by weight. The water sorption ability of the sisal, starch, and their composites was evaluated by varying the reinforcement compositions. The authors reported that sisal fiber absorbed less moisture than the base matrix; however, the diffusion coefficient value was approximately same for all fibers. In addition, the fibers possessed a higher diffusion coefficient than the base matrix. The diffusion coefficient was marginally increased with increasing fiber content, and the moisture content had a harmful effect on the mechanical behavior of the composites. The flexural modulus was reduced with an increase in fiber content.

Liang et al. [308] developed kenaf-reinforced PBS polymer matrix composites using a mixing process. The mechanical behavior, morphology, and crystallization behavior were evaluated. The composition of kenaf was varied from zero to 30% by weight in steps of 10%. The authors reported that the moduli and crystallization rate were increased with the addition of kenaf material to the base matrix. The authors also observed that the tensile strength and storage modulus of the newly developed composite were increased by 53 and 154%, respectively, with the addition of 30% kenaf to the base matrix and the increase in the crystallization temperature from 76.3 to 87.7 °C. SEM result illustrates that further improvement at the PBS and kenaf interface is required to increase the interfacial bonding.

Zhu et al. [309] blended sisal fiber-reinforced PLA composites. Initially, hybrid sisal fibers were prepared by mixing treated and untreated sisal fibers and then introduced into the PLA matrix. The authors reported an enhancement in the crystallinity and the impact strength of the base matrix with the addition of sisal fibers due to β crystals, as observed in XRD patterns. It was also reported that properties of the hybrid sisal reinforced composites were significantly improved, with the treated and untreated sisal-reinforced composite and PLA/HSFs composites possessing 47.1% and 30.8% higher strength and crystallinity, respectively, than the other samples.

Jung et al. [310] investigated the effect of chemical bonding on the mechanical behavior of biopolymer composites. The behavior of these bonds under mechanical deformation was also evaluated. The authors observed that hydrogen bonds enabled high deformation and changed the structure under low loading conditions. In addition, mutation caused a change in the stable structure of the biopolymers.

Bahrami et al. [311] reviewed the mechanical behavior of hybrid biocomposites. The properties reviewed by the authors included the strength, water absorption, and flammability of the biocomposites. It was found that the use of hybrid fibers and the treatment of the fibers, among other factors, have a significant effect on the mechanical behavior of the biocomposites. The authors reported that biocomposites will be superior to numerous alternative engineering materials in the future.

Kremensas et al. [312] developed wood fiber-reinforced biopolymer matrix composites boards. The authors used corn starch and hemp shiv treatment during the production of biocomposites. The results revealed that compressive, tensile, and bending stresses of 3, 0.45, and 6.3 MPa, respectively, were obtained at 10% hemp shiv by weight. In addition, the composite containing 10% corn starch by weight exhibited better a contact zone and increased the product strength.

Aslam Khan et al. [313] presented a review article on the recent progress of biocomposites for tissue engineering and regenerative medicines. Biopolymers are used in numerous applications in wound healing and other medical areas, as reported in Figure 13. Biopolymers are the best alternate among petroleum-based synthetic polymers because they are ecofriendly, environmentally sustainable, and readily available. Biopolymers lack strength and stability, which can be overcome with the use of ceramic-reinforced biopolymer composites. Further study on these materials is required to increase their industrial applications.

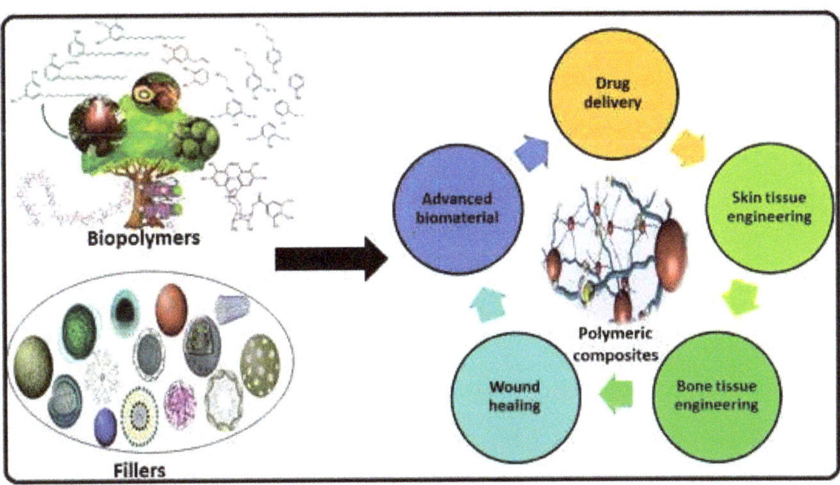

Figure 13. A graphical description highlighting the numerous promising functionalities of biodegradable polymerics with numerous filler particles. Reproduced with permission from [313].

Baldino et al. [314] proposed a new process that improves ESPR atomization due to a mixture of SC-CO$_2$ in which polymers can dissolve. This occurs due to the reduction of surface tension and viscosity, which enables the production of micro- and nanoparticles of controlled dimensions.

The above summary provides insights into synthesized biodegradable and bioactive polymeric composites with biological functionality and remarkable compatibility for biomedical application domains, as shown in Figure 14.

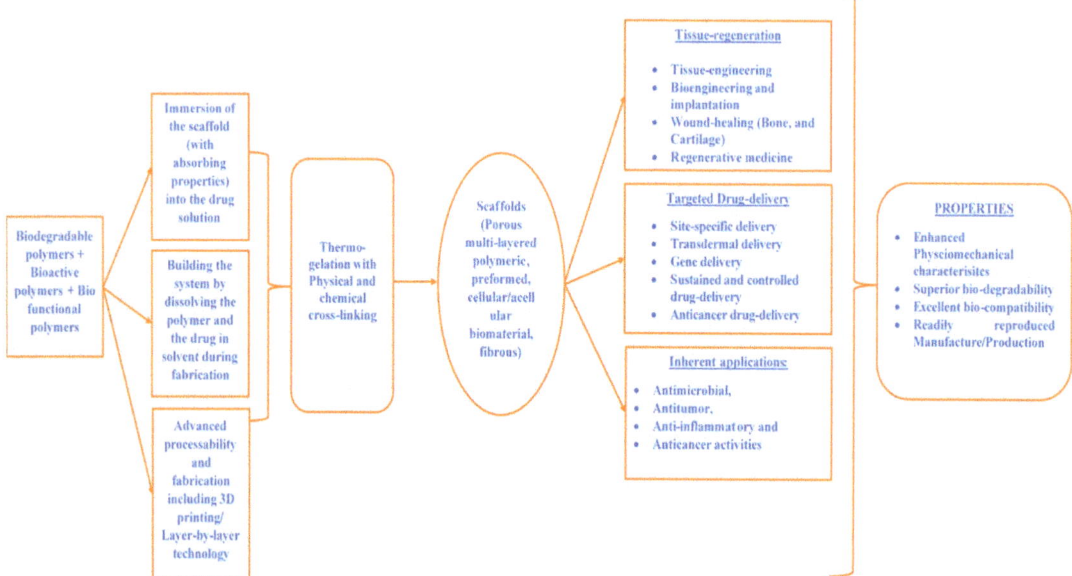

Figure 14. Visualization summarizing the numerous potentially attractive characteristics of biodegradable/bioactive polymerics for biomedical applications.

Considerable work has been carried out on polymeric-blend materials and copolymers to control crucial aspects of biocompatible polymers, such as degradation rates and physicomechanical characteristics. Moreover, cellular films, implantable biomaterials, three-dimension-printed scaffolds, and nano-structured biomaterials have been produced that acquire the benefits of the biopolymeric stimuli-responsive properties of certain polymerics to improve modulation of biologically active molecule delivery, tissue regeneration, regenerative medicine, and wound healing. Consequently, the current review underlines the critical factors that must be considered in biocompatible hydrogels throughout tissue regeneration and drug carrier delivery; discusses the numerous frequently researched natural and synthetic biomimetic polymeric materials; and provides insight into the developed production and manufacturing methods that employ biodegradable polymeric materials for biomedical applications.

Therefore, this summary examines the contemporary status of biodegradable naturally and synthetically derived polymers for biomedical fields, such as tissue engineering, regenerative medicine, bioengineering, targeted drug delivery, implantation, and wound repair and healing. Furthermore, this review presents insights into a small number of the commonly used tissue engineering applications, including drug delivery carrier systems, demonstrated in the recent findings. Due to the inherent remarkable properties of the biodegradable and bioactive polymers, such as their antimicrobial, antitumor, anti-inflammatory, and anticancer activities, certain materials have gained significant interest in recent years. These systems are being actively researched to improve therapeutic activity and to mitigate adverse consequences. This article also presents the main drug delivery systems reported in the literature and the primary methods for impregnating polymeric scaffolds with drugs, their properties, and the respective benefits for tissue engineering.

2. Methodology

The systematic literature review is the most well-known form of literature review and presents a clear picture to the researchers in a more transparent manner. The phases of a systematic review are the identification of articles; screening of the articles according to the established criteria; assessing eligibility according to the content of the articles; and inclusion of the final articles for the analysis.

2.1. Identification

The identification of articles is undertaken in such a manner that it can be reproduced if the search database is given to other researchers. It also allows the review study to be transparent. In this review, we examined the articles related to biodegradable and bioactive polymers, for antimicrobial, antitumor, anti-inflammatory, and anticancer purposes, published during the past 21 years i.e., from 2000 to 2021, using the advanced search option of the Web of Science (WoS) and SCOPUS databases. The searches used for identifying the relevant articles were: "Biodegradable Polymer Nanocomposites for Biomedical applications", "Natural polymeric biomaterials for tissue-engineering and drug-carrier", and "Bioactive polymers for biomedical applications". These specific document searches identified 485 papers that were sent to the screening process.

2.2. Screening and Eligibility According to the Relevancy of the Articles

The identification process was followed by the screening process to select the relevant articles according to the theme.

During the process of screening, duplicate articles were removed. Furthermore, some records were also excluded based on article titles that are not relevant to the theme. Finally, the remaining 88 full-text journal papers were further assessed for eligibility based on the content according to the criteria. All of the articles were thoroughly read to determine if they focused on any one of the biomedical applications of the biodegradable polymers or other natural polymeric biomaterials.

2.3. Inclusions

The research articles selected for the review were focused on the performance characteristics of biodegradable and natural polymeric materials for biomedical applications. All of the articles focusing on the performance of biodegradable and bioactive polymers, for antimicrobial, antitumor, anti-inflammatory, and anticancer purposes, were included. Thus, 88 journal papers were finally included for analysis.

2.4. Analysis of the Articles

The data of the 88 journal articles selected for the analysis were tabulated using Bibliometric Scientific mapping analysis and Microsoft Excel. The collected data included the author's affiliation, year of publication, journal name, and publisher's name. Analysis was then undertaken of the biodegradable polymer nanocomposites and other natural polymeric biomaterials for tissue engineering and drug carrier applications.

3. Results and Discussions

3.1. Publication Trends

In this section, we discuss the publication of articles based on the articles' year of publication, journal, publisher, geographic location of the conducted study, and author affiliation.

3.1.1. Year of Publication

The articles related to biodegradable and natural polymers for biomedical applications published between 2000 and 2021 were taken into consideration. Here, we consider only the publication trend of the relevant publications that were finally included according to the inclusion and exclusion criteria. The annual publication of the articles is shown in Figure 15.

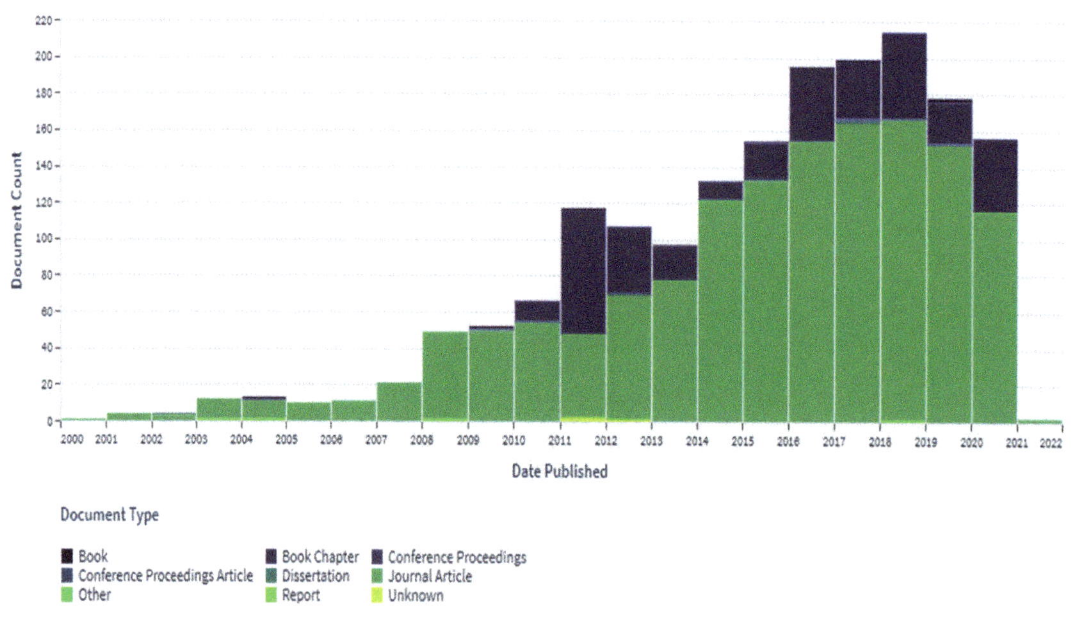

Figure 15. Number of articles published per year.

It can be noted that researchers increased the study of performance characteristics of polymer carriers for drug delivery in tissue engineering, antimicrobial, antitumor, anti-inflammatory, and anticancer applications in 2014. The lowest number of relevant publications was found in the initial years, i.e., between 2000 and 2013. However, from 2014, a noticeable increase in the number of articles can be seen, with articles numbering 214, 199, 195, and 178 in 2018, 2017, 2016, and 2019 respectively. In 2020, 156 relevant articles were published.

3.1.2. Journal and Publisher

This section shows the classification of relevant published journal articles ($n = 88$) according to the journal and publisher. There are wide varieties of journals and publishers with which researchers were associated for the publication of their studies. A total of 25 journals and 41 publishers were reported during the data processing of the 88 identified articles, as shown in Figure 16a,b. From Figure 16a, it can be seen that *Acta Biomaterialia* was the most popular journals for publishing articles related to the performance characteristics of natural polymeric and biodegradable polymers for tissue engineering and drug delivery applications, i.e., 56 of the total number of articles, followed by *Materials* (Basel, Switzerland) with 35 articles, and the *International Journal of Molecular Sciences*, with 33 articles. Figure 16b shows that Elsevier was the most frequent publisher, contributing to 21.1167%, of the articles followed by Intech (13.86%) and Wiley (7.25%).

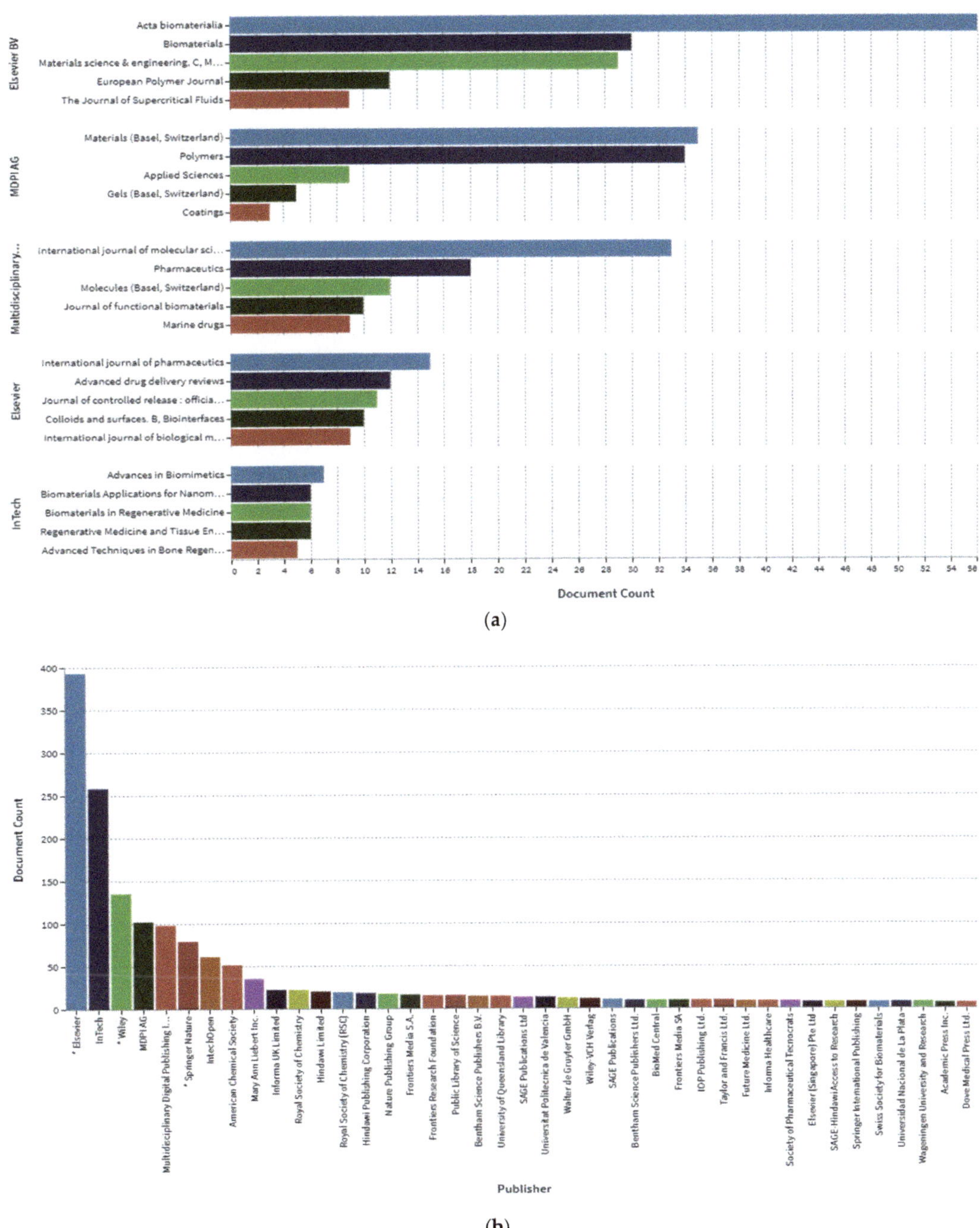

(a)

(b)

Figure 16. (**a**) Number of relevant articles (n = 88) by journal. (**b**) Number of relevant journal articles (n = 88) by publisher.

3.1.3. Author's Affiliation

It was observed during data processing that authors from various countries were interested in the field of biomimetic polymers for biomedical applications. The database used for the present study only gathered the information relating to articles published in the English language. This indicates that the majority of the analyzed articles originated in countries where English language is primarily used for research and technical reports. Thus, the majority of articles was found to come from the UK, followed by Portugal. The distribution of authors, participating institutes, and countries may vary if articles of other languages are considered. The most active institutions in the fields of study related to the identified articles were the University of Minh, followed by Massachusetts Institute of Technology and Polytechnic University of Turin, as illustrated in Figure 17a,b.

The data were tabulated to classify the articles according to the author's affiliation, as shown in Figure 18. It was found that authors from 50 different institutes or universities, and a variety of countries, undertook the studies reported in these 88 articles. The maximum numbers of authors related to the published articles were affiliated with UK (17%), followed by Portugal (11%) and USA (9.7%).

Following the categorization of the articles according to the author's affiliation, we report the distribution of articles according to the country in which the experiments were carried out, as shown in Figure 18. In some cases, it was found that studies involved as many as six or seven authors in a single research paper, and affiliations with two or three countries. Figure 18 shows the distribution of the articles according to the country in which the experiments were conducted. Experiments were also carried out in China, Spain, Australia, Malaysia, USA, Mexico, Oman, China, Philippines, Colombia, Portugal, France, and Indonesia.

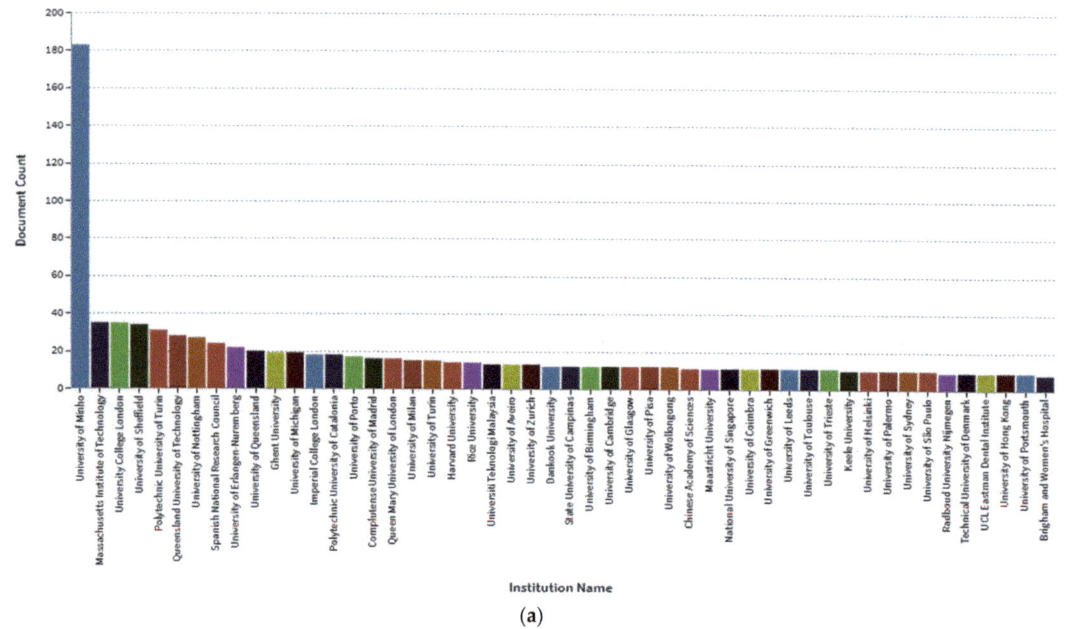

(a)

Figure 17. *Cont.*

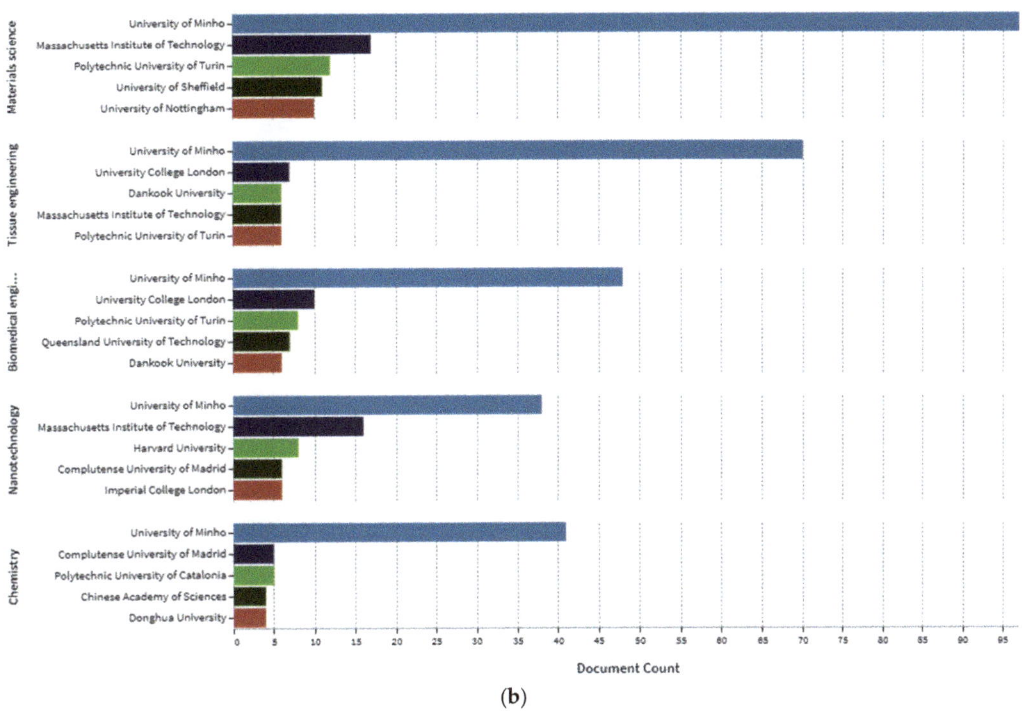

(b)

Figure 17. (**a**) Number of papers published (*n* = 88) according to the author's institute. (**b**) Number of papers published (*n* = 88) according to the author's institute and field of study.

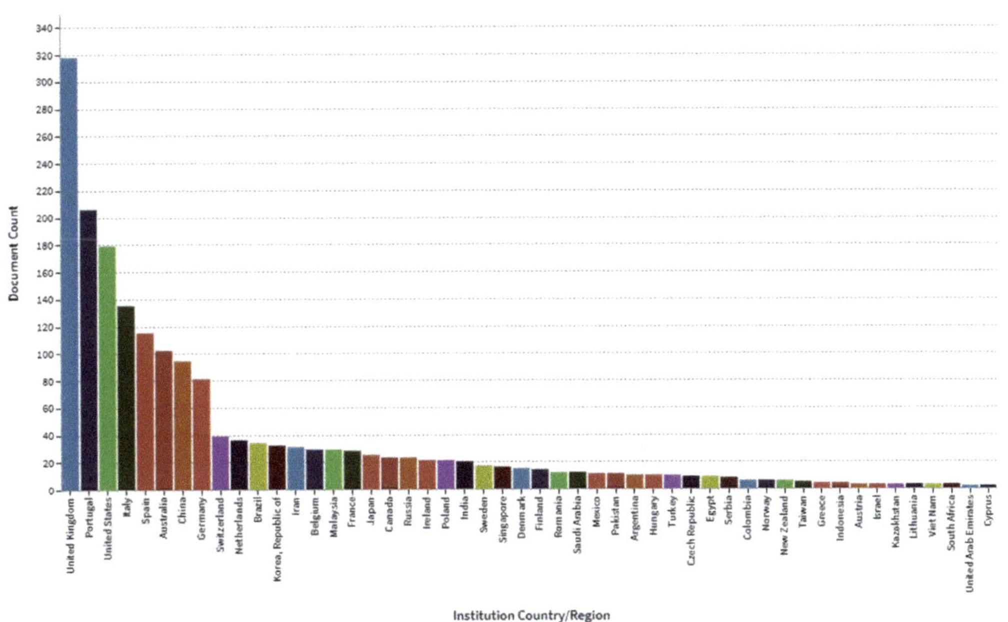

Figure 18. Number of papers published (*n* = 88) according to the geographical location and country affiliation.

The most prolific and dominant author, Rui L. Reis, has the maximum number of articles published, contributing to 30% of the articles, followed by the João F. Mano (13%) and Aldo R. Boccaccini (4.8%), as exhibited in Figure 19.

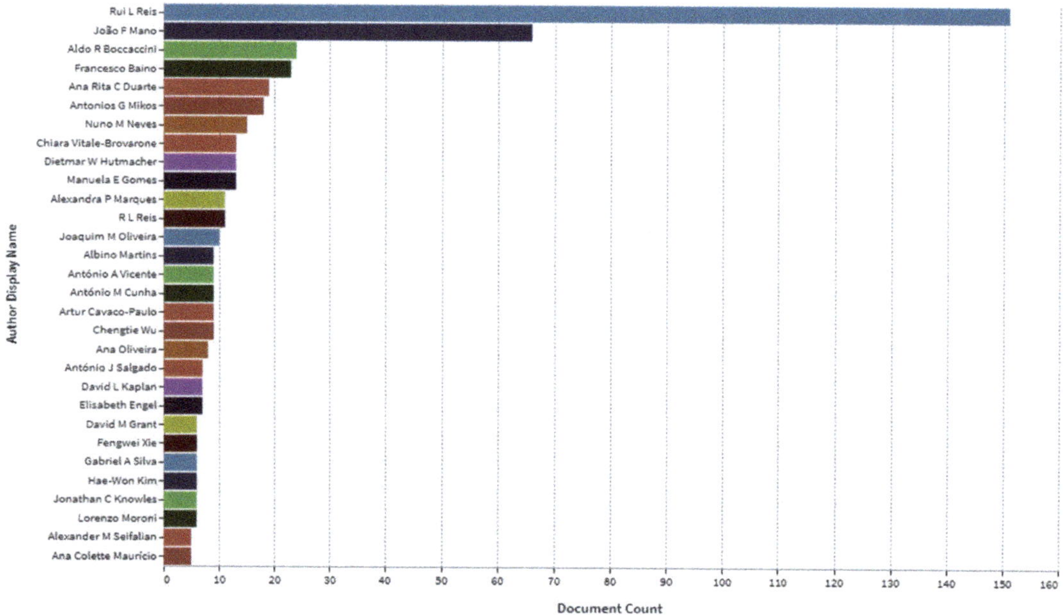

Figure 19. Number of papers published (*n* = 88) according to the most active authors.

3.2. Drug Delivery Systems of Biodegradable and Other Natural Polymeric Biomaterials in Hard Tissue Engineering

In recent years, a new generation of system has appeared as an alternative solution to many cases of trauma or diseases of bone tissue, in which a wide range of synthetic bone substitutes and biomaterials are used as scaffolds (such as chitosan, alginate, collagen, and hydroxyapatite) [315,316]. The following characteristics are required for the successful implementation of a bone scaffold: It must be able to be sterilized;

a. It must provide mechanical support;
b. It must deliver bioactive molecules;
c. It must not cause inflammatory reactions;
d. It must have interconnected pores to facilitate the growth of a new bone;
e. It must to promote the osteogenic differentiation;
f. It must degrade as the new bone forms;
g. It must not create non-toxic degradation products;
h. It must sustain the bone cell migration [5].

Several methods have been reported for impregnating the scaffolds with drugs [1]. The first method is the simplest, and entails the immersion of the scaffold (with absorbing properties) into the drug solution, as shown in Figure 20.

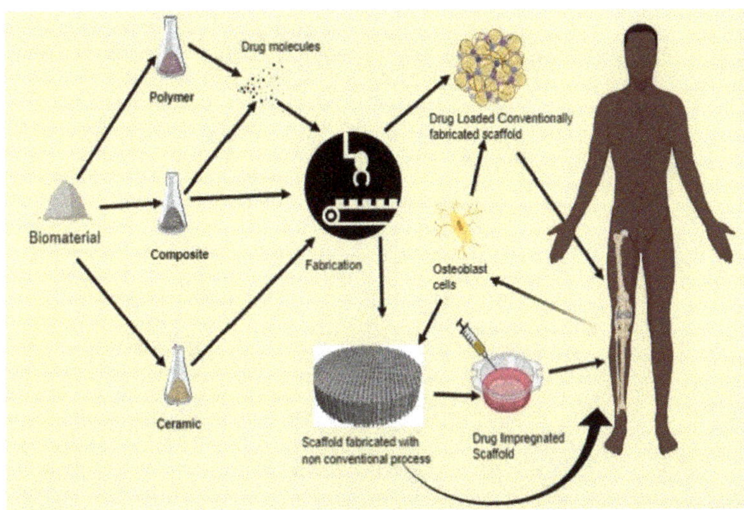

Figure 20. The method in which the scaffold is immersed in the drug solution. Reproduced with permission from [1].

The second method refers to building the system by dissolving the polymer and the drug in solvent during fabrication, as show in Figure 21. It was reported in the literature that, in the case of these two methods, the drug release profiles depend on the scaffold features, such as degradation rate and porosity [1].

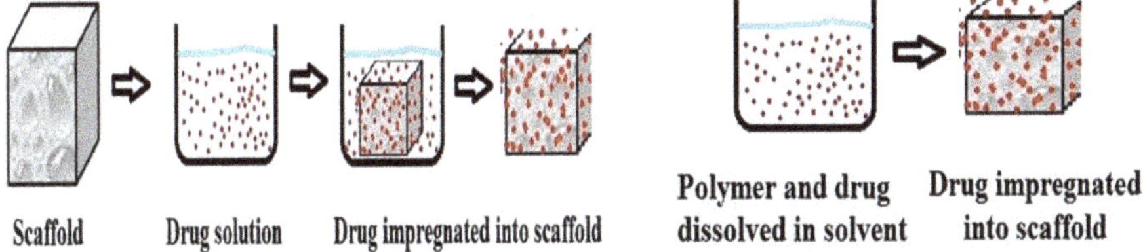

Figure 21. The formation of the system during fabrication. Reproduced with permission from [1].

One new method that has revolutionized medicine is 3D printing. The biggest advantage of this method is the accurate control of the architecture, shape, size, location, and dosage of drugs, as shown in Figure 22 [1].

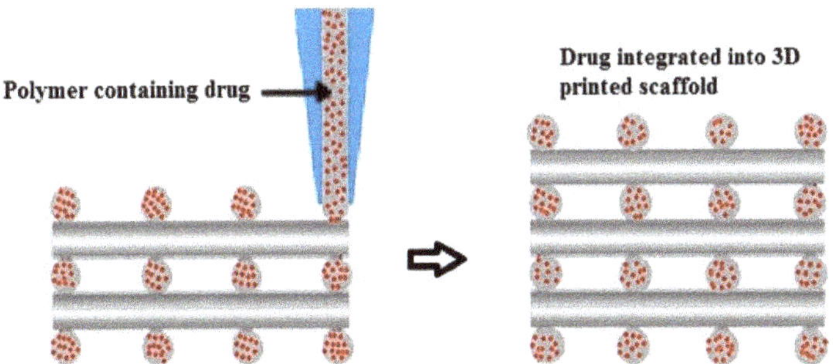

Figure 22. The 3D printing method. Reproduced with permission from [1].

Another widely used method to address the side effects of orthopedic implants is layer-by-layer technology, as exhibited in Figure 23. This method involves covering certain surfaces with different layers, between which the drugs (drugs A and B) are caught. Subsequently, capsules can be formed by decomposition [1].

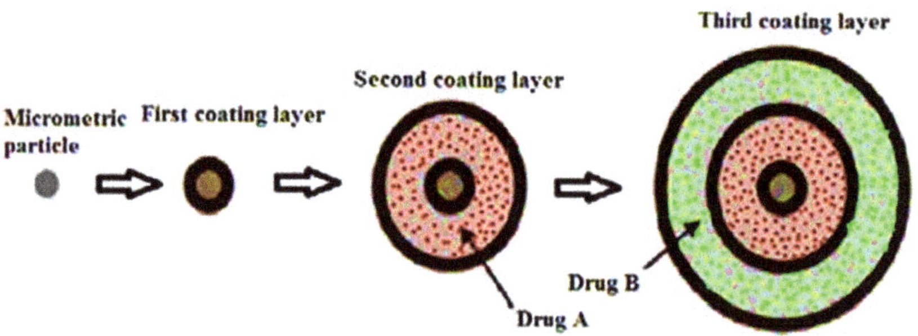

Figure 23. Layer-by-layer method. Reproduced with permission from [1].

One of the most common global diseases at present is osteoporosis. Most drugs used to treat this disease have been shown to be ineffective due to their side effects. Thus, to prevent these side effects, drug delivery technologies have been used with the effects of enhancing the release profile, reducing toxicity, and improving the therapeutic effectiveness of the drugs [317]. A number of drug delivery systems used in hard tissue engineering are presented in Table 3. The main antibiotics used in hard tissue applications are gentamicin, ampicillin, penicillin, oxacillin, kanamycin, and methicillin from bone cement. Currently, for this application, the researchers are investigating a new type of scaffold that contains antibiotic-loaded nanoparticles [318].

The bibliometric mapping analysis shows that the use of biodegradable polymers in biomedical applications has undergone significant advances during the past 80 years, as illustrated in Figure 24. Biodegradable materials have emerged for the development of therapy devices, such as provisional implantable devices and 3D scaffolds for tissue engineering. Additional progress has been made in the use of biodegradable polymers for pharmaceutical applications, including drug delivery carriers with sustained delivery. To ensure efficient treatment, these developments require that the materials have desirable mechanical, biochemical, and decomposition characteristics. As a consequence, a wide spectrum of polymeric materials that enable hydrolytic and enzymatic deterioration are now being explored for biological applications, including tissue regeneration, regenerative medicine, prostheses, temporary implants, tissue repair and regeneration, wound healing, and drug carriers.

3.3. Polymer Nanocomposites and Natural Polymeric Carriers in Drug Delivery and Biomedical Engineering

Jordi Puiggali et al. (2019) highlighted nanocomposites comprising novel materials. Although nanocomposites have traditionally been developed from natural materials, in this article, the author specifically describes novel materials in the formulation of nanocomposites. Nanocomposites composed of hydroxyapatite, bioactive glass, chitosan, collagen, fibrin, gelatin, and silk were thoroughly investigated. The basic compositions of each of the basic materials, in addition to their synthesis techniques and applications in different fields were reported. Most of the polymers were found to be biodegradable and biocompatible, and to have unique advantages in biomedical and pharmaceutical fields. Important areas within the biomedical field include tissue regeneration and bone implants, whereas key pharmaceutical applications include drug delivery and solubility enhancement, as indicated in Figure 25. The mechanical properties of the above-mentioned polymers were also found to be improved [319].

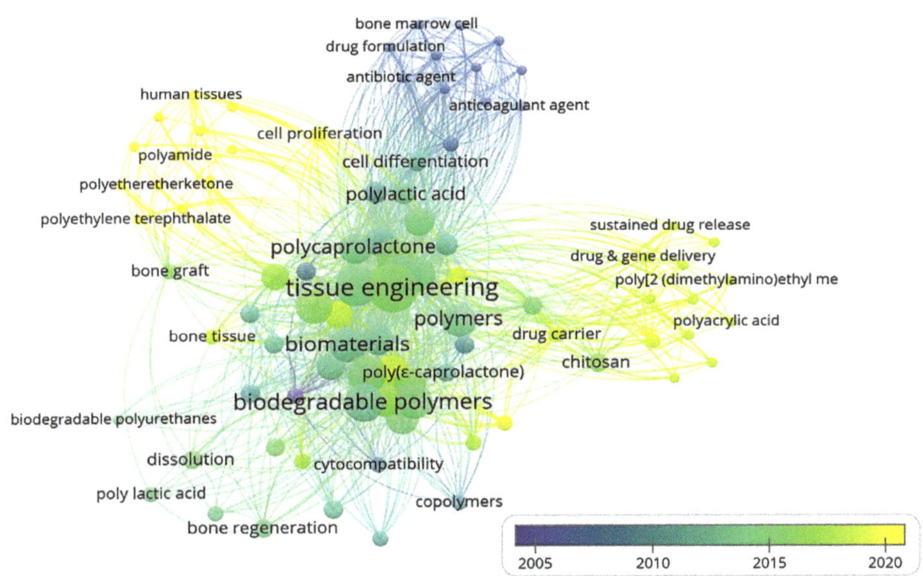

Figure 24. Bibliometric analysis of the utilization of biodegradable and other natural polymeric biomaterials in hard tissue engineering applications.

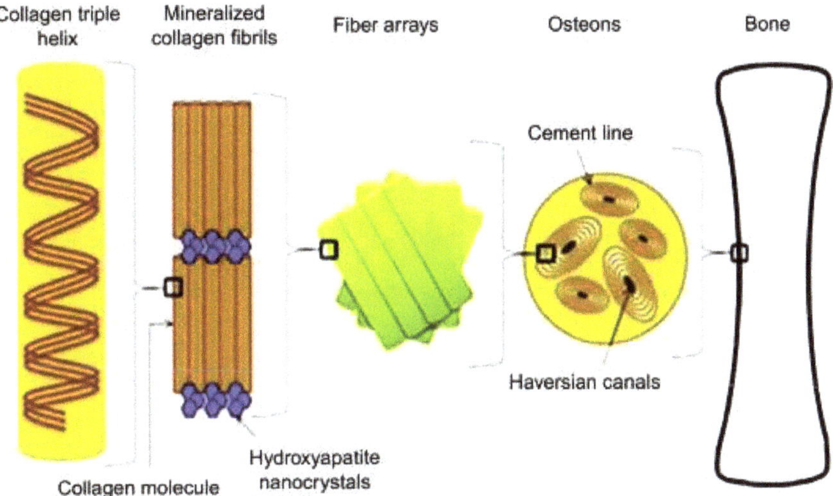

Figure 25. Schematic flow diagram indicating that hierarchical bone arrangement relies on the self-assembly of triple helices of collagen and the accumulation on the surface by HAp-precipitated crystals. The formation of structured and layered fibrous assemblies and osteons (i.e., concentric strands) corresponds to later measures. Reproduced with permission from [319].

Alvarez G.S. et al. (2017) discussed the formulation of biomaterial nanocomposites. Biomaterials are agents that are placed in contact with living tissue to improve or replace the unique functioning of the tissue or a particular organ. Tissue engineering has been found to be an important technology in the formulation of functional substitutes to help regenerate and repair damaged tissues or organs. The study also described hydrogels, which are polymers that have an open core structure and retain a large amount of water in their structure. Hydrogels were found to consist of two types, namely, from natural and synthetic polymers. Due to their distinctive properties, such as biocompatibility, natural polymers have wide applications in the biomedical field, and include polymers such as chitosan, collagen, and gelatin. Synthetic nanocomposites composed of polylactic acid, polyglycolic acid, polymethyl methacrylate, and polyvinyl alcohol were also discussed in details [320].

Priscila Anadao et al. (2012) highlighted the concept and applications of polymer-clay nanocomposites. In 1949, Bower conducted an experiment for DNA absorption using montmorillonite clay. In 1963, Greenland highlighted the use of polyvinyl alcohol montmorillonite nanocomposites. Subsequently, a large amount of research has been carried out on polymer-clay nanocomposites. These are composites that have a polymer matrix, whose dispersed phase is formed by silicate particles that have dimensions in the nanometer range. Based on the interphase forces in clay and polymer, the authors highlighted different thermodynamically accepted morphologies. Intercalated, exfoliated, and flocculated morphologies are described for nanocomposites. In addition, methods of preparation of polymer nanoclay nanocomposites were briefly discussed. Important methods include in situ polymerizations, solution dispersions, and fusion intercalations. In in situ polymerization, the clay particles are dispersed in monomer medium and, using suitable conditions, the polymerization process is carried out. The use of a compatibilizing agent is also highlighted in the literature. The name indicates that the agent is compatible with clay and the polymer. An example is maleinanhydride, which is used as a compatibilizer for polypropylene and polyethylene polymer. The widely used polymers for the preparation of nanocomposites are polyethylene, polypropylene, polyvinyl chloride, polyamide, and polyethylene oxide. The use of biodegradable polymer from natural and synthetic sources was also mentioned. These sources include polyhydroxy butarate, chitosan, polycarbolactone, and polylactic glycolic acid. The application of polymer nanoclay nanocomposites was also reported in literature. These nanocomposites can be used in different areas, such as the biomedical applications of artificial tissues, dental and bone surgery, and medicine and drug delivery. Other applications include automatic energy, packaging, and construction [321].

S. Latha et al. (2013) formulated hydrogels of Captopril nanocomposites. Due to their better diffusion, biocompatibility, and excellent water sorption, hydrogels have wide applicability in the biomedical field. The authors developed hydrogels of nanocomposites using a novel free radical reaction technique of polymerization based on nanoclays of montmorillonite. The formulated hydrogels were evaluated in terms of their swelling behavior, SEM, DSC, TGA, FTIR, drug loading, and in vitro drug release. It was observed that the pH of the external media and the clay addition order had a profound effect on the swelling characteristics of nanocomposite hydrogels, as shown in Figure 26. The targeted release of Captopril was observed in the intestine [322].

Karak et al. (2019), highlighted the fundamental concept of nanocomposites of polymers and nanomaterials. Nanomaterials are defined as materials whose dimensions are less than 100 nm. They also highlighted how shape, surface structure, and size influence the properties of nanocomposites. The developed nanocomposites have unique thermal, biodegradability, and mechanical properties; in addition, their flame retardant ability was found to be enhanced. The historical perspective, classifications, and raw materials required for the formulation of nanocomposites were discussed in the study. In addition, the methods used to prepare polymer nanocomposites, including physical and chemical approaches, were discussed. The study also highlighted the significant properties of nanocomposites, such as mechanical strength, optical activity, toughness, catalytic activity, thermal stability, biological activity, and barrier properties. The improvement in the properties of nanocomposites with the use of a core polymer was also discussed in detail. The various applications of nanocomposites were outlined in the study; these applications include use in the biomedical, pharmaceutical, industrial, agricultural, sports, and electronics fields [323].

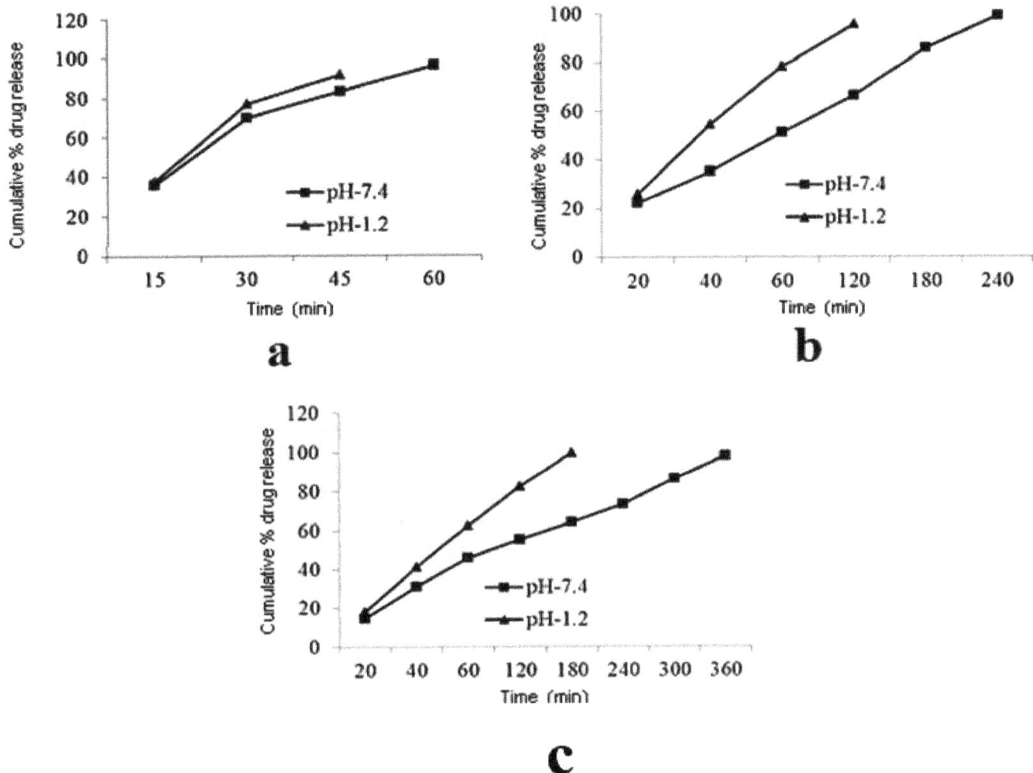

Figure 26. Influence of concentration on the cumulative percentage of drug delivery at (**a**) 0.05%, (**b**) 0.1%, and (**c**) 0.15% of capoten-loaded nanocomposite hydrogels. Reproduced with permission from [322].

Bhat. M. et al. (2015) reported that the Biopharmaceutical Classification System (BCS) Class II drugs have poor aqueous solubility. This affects the drug's release, which is significantly influenced by the aqueous surrounding in the gastro-intestinal (GI) tract, which particularly affects the drug's bioavailability. Bio-nanocomposites are a hybrid form of biopolymer in which two or more components are fused together. These have numerous applications in different dosage forms. The review highlighted the simple and convenient method of preparation of bio-nanocomposites using the microwave irradiation method by means of carriers of natural origin, such as acacia, ghatti gum, cassia, and gelatin, to enhance the solubility of BCS Class II drugs and improve the rate of dissolution for these drug entities, thereby affecting their bioavailability. The reported fusion method (MW) offers numerous advantages, such as its simplicity, time saving, and cost effectiveness. The MW technique is a recent and advanced technology in materials processing and chemical manufacturing, and presents promising advantages compared to the traditional thermal treatments. The mixture was heated to form a molten mass which was cooled and solidified. The final product was crushed and sieved. The developed nanocomposites can be characterized by different analytical techniques, such as SEM, FTIR, DSC, and XRD. At present, materials processed via the MW approach are extensively used in pharmaceutics that are developed specifically to enhance the speed and efficiency of the extraction process of polar solvents [324].

Paul D.R. et al. (2008) synthesized a polymer matrix nanocomposite and suggested it may be used in targeted drug delivery applications. The addition of nanoparticles to the drugs provided significant advantages, and resulted in a slower and steadier drug delivery process that enabled more control of the release, enhanced the mechanical integrity of the hydrogel-based nanocomposites, and reduced swelling. In the literature, iron oxide nanoparticles have been examined for a wide range of applications, such as immunoassays, cellular therapies, magnetic resonance imaging contrast enhancement, and drug delivery. These experimentations usually deploy magnetic dispersal in a polymeric microsphere or microcapsule, involving biodegradable and/or natural polymers of poly(L-lysine) microspheres containing magnetic nanoparticles. These were manufactured through coacervation and were characterized for further potential use, such as in targeted drug delivery systems. The Fe and Co nanoparticles encapsulated in polydimethylsiloxane were tested for the treatment of retinal detachment disorder [325].

Kushare S. et al. (2013) synthesized bio-nanocomposites using microwaves to improve the dissolution and solubility of the poorly water-soluble drug, Glipizide. However, when the material was characterized, it was confirmed that there no interactions occurred between the polymers and the drug. The researchers then concluded that Glipizide was converted into nanocrystals in the composites, to which the improvement in solubility was solely attributed. The use of microwave irradiation generated by a microwave oven caused the breakage of the internal structure of the drug particles, resulting in the formation of nanoparticles, and ultimately leading to an enhancement in solubility. The in vitro and in vivo evaluations conducted for the optimized formulation reaffirmed the use of BNCs to improve the dissolution and solubility using natural carriers. In stability studies conducted by the researchers, the BNC-containing formulations were found to be stable. The microwave irradiation method is a novel method for the improvement in drug solubility [326].

Mukhija Umesh et al. (2012) emphasized the challenge of the poor aqueous solubility of drugs in their formulation. They also stressed that the drug must be eventuality converted into a bio-nanocomposite using natural carriers. The attempts to enhance the solubility rate of Meloxicam were reported because this drug has poor water solubility, and its solubility–dissolution was improved using different solid dispersion techniques based on Poloxamer 188. The solid dispersions of Meloxicam for in vitro dissolution were prepared by two methods, namely, the microwave-assisted method and the hot melt method. From the results obtained from the formulation, it was found that the model with the best fit for the mechanism of dissolution was the Higuchi matrix release, which had the highest release. When compared with the melting method, the microwave-assisted method proved to be highly compatible for producing better solubility in the preparation of a solid dispersion, as exhibited in Figure 27. Microwave irradiation is an effective and efficient instrument in the generation of molecular dispersion [327].

D. Bikiaris et al. (2008) synthesized a solid dispersion of Tibolone from polyethylene glycol to increase the dissolution rate of drugs with poor water solubility using microwave irradiation. The researchers found that the time required for dissolution of Tibolone in PEG melting under microwave irradiation was significantly less than that of the traditional melting method using heat application (>15 min). Thus, they determined that microwave-induced synthesis of solid drug dispersions is a simple and efficient process compared to traditional methods of preparing solid dispersions, such as melt mixing or solvent evaporation through heat application. The application of microwaves at the instance at which Tibolone solid dispersions were synthesized in PEG resulted in the formation of various-sized drug crystals in the dispersion, compared to the solid dispersions manufactured using traditional methods. Furthermore, the researchers also noted and compared the results obtained in terms of various drug properties. It was mentioned that the drug's dissolution rate appeared to be higher when the dispersions were formulated using microwave irradiation [328].

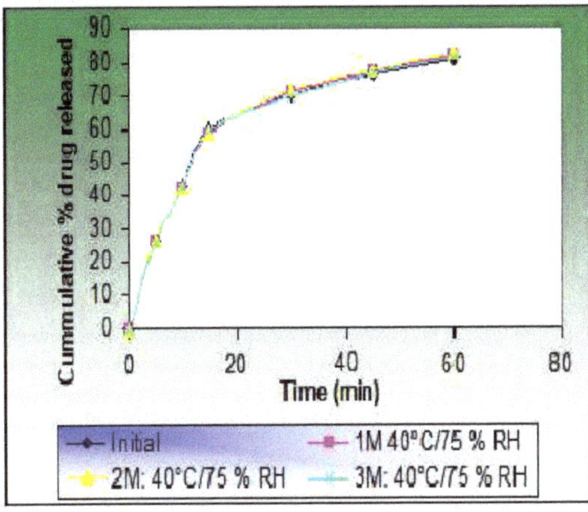
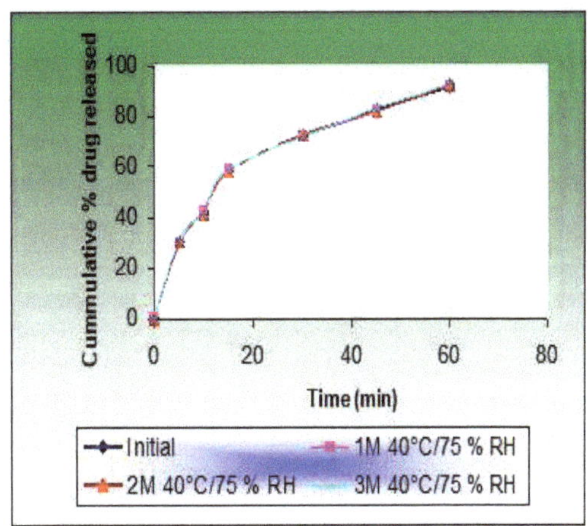

a **b**

Figure 27. Comparison of in vitro dissolution characteristics of Meloxicam/Poloxamer solid dispersions fabricated using (**a**) the melting process and (**b**) the microwave method. Reproduced with permission from [327].

M. Moneghini et al. (2008) synthesized solid dispersions of ibuprofen using microwaves. For this purpose, the active system was prepared with PVP/VA/60/40, whereas HP-Cyclodextrin was designated as the carrier. This significantly increased the dissolution profile of ibuprofen, which is usually considered to be a poorly soluble drug. Furthermore, the researchers analyzed the physical characteristics of the microwave-activated system. They found that the drug was completely amorphized and no polymorphic forms were found. This study deployed the microwave technique to prepare SD. The study results indicated that this is a viable technique and a suitable alternative in the preparation of solvent-free binary systems. In contrast to the conventional heating process, in which heating is confined to the surface, heating produced by this technique is uniform across the material. Unwanted side reactions are also mitigated (reaction quenching). In this study, the active system was prepared with PVP/VA 60/40, whereas HP-Cyclodextrin was designated as the carrier [329].

P. Bergese et al. (2003) synthesized nanocomposites using microwaves to increase the solubility of drugs. Microwave irradiation is a widely used technique that offers numerous advantages compared to traditional thermal processing. Ibuprofen, nimesulide, and nifedipine were the model drugs utilized from BCS Class II, whereas polyvinyl pyrrolidone and βCyclodextrin were the polymers used. When characterized, the materials confirmed that there was no interaction between the polymer and the drug. The study concluded that microwave processing has a positive and significant impact on the drug, i.e., transformation from a microcrystal to a (matrix embedded) molecular cluster, as shown in Figure 28 [330]. Such (thermally activated endothermic) transition, as confirmed by DSC & Thermo-gravimetric analyses (Figure 3), seems to be a solid/solid shift transformation in which the hydrates β-CD ends up losing the water of crystallisation (de-hydration).

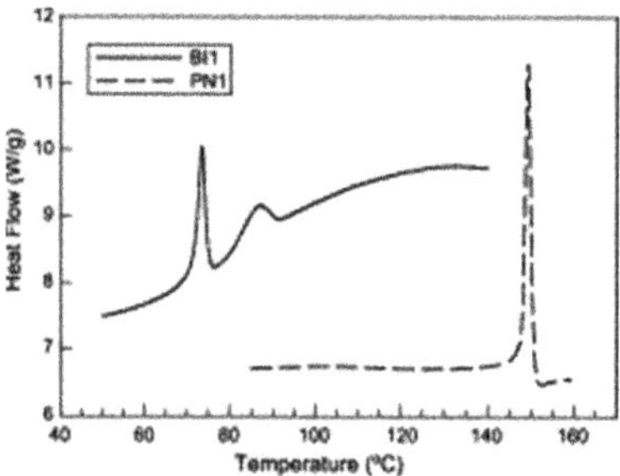

Figure 28. DSC thermogram. Solid line: Composite BI1; the 75 °C peak conforms to the melting point of the microcrystalline ibuprofen stage, whereas the β-CD dehydration is correlated with the highest point at 85 °C. Broken/dotted line: Composite PN1; the maximum point is correlated with the melting of the microcrystalline nimesulide stage. Reproduced with permission from [330].

D. Maurya et al. (2010) synthesized poorly water-soluble atorvastatin calcium using microwaves, which induced its solubility. Their solid dispersions of atorvastatin and PEG 6000, generated using the microwave-induced fusion method, notably enhanced the rate of dissolution ($p < 0.05$). The researchers found that the reasons for the improved dissolution rate were the solubilizing effect of PEG 6000, increased wetting, alteration of the drug surface properties, and the molecular dispersion of the drug in solid dispersions. When the material was characterized, the researchers confirmed that there was no interaction between the polymers and the drug. They also concluded that the results of solid dispersions prepared by the MIND method under in vivo conditions showed increased solubility [331].

Yuen M. et al. (2017) synthesized an advanced microcapsule that could be further applied via bandages or socks to release antifungal drugs to treat fungal skin diseases in a controller manner under pressure, as shown in Figure 29. Chitosan/miconazole nitrate and chitosan/clotrimazole microcapsules were the two kinds of microcapsules prepared for the study. The mean particle size was 2.6 μm for the chitosan/miconazole nitrate microcapsules and 4.1 μm for the chitosan/clotrimazole microcapsules. High-performance liquid chromatography (HPLC) was used to determine the drug loading and encapsulation efficiency. The above-prepared microcapsules, loaded with the drug, can be directly applied to socks or bandages that release antifungal drugs in a controlled manner under pressure. Patients who suffer from tinea pedis or other fungal infections can administer this medical treatment by wrapping a bandage around the body area or putting on socks [332].

Salomy M. et al. (2015) created a topical gel for application to the skin or specific mucosal surfaces. The aim of the gel was to perform local actions or transdermal penetration of the medicament, or for its emollient or protective actions. The study attempted topical delivery of the drugs directly upon the hydrogel matrix, so that the drugs were effectively delivered at the required site and, simultaneously, avoided first pass metabolism, enhanced local action in pain management, and treated skin diseases. Hydrophilic polymers, such as guar gum and Carbopol 940, of varying concentrations, were used to develop a topical hydrogel formulation of the drugs, as exhibited in Figure 30a–e [333].

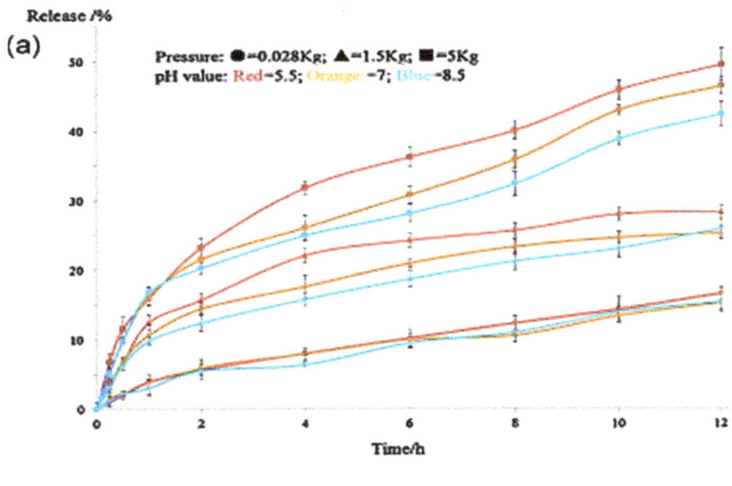

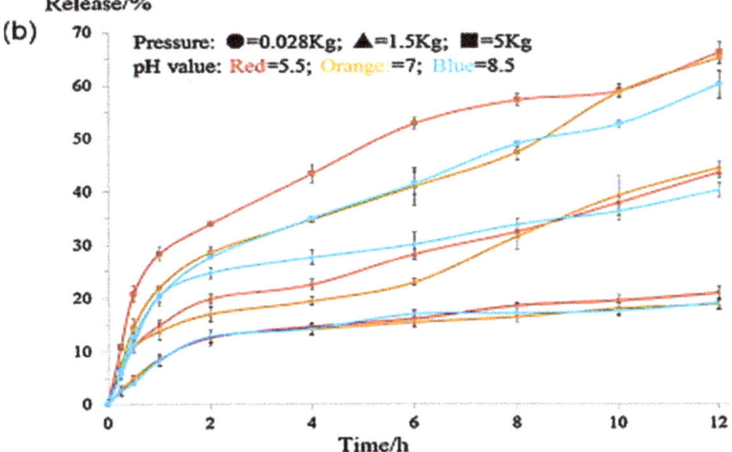

Figure 29. Quantity of (**a**) chitosan/miconazole nitrate and (**b**) chitosan/clotrimazole microspheres released in vitro from the drug at diverse pressures at various concentrations. Reproduced with permission from [332].

Akhilesh K. Gaharwar et al. (2014) specifically examined the updated information on nanocomposite hydrogels, with a particular focus on biomedical and pharmaceutical applications. These are hybrid hydrogels having a hydrated network of polymers that are cross-linked with each other or nanostructures. The researchers highlighted the advances in the field of nanocomposite hydrogels in terms of their physical properties and applications. Two-phase and multi-phase systems were also discussed. Due to their porous and hydrated molecular structure, the nanocomposite hydrogels usually stimulate the native tissue microenvironment. The factors and challenges associated with the fabrication and design of nanocomposite hydrogels were also discussed. This study provided a novel approach to reinforcing polymeric hydrogels, including different functionalities that were focused on the implementation of nanoparticles within the hydrogel network [334].

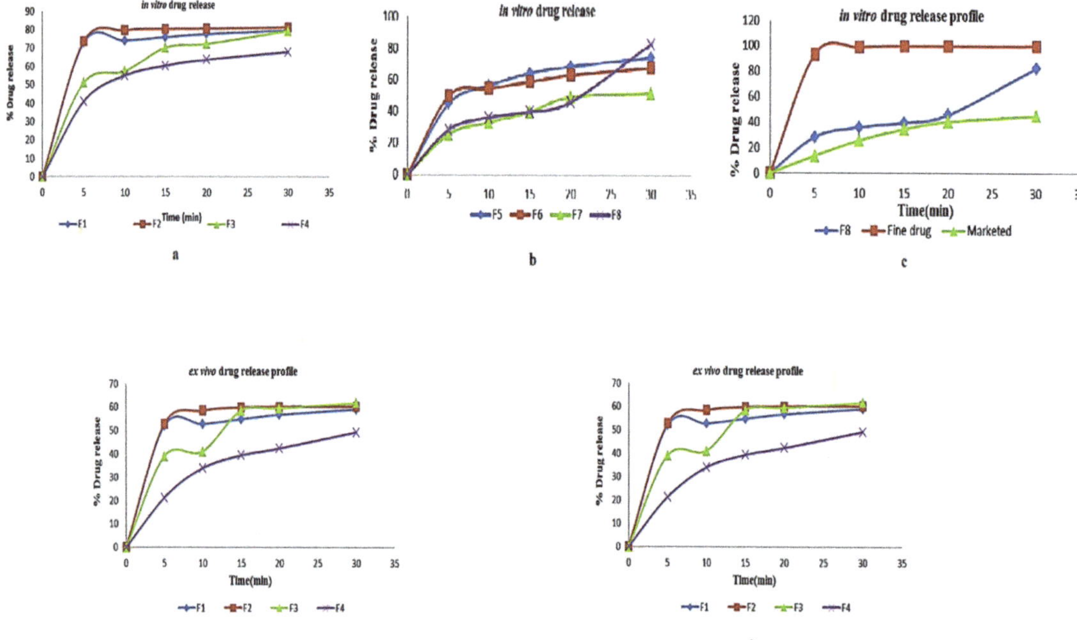

Figure 30. (**a,b**) In vitro drug release of synthesized hydrophilic polymer–gel in various formulations; (**c**) in vitro drug release of synthesized hydrophilic polymer–gel in the marketed product formulation; and (**d,e**) ex vivo drug release profile of developed gels in different formulations. Reproduced with permission from [333].

Surendra G. Gattani et al. (2016) developed bio-nanocomposites using the microwave-induced diffusion technique (MIND) to enhance the solubility of the drug ketoprofen. They highlighted the importance of solubility in achieving the concentration of the drug in the systemic circulation. Among all of the available drug molecules, only 8% have sufficient solubility. Different formulations were developed using the microwave-induced diffusion technique. The MIND process is an effective and simple technique for the enhancement of the solubility of molecules. It is an advanced and current technology in materials processing and the manufacture of chemicals, and presents promising advantages compared to the traditional thermal treatments. Heating is a significant component in the energy exchange. The solubility enhancement of bio-nanocomposites was investigated by dissolution and an in vitro solubility study, as shown in Figure 31. The polymers were selected based on the surfactant and the wetting properties. The solubility of ketoprofen was enhanced using the microwave-induced diffusion technique. The microwave technique enables rapid and uniform heating of materials with low heat conductivity because energy can easily be converted into heat within the material and most of the materials comprise polymers. The enhancement in solubility may be attributed to the drug dispersion at micro- and nanoscales. The in vivo study of optimized bio-nanocomposites was also conducted using the rat paw edema model [335].

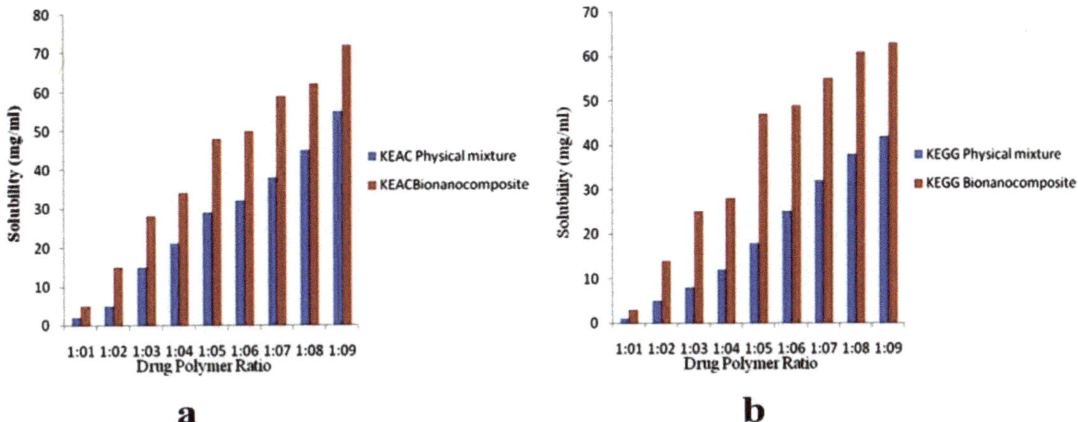

Figure 31. (**a**) Comparison of the solubility between the physical KEAC combination and KENC. (**b**) Correlations of solubility in KEGG physical blend and KEGG bio-nanocomposite. The information is the average ± SD, *n* = 3. The results are expressed in terms of the solubility percentage of virgin ketoprofen, where: KE–ketoprofen, AC–acacia, GG–ghatti gum, SD–standard deviation. Reproduced with permission from [335].

Xiao H. et.al. (2000) analyzed the mechanisms and the important role of the burst release in drug delivery systems under the control of the matrix. In this research, the authors reviewed the burst release experiments on monolithic polymer-controlled drug delivery systems. Furthermore, they reviewed the theories regarding the physical mechanisms that cause bursting, and presented novel ideas to prevent bursting and to treat burst release under controlled release models. This article also discussed the significance of burst release and suggested that burst release may be applied in the treatment of wound and bacterial bone infection, in which an initially high concentration of antibiotics is needed for effective eradication of the infection, as illustrated in Figure 32. The burst release profile is also useful in the case of targeted drug delivery and pulsatile drug delivery systems [336].

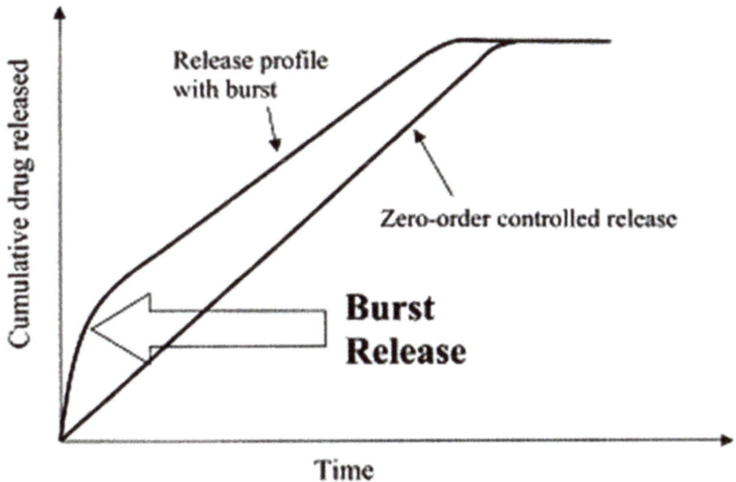

Figure 32. Schematic representing the bursting effect in a zero-ordered drug delivery system. Reproduced with permission from [336].

Nanostructured polymers have attracted increased attention as a promising class with functional materials, and in the design of biocompatible frameworks for biomedical applications, as shown in the bibliometric mapping analysis in Figure 33. The combined effect of multiple classes of nanomaterials enhances not only the inherent characteristics of composite materials, but also their shape, and their stereochemical, biological, functional, and compositional resemblance to organic and inorganic body parts. A broad spectrum of formulations, mixtures, and nanofillers is comprehensively used in medical applications, primarily in the role of a drug carrier. The key obstacle for polymeric nanocomposites is to mimic (biologically, synthetically, and functionally) the extracellular matrices of numerous body parts to facilitate tissue regeneration.

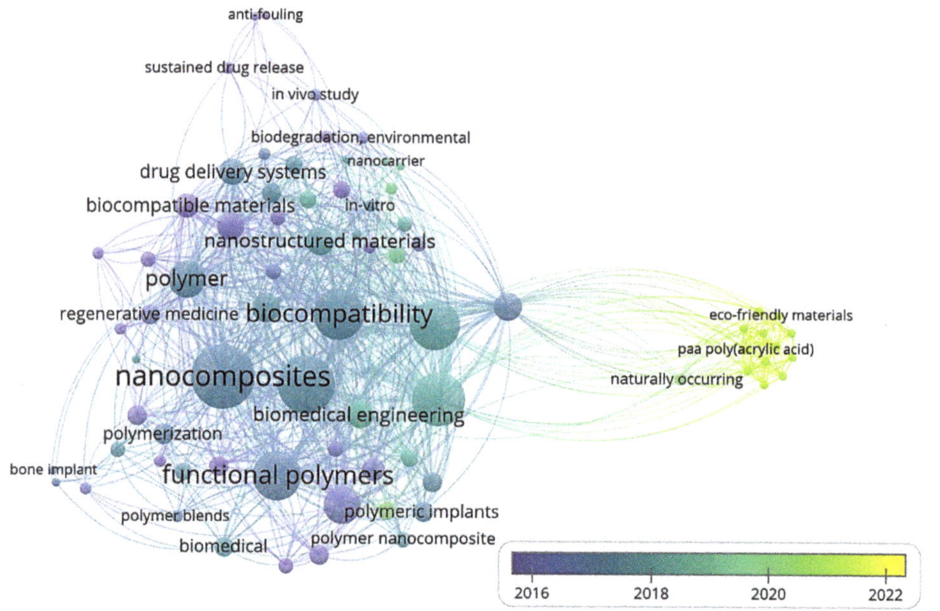

Figure 33. Bibliometric analysis of the utilization of natural polymeric biomaterials and nanopolymers in drug carriers and tissue engineering applications.

3.4. Antimicrobial Drug Delivery of Biodegradable and Other Natural Polymeric Biomaterials in Hard Tissue Engineering

Antibiotics are frequently used in the case of bone implants to prevent postsurgical infection or in the case when an infection has been diagnosed [3]. For example, in the case of osteomyelitis, the classic treatment entails the surgical removal of the diseased bone followed by the administration of antibiotics. This method is complicated because it causes weakening of the musculoskeletal support and the effectiveness of the antibiotics decreases. Thus, this problem can be resolved using systems capable of local delivery of antimicrobial agents [2]. Gomes D. and co-workers reported that composite nanostructures, such as hydroxyapatite and PLGA, are used for the treatment of osteomyelitis and the delivery of antibiotics to the infected bone [337]. Examples of drug delivery systems for osteomyelitis are presented in Table 4.

Logith Kumar R and co-workers reported that the antimicrobial activity of chitosan can be improved by modifying its structure, thereby increasing its suitability for use in hard tissue engineering [338]. In addition, the team of researchers investigated the antimicrobial activity of the incorporated vancomycin-loaded liposomes into a nano-hydroxyapatite-chitosan-konjac glucomannan scaffold.

It was reported that the scaffold was biocompatible and biodegradable, and enabled modification of the release profile of the drug by adjusting the ratio between the chitosan and konjac glucomannan. The study used a scaffold with a content of 60–70% nano-hydroxyapatite, and the content of chitosan and konjac glucomannan was varied to allow the differences to be observed. A slower release of the drug was observed at the largest amounts of chitosan and konjac glucomannan used. The in vitro tests confirmed that the system comprising vancomycin-loaded liposomes and scaffolds resulted in greater inhibition of the formation of *S. aureus* biofilms than the drug-loaded scaffold [339].

In another study, Hornyák I and co-workers studied the antimicrobial activity of human bone allografts incubated with antibiotic solution (vancomycin) and coated with chitosan. Sustained release of the drug for 50 days was observed. It was reported that the MIC for Enterococcus faecalis was 0.2 μg/mL of vancomycin, and for methicillin-resistant S. aureus, was 2 μg/mL of vancomycin [340]. Another advantage is that the alginate and allograft are biodegradable, which makes the development of the biofilm more difficult [341]. Another study reported on a system formed from a chitosan scaffold with bactericidal agents coated with a nano-hydroxyapatite-poly(amide). The study reported the continued release of the bactericidal agents for over 150 h, the decrease in the extent of bacterial growth, and cell adhesion. In addition, it was reported that scaffolds made of chitosan/nano-hydroxyapatite/nano-silver particles showed good antimicrobial activity against Gram-negative and Gram-positive bacterial strains. Furthermore, it was observed that these scaffolds are not toxic to rat osteoprogenitor cells or human osteosarcoma cell lines [342].

González-Sánchez MI and co-workers attempted to maximize the antimicrobial activity of osteoconductive acrylate hydrogels against *Staphylococcus epidermidis* and methicillin-resistant *Staphylococcus aureus* by charging silver nanoparticles using three methods. The first method encapsulated the silver nanoparticles during the synthesis. It was observed that the hydrogels with different cross-linking degrees containing silver nanoparticles showed no changes in antimicrobial activity compared with the control (Ag 0%) against *Staphylococcus epidermidis* and methicillin-resistant *Staphylococcus aureus*. The second method diffused the nanoparticles into the composite by diluting the sodium dihydrogen phosphate in the silver nanoparticle suspension. A slightly higher antibacterial activity was observed compared to the control, but it was reported that these results were not statistically significant. The third method used the adsorption of silver nanoparticles into the scaffold by placing the silver nanoparticle suspension in contact with the mineralized hydrogel for a period of between 1 and 6 days. It was observed that the samples that were in contact with the 1 mM silver nanoparticle suspension showed significantly higher antimicrobial activity compared to the samples that were in contact with the 0.5 mM silver nanoparticle suspension, against both *Staphylococcus epidermidis* and methicillin-resistant *Staphylococcus aureus*. In addition, it was reported that the greatest antimicrobial activity of the scaffolds was achieved for *Staphylococcus epidermidis* and, for the samples that were in contact with silver nanoparticles for 2 days, the antimicrobial activity decreased thereafter. It was reported that this method does not have a negative impact on osteoblasts. This is one of the few studies performed on the acrylate hydrogel with antimicrobial activity using a non-antibiotic-based antibacterial [343].

Polymeric biomaterials have had a substantial influence for a sustained period. During the past several centuries, biomimetic and biodegradable polymeric materials have emerged, and have promised exceptional advances in a diverse variety of diagnostic and therapeutic medical devices. Awareness of the interfacial interrelations of polymeric biomaterials, and controlling these materials with biological components such as water, ions, peptides, enzymes, microorganisms, microbes, and cell types, appears to be crucial for their productive use in biomedical fields. This is shown in the bibliometric mapping analysis in Figure 34.

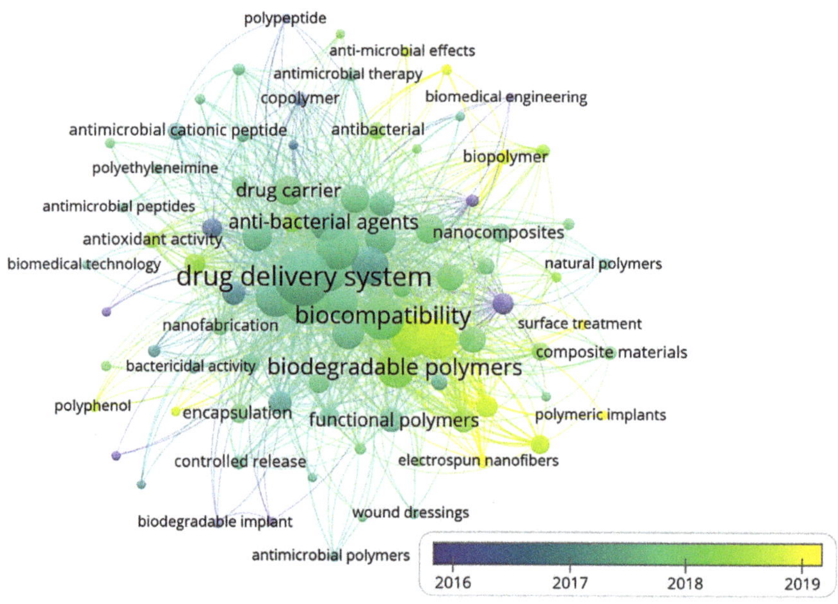

Figure 34. Bibliometric mapping of the antimicrobial drug delivery of biodegradable and natural polymeric biomaterials in hard tissue engineering.

3.5. Antitumor Drug Delivery of Biodegradable and Other Natural Polymeric Biomaterials in Hard Tissue Engineering

Gu W and co-workers reported that the skeleton is the organ that has the highest mortality percentage and is most affected by metastatic cancer [344]. To overcome the limitations of chemotherapy (nonspecific biodistribution and targeting) in the case of cancer, research attention is rapidly focusing on drug delivery systems [345]. For example, El-Kady and co-workers synthetized lithium-modified bioactive glass nanoparticles using the sol-gel method, in which the nanoparticles were loaded with 5-fluorouracil. The release profile of the drug was in two phases: rapid release in the first 24 h, followed by slow release for 32 days. It was reported that the in vitro bioactivity assessment in SBF indicated that this system can be used for bone engineering, and that the controlled release of lithium ions accelerates bone regeneration [346]. In another study, mesoporous silica nanoparticles were synthesized with a dimension of 40 nm, anchored by zoledronic acid, and loaded with doxorubicin for bone cancer therapy. It was reported that the system had improved bone-targeting ability compared with the mesoporous silica nanoparticles. Although the mesoporous silica nanoparticles presented a maximum loading capacity of 1671 mg/g and a loading efficiency of 83.56%, compared to the DOX@MSNs4ZOL system, which presented a maximum loading capacity of 1547 mg/g and a loading efficiency of 77.34%, it was reported that DOX@MSNs4ZOL offered better cytotoxicity against A549 cells and decreased cell migration in vitro [347].

Nanotech advances have contributed to the emergence of novel polymeric compositions that enable the modulation of the biotech and biomedical characteristic rates of compounds. The unique physicochemical and technical attributes of polymeric nanocomposite-based therapeutic agents have resulted in numerous promising therapeutic applications. The utilization of polymer–nanomaterials as anti-cancer compound drug carriers, their physical characteristics, and their ability to be effectively concentrated in particular tumors, were portrayed in this review. The nano-encapsulation of antitumor productive substances in biocompatible polymers is a viable strategy for increasing the effectiveness of numerous tumor treatment options, as depicted in the bibliometric mapping analysis in Figure 35.

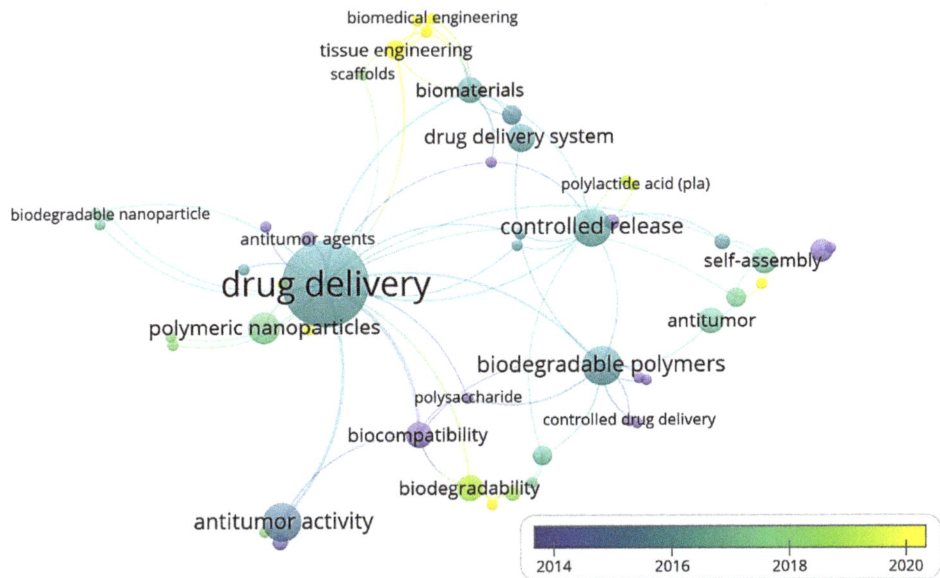

Figure 35. Antitumor drug delivery of biodegradable and natural polymeric biomaterials in hard tissue engineering.

Yang L. and co-workers reported that selenium nanoparticles present biocompatibility and anticancer activities, and, when grown on titanium, have the ability to inhibit the growth of cancerous osteoblasts and increase the growth of healthy osteoblasts [348]. Another study investigated new drug delivery systems formed from calcium phosphate cement, calcium phosphate cement containing caffeine or cisplatin, and solely caffeine and cisplatin. The in vitro tests on SOSN2 cells demonstrated that the system formed from calcium phosphate cement, caffeine, and cisplatin released a greater quantity of the drug. In addition, in vivo tests on male Fischer 344/NSlc 7 week old rats demonstrated greater tumor growth inhibition when the calcium phosphate cement, caffeine, and cisplatin system was used. Based on these studies, the authors reported that this system possesses suitable antitumor effects [349].

A new class that treats cancer bone metastasis is represented by the bisphosphonates, which show an affinity for bone tissue and can be used to deliver other anticancer drugs. Figure 36 presents a drug delivery system for bone cancer. The system involves mesoporous silica nanoparticles loaded with anticancer drugs and coated with bisphosphonates. The positive charge of the nanoparticles can be transported with siRNAs. At the moment of administration, the nanoparticles remain attached to the bone cells, kill the cancer cells, and release drugs or siRNAs. It was reported that poly-l-lysine grafted with beta-cyclodextrin for RIS delivery warned of the induction of metastatic cancer in animal models [344].

Wang F and co-workers created a liposomal system conjugated with cyclic arginine-glycine-aspartic acid-tyrosine-lysine peptide (cRGDyk)-loaded cisplatin. It was reported, after in vivo tests, that this system presents low organ toxicity and high therapeutic efficacy, and can be successfully used for therapy of bone metastases [350]. Another study developed an anti-tumoral-loaded bone graft material for the treatment of bone cancer. The system consisted of collagen, hydroxyapatite, and cisplatin, and was tested on the osteosarcoma G292 cell line. It was observed that the cytotoxic, anti-proliferative, and anti-invasive activities depend on the released concentration of cisplatin [351].

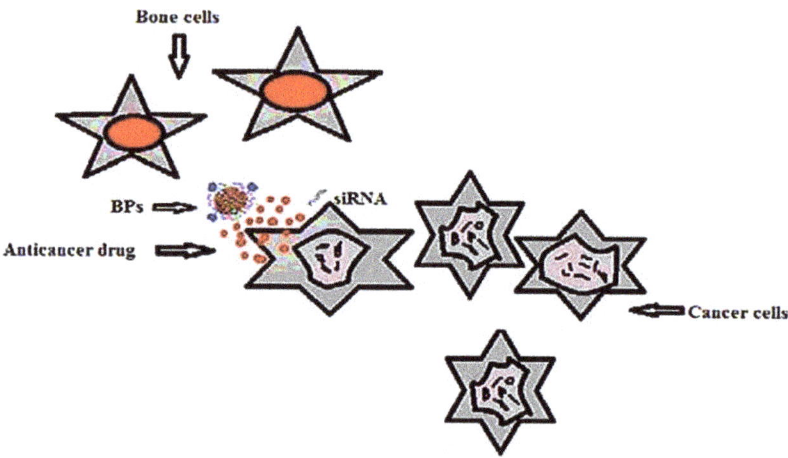

Figure 36. Drug delivery system for bone cancer.

Hyperthermia can be used for the destruction of cancer cells. For this purpose, magnetic nanoparticles are loaded onto the scaffold and exposed to alternating magnetic fields, in combination with anticancer drugs. This method is widely used in hard tissue engineering [1]. In addition, Zhang and co-workers created a scaffold consisting of Fe_3O_4 nanoparticles, mesoporous bioactive glass, and polycaprolactone produced using the 3D printing technique. It was reported that this system presents excellent apatite-forming bioactivity, good magnetic heating properties, and can be used for the treatment of bone tumors [352].

3.6. Anti-Inflammatory Drug Delivery of Biodegradable and Other Natural Polymeric Biomaterials in Hard Tissue Engineering

Conventional nanocarriers have been replaced by nanotherapeutics due to their advantages, such as simultaneous delivery of multiple drugs and the presence of the targeting agents on the surface. These systems can be used to treat different pathologies, such as inflammatory diseases, and can be adjusted depending on the patient. It was reported that chitin dressings accelerate wound repair and can regulate the secretion of inflammatory mediators, such as prostaglandin E, IL-8, and IL-1 β [353].

In the case of anti-inflammatory applications, steroids and non-steroids (ibuprofen) are commonly used. For example, Paris and co-workers created a scaffold consisting of apatite and agarose polymer loaded with two drugs (ibuprofen and zoledronic acid) during the scaffold fabrication and after consolidation. In the first step, the agarose polymer was introduced into deionized water and was subjected to magnetic stirring under heating to 90 °C. Then, the temperature was gradually reduced to 45 °C, and the apatite and drug 1 was added. The scaffold was then shaped and freeze dried, and drug 2 was injected into the scaffold. It was observed that this system provides a very fast delivery of ibuprofen (to reduce the inflammation after implantation) and zoledronic acid (to promote bone regeneration). Due to its rapid release, the authors encapsulated the ibuprofen into chitosan spheres. It was reported that, as a result of this change, a release profile was obtained that is suitable for clinical application [354]. Another study developed an anti-inflammatory delivery system for bone applications formed from porous β-TCP pellets loaded with ibuprofen by physisorption. It was reported that the interaction between porous β-TCP pellets and ibuprofen is weak. In vitro tests showed the complete release (100%) of ibuprofen due to Van der Waals forces [355]. Xiao and co-workers reported that an asymmetric coating formed from hydroxyapatite and gelatin on a Ti6Al4V alloy implant released ibuprofen for a minimum of 30 days. In addition, it was reported that in vitro studies in SBF led to the formation of apatite and the implant was fully covered after

14 days [356]. By comparison, Lin and co-workers released aspirin from a composite formed from PMMA and silica with various 3-(trimethoxysilyl) propyl methacrylate proportions and silica contents. It was observed that the release of the drug in PBS decreased with the increase in the 3-(trimethoxysilyl) propyl methacrylate content and increased with the silica content in the composites [357].

Both non-enzymatic and enzymatic decomposition of biopolymers tend to produce an innocuous, functionalized biomimetic co-product [358–360]. Within the framework of bioactive novel genetic materials in specific targeted drug delivery applications, in particular, biopolymeric materials place a considerable accent on science, as illustrated in the bibliometric mapping analysis in Figure 37 [361–368]. Utilization with biocompatible polymers minimizes a drug's adverse effects and negative consequences. Biopolymers, including biodegradable biopolymers, do not have a persistent inflammatory influence, and are characterized by high porosity and permeability, and outstanding therapeutic properties [369–371].

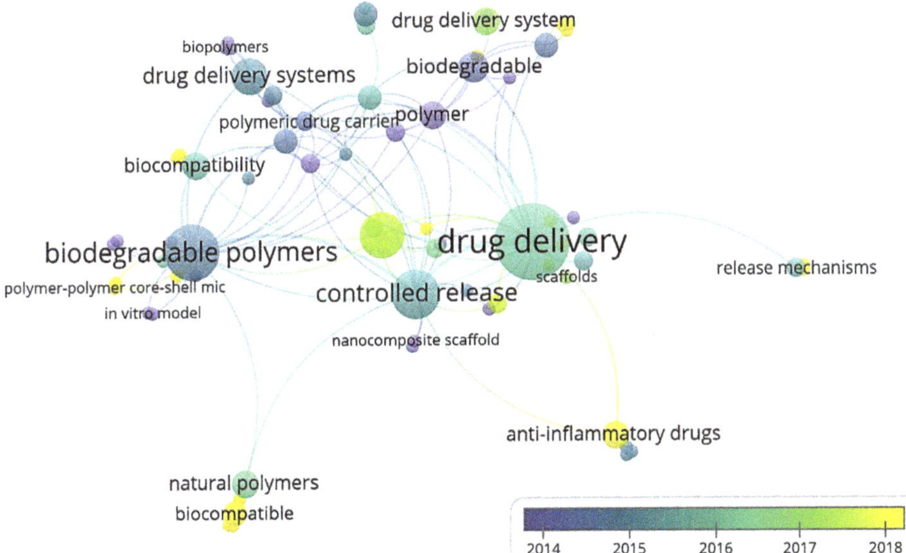

Figure 37. Scientometric bio-informatic mapping of anti-inflammatory drug delivery of biodegradable and natural polymeric biomaterials in hard tissue engineering.

Table 3. Examples of systems used in hard tissue engineering [360].

Type	Fabrication Method	Materials		Applications
		Core	**Shell**	
Nanofiber	Co-axial electrospinning	PLGA	Collagen	Dual drug delivery systems for hard tissue engineering
	Co-axial electrospinning	PEO	PCL-PEG	Drug delivery systems for hard tissue engineering
	Co-axial electrospinning	PLLC	Collagen	Dual drug delivery systems for hard tissue engineering

Table 3. *Cont.*

Type	Fabrication Method	Materials		Applications
		Core	**Shell**	
Microfiber	Co-concentric extrusion	Tricalcium Phosphate and alginate	Alginate	Dual drug delivery systems for bone regeneration
	Droplet coating	Alginate	Calcium silicate	Protein delivery control for hard tissue engineering
Micropheres	Co-axial electrodropping	PLGA	Alginate	Dual drug delivery systems for hard tissue engineering (Dexamethasone and BMP2)
	Biomimetic approach	Gelatin	Calcium phosphate	Drug delivery systems for hard tissue engineering

Table 4. Examples of studies of drug delivery systems for osteomyelitis [337].

Class	Material	Antibiotic	Tested on Microorganism	Animal Model
Bioceramic	Calcium phosphate	Gentamicin	*S. aureus*	Rabbits
	Calcium sulphate	Moxifloxacin	Methicillin resistant *S. aureus*	Rabbits
Polymer	Hydroxyapatite	Vancomycin	*S. aureus*	Rabbits
	Collagen	Gentamicin	*S. aureus*	Rabbits
	PEG, PLGA	Tobramycin, Cefazolin	*S. aureus*	Rabbits
	Polylactide/polyglycolide	Gentamicin	*S. aureus*	Dogs
Bioactive glass	Borate	Vancomycin	Methicillin resistant *S. aureus*	Rabbits
	Boro-silicate	Ceftriaxone–sulbactam	*S. aureus*	Rabbits
Polymer composite	Chitosan, borate glass	Teicoplanin	*S. aureus*	Rabbits
	PLGA, bioactive glass	Ciprofloxacin	*S. aureus*	Rabbits

4. Concluding Remarks and Future Outlook

Biodegradable polymer nanocomposites and other natural polymeric biomaterials (which are usually a combination of two or more materials) possess distinctive characteristics fused with sufficient energy to ensure that the outcome benefits from the best properties of both materials. In contrast to an individual material, a polymer nanocomposite is composed of two materials, and thus combines two sets of properties. A polymer bio-nanocomposite is the combination of a drug with a natural- or bio-carrier using nanotechnology. The parameters of bio-nanocomposites are evaluated via their drug release profile determined in vivo and in vitro, and their bioavailability in a biological system. Bio-nanocomposites are a class of materials comprised of nanosized particles within a composition of other materials. Drug delivery systems represent an emerging area that is essential for the treatment of numerous diseases. These systems can be synthesized using various methods, depending on the applications for which they are required, such as anticancer or anti-inflammatory applications. Research is currently underway to develop controlled release systems loaded with natural products, such as medicinal plants or phenolic compounds, to treat different pathologies. It has been reported that the biggest challenge to the future development of nanotherapeutics is advancing the research on systems based on natural products that are capable of enabling a targeted release. Another research challenge is the design and testing of novel methods of controlling the interaction of nanomaterials with the body. This paper also emphasized that current methods aim to target the disadvantage of polymeric nanomaterials when applied to certain organs, such as the spleen and the liver. The employment of biodegradable and biorelated co-polymeric materials in the treatment of cancer, and particularly the utilization of these materials as processing methods and modes of delivery for efficacious anticancer medications, has played a pioneering role. The consolidation of insights from synthetic and biological domains indicates that a paradigm shift for the development of both biopolymeric drug and genetic delivery systems is required. Substantial technological breakthroughs relating to the fabrication of relatively new biopolymers, in addition to the comprehension of biological processes, have laid the path for this barrier to be overcome. Targeted polymeric drug delivery systems that rely on bacterial pathogens and viruses may have a virulent immunosuppressive effect on the body. In the near future, attempting to combine viewpoints from synthetic and biological areas will offer a novel framework for the development of biopolymeric targeted drug delivery applications.

Author Contributions: Conceptualization, S.S.; methodology, S.S.; formal analysis, S.S., P.S., J.S.; investigation, S.S., P.S., J.S.; resources, S.S., P.S., J.S., R.A.I.; writing—original draft preparation, S.S.; writing—review and editing, S.S., P.S., J.S., R.A.I., M.R.M.A., M.R.R.; supervision, S.S., P.S., J.S., R.A.I.; project administration, S.S., P.S., J.S.; funding acquisition, S.S., R.A.I., M.R.M.A., M.R.R. All authors have read and agreed to the published version of the manuscript.

Funding: This research was supported by Universiti Kebangsaan Malaysia, PP/LESTARI/2021 and XX-2018-008.

Institutional Review Board Statement: Not applicable.

Informed Consent Statement: Not applicable.

Data Availability Statement: No data were used to support this study.

Acknowledgments: This research was supported by Universiti Kebangsaan Malaysia, PP/LESTARI/2021 and XX-2018-008. The author Shubham Sharma wishes to acknowledge the Department of RIC, IKGPTU, Kapurthala, Punjab, India for providing the opportunity to conduct this research task. The author P. Sudhakara gratefully acknowledges the support from the Science and Engineering Research Board (SERB-YSS/2015/001294)), New Delhi, India.

Conflicts of Interest: The authors declare no conflict of interest.

References

1. Bagde, A.; Kuthe, A.M.; Quazi, S.; Gupta, V.; Jaiswal, S.; Jyothilal, S.; Lande, N.; Nagdeve, S. State of the Art Technology for Bone Tissue Engineering and Drug Delivery. *IRBM* **2019**, *40*, 133–144. [CrossRef]
2. Uskoković, V.; Desai, T.A. Nanoparticulate drug delivery platforms for advancing bone infection therapies. *Expert Opin. Drug Deliv.* **2014**, *11*, 1899–1912. [CrossRef]
3. Uskoković, V. When 1 + 1 > 2: Nanostructured composites for hard tissue engineering applications. *Mater. Sci. Eng. C* **2015**, *57*, 434–451. [CrossRef] [PubMed]
4. Th, S.; Reis, R. Drug delivery systems and cartilage tissue engineering scaffolding using marine-derived products. In *Functional Marine Biomaterials: Properties and Applications*; Woodhead Publishing: Sawston, Cambridge, UK, 2015; pp. 123–136.
5. Romagnoli, C.; D'Asta, F.; Brandi, M.L. Drug delivery using composite scaffolds in the context of bone tissue engineering. *Clin. Cases Miner. Bone Metab.* **2013**, *10*, 155–161. [PubMed]
6. Hoffman, A.S. Hydrogels for biomedical applications. *Adv. Drug Deliv. Rev.* **2012**, *64*, 18–23. [CrossRef]
7. Shi, J. Nanotechnology in Drug Delivery and Tissue Engineering: From Discovery to Applications. *Nano Lett.* **2010**, *10*, 3223–3230. [CrossRef]
8. Porter, J.R.; Ruckh, T.T.; Popat, K.C. Bone tissue engineering: A review in bone biomimetics and drug delivery strategies. *Biotechnol. Prog.* **2009**, *25*, 1539–1560. [CrossRef]
9. Chu, P.K.; Liu, X. *Biomaterial's Fabrication and Processing Handbook*; CRC Press: Boca Raton, FL, USA, 2008; ISBN 9780849379734.
10. Rezwan, K.; Chen, Q.Z.; Blaker, J.J.; Boccaccini, A.R. Biodegradable and bioactive porous polymer/inorganic composite scaffolds for bone tissue engineering. *Biomaterials* **2006**, *27*, 3413–3431. [CrossRef]
11. Stamatialis, D.F.; Papenburg, B.J.; Gironés, M.; Saiful, S.; Bettahalli, S.N.; Schmitmeier, S.; Wessling, M. Medical applications of membranes: Drug delivery, artificial organs and tissue engineering. *J. Membr. Sci.* **2008**, *308*, 1–34. [CrossRef]
12. Gaikwad, V.V.; Patil, A.B.; Gaikwad, M.V. Scaffolds for Drug Delivery in Tissue Engineering. *Int. J. Pharm. Sci. Nanotechnol.* **2008**, *1*, 113–123. [CrossRef]
13. Goldberg, M.; Langer, R.; Jia, X. Nanostructured materials for applications in drug delivery and tissue engineering. *J. Biomater. Sci. Polym. Ed.* **2007**, *18*, 241–268. [CrossRef]
14. Sokolsky-Papkov, M.; Agashi, K.; Olaye, A.; Shakesheff, K.; Domb, A.J. Polymer carriers for drug delivery in tissue engineering. *Adv. Drug Deliv. Rev.* **2007**, *59*, 187–206. [CrossRef]

15. Gayathri, S.; Ghosh, O.; Sudhakara, P.; Viswanath, K. Chitosan conjugation: A facile approach to enhance the cell viability of LaF3: Yb, Er upconverting nanotransducers in human breast cancer cells. *Carbohydr. Polym.* **2015**, *121*, 302–308. [CrossRef]
16. Prasad, C.V.; Swamy, B.Y.; Reddy, C.L.N.; Prasad, K.V.; Sudhakara, P.; Subha, M.C.S.; Il, S.J.; Rao, K.C. Formulation and Characterization of Sodium Alginate g-Hydroxy Ethylacrylate Bio-Degradable Polymeric Beads: In Vitro Release Studies. *J. Polym. Environ.* **2012**, *20*, 344–352. [CrossRef]
17. Holland, T.A.; Mikos, A.G. Biodegradable polymeric scaffolds. Improvements in bone tissue engineering through controlled drug delivery. *Tissue Eng. I* **2005**, *102*, 161–185.
18. Dhandayuthapani, B.; Yoshida, Y.; Maekawa, T.; Kumar, D.S. Polymeric Scaffolds in Tissue Engineering Application: A Review. *Int. J. Polym. Sci.* **2011**, *2011*, 290602. [CrossRef]
19. Holzwarth, J.M.; Ma, P.X. Biomimetic nanofibrous scaffolds for bone tissue engineering. *Biomaterials* **2011**, *32*, 9622–9629. [CrossRef]
20. Zhang, R.; Ma, P.X. Poly(alpha-hydroxyl acids)/hydroxyapatite porous composites for bone tissue engineering. I. Preparation and morphology. *J. Biomed. Mater. Res.* **1999**, *44*, 446–455. [CrossRef]
21. Boccaccini, A.R.; Blaker, J.J. Bioactive composite materials for tissue engineering scaffolds. *Expert Rev. Med. Devices* **2005**, *2*, 303–317. [CrossRef]
22. Dorozhkin, S.V. Nanosized and nanocrystalline calcium orthophosphates. *Acta Biomater.* **2010**, *6*, 715–734. [CrossRef]
23. Chou, Y.F.; Dunn, J.C.Y.; Wu, B.M. In vitro response of MC3T3-E1 preosteoblasts within three-dimensional apatite-coated PLGA scaffolds. *J. Biomed. Mater. Res. Part B Appl. Biomater.* **2005**, *75B*, 81–90. [CrossRef] [PubMed]
24. Zhang, R.; Ma, P.X. Porous poly(L-lactic acid)/apatite composites created by biomimetic process. *J. Biomed. Mater. Res.* **1999**, *45*, 285–293. [CrossRef]
25. Pollok, J.M.; Vacanti, J.P. Tissue engineering. *Semin Pediatr. Surg.* **1996**, *5*, 191–196. [PubMed]
26. Liao, S.; Watari, F.; Zhu, Y.; Uo, M.; Akasaka, T.; Wang, W.; Xu, G.; Cui, F. The degradation of the three layered nano-carbonated hydroxyapatite/collagen/PLGA composite membrane in vitro. *Dent. Mater.* **2007**, *23*, 1120–1128. [CrossRef]
27. Ngiam, M.; Liao, S.; Patil, A.J.; Cheng, Z.; Chan, C.K.; Ramakrishna, S. The fabrication of nano-hydroxyapatite on PLGA and PLGA/collagen nanofibrous composite scaffolds and their effects in osteoblastic behavior for bone tissue engineering. *Bone* **2009**, *45*, 4–16. [CrossRef]
28. Wei, G.; Ma, P.X. Structure and properties of nano-hydroxyapatite/polymer composite scaffolds for bone tissue engineering. *Biomaterials* **2004**, *25*, 4749–4757. [CrossRef]
29. Lei, B.; Shin, K.H.; Noh, D.Y.; Jo, I.H.; Koh, Y.H.; Choi, W.Y.; Kim, H.E. Nanofibrous gelatin-silica hybrid scaffolds mimicking the native extracellular matrix (ECM) using thermally induced phase separation. *J. Mater. Chem.* **2012**, *22*, 14133–14140. [CrossRef]
30. Wei, G.; Ma, P.X. Macroporous and nanofibrous polymer scaffolds and polymer/bone-like apatite composite scaffolds generated by sugar spheres. *J. Biomed. Mater. Res. Part. A* **2006**, *78A*, 306–315. [CrossRef] [PubMed]
31. Liu, X.; Smith, L.A.; Hu, J.; Ma, P.X. Biomimetic nanofibrous gelatin/apatite composite scaffolds for bone tissue engineering. *Biomaterials* **2009**, *30*, 2252–2258. [CrossRef] [PubMed]
32. He, C.L.; Xiao, G.Y.; Jin, X.B.; Sun, C.H.; Ma, P.X. Electrodeposition on nanofibrous polymer scaffolds: Rapid mineralization, tunable calcium phosphate composition and topography. *Adv. Funct. Mater.* **2010**, *20*, 3568–3576. [CrossRef] [PubMed]
33. He, C.; Jin, X.; Ma, P.X. Calcium phosphate deposition rate, structure and osteoconductivity on electrospun poly(l-lactic acid) matrix using electrodeposition or simulated body fluid incubation. *Acta Biomater.* **2014**, *10*, 419–427. [CrossRef]
34. Mamidi, N.; Delgadillo, R.M.V. Design, fabrication and drug release potential of dual stimuli-responsive composite hydrogel nanoparticle interfaces. *Colloids Surf. B Biointerfaces* **2021**, *204*, 111819. [CrossRef] [PubMed]
35. Mamidi, N.; Delgadillo, R.V.; Ortiz, A.G.; Barrera, E. Carbon Nano-Onions Reinforced Multilayered Thin Film System for Stimuli-Responsive Drug Release. *Pharmaceutics* **2020**, *12*, 1208. [CrossRef] [PubMed]
36. Nair, L.S.; Laurencin, C.T. Polymers as biomaterials for tissue engineering and controlled drug delivery. In *Tissue engineering I. Advances in Biochemical Engineering/Biotechnology*; Lee, K., Kaplan, D., Eds.; Springer: Berlin/Heidelberg, Germany, 2006; pp. 47–90.
37. Tian, H.Y.; Tang, Z.H.; Zhuang, X.L.; Chen, X.S.; Jing, X.B. Biodegradable synthetic polymers: Preparation, functionalization and biomedical application. *Prog. Polym. Sci.* **2012**, *37*, 237–280. [CrossRef]
38. Chandra, R. Biodegradable polymers. *Prog. Polym. Sci.* **1998**, *23*, 1273–1335. [CrossRef]
39. Dawson, E.; Mapili, G.; Erickson, K.; Taqvi, S.; Roy, K. Biomaterials for stem cell differentiation. *Adv. Drug Deliv. Rev.* **2008**, *60*, 215–228. [CrossRef]
40. Del Valle, E.M.M.; Galán, M.A.; Carbonell, R.G. Drug Delivery Technologies: The Way Forward in the New Decade. *Ind. Eng. Chem. Res.* **2009**, *48*, 2475–2486. [CrossRef]
41. Schmaljohann, D. Thermo- and pH-responsive polymers in drug delivery. *Adv. Drug Deliv. Rev.* **2006**, *58*, 1655–1670. [CrossRef]
42. Ha, C.S.; Gardella, J.A. Surface Chemistry of Biodegradable Polymers for Drug Delivery Systems. *Chem. Rev.* **2005**, *105*, 4205–4232. [CrossRef]
43. Siepmann, J.; Siepmann, F. Microparticles Used as Drug Delivery Systems. In *Smart Colloidal Materials*; Progress in Colloid and Polymer Science; Richtering, W., Ed.; Springer: Berlin/Heidelberg, Germany, 2006; Volume 133, p. 15.
44. Kretlow, J.D.; Klouda, L.; Mikos, A.G. Injectable matrices and scaffolds for drug delivery in tissue engineering. *Adv. Drug Deliv. Rev.* **2007**, *59*, 263–273. [CrossRef] [PubMed]

45. Langer, R.; Vacanti, J. Tissue engineering. *Science* **1993**, *260*, 920–926. [CrossRef] [PubMed]
46. Hutmacher, D.W. Scaffold design and fabrication technologies for engineering tissues—State of the art and future perspectives. *J. Biomater. Sci. Polym. Ed.* **2001**, *12*, 107–124. [CrossRef]
47. Hutmacher, D.W.; Schantz, T.; Zein, I.; Ng, K.W.; Teoh, S.H.; Tan, K.C. Mechanical properties and cell cultural response of polycaprolactone scaffolds designed and fabricated via fused deposition modeling. *J. Biomed. Mater. Res.* **2001**, *55*, 203–216. [CrossRef]
48. Xia, Y.; Whitesides, G.M. Soft Lithography. *Annu. Rev. Mater. Sci.* **1998**, *28*, 153–184. [CrossRef]
49. Xia, Y.; Rogers, J.A.; Paul, K.E.; Whitesides, G.M. Unconventional Methods for Fabricating and Patterning Nanostructures. *Chem. Rev.* **1999**, *99*, 1823–1848. [CrossRef]
50. Gates, B.D.; Xu, Q.; Stewart, M.; Ryan, D.; Willson, C.G.; Whitesides, G.M. New Approaches to Nanofabrication: Molding, Printing, and Other Techniques. *Chem. Rev.* **2005**, *105*, 1171–1196. [CrossRef]
51. Tessmar, J.K.; Göpferich, A.M. Matrices and scaffolds for protein delivery in tissue engineering. *Adv. Drug Deliv. Rev.* **2007**, *59*, 274–291. [CrossRef]
52. Clapper, J.D.; Skeie, J.M.; Mullins, R.F.; Guymon, C.A. Development and characterization of photopolymerizable biodegradable materials from PEG–PLA–PEG block macromonomers. *Polymer* **2007**, *48*, 6554–6564. [CrossRef]
53. Chan-Park, M.B.; Zhu, A.P.; Shen, J.Y.; Fan, A.L. Novel Photopolymerizable Biodegradable Triblock Polymers for Tissue Engineering Scaffolds: Synthesis and Characterization. *Macromol. Biosci.* **2004**, *4*, 621. [CrossRef]
54. Zhu, A.; Chen, R.; Chan-Park, M.B. Patterning of a Random Copolymer of Poly[lactide-co-glycotide-co-(ε-caprolactone)] by UV Embossing for Tissue Engineering. *Macromol. Biosci.* **2006**, *6*, 51–57. [CrossRef] [PubMed]
55. He, B.; Chan-Park, M.B. Synthesis and Characterization of Functionalizable and Photopatternable Poly(ε-caprolactone-co-RS-β-malic acid). *Macromolecules* **2005**, *38*, 8227–8234. [CrossRef]
56. Shen, J.Y.; Chan-Park, M.B.E.; Feng, Z.Q.; Chan, V.; Feng, Z.W. UV-embossed microchannel in biocompatible polymeric film: Application to control of cell shape and orientation of muscle cells. *J. Biomed. Mater. Res. Part. B Appl. Biomater.* **2006**, *77B*, 423–430. [CrossRef] [PubMed]
57. Brey, D.M.; Ifkovits, J.L.; Mozia, R.I.; Katz, J.S.; Burdick, J.A. Controlling poly(β-amino ester) network properties through macromer branching. *Acta Biomater.* **2008**, *4*, 207–217. [CrossRef]
58. Doppalapudi, S.; Jain, A.; Khan, W.; Domb, A.J. Biodegradable polymers-an overview. *Polym. Adv. Technol.* **2014**, *25*, 427–435. [CrossRef]
59. Nair, L.S.; Laurencin, C.T. Biodegradable polymers as biomaterials. *Prog. Polym. Sci.* **2007**, *32*, 762–798. [CrossRef]
60. Briassoulis, D.; Dejean, C. Critical Review of Norms and Standards for Biodegradable Agricultural Plastics Part I. Biodegradation in Soil. *J. Polym. Environ.* **2010**, *18*, 384–400. [CrossRef]
61. Briassoulis, D.; Dejean, C.; Picuno, P. Critical Review of Norms and Standards for Biodegradable Agricultural Plastics Part II: Composting. *J. Polym. Environ.* **2010**, *18*, 364–383. [CrossRef]
62. Gross, R.A. Biodegradable Polymers for the Environment. *Science* **2002**, *297*, 803–807. [CrossRef]
63. Luckachan, G.E.; Pillai, C.K.S. Biodegradable Polymers—A Review on Recent Trends and Emerging Perspectives. *J. Polym. Environ.* **2011**, *19*, 637–676. [CrossRef]
64. Gliding, D.K.; Reed, A.M. Biodegradable polymers for use in surgery: Poly (glycolic)/poly (lactic acid) homo and co-polymers. *Polymer* **1979**, *20*, 1459–1464. [CrossRef]
65. Asghari, F.; Samiei, M.; Adibkia, K.; Akbarzadeh, A.; Davaran, S. Biodegradable and biocompatible polymers for tissue engineering application: A review. *Artif. Cells Nanomed. Biotechnol.* **2017**, *45*, 185–192. [CrossRef] [PubMed]
66. Datta, R.; Henry, M. Lactic acid: Recent advances in products, processes and technologies—A review. *J. Chem. Technol. Biotechnol.* **2006**, *81*, 1119–1129. [CrossRef]
67. Hamad, K.; Kaseem, M.; Yang, H.W.; Deri, F.; Ko, Y.G. Properties and medical applications of polylactic acid: A review. *Express Polym. Lett.* **2015**, *9*, 435–455. [CrossRef]
68. Tokiwa, Y.; Calabia, B.P. Biodegradability and biodegradation of poly(lactide). *Appl. Microbiol. Biotechnol.* **2006**, *72*, 244–251. [CrossRef] [PubMed]
69. Arrieta, M.P.; Samper, M.D.; López, J.; Jiménez, A. Combined Effect of Poly(hydroxybutyrate) and Plasticizers on Polylactic acid Properties for Film Intended for Food Packaging. *J. Polym. Environ.* **2014**, *22*, 460–470. [CrossRef]
70. Lasprilla, A.J.; Martinez, G.A.; Lunelli, B.H.; Jardini, A.L.; Filho, R.M. Poly-lactic acid synthesis for application in biomedical devices—A review. *Biotechnol. Adv.* **2012**, *30*, 321–328. [CrossRef] [PubMed]
71. Hamad, K.; Kaseem, M.; Deri, F. Melt Rheology of Poly(Lactic Acid)/Low Density Polyethylene Polymer Blends. *Adv. Chem. Eng. Sci.* **2011**, *1*, 208–214. [CrossRef]
72. Omura, M.; Tsukegi, T.; Shirai, Y.; Nishida, H.; Endo, T. Thermal Degradation Behavior of Poly(Lactic Acid) in a Blend with Polyethylene. *Ind. Eng. Chem. Res.* **2006**, *45*, 2949–2953. [CrossRef]
73. Ren, Z.; Dong, L.; Yang, Y. Dynamic mechanical and thermal properties of plasticized Poly(lactic acid). *J. Appl. Polym. Sci.* **2006**, *101*, 1583–1590. [CrossRef]
74. Mohamed, R.M.; Yusoh, K. A Review on the Recent Research of Polycaprolactone (PCL). *Adv. Mater. Res.* **2015**, *1134*, 249–255. [CrossRef]

75. Kunioka, M.; Ninomiya, F.; Funabashi, M. Novel evaluation method of biodegradabilities for oil-based polycaprolactone by naturally occurring radiocarbon-14 concentration using accelerator mass spectrometry based on ISO 14855-2 in controlled compost. *Polym. Degrad. Stab.* **2007**, *92*, 1279–1288. [CrossRef]

76. Sarasam, A.R.; Krishnaswamy, R.K.; Madihally, S.V. Blending Chitosan with Polycaprolactone: Effects on Physicochemical and Antibacterial Properties. *Biomacromolecules* **2006**, *7*, 1131–1138. [CrossRef] [PubMed]

77. Liu, L.; Li, S.; Garreau, H.; Vert, M. Selective Enzymatic Degradations of Poly(l-lactide) and Poly(ε-caprolactone) Blend Films. *Biomacromolecules* **2000**, *1*, 350–359. [CrossRef]

78. Fukuzaki, H.; Yoshida, M.; Asano, M.; Kumakura, M.; Mashimo, T.; Yuasa, H.; Imai, K.; Hidetoshi, Y. Synthesis of low- molecular-weight copoly(l-lactic acid/ε-caprolactone) by direct copolycondensation in the absence of catalysts, and enzymatic degradation of the polymers. *Polymer* **1990**, *31*, 2006–2014. [CrossRef]

79. Mochizuki, M.; Hirano, M.; Kanmuri, Y.; Kudo, K.; Tokiwa, Y. Hydrolysis of polycaprolactone fibers by lipase: Effects of draw ratio on enzymatic degradation. *J. Appl. Polym. Sci.* **1995**, *55*, 289–296. [CrossRef]

80. Gan, Z.; Liang, Q.; Zhang, J.; Jing, X. Enzymatic degradation of poly(ε-caprolactone) film in phosphate buffer solution containing lipases. *Polym. Degrad. Stab.* **1997**, *56*, 209–213. [CrossRef]

81. Gan, Z.; Yu, D.; Zhong, Z.; Liang, Q.; Jing, X. Enzymatic degradation of poly(ε-caprolactone)/poly(dl-lactide) blends in phosphate buffer solution. *Polymer* **1999**, *40*, 2859–2862. [CrossRef]

82. Khatiwala, V.K.; Shekhar, N.; Aggarwal, S.; Mandal, U.K. Biodegradation of Poly(ε-caprolactone) (PCL) Film by Alcaligenes faecalis. *J. Polym. Environ.* **2008**, *16*, 61–67. [CrossRef]

83. Chen, D.R.; Bei, J.Z.; Wang, S.G. Polycaprolactone microparticles and their biodegradation. *Polym. Degrad. Stab.* **2000**, *67*, 455–459. [CrossRef]

84. Murphy, C.A.; Cameron, J.A.; Huang, S.J.; Vinopal, R.T. Fusarium polycaprolactone depolymerase is cutinase. *Appl. Environ. Microbiol.* **1996**, *62*, 456–460. [CrossRef]

85. Oda, Y.; Asari, H.; Urakami, T.; Tonomura, K. Microbial degradation of poly(3-hydroxybutyrate) and polycaprolactone by filamentous fungi. *J. Ferment. Bioeng.* **1995**, *80*, 265–269. [CrossRef]

86. Oda, Y.; Oida, N.; Urakami, T.; Tonomura, K. Polycaprolactone depolymerase produced by the bacterium Alcaligenes faecalis. *FEMS Microbiol. Lett.* **1997**, *52*, 339–382. [CrossRef]

87. Abdel-Motaal, F.F.; El-Sayed, M.A.; El-Zayat, S.A.; Ichi Ito, S. Biodegradation of poly (ε-caprolactone) (PCL) film and foam plastic by Pseudozyma japonica sp. nov., a novel cutinolytic ustilaginomycetous yeast species. *3 Biotech.* **2014**, *4*, 507–512. [CrossRef]

88. Azimi, B.; Nourpanah, P.; Rabiee, M.; Arbab, S. Poly(ε-caprolactone) Fiber: An Overview. *J. Eng. Fibers Fabr.* **2014**, *9*, 155892501400900.

89. Flieger, M.; Kantorová, M.; Prell, A.; Řezanka, T.; Votruba, J. Biodegradable plastics from renewable sources. *Folia Microbiol.* **2003**, *48*, 27–44. [CrossRef]

90. Mitrus, M.; Wojtowicz, A.; Moscicki, L. Biodegradable polymers and their practical utility. In *Thermoplastic Starch: A Green Material for Various Industries*; Wiley-VCH Verlag GmbH & Co. KGaA: Weinheim, Germany, 2009; pp. 1–33.

91. Fields, R.D.; Rodriguez, F.; Finn, R.K. Microbial degradation of polyesters: Polycaprolactone degraded by P. pullulans. *J. Appl. Polym. Sci.* **1974**, *18*, 3571–3579. [CrossRef]

92. Chen, G.Q. Plastics completely synthesized by bacteria: *Polyhydroxyalkanoates. Microbiol. Monogr.* **2010**, *14*, 17–37.

93. Salehizadeh, H.; Loosdrecht, M.C.M.V. Production of polyhydroxyalkanoates by mixed culture: Recent trends and biotechnological importance. *Biotechnol. Adv.* **2004**, *22*, 261–279. [CrossRef] [PubMed]

94. Verlinden, R.A.J.; Hill, D.J.; Kenward, M.A.; Williams, C.D.; Radecka, I. Bacterial synthesis of biodegradable polyhydroxyalkanoates. *J. Appl. Microbiol.* **2007**, *102*, 1437–1449. [CrossRef]

95. Poirier, Y. Green chemistry yields a better plastic. *Nat. Biotechnol.* **1999**, *17*, 960–961. [CrossRef]

96. Zinn, M.; Witholt, B.; Egli, T. Occurrence, synthesis and medical application of bacterial polyhydroxyalkanoate. *Adv. Drug Deliv. Rev.* **2001**, *53*, 5–21. [CrossRef]

97. Chee, J.Y.; Yoga, S.S.; Lau, N.S.; Ling, S.C.; Abed, R.M.; Sudesh, K. Bacterially produced Polyhydroxyalkanoate (PHA): Converting renewable resources into bioplastics. *Curr. Res.* **2010**, *2*, 1395–1404.

98. Keshavarz, T.; Roy, I. Polyhydroxyalkanoates: Bioplastics with a green agenda. *Curr. Opin. Microbiol.* **2010**, *13*, 321–326. [CrossRef] [PubMed]

99. Leong, Y.K.; Show, P.L.; Ooi, C.W.; Ling, T.C.; Lan, J.C.W. Current trends in polyhydroxyalkanoates (PHAs) biosynthesis: Insights from the recombinant Escherichia coli. *J. Biotechnol.* **2014**, *180*, 52–65. [CrossRef] [PubMed]

100. Laycock, B.; Halley, P.; Pratt, S.; Werker, A.; Lant, P. The chemomechanical properties of microbial polyhydroxyalkanoates. *Prog. Polym. Sci.* **2014**, *39*, 397–442. [CrossRef]

101. Bugnicourt, E.; Cinelli, P.; Lazzeri, A.; Alvarez, V. Polyhydroxyalkanoate (PHA): Review of synthesis, characteristics, processing and potential applications in packaging. *Express Polym. Lett.* **2014**, *8*, 791–808. [CrossRef]

102. Chen, G.Q. A microbial polyhydroxyalkanoates (PHA) based bio- and materials industry. *Chem. Soc. Rev.* **2009**, *38*, 2434–2446. [CrossRef]

103. Roy, P.K.; Titus, S.; Surekha, P.; Tulsi, E.; Deshmukh, C.; Rajagopal, C. Degradation of abiotically aged LDPE films containing pro-oxidant by bacterial consortium. *Polym. Degrad. Stab.* **2008**, *93*, 1917–1922. [CrossRef]

104. Reddy, N.; Nama, D.; Yang, Y. Polylactic acid/polypropylene polyblend fibers for better resistance to degradation. *Polym. Degrad. Stab.* **2008**, *93*, 233–241. [CrossRef]
105. Nishida, H.; Arazoe, Y.; Tsukegi, T.; Yan, W.; Shirai, Y. Selective Depolymerization and Effects of Homolysis of Poly(L-lactic acid) in a Blend with Polypropylene. *Int. J. Polym. Sci.* **2009**, *2009*, 287547. [CrossRef]
106. Hamad, K.; Kaseem, M.; Deri, F. Rheological and mechanical characterization of poly(lactic acid)/polypropylene polymer blends. *J. Polym. Res.* **2011**, *18*, 1799–1806. [CrossRef]
107. Choudhary, P.; Mohanty, S.; Nayak, S.K.; Unnikrishnan, L. Poly(L-lactide)/polypropylene blends: Evaluation of mechanical, thermal, and morphological characteristics. *J. Appl. Polym. Sci.* **2011**, *121*, 3223–3237. [CrossRef]
108. Grund, S.; Bauer, M.; Fischer, D. Polymers in Drug Delivery-State of the Art and Future Trends. *Adv. Eng. Mater.* **2001**, *13*, 61–87. [CrossRef]
109. Stewart, S.; Domínguez-Robles, J.; Donnelly, R.; Larrañeta, E. Implantable Polymeric Drug Delivery Devices: Classification, Manufacture, Materials, and Clinical Applications. *Polymers* **2018**, *10*, 1379. [CrossRef]
110. Saltzman, M.W. *Drug Delivery, Engineering Principles for Drug Therapy*; Oxford University Press: New York, NY, USA, 2001.
111. Akash, M.S.H.; Rehman, K.; Tariq, M.; Chen, S. Development of therapeutic proteins: Advances and challenges. *Turk. J. Biol.* **2015**, *39*, 343–358. [CrossRef]
112. Langer, R.; Peppas, N.A. Advances in biomaterials, drug delivery, and bionanotechnology. *AIChE J.* **2003**, *49*, 2990–3006. [CrossRef]
113. Uhrich, K.E.; Cannizzaro, S.M.; Langer, R.S.; Shakesheff, K.M. Polymeric Systems for Controlled Drug Release. *Chem. Rev.* **1999**, *99*, 3181–3198. [CrossRef]
114. Bunker, A. Poly(Ethylene Glycol) in Drug Delivery, Why Does it Work, and Can We do Better? All Atom Molecular Dynamics Simulation Provides Some Answers. *Phys. Procedia* **2012**, *34*, 24–33. [CrossRef]
115. Knop, K.; Hoogenboom, R.; Fischer, D.; Schubert, U. Poly(ethylene glycol) in Drug Delivery: Pros and Cons as Well as Potential Alternatives. *Angew. Chem. Int. Ed.* **2010**, *49*, 6288–6308. [CrossRef] [PubMed]
116. Zhong, H.; Chan, G.; Hu, Y.; Hu, H.; Ouyang, D. A Comprehensive Map of FDA-Approved Pharmaceutical Products. *Pharmaceutics* **2018**, *10*, 263. [CrossRef]
117. Danhier, F.; Ansorena, E.; Silva, J.M.; Coco, R.; Breton, A.L.; Préat, V. PLGA-based nanoparticles: An overview of biomedical applications. *J. Control. Release* **2012**, *161*, 505–522. [CrossRef] [PubMed]
118. Makadia, H.K.; Siegel, S.J. Poly Lactic-co-Glycolic Acid (PLGA) as Biodegradable Controlled Drug Delivery Carrier. *Polymers* **2011**, *3*, 1377–1397. [CrossRef]
119. Mahboubian, A.; Hashemein, S.K.; Moghadam, S.; Atyabi, F.; Dinarvand, R. Preparation and In-vitro Evaluation of Controlled Release PLGA Microparticles Containing Triptoreline. *Iran. J. Pharm. Res.* **2010**, *9*, 369–378.
120. Xie, J.; Lei, C.; Hu, Y.; Gay, G.K.; Jamali, N.H.B.; Wang, C.H. Nanoparticulate Formulations for Paclitaxel Delivery Across MDCK Cell Monolayer. *Curr. Pharm. Des.* **2010**, *16*, 2331–2340. [CrossRef]
121. Averineni, R.K.; Shavi, G.V.; Gurram, A.K.; Deshpande, P.B.; Arumugam, K.; Maliyakkal, N.; Meka, S.R.; Nayanabhirama, U. PLGA 50:50 nanoparticles of paclitaxel: Development, *in vitro* anti-tumor activity in BT-549 cells and in vivo evaluation. *Bull. Mater. Sci.* **2012**, *35*, 319–326. [CrossRef]
122. Şengel-Türk, C.T.; Hasçiçek, C.; Dogan, A.L.; Esendagli, G.; Guc, D.; Gönül, N. Preparation andin vitroevaluation of meloxicam-loaded PLGA nanoparticles on HT-29 human colon adenocarcinoma cells. *Drug Dev. Ind. Pharm.* **2012**, *38*, 1107–1116.
123. Schleich, N.; Sibret, P.; Danhier, P.; Ucakar, B.; Laurent, S.; Muller, R.N.; Jérôme, C.; Gallez, B.; Préat, V.; Danhier, F. Dual anticancer drug/superparamagnetic iron oxide-loaded PLGA-based nanoparticles for cancer therapy and magnetic resonance imaging. *Int. J. Pharm.* **2013**, *447*, 94–101. [CrossRef]
124. Cooper, D.L.; Harirforoosh, S. Design and Optimization of PLGA-Based Diclofenac Loaded Nanoparticles. *PLoS ONE* **2014**, *9*, e87326. [CrossRef]
125. Rafiei, P.; Haddadi, A. Docetaxel-loaded PLGA and PLGA-PEG nanoparticles for intravenous application: Pharmacokinetics and biodistribution profile. *Int. J. Nanomed.* **2017**, *12*, 935–947. [CrossRef]
126. Afrooz, H.; Ahmadi, F.; Fallahzadeh, F.; Mousavi-Fard, S.H.; Alipour, S. Design and characterization of paclitaxel-verapamil co-encapsulated PLGA nanoparticles: Potential system for overcoming P-glycoprotein mediated MDR. *J. Drug Deliv. Sci. Technol.* **2017**, *41*, 174–181. [CrossRef]
127. Ahmadi, F.; Bahmyari, M.; Akbarizadeh, A.; Alipour, S. Doxorubicin-verapamil dual loaded PLGA nanoparticles for overcoming P-glycoprotein mediated resistance in cancer: Effect of verapamil concentration. *J. Drug Deliv. Sci. Technol.* **2019**, *53*, 101206. [CrossRef]
128. Vakilinezhad, M.A.; Amini, A.; Dara, T.; Alipour, S. Methotrexate and Curcumin co-encapsulated PLGA nanoparticles as a potential breast cancer therapeutic system: In vitro and in vivo evaluation. *Colloids Surf. B Biointerfaces* **2019**, *184*, 110515. [CrossRef] [PubMed]
129. Hami, Z.; Amini, M.; Ghazi-Khansari, M.; Rezayat, S.M.; Gilani, K. Doxorubicin-conjugated PLA-PEG-Folate based polymeric micelle for tumor-targeted delivery: Synthesis and in vitro evaluation. *DARU J. Pharm. Sci.* **2014**, *22*, 22–30. [CrossRef] [PubMed]
130. Conde, J.; Dias, J.T.; Grazú, V.; Moros, M.; Baptista, P.V.; de la Fuente, J.M. Revisiting 30 years of biofunctionalization and surface chemistry of inorganic nanoparticles for nanomedicine. *Front. Chem.* **2014**, *2*, 48. [CrossRef]

131. Diamandis, E.P.; Christopoulos, T.K. The biotin-(strept)avidin system: Principles and applications in biotechnology. *Clin. Chem.* **1991**, *37*, 625–636. [CrossRef]

132. Park, J.; Mattessich, T.; Jay, S.M.; Agawu, A.; Saltzman, W.M.; Fahmy, T.M. Enhancement of surface ligand display on PLGA nanoparticles with amphiphilic ligand conjugates. *J. Control. Release* **2011**, *156*, 109–115. [CrossRef]

133. Liu, K.C.; Yeo, Y. Zwitterionic Chitosan–Polyamidoamine Dendrimer Complex Nanoparticles as a pH-Sensitive Drug Carrier. *Mol. Pharm.* **2013**, *10*, 1695–1704. [CrossRef]

134. Poon, Z.; Lee, J.B.; Morton, S.W.; Hammond, P.T. Controlling in Vivo Stability and Biodistribution in Electrostatically Assembled Nanoparticles for Systemic Delivery. *Nano Lett.* **2011**, *11*, 2096–2103. [CrossRef]

135. Fang, R.H.; Hu, C.M.J.; Chen, K.N.H.; Luk, B.T.; Carpenter, C.W.; Gao, W.; Li, S.; Zhang, D.E.; Lu, W.; Zhang, L. Lipid-insertion enables targeting functionalization of erythrocyte membrane-cloaked nanoparticles. *Nanoscale* **2013**, *5*, 8884. [CrossRef]

136. Fang, Y.; Xue, J.; Gao, S.; Lu, A.; Yang, D.; Jiang, H.; He, Y.; Shi, K. Cleavable PEGylation: A strategy for overcoming the "PEG dilemma" in efficient drug delivery. *Drug Deliv.* **2017**, *24*, 22–32. [CrossRef]

137. Ruan, G.; Feng, S.S. Preparation and characterization of poly(lactic acid)–poly(ethylene glycol)–poly(lactic acid) (PLA–PEG–PLA) microspheres for controlled release of paclitaxel. *Biomaterials* **2003**, *24*, 5037–5044. [CrossRef]

138. Danafar, H.; Rostamizadeh, K.; Davaran, S.; Hamidi, M. PLA-PEG-PLA copolymer-based polymersomes as nanocarriers for delivery of hydrophilic and hydrophobic drugs: Preparation and evaluation with atorvastatin and lisinopril. *Drug Dev. Ind. Pharm.* **2014**, *40*, 1411–1420. [CrossRef] [PubMed]

139. Dong, Y.; Feng, S.S. Methoxy poly(ethylene glycol)-poly(lactide) (MPEG-PLA) nanoparticles for controlled delivery of anticancer drugs. *Biomaterials* **2004**, *25*, 2843–2849. [CrossRef] [PubMed]

140. Danhier, F.; Lecouturier, N.; Vroman, B.; Jérôme, C.; Marchand-Brynaert, J.; Feron, O.; Préat, V. Paclitaxel-loaded PEGylated PLGA-based nanoparticles: In vitro and in vivo evaluation. *J. Control. Release* **2009**, *133*, 11–17. [CrossRef]

141. Rafiei, P.; Michel, D.; Haddadi, A. Application of a rapid ESI–MS/MS method for quantitative analysis of docetaxel in polymeric matrices of PLGA and PLGA–PEG nanoparticles through direct injection to mass spectrometer. *Am. J. Anal. Chem.* **2015**, *6*, 164–175. [CrossRef]

142. Sims, L.B.; Curtis, L.T.; Frieboes, H.B.; Steinbach-Rankins, J.M. Enhanced uptake and transport of PLGA-modified nanoparticles in cervical cancer. *J. Nanobiotechnol.* **2016**, *14*, 33. [CrossRef] [PubMed]

143. Khalil, N.M.; Nascimento, T.C.D.; Casa, D.M. Pharmacokinetics of curcum in loaded PLGA and PLGA-PEG blend nanoparticles after oral administration in rats. *Colloids Surf. B Biointerfaces* **2013**, *101*, 353–360. [CrossRef]

144. Mahou, R.; Wandrey, C. Versatile Route to Synthesize Heterobifunctional Poly(ethylene glycol) of Variable Functionality for Subsequent Pegylation. *Polymers* **2012**, *4*, 561–589. [CrossRef]

145. Gref, R.; Couvreur, P.; Barratt, G.; Mysiakine, E. Surface-engineered nanoparticles for multiple ligand coupling. *Biomaterials* **2003**, *24*, 4529–4537. [CrossRef]

146. Popielarski, S.R.; Pun, S.H.; Davis, M.E. A Nanoparticle-Based Model Delivery System To Guide the Rational Design of Gene Delivery to the Liver. 1. Synthesis and Characterization. *Bioconjug. Chem.* **2005**, *16*, 1063–1070. [CrossRef]

147. Patil, Y.B.; Toti, U.S.; Khdair, A.; Ma, L.; Panyam, J. Single-step surface functionalization of polymeric nanoparticles for targeted drug delivery. *Biomaterials* **2009**, *30*, 859–866. [CrossRef]

148. Toti, U.S.; Guru, B.R.; Grill, A.E.; Panyam, J. Interfacial Activity Assisted Surface Functionalization: A Novel Approach to Incorporate Maleimide Functional Groups and cRGD Peptide on Polymeric Nanoparticles for Targeted Drug Delivery. *Mol. Pharm.* **2010**, *7*, 1108–1117. [CrossRef]

149. Roger, E.; Kalscheuer, S.; Kirtane, A.; Guru, B.R.; Grill, A.E.; Whittum-Hudson, J.; Panyam, J. Folic Acid Functionalized Nanoparticles for Enhanced Oral Drug Delivery. *Mol. Pharm.* **2012**, *9*, 2103–2110. [CrossRef]

150. Dhoke, D.M.; Basaiyye, S.S.; Khedekar, P.B. Development and characterization of L-HSA conjugated PLGA nanoparticle for hepatocyte targeted delivery of antiviral drug. *J. Drug Deliv. Sci. Technol.* **2018**, *47*, 77–94. [CrossRef]

151. Turk, M.J.; Breur, G.J.; Widmer, W.R.; Paulos, C.M.; Xu, L.C.; Grote, L.A.; Low, P.S. Folate-targeted imaging of activated macrophages in rats with adjuvant-induced arthritis. *Arthritis Rheum.* **2002**, *46*, 1947–1955. [CrossRef]

152. Fernández, M.; Javaid, F.; Chudasama, V. Advances in targeting the folate receptor in the treatment/imaging of cancers. *Chem. Sci.* **2018**, *9*, 790–810. [CrossRef]

153. Vortherms, A.R.; Doyle, R.P.; Gao, D.; Debrah, O.; Sinko, P.J. Synthesis, characterization, and in vitro assay of folic acid conjugates of 3′-azido-3′-deoxythymidine (AZT): Toward targeted AZT based anticancer therapeutics. *Nucleosides Nucleotides Nucleic Acids* **2008**, *27*, 173–185. [CrossRef] [PubMed]

154. Xiong, J.; Meng, F.; Wang, C.; Cheng, R.; Liu, Z.; Zhong, Z. Folate-conjugated crosslinked biodegradable micelles for receptor-mediated delivery of paclitaxel. *J. Mater. Chem.* **2011**, *21*, 5786–5794. [CrossRef]

155. Goren, D.; Horowitz, A.T.; Tzemach, D.; Tarshish, M.; Zalipsky, S.; Gabizon, A. Nuclear delivery of doxorubicin via folate-targeted liposomes with bypass of multidrugresistance efflux pump. *Clin. Cancer Res.* **2000**, *6*, 1949–1957. [PubMed]

156. Chandrasekar, D.; Sistla, R.; Ahmad, F.J.; Khar, R.K.; Diwan, P.V. The development of folate-PAMAM dendrimer conjugates for targeted delivery of anti-arthritic drugs and their pharmacokinetics and biodistribution in arthritic rats. *Biomaterials* **2007**, *28*, 504–512. [CrossRef] [PubMed]

157. Zhang, Z.; Lee, S.H.; Feng, S.S. Folate-decorated poly(lactide-co-glycolide)-vitamin E TPGS nanoparticles for targeted drug delivery. *Biomaterials* **2007**, *28*, 1889–1899. [CrossRef]

158. Pan, J.; Feng, S.S. Targeted delivery of paclitaxel using folate-decorated poly(lactide)-vitamin E TPGS nanoparticles. *Biomaterials* **2008**, *29*, 2663–2672. [CrossRef]
159. Lühder, F.; Reichardt, H. Novel Drug Delivery Systems Tailored for Improved Administration of Glucocorticoids. *Int. J. Mol. Sci.* **2017**, *18*, 1836. [CrossRef] [PubMed]
160. Morita, K.; Ishimura, K.; Tsuruo, Y.; Wong, D.L. Dexamethasone enhances serum deprivation-induced necrotic death of rat C6 glioma cells through activation of glucocorticoid receptors. *Brain Res.* **1999**, *816*, 309–316. [CrossRef]
161. Shapiro, W.R.; Posner, J.B. Corticosteroid Hormones: Effects in an Experimental Brain Tumor. *Arch. Neurol.* **1974**, *30*, 217–221. [CrossRef]
162. Grasso, R.J. Transient Inhibition of Cell Proliferation by Glucocorticoids in Rat Glioma Monolayer Cultures. *Cancer Res.* **1976**, *36*, 2408–2414. [PubMed]
163. Grasso, R.J.; Johnson, C.E. Dose-Response Relationships between Glucocorticoids and Growth Inhibition in Rat Glioma Monolayer Cultures. *Exp. Biol. Med.* **1977**, *154*, 238–241. [CrossRef]
164. Gurcay, O.; Wilson, C.; Barker, M.; Eliason, J. Corticosteroid Effect on Transplantable Rat Glioma. *Arch. Neurol.* **1971**, *24*, 266–269. [CrossRef] [PubMed]
165. Kaup, B.; Schindler, I.; Knüpfer, H.; Schlenzka, A.; Preiss, R.; Knüpfer, M.M. Timedependent inhibition of glioblastoma cell proliferation by dexamethasone. *J. Neurooncol.* **2001**, *51*, 105–110. [CrossRef]
166. Fan, Z.; Sehm, T.; Rauh, M.; Buchfelder, M.; Eyupoglu, I.Y.; Savaskan, N.E. Dexamethasone Alleviates Tumor-Associated Brain Damage and Angiogenesis. *PLoS ONE* **2014**, *9*, e93264.
167. Sand, W. Microbial mechanisms of deterioration of inorganic substrates—A general mechanistic overview. *Int. Biodeterior. Biodegrad.* **1997**, *40*, 183–190. [CrossRef]
168. Colin, G.; Mitton, M.T.; Carlsson, D.J.; Wiles, D.M. The effect of soil burial exposure on some geotechnical fabrics. *Geotext. Geomembr.* **1986**, *4*, 1–8. [CrossRef]
169. Inonesuen, A.; Kiss, S.; Dragan-Bularda, M.; Redulescu, D.; Kolozski, E.; Pintea, H.; Crisan, R. *Proceedings of the 2nd International Conference on Geotextiles—2nd ICG, Las Vegas, NV, USA, 1–6 August 1982*; Industrial Fabrics Association International: Las Vegas, NV, USA, 1982; p. 547.
170. Darby, R.T.; Kaplan, A.M. Fungal susceptibility of polyurethanes. *Appl. Microbiol.* **1968**, *16*, 900–905. [CrossRef]
171. Griffin, G.J.L. Preprints. In Proceedings of the 15th Prague Microsymposium on Degradation and stabilization of Polyolefins, Prague, Czech Republic, 21–24 July 1974.
172. Dey, A.S.; Bose, H.; Mohapatra, B.; Sar, P. Biodegradation of Unpretreated Low-Density Polyethylene (LDPE) by Stenotrophomonas sp. and Achromobacter sp., Isolated from Waste Dumpsite and Drilling Fluid. *Front. Microbiol.* **2020**. [CrossRef] [PubMed]
173. Potts, J.E. *Kirk- Encyclopedia of Chemical Technology*; Wiley-Interscience: New York, NY, USA, 1984; p. 629.
174. Potts, W.T.; Fleming, W.R. The Effect of Environmental Calcium and Ovine Prolactin on Sodium Balance in Fundulus Kansae. *J. Exp. Biol.* **1971**, *55*, 63–75. [CrossRef]
175. Potts, J.E.; Clendining, R.A.; Ackart, W.B.; Niegisch, W.D. Evaluation of polypropylene degradation with commercial additives in different media of exposure. *Rev. Eletrônica Gestão, Educ. Tecnol. Ambient.* **1973**, *24*. [CrossRef]
176. Potts, J.E.; Clendining, R.A.; Ackart, W.B.; Niegisch, W.D. *Polymer Science and Technology*; The Biodegradability of Synthetic Polymers; Springer: Boston, MA, USA, 1973; pp. 61–79.
177. Weiland, M.; Daro, A.; David, C. Biodegradation of thermally oxidized polyethylene. *Polym. Degrad. Stab.* **1995**, *48*, 275–289. [CrossRef]
178. Khabhaz, F.; Albersson, A.; Karlsson, S. Chemical and morphological changes of environmentally degradable polyethylene films exposed to thermo-oxidation. *Polym. Degrd. Stab.* **1999**, *63*, 127–138. [CrossRef]
179. Potts, J.E. *Aspect of Degradation and Stabilization of Polymers*; Elsevier: Amsterdam, The Netherlands, 1978; p. 617.
180. Bhatt, N.B.; Bhatt, N.B. Book Reviews. In *IETE Journal of Research*; Informa UK Limited: London, UK, 1972; Volume 18, p. 297.
181. Albertsson, A.C.; Bánhidi, Z.G. Microbial and oxidative effects in degradation of polyethene. *J. Appl. Polym. Sci.* **1980**, *25*, 1655–1671. [CrossRef]
182. Fields, R.D.; Rodriguez, F. Microbial degradation of aliphatic polyesters. In Proceedings of the 3rd International Biodegradation Symposium Applied Science, New York, NY, USA, 25–28 June 1976; p. 775.
183. Tokiwa, Y.; Suzuki, T. Hydrolysis of Polyesters by Rhizopus delemar Lipase. *Agric. Biol. Chem.* **1978**, *42*, 1071–1072. [CrossRef]
184. Guillet, J.E. Polymer and Ecological Problems. *Polym. Sci. Technol.* **1973**, *3*, 1.
185. Lee, B.; Pometto, A.L.; Fratzke, A.; Bailey, T.B. Biodegradation of Degradable Plastic Polyethylene by Phanerochaete and Streptomyces Species. *Appl. Environ. Microbiol.* **1991**, *57*, 678–685. [CrossRef] [PubMed]
186. Albertsson, A.C.; Karlsson, S. Degradable polyethylene-starch complex. *Makromol. Chem. Macromol. Symp.* **1991**, *48–49*, 395–402. [CrossRef]
187. Evangelista, R.L.; Sung, W.; Jane, J.L.; Gelina, R.J.; Nikolov, Z.L. Effect of compounding and starch modification on properties of starch-filled low-density polyethylene. *Ind. Eng. Chem. Res.* **1991**, *30*, 1841–1846. [CrossRef]
188. Griffin, G.J.L. Biodegradable fillers in thermoplastics. In *Fillers and Reinforcements for Plastics*; Deanin, R.D., Schott, N.R., Eds.; American Chemical Society: Washington, WA, USA.
189. Griffin, G.J.L. Biodegradable fillers in thermoplastics. *Adv. Chem. Ser.* **1974**, *134*, 159.

190. Griffin, G.J.L. Degradation of polyethylene in compost burial. *J. Polym. Sci. Polym. Symp.* **1976**, *57*, 281–286. [CrossRef]
191. Griffin, G.J.L. Fillers and reinforcement for plastics. *ACS Adv. Chem. Ser.* **1976**, *57*, 281.
192. Griffins, G.J.L. Synthetic Resin Sheet Material. U.S. Patent US4021388A, 3 May 1977.
193. Griffins, G.J.L. Biodegradable Synthetic Resin Sheet Material Containing Starch and a Fatty Materia. U.S. Patent US4016117A, 5 April 1977.
194. Nakashima, T.; Matsu, M.J. Development of high-strength and high-modulus polyethylene-starch composite films and biodegradation of the composite films. *Macromol. Sci. Phys.* **1996**, *25*, 659–679. [CrossRef]
195. Tsao, R.; Anderson, T.A.; Coats, J.R. The influence of soil macroinvertebrates on primary biodegradation of starch-containing polyethylene films. *J. Environ. Polym. Degrad.* **1993**, *1*, 301–306. [CrossRef]
196. Arvalo-Nio, K.; Sandoval, C.F.; Galan, L.J.; Imam, S.H.; Gordon, S.H.; Greene, R.V. Starch-based extruded plastic films and evaluation of their biodegradable properties. *Biodegradation* **1996**, *7*, 231–237. [CrossRef]
197. Psomiadou, E.; Arvanitoyannis, I.; Biliaderis, C.G.; Ogawa, H.; Kawasaki, N. Biodegradable films made from low density polyethylene (LDPE), wheat starch and soluble starch for food packaging applications. Part 2. *Carbohydr. Polym.* **1997**, *33*, 227–242. [CrossRef]
198. Marjonavic, M.; Pospisil, J.; Cikovic, N. Gas permeability of starch modified low density polyethylene film. *Kemija u Industriji* **1994**, *43*, 395–398.
199. Feng, F.; Zhao, X.; Ye, L. Structure and properties of ultradrawn polylactide/thermoplastic polyurethane elastomer blends. *J. Macromol. Sci. Part B* **2011**, *50*, 1500–1507. [CrossRef]
200. Nakada, K.; Hanawa, T.; Takei, N.; Takemura, Y. Enhancement of Dissolution Rate of Mefenamic Acid-Coating of Mefenamic Acid on the Potato Starch by Use of Wet Coating Method in High Speed Jet Stream. *PDA J. Pharm. Sci. Technol.* **1993**, *53*, 109.
201. Kiatkamjornwong, S.U.; Pabunruang, T.H.; Wongvisetsirikul, N.I.; Prasassarakich, P.A. Degradation of cassava starch-polyethylene blends. *J. Sci. Soc.* **1997**, *23*, 135.
202. Sastry, P.K.; Satyanarayana, D.; Rao, D.V.M. Accelerated and environmental weathering studies on polyethylene-starch blend films. *J. Appl. Polym. Sci.* **1998**, *70*, 2251–2257. [CrossRef]
203. Pometto, A.L.; Lee, B.T.; Johnson, K.E. Production of an extracellular polyethylene-degrading enzyme(s) by Streptomyces species. *Appl. Environ. Microbiol.* **1992**, *58*, 731–733. [CrossRef]
204. Goheen, S.M.; Wool, R.P. Degradation of polyethylene–starch blends in soil. *J. Appl. Polym. Sci.* **1991**, *42*, 2691–2701. [CrossRef]
205. Dave, H.; Rao, P.V.C.; Desai, J.D. Biodegradation of starch-polyethylene films in soil and by microbial cultures. *World J. Microbiol. Biotechnol.* **1997**, *13*, 655–658. [CrossRef]
206. Breslin, V.T.; Li, B. Weathering of starch–polyethylene composite films in the marine environment. *J. Appl. Polym. Sci.* **1993**, *48*, 2063–2079. [CrossRef]
207. Peanansky, J.S.; Long, J.M.; Wool, R.P. Percolation effects in degradable polyethylene-starch blends. *J. Polym. Sci. Part B Polym. Phys.* **1992**, *29*, 565–579. [CrossRef]
208. Griffin, G.J.L. *Chemistry and Technology of Biodegradable Polymers*; Blackie Academic & Professional: London, UK, 1994; pp. 135–150.
209. Chiquet, A. Degradable Synthetic Compositions. Google Patents EP0305413A1, 12 May 1993.
210. Chandra, R.; Rustgi, R. Biodegradation of maleated linear low-density polyethylene and starch blends. *Polym. Degrad. Stab.* **1997**, *56*, 185–202. [CrossRef]
211. Shogren, R.L.; Thompson, A.R.; Felker, F.C.; Harry-O'kuru, R.E.; Greene, R.V.; Gordon, S.H.; Gould, J.M. Polymer compatibility and biodegradation of starch–poly (ethylene-co-acrylic acid)–polyethylene blends. *J. Appl. Polym. Sci.* **1971**, *44*, 1971–1978. [CrossRef]
212. Willett, J.L. Mechanical properties of LDPE/granular starch composites. *J. Appl. Polym. Sci.* **1994**, *54*, 1685–1695. [CrossRef]
213. Jasberg, B.K.; Swanson, C.L.; Shogren, R.L.; Doane, W.M. Effect of moisture on injection molded starch-EAA-HDPE composites. *J. Polym. Mater.* **1992**, *9*, 163–170.
214. Otey, F.H.; Westhoff, R.P.; Doane, W.M. Starch-Based Blown Films. *Ind. Eng. Chem. Prod. Res. Dev.* **1980**, *19*, 592–595. [CrossRef]
215. Otey, F.H.; Westhoff, R.P.; Doane, W.M. Starch-Based Blown Films. *Ind. Eng. Chem. Res.* **1987**, *26*, 1659–1663. [CrossRef]
216. Otey, F.H.; Westhoff, R.P. Biodegradable Starch-Based Blown Films. U.S. Patent US4337181A, 29 June 1982.
217. Li, X.; Yuan, H.; Tang, S.; Zhou, Q.; Pan, R. Study on grafted modified starch as compatibilizer of LLDPE/starch blend system. *Hecheng Shuzki Ji Suliao* **1996**, *13*, 5.
218. Okieimen, E.F.; Egharevba, F. Graft copolymerization of ethyl acrylate on starch. *Eur. Polym. J.* **1989**, *25*, 99–102. [CrossRef]
219. Athawale, V.D.; Rathi, S.C. Graft polymerisation of N-methylol acrylamide onto starch using Ce^{4+} as initiator. *J. Polym. Mat.* **1996**, *13*, 335–336.
220. Swanson, C.L.; Westhoff, R.; Doane, W.P. Modified Starches in plastic films. In *Proceedings of the com Utilization Conference II*; National Corn Growers Association: Columbus, OH, USA, 1988.
221. Takigawa, M.; Sakurai, N.; Takagi, T.; Yanaka, N.; Horikiri, Y.; Tamura, T. A Composition for Regenerative Treatment of Cartilage Disease. Google Patent US20070155652A1, 28 August 2012.
222. Nakajima, T.; Takagi, S.; Ueda, T. Jpn. Kokai Tokkyo Koho. Japan Patent JP 92202543, 1992.
223. Nakajima, T.; Takagi, S.; Ueda, T. Jpn. Kokai Tokkyo Koho. Japan Patent JP 92202544, 1992.
224. Nakajima, T.; Takagi, S.; Ueda, T. Jpn. Kokai Tokkyo Koho. Japan Patent JP 92202567, 1992.
225. Nakajima, H.; Yamamoto, N.; Kaizu, T.; Kino, T. Jpn. Kokai Tokkyo Koho. Japan Patent JP 07206668, 1995.

226. Favis, B.D.; Yuan, Z.; Riscanu, D. Thermoplastic Starch and Synthetic Polymer Blends and Method of Making. U.S. Patent US8841362B2 2006, 23 September 2014.

227. Shah, P.B.; Bandopadhyay, S.; Bellare, J.R. Environmentally degradable starch filled low density polyethylene. *Polym. Degrad. Stab.* **1995**, *47*, 165–173. [CrossRef]

228. Muller, J.; González-Martínez, C.; Chiralt, A. Combination of Poly(lactic) Acid and Starch for Biodegradable Food Packaging. *Materials* **2017**, *10*, 952. [CrossRef] [PubMed]

229. St-Pierre, N.; Favis, B.D.; Ramsay, B.A.; Ramsay, J.A.; Verhoogt, H. Processing and characterization of thermoplastic starch/polyethylene blends. *Polymer* **1997**, *38*, 647–655. [CrossRef]

230. Gould, J.M.; Imam, S.H.; Gordon, S.H. Biodegradation of starch-containing plastics in aqueous environments. *Proc. ACS Div. Polym. Mater. Sci. Eng.* **1990**, *63*, 853.

231. Shogren, R.L.; Gordon, S.H.; Harry-O'Kuru, R.E.; Gould, J.M. Biodegradation of starch-containing plastics films. Structure-function relationships. Polymeric Materials Science and Engineering. In Proceedings of the ACS Division of Polymeric Materials Science and Engineering, Washington DC, WA, USA, 26–31 August 1990.

232. Tian, R.; Yu, J.; Duan, M.; Chen, Y.; Xu, Z.; Xu, X. Tianjin Daxue Xuebao. *J. Tianjin Univ. Sci. Technol.* **1990**, *2*, 23.

233. Rivard, C.; Moens, L.; Roberts, K.; Brigham, J.; Kelley, S. Starch esters as biodegradable plastics: Effects of ester group chain length and degree of substitution on anaerobic biodegradation. *Enz. Microbial. Tech.* **1995**, *17*, 848–852. [CrossRef]

234. Bikiaris, D.; Prinos, J.; Panayiotou, C. Effect of methyl methacrylate-butadiene-styrene copolymer on the thermooxidation and biodegradation of LDPE/plasticized starch blends. *Polym. Degrad. Stab.* **1997**, *58*, 215–228. [CrossRef]

235. Bikiaris, D.; Aburto, J.; Alric, I.; Borredon, E.; Botev, M.; Betchev, C.; Panayiotou, C. Mechanical properties and biodegradability of LDPE blends with fatty-acid esters of amylose and starch. *J. Appl. Polym. Sci.* **1999**, *71*, 1089–1100. [CrossRef]

236. Thiebaud, S.; Aburto, J.; Alric, I.; Borredon, E.; Bikiaris, D.; Prinos, J.; Panayiotou, C. Properties of fatty-acid esters of starch and their blends with LDPE. *J. Appl. Polym. Sci.* **1997**, *65*, 705–721. [CrossRef]

237. Aburto, J.; Thiebaud, S.; Alric, I.; Borredon, E.; Bikiaris, D.; Prinos, J.; Panayiotou, C. Properties of octanoated starch and its blends with polyethylene. *Carbohydr. Polym.* **1997**, *34*, 101–112. [CrossRef]

238. Albertsson, A.C.; Huang, S.J. *Degradable Polymers, Recycling and Plastics Waste Management*; Marcel Drekker: New York, NY, USA, 1995.

239. Kshirsagar, N.J.; Rangaprasad, R.; Naik, V.G.; Kale, D.D. Rheology and Permeability Characteristics of Blends of Starch (and Starch Acetate) with LDPE. *J. Polym. Mater.* **1992**, *9*, 17.

240. Sailaja, R.R.N.; Chanda, M. Use of maleic anhydride-grafted polyethylene as compatibilizer for HDPE-tapioca starch blends: Effects on mechanical properties. *J. Appl. Polym. Sci.* **2001**, *80*, 863–872. [CrossRef]

241. Griffith, G.J.L. *SPI Symposium on Degradable Plastics*; Society of Plastics Industries: Washington, DC, USA, 1987; pp. 47–50.

242. Otey, F.H.; Mark, A.M.; Mehltretter, C.L.; Russell, C.R. Starch-Based Film for Degradable Agricultural Mulch. *Prod. RD* **1974**, *13*, 90–92. [CrossRef]

243. Otey, F.H.; Sloane, W.M. *SPI Symposium on Degradable Plastics*; Society of Plastics Industries: Washington, DC, USA, 1987; pp. 39–40.

244. Vaidya, U.R.; Bhattacharya, M. Properties of blends of starch and synthetic polymers containing anhydride groups. *J. Appl. Polym. Sci.* **1994**, *52*, 617–628. [CrossRef]

245. Ramkumar, D.; Vaidya, U.R.; Bhattacharya, M.; Hakkarainen, M.; Albertsson, A.C.; Karlsson, S. Properties of injection moulded starch/synthetic polymer blends—I. Effect of processing parameters on physical properties. *Eur. Polym. J.* **1996**, *32*, 999–1010. [CrossRef]

246. Seethamraju, K.; Bhattacharya, M.; Vaidya, U.R.; Fulcher, R.G. Rheology and morphology of starch/synthetic polymer blends. *Rheol. Acta* **1994**, *33*, 553–567. [CrossRef]

247. Otey, F.H.; Doane, W.M. Starch-based Degradable Plastics Film. In Proceedings of the Society of the Plastic Industry, Symposium on Degradable Plastics, Washington, DC, WA, USA, 10 June 1987; pp. 39–40.

248. Fanta, F.G.; Swanson, C.L.; Shogren, R.L. Starch–poly(ethylene-co-acrylic acid) composite films. Effect of processing conditions on morphology and properties. *J. Appl. Polym. Sci.* **1992**, *44*, 2037–2042. [CrossRef]

249. Reis, R.L.; Mendes, S.C.; Cunha, A.M.; Bevis, M.J. Processing andin vitro Degradation of Starch/EVOH Thermoplastic Blends. *Polym. Int.* **1997**, *43*, 347–352. [CrossRef]

250. Bastioli, C.; Bellotti, V.; Giudice, L.; Gilli, G. Mater-Bi: Properties and biodegradability. *J. Environ. Polymer Degrad.* **1993**, *1*, 181–191. [CrossRef]

251. Bastioli, C.; Degli Innocenti, F.; Guanella, I.; Romano, G.C. Compostable Films of Mater-Bi Z Grades. *J. Macromol. Sci. Part A Pure Appl. Chem.* **1995**, *32*, 839–842. [CrossRef]

252. Thomas, H.R.; Li, C.L. A parallel computing solution of heat and moisture transfer in unsaturated soil. *Int. J. Numer. Methods Eng.* **1996**, *39*, 3793–3808. [CrossRef]

253. Bastioli, C.; Cerutti, A.; Guanella, I.; Romano, G.; Tosin, M. Physical state and biodegradation behavior of starch-polycaprolactone systems. *J. Environ. Polym. Degrad.* **1995**, *3*, 81–95. [CrossRef]

254. Degli Innocenti, F.; Tosin, M.; Bastioli, C. Evaluation of the Biodegradation of Starch and Cellulose Under Controlled Composting Conditions. *J. Polym. Environ.* **1997**, *6*, 197–202. [CrossRef]

255. Funke, U.; Bergthaller, W.; Lindhauer, M.G. Processing and characterization of biodegradable products based on starch. *Polym. Degrad. Stab.* **1998**, *59*, 293–296. [CrossRef]
256. Dake, I.; Borchers, G.; Zdrahala, R.; Dreiblatt, A.; Rathmer, P. Biodegradable Compositions Comprising Starch Derivatives. Google Patents CA2105479A, 20 September 1992.
257. Chen, L.; Imam, S.H.; Gordon, S.H.; Greene, R.V. Starch- polyvinyl alcohol crosslinked film: Performance and biodegradation. *J. Environ. Polym. Degrad.* **1997**, *5*, 111–117. [CrossRef]
258. Ahamed, T.N.; Snghal, R.S.; Kulkarni, P.R.; Kale, D.D.; Pal, M. Studies on Chenopodium quinoa and Amaranthus paniculatas starch as biodegradable fillers in LDPE films. *Carbohydr. Polym.* **1996**, *31*, 157–160. [CrossRef]
259. Doane, W.M.; Swanson, E.L.; Fanta, G.F. Emerging polymeric materials based on starch. In *Emerging Technologies/or Materials and Chemicals/i'om Biomass*; Rowell, R.M., Schultz, T.P., Narayan, R., Eds.; American Chemical Society: Washington, DC, USA, 1992; pp. 197–230.
260. Neumann, P.; Seib, P.A. Starch-Based, Biodegradable Packing Filler and Method of Preparing Same. Google Patents US5208267A, 4 May 1993.
261. Xiong, H.G.; Tang, S.W.; Tang, H.L.; Zou, P. The structure and properties of a starch-based biodegradable film. *Carbohydr. Polym.* **2008**, *71*, 263–268. [CrossRef]
262. Otey, F.H. Current and Potential Uses of Starch Products in Plastics. *Polym. Plast. Technol. Eng.* **1976**, *7*, 221–234. [CrossRef]
263. Ritter, W.; Beck, M.; Schaefer, M. Thermoplastic Composite Materials Based on Starch. Google Patents DE4333858A1, 1995.
264. Conville, C.; Major, I.; Friend, D.R.; Clark, M.R.; Woolfson, A.D.; Malcolm, R.K. Development of polylactide and polyethylene vinyl acetate blends for the manufacture of vaginal rings. *J. Biomed. Mater. Res. B Appl Biomater.* **2012**, *100*, 891–895. [CrossRef]
265. Muranaka, M.; Ueno, H.; Hirai, K.; Nakajima, K.; Ishida, F. Classification and immunochemical basis of allergic drug reactions. *Arerugi* **1996**, *45*, 1219–1230.
266. Otey, F.H.; Westhoff, R.P.; Russell, C.R. Biodegradable Films from Starch and Ethylene-Acrylic Acid Copolymer. *Prod. RD* **1977**, *16*, 305–308. [CrossRef]
267. Koller, I.; Owen, A.J. Starch-Filled PHB and PHB/HV Copolymer. *Polym. Int.* **1996**, *39*, 175–181. [CrossRef]
268. Kotnis, M.A.; O'Brien, G.S.; Willett, J.L. Processing and mechanical properties of biodegradable Poly(hydroxybutyrate-co-valerate)-starch compositions. *J. Environ. Polym. Degrad.* **1995**, *3*, 97–105. [CrossRef]
269. Imam, S.H.; Gordon, S.H.; Shogren, R.L.; Greene, R.V. Biodegradation of starch-poly(β-hydroxybutyrate-co-valerate) composites in municipal activated sludge. *J. Environ. Polym. Degrad.* **1995**, *3*, 205–213. [CrossRef]
270. Ajioka, M.; Enomoto, K.; Suzuki, K.; Yamaguchi, A. The basic properties of poly(lactic acid) produced by the direct condensation polymerization of lactic acid. *J. Environ. Polym. Degrad.* **1995**, *3*, 225–234. [CrossRef]
271. U.S. National Committee on Tunneling Technology Subcommittee on Contracting Practices for the Superconducting Super Collider. Contracting practices for the underground construction of the superconducting super collider. *Tunn. Undergr. Space Technol.* **1990**, *5*, 385–412. [CrossRef]
272. Chiang, C.-R.; Chang, F. Polymer blends of polyamide-6 (PA6) and poly(phenylene oxide) (PPO) compatibilized by styrene-maleic anhydride (SMA) copolymer. *Polymer* **1997**, *38*, 4807–4817. [CrossRef]
273. Nakajima, K.; Ohashi, M.; Miyajima, Y. Four-wave-mixing suppression effect of dispersion varying fiber. *IEEE Photonics Technol. Lett.* **1998**, *10*, 537–539. [CrossRef]
274. Gonsalves, K.; Wong, T.K.; Patel, S.H.; Chen, X.; Trivedi, D. Degradable materials for the marine environment. *Polym. Mater. Sci. Eng.* **1990**, *63*, 854.
275. Boggs, W.H. Carcinoma of the colon, rectum and anal canal in Negro patients: A review of 96 cases. *Dis. Colon Rectum* **1959**, *2*, 205–209. [CrossRef]
276. Dosmann, L.P.; Steel, R.N. To United States Rubber Co. U.S. Patent 3,004,934, 17 October 1961.
277. Bennett, F.L.; Otey, F.H.; Mehltretter, C.L. Rigid Urethane Foam Extended with Starch. *J. Cell. Plast.* **1967**, *3*, 369–373. [CrossRef]
278. Maddever, W.J.; Chapman, G. Modified starch-based biodegradable plastics. *Plast. Eng.* **1989**, *45*, 31–34.
279. Westhoff, R.P.; Otey, F.H.; Mehltretter, C.L.; Russell, C.R. Starch-Filled Polyvinyl Chloride Plastics-Preparation and Evaluation. *Prod. RD* **1974**, *13*, 123–125. [CrossRef]
280. Otey, F.H.; Westhoff, P.; Russell, C.R. Starch Based Plastics and Films. In Proceedings of the Technical Symposium Nonwoven Product Technology. International Nonwoven Disposables Association, Miami Beach, FL, USA, March 1975.
281. Yingfeng, Z.; Liu, W.; Wu, Y.; Li, P.; Xiao, J. Dialdehyde Starch and Preparation Method Thereof. Google Patents CN105646723A, 7 September 2018.
282. Paul, J.W.; Beauchamp, E.G. Nitrogen Flow on Two Livestock Farms in Ontario: A simple model to evaluate strategies to improve N utilization. *J. Sustain. Agric.* **1995**, *5*, 35–50. [CrossRef]
283. Ueda, T. Japan Patent 04 363 301 (92 363 301), 1992.
284. Watanabe, T.; Kuwabara, M.; Koshijima, T.; Ueda, M.; Nakajima, M. Preparation of oligosaccharides by enzymic transglycosidation of cellooligosaccharides with starch, etc. *Jpn. Kokai Tokkyo Koho.* **1994**, 8.
285. Jeremić, K.; Dunjić, B.; Djonlagić, J.; Jovanović, S. Blends of thermoplastic starch and some thermoplastic polymers. *J. Serb. Chem. Soc.* **1998**, *63*, 753–762.
286. Scott, G.; Gilead, D. *Degradable Polymers in the Waste and Litter Control, Degradable Polymers*, 1st ed.; Scott, G., Gilead, D., Eds.; Chapman & Hall: London, UK, 1995; Chapter 13; pp. 247–258.

287. Jane, J.-L.; Gelina, R.J.; Nikolov, Z.L.; Evangelista, R.L. Degradable Plastics from Octenyl Succinate Starch. U.S. Patent US5059642A, 22 October 1991.

288. Chinnaswami, R.; Hanna, M.A. Biodegradable Polymers. Google Patents EP0438585A1, 10 August 1989.

289. Zhang, S.; Wang, B.; Tang, J.Y. Processing and characterisation of diamondlike carbon films. *Surf. Eng.* **1997**, *13*, 303–309. [CrossRef]

290. Imam, S.H.; Chen, L.; Gordon, S.H.; Shogren, R.L.; Weisleder, D.; Greene, R.V. Biodegradation of injection molded starch-poly(3-hydroxybutyrate-co-3-hydroxyvalerate) blends in a natural compost environment. *J. Environ. Polym. Degrad.* **1998**, *6*, 91–98. [CrossRef]

291. Jeffs, H.J. Biodegradable Packaging Foam and Method of Preparation. U.S. Patent 5252271, 12 October 1993.

292. Bastioli, C.; Lombi, R.; Tredici, G.D.; Guanolla, I. Biodegradable Material for Injection Molding and Articles Obtained Therewith. European Patent WO2008071712A1, 19 June 2008.

293. Bastioli, C.; Giovanni, F.; Tredici, G.D. Biodegradable Compositions Based on Nanoparticulate Starch. Wo. Patent CA2662446A1, 29 July 2014.

294. Bastioli, C.; Bellotti, V.; Giudice, L.D.; Gilli, G. Microstructure and Biodegradability of Mater-bi Products. Biodegradable Polymers and Plastics. Wo. Patent 15 June 1992.

295. Buehler, F.S.; Schmid, E.; Schultze, H.J. Starch/Polymer Mixture, Process for the Preparation Thereof, and Products Obtainable Therefrom. U.S. Patent US5346936A, 13 September 1994.

296. Berruezo, M.; Luduena, L.N.; Rodriguez, E.; Alvarez, V.A. Preparation and characterization of polystyrene/starch blends for packaging applications. *J. Plast Film Sheet* **2014**, *30*, 141–161. [CrossRef]

297. Mamidi, N.; Zuníga, A.E.; Villela-Castrejón, J. Engineering and evaluation of forcespun functionalized carbon nano-onions reinforced poly (ε-caprolactone) composite nanofibers for pH-responsive drug release. *Mater. Sci. Eng. C* **2020**, *112*, 110928. [CrossRef]

298. Mamidi, N.; Delgadillo, R.M.V.; González-Ortiz, A. Engineering of carbon nano-onion bioconjugates for biomedical applications. *Mater. Sci. Eng. C* **2021**, *120*, 111698. [CrossRef]

299. Mamidi, N.; Delgadillo, R.M.V.; Castrejón, J.V. Unconventional and facile production of a stimuli-responsive multifunctional system for simultaneous drug delivery and environmental remediation. *Environ. Sci. Nano* **2021**, *8*, 2081–2097. [CrossRef]

300. Mamidi, N.; Romo, I.L.; Gutiérrez, H.M.L.; Barrera, E.V.; Elías-Zúñiga, A. Development of forcespun fiber-aligned scaffolds from gelatin-zein composites for potential use in tissue engineering and drug release. *MRS Commun.* **2018**, *8*, 885–892. [CrossRef]

301. Mamidi, N.; Delgadillo, R.M.V.; Barrera, E.V. Covalently Functionalized Carbon Nano-Onions Integrated Gelatin Methacryloyl Nanocomposite Hydrogel Containing γ-Cyclodextrin as Drug Carrier for High-Performance pH-Triggered Drug Release. *Pharmaceuticals* **2021**, *14*, 291. [CrossRef] [PubMed]

302. Cole, D.M.; Ringelberg, D.B.; Reynolds, C.M. Small-Scale Mechanical Properties of Biopolymers. *J. Geotech. Geoenviron. Eng.* **2012**, *138*, 1063–1074. [CrossRef]

303. Nam, T.H.; Ogihara, S.; Tung, N.H.; Kobayashi, S. Effect of alkali treatment on interfacial and mechanical properties of coir fiber reinforced poly(butylene succinate) biodegradable composites. *Compos. Part B Eng.* **2011**, *42*, 1648–1656. [CrossRef]

304. Gallo, E.; Schartel, B.; Acierno, D.; Cimino, F.; Russo, P. Tailoring the flame retardant and mechanical performances of natural fiber-reinforced biopolymer by multi-component laminate. *Compos. Part B Eng.* **2013**, *44*, 112–119. [CrossRef]

305. Motru, S.; Adithyakrishna, V.H.; Bharath, J.; Guruprasad, R. Development and Evaluation of Mechanical Properties of Biodegradable PLA/Flax Fiber Green Composite Laminates. *Mater. Today Proc.* **2020**, *24*, 641–649.

306. Russo, P.; Carfagna, C.; Cimino, F.; Acierno, D.; Persico, P. Biodegradable Composites Reinforced with Kenaf Fibers: Thermal, Mechanical, and Morphological Issues. *Adv. Polym. Technol.* **2013**, *32*, E313–E322. [CrossRef]

307. Alvarez, V.A.; Fraga, A.N.; Vázquez, A. Effects of the moisture and fiber content on the mechanical properties of biodegradable polymer-sisal fiber biocomposites. *J. Appl. Polym. Sci.* **2004**, *91*, 4007–4016. [CrossRef]

308. Liang, Z.; Pan, P.; Zhu, B.; Dong, T.; Inoue, Y. Mechanical and thermal properties of poly(butylene succinate)/plant fiber biodegradable composite. *J. Appl. Polym. Sci.* **2010**, *115*, 3559–3567. [CrossRef]

309. Zhu, Z.; Wu, H.; Ye, C.; Fu, W. Enhancement on Mechanical and Thermal Properties of PLA Biocomposites Due to the Addition of Hybrid Sisal Fibers. *J. Nat. Fibers* **2017**, *14*, 875–886.

310. Jung, G.; Qin, Z.; Buehler, M.J. Mechanical properties and failure of biopolymers: Atomistic reactions to macroscale response. *Polym. Mechanochem.* **2015**, *369*, 317–343.

311. Bahrami, M.; Abenojar, J.; Ángel Martínez, M. Recent Progress in Hybrid Biocomposites: Mechanical Properties, Water Absorption, and Flame Retardancy. *Materials* **2020**, *13*, 5145.

312. Kremensas, A.; Kairytė, A.; Vaitkus, S.; Vėjelis, S.; Balčiūnas, G. Mechanical Performance of Biodegradable Thermoplastic Polymer-Based Biocomposite Boards from Hemp Shivs and Corn Starch for the Building Industry. *Materials* **2019**, *12*, 845. [CrossRef]

313. Khan, M.U.A.; Razak, S.I.A.; Arjan, W.S.A.; Nazir, S.; Anand, T.J.S.; Mehboob, H.; Amin, R. Recent Advances in Biopolymeric Composite Materials for Tissue Engineering and Regenerative Medicines: A Review. *Molecules* **2021**, *26*, 619. [CrossRef]

314. Baldino, L.; Cardea, S.; Reverchon, E. Supercritical Assisted Electrospray: An Improved Micronization Process. *Polymers* **2019**, *11*, 244.

315. Sharma, C. Fabrication and characterization of novel nano- biocomposite scaffold of chitosan-gelatin-alginate-hydroxyapatite for bone tissue engineering. *Mater. Sci. Eng. C* **2016**, *64*, 416–427. [CrossRef] [PubMed]

316. Kim, H.L. Preparation and characterization of nano-sized hydroxyapatite/alginate/chitosan composite scaffolds for bone tissue engineering. *Mater. Sci. Eng. C* **2015**, *54*, 20–25.

317. Xinluan, W.; Yuxiao, L.; Helena, N.; Zhijun, Y.; Ling, Q. Systemic Drug Delivery Systems for Bone Tissue Regeneration—A Mini Review. *Curr. Pharm. Des.* **2015**, *21*, 1575–1583. [CrossRef]

318. Grumezescu, A. *Nanobiomaterials in Hard Tissue Engineering: Applications of Nanobiomaterials*; William Andrew: Norwich, NY, USA, 2016.

319. Puiggali, J. Other Miscellaneous Materials and Their Nanocomposites. In *Nanomaterials and Polymer Nanocomposites- Raw Materials to Applications*; Elsevier Publications: Philadelphia, PA, USA, 2019.

320. Gisela, S.A. *Bionanocomposites: Integrating Biological Processes for Bioinspired Nanotechnologies*; Wiley & Sons, Inc.: Paris, France, 2017.

321. Anadao, P. Polymer/Clay Nanocomposites: Concepts, Researches, Applications and Trends for the Future. In *Nanocomposites-New Trends and Developments*; Intech Publications: Rijeka, Croatia, 2012.

322. Latha, S.; Selvamani, P.; Prabu, C. Preparation of Nanocomposite Hydrogels and it's in vitro Release Behaviour on Captopril. *Asian J. Chem.* **2013**, *25*, 7211–7215. [CrossRef]

323. Karak, N. Fundamentals of Nanomaterials and Polymer Nanocomposites. In *Nanomaterials and Polymer Nanocomposites- Raw Materials to Applications*; Elsevier Publications: Philadelphia, PA, USA, 2019.

324. Bhat, M.R. Bionanocomposites: Technique towards Enhancement of Solubility, Drug Release and Bioavailability. *J. Med. Pharm. Innov.* **2015**, *2*, 6–18.

325. Paul, D.R.; Robeson, L.M. Polymer nanotechnology: Nanocomposites. *Polymer* **2008**, *49*, 3187–3204. [CrossRef]

326. Kushare, S.S.; Gattani, S.G. Microwave-generated bionanocomposites for solubility and dissolution enhancement of poorly water-soluble drug glipizide: In-vitro and in-vivo studies. *J. Pharm. Pharmacol.* **2012**, *65*, 79–93. [CrossRef] [PubMed]

327. Mukhija, U.; Soni, N.; Chawla, A.; Bhatt, D.C. Physical Properties and Dissolution Behaviour of Meloxicam/Poloxamer Solid Dispersions Prepared by Hot Melt Method and Microwave Assisted Method. *Int. J. Res. Pharm. Sci.* **2012**, *2*, 64–74.

328. Papadimitriou, S.A.; Bikiaris, D.; Avgoustakis, K. Microwave-induced enhancement of the dissolution rate of poorly water-soluble tibolone from poly(ethylene glycol) solid dispersions. *J. Appl. Polym. Sci.* **2008**, *108*, 1249–1258. [CrossRef]

329. Moneghini, M.; Bellich, B.; Baxa, P.; Princivalle, F. Microwave generated solid dispersions containing Ibuprofen. *Int. J. Pharm.* **2008**, *361*, 125–130. [CrossRef]

330. Bergese, P.; Colombo, I.; Gervasoni, D.; Depero, L.E. Microwave generated nanocomposites for making insoluble drugs soluble. *Mater. Sci. Eng. C.* **2003**, *23*, 791–795. [CrossRef]

331. Maurya, D.; Belgamwar, V.; Tekade, A. Microwave induced solubility enhancement of poorly water-soluble atorvastatin calcium. *J. Pharm. Pharmacol.* **2010**, *62*, 1599–1606. [CrossRef]

332. Yuen, M.; Yip, J. Chitosan microcapsules loaded with either Miconazole nitrate or Clotrimazole, prepared via emulsion technique. *Carbohydr. Polym.* **2012**, *89*, 795–801. [CrossRef]

333. Salomy, M.; Gautami, J. Design and Evaluation of topical formulation of diclofenac sodium for improved therapy. *Int. J. Pharm. Sci. Res.* **2015**, *5*, 1973.

334. Carrow, J.K.; Gaharwar, A.K. Bioinspired Polymeric Nanocomposites for Regenerative Medicine. *Macromol. Chem. Phys.* **2015**, *216*, 248–264. [CrossRef]

335. Gattani, S.G.; Patwekar, S.L. Solubility and Dissolution Enhancement of Poorly Water-soluble Ketoprofen by Microwave-assisted Bionanocomposites: In Vitro and In Vivo Study. *Asian J. Pharm.* **2016**, *10*, 601–611.

336. Huang, X.; Brazel, C.S. On the importance and mechanisms of burst release in matrix-controlled drug delivery systems. *J. Control. Release* **2001**, *73*, 121–136. [CrossRef]

337. Gomes, D.; Pereira, M.; Bettencourt, A.F. Osteomyelitis: An overview of antimicrobial therapy. *Braz. J. Pharm. Sci.* **2013**, *49*, 13–27. [CrossRef]

338. LogithKumar, R.; KeshavNarayan, A.; Dhivya, S.; Chawla, A.; Saravanan, S.; Selvamurugan, N. A review of chitosan and its derivatives in bone tissue engineering. *Carbohydr. Polym.* **2016**, *151*, 172–188. [CrossRef]

339. Rukavina, Z.; Vanić, Ž. Current Trends in Development of Liposomes for Targeting Bacterial Biofilms. *Pharmaceutics* **2016**, *8*, 18. [CrossRef]

340. Hornyák, I.; Madácsi, E.; Kalugyer, P.; Vácz, G.; Horváthy, D.B.; Szendrői, M.; Han, W.; Lacza, Z. Increased Release Time of Antibiotics from Bone Allografts through a Novel Biodegradable Coating. *BioMed Res. Int.* **2014**, *2014*, 459867. [CrossRef]

341. Venkatesan, J. Alginate composites for bone tissue engineering: A review. *Int. J. Biol. Macromol.* **2015**, *72*, 269–281. [CrossRef]

342. Ivanova, E.; Bazaka, K.; Crawford, R. Natural polymer biomaterials: Advanced applications. *New Funct. Biomater. Med. Healthc.* **2014**, *1*, 32–70.

343. González-Sánchez, M.I. Silver nanoparticle based antibacterial methacrylate hydrogels potential for bone graft applications. *Mater. Sci. Eng. C* **2015**, *50*, 332–340. [CrossRef] [PubMed]

344. Gu, W. Nanotechnology in the targeted drug delivery for bone diseases and bone regeneration. *Int. J. Nanomed.* **2013**, *8*, 2305–2317. [CrossRef]

345. Subia, B.; Dey, T.; Sharma, S.; Kundu, S.C. Target Specific Delivery of Anticancer Drug in Silk Fibroin Based 3D Distribution Model of Bone–Breast Cancer Cells. *ACS Appl. Mater. Interfaces* **2015**, *7*, 2269–2279. [CrossRef]

346. El-Kady, A.M.; Farag, M.M.; El-Rashedi, A.M. Bioactive glass nanoparticles designed for multiple deliveries of lithium ions and drugs: Curative and restorative bone treatment. *Eur. J. Pharm. Sci.* **2016**, *91*, 243–250. [CrossRef]

347. Sun, W. Bone targeted mesoporous silica nanocarrier anchored by zoledronate for cancer bone metastasis. *Langmuir* **2016**, *32*, 9237–9244. [CrossRef]

348. Yang, L. *Nanotechnology-Enhanced Orthopedic Materials: Fabrications. Applications and Future Trends*; Woodhead Publishing: Sawston, Cambridge, UK, 2015.

349. Tanzawa, Y. Potentiation of the antitumor effect of calcium phosphate cement containing anticancer drug and caffeine on rat osteosarcoma. *J. Orthop. Sci.* **2011**, *16*, 77–84. [CrossRef]

350. Wang, F. RGD peptide conjugated liposomal drug delivery system for enhance therapeutic efficacy in treating bone metastasis from prostate cancer. *J. Control. Release* **2014**, *196*, 222–233. [CrossRef]

351. Andronescu, E. Collagen-hydroxyapatite/cisplatin drug delivery systems for locoregional treatment of bone cancer. *Technol. Cancer Res. Treat.* **2013**, *12*, 275–284. [CrossRef]

352. Zhang, J. 3D-printed magnetic Fe 3 O 4/MBG/PCL composite scaffolds with multifunctionality of bone regeneration, local anticancer drug delivery and hyperthermia. *J. Mater. Chem. B* **2014**, *2*, 7583–7595. [CrossRef] [PubMed]

353. Deepthi, S. An overview of chitin or chitosan/nano ceramic composite scaffolds for bone tissue engineering. *Int. J. Biol. Macromol.* **2016**, *93*, 1338–1353. [CrossRef]

354. Paris, J. Tuning dual-drug release from composite scaffolds for bone regeneration. *Int. J. Pharm.* **2015**, *486*, 30–37. [CrossRef]

355. Baradari, H.; Damia, C.; Dutreih-Colas, M.; Champion, E.; Chulia, D.; Viana, M. β-TCP porous pellets as an orthopaedic drug delivery system: Ibuprofen/carrier physicochemical interactions. *Sci. Technol. Adv. Mater.* **2011**, *12*, 055008. [CrossRef]

356. Xiao, J. An asymmetric coating composed of gelatin and hydroxyapatite for the delivery of water insoluble drug. *J. Mater. Sci. Mater. Med.* **2009**, *20*, 889–896. [CrossRef]

357. Lin, M. Structure and release behavior of PMMA/silica composite drug delivery system. *J. Pharm. Sci.* **2007**, *96*, 1518–1526. [CrossRef] [PubMed]

358. Di Marzio, L.; Ventura, C.A.; Cosco, D.; Paolino, D.; Di Stefano, A.; Stancanelli, R.; Tommasini, S.; Cannava, C.; Celia, C.; Fresta, M. Nanotherapeutics for anti-inflammatory delivery. *J. Drug Deliv. Sci. Technol.* **2016**, *32*, 174–191. [CrossRef]

359. Safari, J.; Zarnegar, Z. Advanced drug delivery systems: Nanotechnology of health design A review. *J. Saudi Chem. Soc.* **2014**, *18*, 85–99. [CrossRef]

360. Perez, R.A.; Kim, H.W. Core–shell designed scaffolds for drug delivery and tissue engineering. *Acta Biomater.* **2015**, *21*, 2–19. [CrossRef]

361. Ali, S.S.S.; Razman, M.R.; Awang, A.; Asyraf, M.R.M.; Ishak, M.R.; Ilyas, R.A.; Lawrence, R.J. Critical Determinants of Household Electricity Consumption in a Rapidly Growing City. *Sustainability* **2021**, *13*, 4441. [CrossRef]

362. Asyraf, M.R.; Rafidah, M.; Azrina, A.; Razman, M.R. Dynamic mechanical behaviour of kenaf cellulosic fibre biocomposites: A comprehensive review on chemical treatments. *Cellulose* **2021**, *28*, 2675–2695. [CrossRef]

363. Ilyas, R.A.; Sapuan, S.M.; Asyraf, M.R.M.; Dayana, D.A.Z.N.; Amelia, J.J.N.; Rani, M.S.A.; Norrrahim, M.N.F.; Nurazzi, N.M.; Aisyah, H.A.; Sharma, S.; et al. Polymer Composites Filled with Metal Derivatives: A Review of Flame Retardants. *Polymers* **2021**, *13*, 1701. [CrossRef]

364. Ilyas, R.A.; Sapuan, S.M.; Harussani, M.M.; Hakimi, M.Y.A.Y.; Haziq, M.Z.M.; Atikah, M.S.N.; Asyraf, M.R.M.; Ishak, M.R.; Razman, M.R.; Nurazzi, N.M.; et al. Polylactic Acid (PLA) Biocomposite: Processing, Additive Manufacturing and Advanced Applications. *Polymers* **2021**, *13*, 1326. [CrossRef]

365. Asyraf, M.R.; Ishak, M.R.; Sapuan, S.M.; Yidris, N.; Ilyas, R.A.; Rafidah, M.; Razman, M.R. Potential Application of Green Composites for Cross Arm Component in Transmission Tower: A Brief Review. *Int. J. Polym. Sci.* **2020**, *2020*, 8878300. [CrossRef]

366. Chohan, J.S.; Mittal, N.; Kumar, R.; Singh, S.; Sharma, S.; Dwivedi, S.P.; Saxena, A.; Chattopadhyaya, S.; Ilyas, R.A.; Le, C.H.; et al. Optimization of FFF Process Parameters by Naked Mole-Rat Algorithms with Enhanced Exploration and Exploitation Capabilities. *Polymers* **2021**, *13*, 1702. [CrossRef] [PubMed]

367. Chohan, J.S.; Mittal, N.; Kumar, R.; Singh, S.; Sharma, S.; Singh, J.; Rao, K.V.; Mia, M.; Pimenov, D.Y.; Dwivedi, S.P. Mechanical Strength Enhancement of 3D Printed Acrylonitrile Butadiene Styrene Polymer Components Using Neural Network Optimization Algorithm. *Polymers* **2020**, *12*, 2250. [CrossRef] [PubMed]

368. Singh, Y.; Singh, J.; Sharma, S. Multi-objective Optimization of Kerf-taper and Surface-roughness Quality Characteristics for Cutting-operation On Coir and Carbon Fibre Reinforced Epoxy Hybrid Polymeric Composites During CO_2-Pulsed Laser-cutting Using RSM. *Lasers Manuf. Mater. Process.* **2021**, *8*, 157–182. [CrossRef]

369. Sharma, S.; Singh, J.; Kumar, H.; Sharma, A.; Aggarwal, V.; Gill, A.; Jayarambabu, N.; Kailasa, S.; Rao, K.V. Utilization of rapid prototyping technology for the fabrication of an orthopedic shoe inserts for foot pain reprieve using thermo-softening viscoelastic polymers: A novel experimental approach. *Meas. Control.* **2020**, *53*, 519–530. [CrossRef]

370. Singh, Y.; Singh, J.; Sharma, S.; Sharma, A.; Chohan, J. Process Parameter Optimization in Laser Cutting of Coir Fiber Reinforced Epoxy Composite—A Review. *Mater. Today Proc.* **2021**, in press. [CrossRef]

371. Chohan, J.S.; Kumar, R.; Singh, T.B.; Singh, S.; Sharma, S.; Singh, J.; Mia, M.; Pimenov, D.Y.; Chattopadhyaya, S.; Dwivedi, S.P.; et al. Taguchi S/N and TOPSIS Based Optimization of Fused Deposition Modelling and Vapor Finishing Process for Manufacturing of ABS Plastic Parts. *Materials* **2020**, *13*, 5176. [CrossRef]

 *polymers*

Article

High-Density Bio-PE and Pozzolan Based Composites: Formulation and Prototype Design for Control of Low Water Flow

Nicola Schiavone, Vincent Verney and Haroutioun Askanian *

CNRS, Clermont Auvergne INP, ICCF, Université Clermont Auvergne, F-63000 Clermont-Ferrand, France; nicola.schiavone@sigma-clermont.fr (N.S.); vincent.verney@uca.fr (V.V.)
* Correspondence: haroutioun.askanian@sigma-clermont.fr

Abstract: An eco-friendly solution to produce new material for the material extrusion process is to use quarry waste as filler for biopolymer composites. A quarry waste that is still studied little as a filler for polymer composites is pozzolan. In this study, the optimization of the formulations and processing parameters of composites produced with pozzolan and bio-based polyethylene for 3D printing technology was performed. Furthermore, a precision irrigation system in the form of a drip watering cup was designed, printed, and characterized. The results showed that the presence of the pozzolan acted as a reinforcement for the composite material and improved the cohesion between the layers of the 3D printed objects. Furthermore, the optimization of the process conditions made it possible to print pieces of complex geometry and permeable parts for the control of the water flow rates with an order of magnitude in the range from mL/h to mL/day.

Keywords: 3D printing; bio-based polyethylene composite; X-ray tomography

check for
updates

Citation: Schiavone, N.; Verney, V.; Askanian, H. High-Density Bio-PE and Pozzolan Based Composites: Formulation and Prototype Design for Control of Low Water Flow. *Polymers* **2021**, *13*, 1908. https://doi.org/10.3390/polym13121908

Academic Editors: Emin Bayraktar, S. M. Sapuan and R. A. Ilyas

Received: 5 May 2021
Accepted: 5 June 2021
Published: 8 June 2021

Publisher's Note: MDPI stays neutral with regard to jurisdictional claims in published maps and institutional affiliations.

1. Introduction

The study and development of the material extrusion process as an additive manufacturing technique has never stopped since the first version presented by Scott Crump at Stratasys Inc. in 1989 [1]. Due to its low cost, ease of use, and the possibility to produce very complex customizable parts, the material extrusion process has consolidated its presence in the manufacturing market [2].

There are many domains where additive manufacturing can be used, such as aeronautics, automotive and medical applications. Among these domains, the field of agriculture equipment has a lot of potential, and few studies have been conducted in relation to it. In this field, precision irrigation (a subgroup of precision agriculture) creates and optimizes systems to control the irrigation of the plants, depending on the environmental and operational context, and to avoid water and energy waste [3–10].

In terms of 3D printing filament materials, acrylonitrile butadiene styrene (ABS) and polylactic acid (PLA) are the polymers filaments that dominate the market of 3D printing due to their availability and suitability from the point of view of adequate mechanical properties and dimensional accuracy of the final printed object [11,12]. Other common polymer filaments used for 3D printing fabrication are polyamide (PA), high-impact polystyrene (HIPS), polycarbonate (PC), and polyether ether ketone (PEEK), which are adopted for final application where higher mechanical properties and thermal stability are needed [13–16].

Compared to common polymer filaments for 3D printing, a polymeric matrix with competitive properties such as excellent impact properties, high chemical stability and excellent electrical insulation is the polyethylene (PE) matrix [17].

While PE dominated the polymers market with a share of 25.7% in 2019, and despite the great interest by both industries and researchers in improving the performance and functionality of 3D printed objects, there is a lack of information about the 3D printing of polyethylene, with only a few studies conducted on polyethylene used as filaments for the material extrusion process [18–23]. An interesting grade of PE that can be used as a 3D printing polymer filament is bio-based polyethylene (BioPE). This type of polyethylene can be synthesized from biomass (e.g., sugar cane, sugar beet and wheat grain). Its chemical structure is identical to petroleum-based polyethylene. It is industrially available and has a great chance to be used for eco-friendly future plastic products [24]. The characteristics of polymer materials which are required for proper 3D printing can be summarized as follows: sufficient stiffness of the filament to avoid Euler buckling, a melting temperature below the upper limit of the 3D printing extruder, adequate thermorheological properties to maintain the shape during cooling and having optimal cohesion between the layers of the printed parts [25]. Concerning these aspects, polyethylene presents some issues regarding suitability for the material extrusion process. For example, as a semi-crystalline polymer, PE is characterized by a strong volume change during cooling, which can lead to warpage of the final object and losing the correct shape with weak interlayer welding. Moreover, this polymer has low adhesion with most of the materials that characterize the printer beds, causing detachment of the object from the printer bed during printing [26–28]. Consequently, it is still a challenge to properly print a 3D final object with polyethylene, and more knowledge on the optimal printing conditions is needed to obtain a comparable final piece as a common 3D printing polymer filament. A possible way to improve the dimensional accuracy and welding layers of printed parts is the use of inorganic fillers by developing new composites. The presence of the fillers affects the viscoelastic behavior and thermal expansion, reducing the mobility of the polymeric chains during cooling [29]. The optimization of a polymer composite also has the objective of maximizing the filler content, limiting the use of the polymeric matrix and therefore of plastic waste. However, the use of a composite with a high filler content makes it challenging for the extrusion process to obtain a uniform diameter filament and for the printing, since it is possible for the nozzle to become clogged [30–32].

Pozzolan powder is an inorganic filler not commonly used for polymer composites. It is a pyroclastic rock extensively used in different sectors such as construction, buildings and roads, sanitation, and agriculture. In the agriculture field, pozzolan is used for drainage, soil amendments, substrate crops, and the restoration of soil. This rock has a natural or artificial origin, and the composition is rich in silicon dioxide and aluminum oxide. It has a porous form and high thermal stability [33–37].

The pozzolan separation process generates a huge amount of a lateral fine fraction of pozzolan. The generated quantity could achieve 50% of the total treated pozzolan mass. Most generated fine fractions are considered by-products and have a very limited market. Each year, there is an accumulation of many thousand tons in each carrier [38]. At the same time, the directives of the quarries department specify that all the extracted pozzolan must be destined for a specific use [39]. Currently, for environmental issues, the world is heading toward zero-waste production in every field and zero or even a positive environmental impact. Pozzolan is a nontoxic material, and it could be used as a fertilizer in soil [40]. Consequently, the valorization of this by-product as composite filler can be a solution to developing new eco-friendly materials. In this domain, only a few studies have focused on the valorization of natural inorganic material by-products as filler for bio-based polymers intended for 3D printing [35,41–43].

In this work, a new composite based on high-density biopolyethylene and pozzolan by-product is studied. Four formulations are produced in the form of 3D printing filaments for the material extrusion process and analyzed through thermal and rheological characterization. Subsequently, the obtained composite filaments are used for the printing of mechanical specimens and the printing of a drip watering close system prototype. The prototype is characterized by X-ray tomography for morphological analysis, and the results are compared with the water flow measurements obtained by a simulation test in real conditions.

2. Materials and Methods

2.1. Raw Materials

Pozzolan waste was obtained from Pouzzolanes des domes S.A.S (Le Vauriat, St Ours des Roches, France). It is an industrial by-product and, in accordance with the Total Alkali Silica (TAS) classification, was placed at the lower limit of trachybasalts [39,40]. The powder was red pozzolan and was part of storage on the quarry site. It was obtained by process separation after heating for 15 min at a temperature of 600 °C. The De Brouckere mean diameter of the powder was 56 μm, and the specific surface was 2939 cm^2/g. The bio-based bimodal high-density polyethylene (HDPE) SGE 7252 was used as a polymer matrix and purchased by Braskem (São Paulo, Brazil). The density and melt flow index (190 °C/2.16 kg) values were 0.952 g/cm^3 and 2 g/10 min, respectively.

2.2. Preparation of Composite Pellets and the Filaments for 3D Printing

The preparation of composite pellets and 3D printing filaments was performed with a Thermo Fisher Scientific Pharma 11 twin-screw extruder (7 heat/cool zones plus 1 heating zone for the die; L/D ratio = 40:1) (Figure 1) (Waltham, MA, USA). The process conditions are reported in Table 1. For production of the composites pellets, the feed for the pozzolan was placed between Zones 6 and 7 of the extruder, and the mass flow rate was set up to attain different mass ratios of pozzolan equal to 0%, 20%, 40%, and 60%. To produce a 3D printing filament with a regular diameter of 1.75 mm, a second extrusion was performed using the composite pellets obtained previously. A second extrusion was performed in order to avoid a second feeding (powder feeding), which would lead to flow instability and less regularity of the polymeric melt at the exit of the extrusion die. The sample code was resin type-%PR, where %PR was the value of the pozzolan mass percentage.

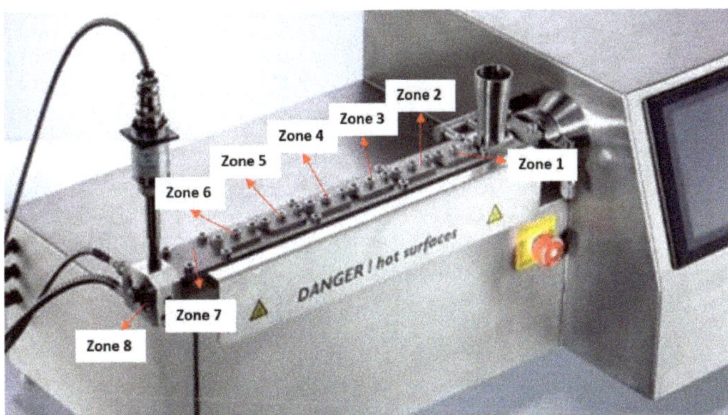

Figure 1. Schematic illustration of the extruder zones.

Table 1. Extrusion conditions for production of the filaments.

Heating Zone	Z1	Z2	Z3	Z4	Z5	Z6	Z7	Z8
Temperature (°C)	180	220	220	220	230	240	240	220
Screw Rate (rpm)					200			
Total Mass Flow (kg/h)					0.6			

2.3. Thermal Characterization

To verify the amount of pozzolan and analyze the effect of the fillers on the thermal stability, thermogravimetric analysis was carried out for all samples. For this purpose, a PerkinElmer TGA 4000 (Waltham, MA, USA) in the range from 25 to 600 °C under an N$_2$ atmosphere at a rate of 10 °C/min was utilized.

The crystallinity and thermal behavior of the samples studied were investigated by a METTLER TOLEDO DSC 3+ differential scanning calorimetry (Columbus, OH, USA) under a nitrogen atmosphere. The DSC thermograms were obtained by using 8–10 mg of material for each sample. A heat–cool–heat cycle experimental method was, used and the first and second heating stages were set from room temperature to 180 °C at 10 °C/min, while the cooling stage went from 180 °C to room temperature at 10 °C/min. The percentage of crystallinity X_c of the samples was calculated using the following equation:

$$X_c = (\Delta H_m - \Delta H_{cc})/(\Delta H_{m0} - (w/w)_p) \times 100 \tag{1}$$

where ΔH_m, ΔH_{cc}, ΔH_{m0}, and $(w/w)_p$ are the melting enthalpy, the cold crystallization enthalpy, the melting enthalpy of pure crystalline polymers, and the mass fraction of the polymer in the matrix, respectively. The ΔH_{m0} taken was 286.7 J/g for the HDPE [44].

In order to compare the crystallization kinetics of the composites as a function of the pozzolan contents, Avrami analysis was performed, and the half-crystallization time $(t_{1/2})$ was measured for each sample. The thermal cycle was as follows: heating from 30 °C to 180 °C at 40 °C/min, remaining for 3 min at 180 °C to uniform the temperature in the sample, and then cooling down to 118 °C (crystallization temperature) at 40 °C/min [45].

2.4. Rheological Measurements

The loss viscosity (η') and storage viscosity (η'') of the melted composites were measured using a shear dynamic experiment. The tests were carried out with an ARES rheometer manufactured by TA Instruments/Waters Corporation (Milford, MA, USA). The frequency sweeps method was chosen, using parallel plate geometry with a diameter of 8 mm and a gap height of 1 mm. The deformation value was chosen following the verification of the linear viscoelasticity range, which was performed with the strain sweep method at a frequency of 10 rad/s. The temperatures used for the test were 140 °C, 150 °C, 170 °C, and 180 °C, while the strain and frequency range were set at 10% and between 0.1 and 100 rad/s, respectively. The Newtonian zero shear viscosity η_0 could be determined from the extrapolation of the arc of a circle plotted from the experimental data, which is characteristic for a Cole–Cole distribution [46].

The effect of the pozzolan on the volumetric contraction stress of the composites was analyzed with the force gap measurement as a function of the temperature and time with a force gap test in a sequential configuration. The gap was fixed at 1 mm, and the temperature profile was characterized by a series of steps of 5 °C each. The initial and final temperatures were 200 °C and 100 °C, respectively, and a duration of 2 min was set for each step [47].

2.5. 3D-Printed Specimens and Drip Watering Close System Prototype Preparation

The 3D printing filament composites were used for printing the specimens for the mechanical tests and the prototypes of close watering systems. The printer used was the Prusa i3 MK3S (Prague, Czech Republic), and the g-code files were elaborated through PrusaSliser software. The geometry of the specimens was selected according to an ASTM D638 for the tensile test and an ASTM D256 for the Charpy impact test. All the specimens were printed on a polypropylene plate. The filling rate and the infill pattern used for the mechanical test samples were set to 100% and linear at an angle of ±45° to the longitudinal axis, respectively. The drip watering prototypes were printed starting from the models developed with the 3D CAD software Autodesk Fusion 360 (Mill Valley, CA, USA). The system was conceived as a bottle cap having a geometrical structure that facilitated insertion into the soil, anchoring the system. Moreover, a permeable porous structure was obtained by using a gyroid 3D printing infill pattern without walls. For the analysis of the porous structure effect on the liquid water flow, three print sets with different infill densities (IDs) of the infill pattern were used, which were equal to 50%, 60%, and 70%.

The printing temperature of the nozzle used in this work was identified to be in the range of 250–270 °C. The used temperature value was set to ensure a suitable viscosity for the extrusion of the HDPE composites and to avoid the material clogging at the nozzle. Polyethylene required particular attention due to the poor adhesion with the materials used for the commercial printer plates and the evident shrinkage during cooling. An atactic polypropylene film was used as a printer bed in order to ensure adhesion with the printed HDPE through the interdiffusion of macromolecular chains, which occurs mainly in the amorphous phase [48,49]. A 3D-printed specimen is shown in Figure 2a.

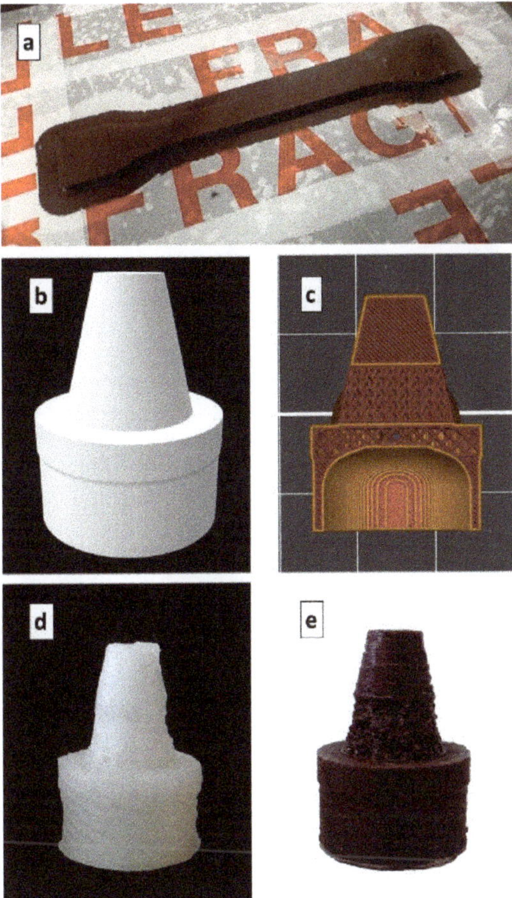

Figure 2. (**a**) Tensile specimen printed with the pozzolan composite. (**b**) 3D model of the prototype for precision irrigation. (**c**) Longitudinal section of the sliced prototype. (**d**) Prototype sample 3D printed with neat HDPE. (**e**) Prototype sample 3D printed with composite at 20% pozzolan.

Another aspect to consider was the cooling rate of the material for the solidification control during printing. The printing of small parts is a critical point, since the added new layers may not have sufficient time to solidify and retain their shape. In the case of HDPE composites, due to the low glass transition temperature, to reach an adequate viscosity of the deposited material, it is important to reach temperatures near crystallization rapidly; otherwise, the geometry is not preserved. In terms of process conditions, the use of an air fan can control the cooling of the material by setting a correct fan rotation speed. For the samples produced in this work, a fan was set 0% of its maximum speed for the first 20 layers and 90% for the rest of the layers.

The 3D model of the system is shown in Figure 2b, and the process conditions used for printing are reported in Table 2.

Table 2. 3D printing process parameters.

Parameter	Value
Nozzle diameter (mm)	0.6
Nozzle temperature (°C)	265
Layer thickness (mm)	0.15
Bed temperature (°C)	35
Printing speed for the first layer (mm/s)	20
Printing speed for the other layers (mm/s)	20

2.6. Mechanical Characterization

The tensile properties were evaluated according to an ASTM D638 IV (West Conshohocken, PA, USA), using a Lloyd EZ50 mechanical test machine (Bognor Regis, UK) at a cross-head speed of 30 mm/min. Tests were carried out at room temperature, and at least five specimens were tested for each sample. The size of the specimens was 115 ± 0.2 mm in length, 10 ± 0.1 mm wide, and 4 ± 0.05 mm thick. The impact properties, according to an ASTM D256 (West Conshohocken, PA, USA)) using Zwick/Roell HIT pendulum impact testers (Ulma, Germany) with a pendulum of 50J in the Charpy configuration, were also evaluated. The size of the impact specimens was 55 ± 0.1 mm in length, 10 ± 0.1 mm wide, and 4 ± 0.06 mm thick. All the results were averaged to obtain a mean value.

2.7. Characterization of the Prototype System

2.7.1. X-ray Tomography

To analyze the morphological and lactic structure of the printed prototype, a Skycan 1174 (Edinburgh, UK) was used for X-ray radiography and tomography. Each sample was placed on a rotating plate while the X-ray beam passed through. The images were recorded by a CCD camera with a resolution of 1024×1024 pixels, which revealed the different levels of X-ray absorption of the sample. The total exposure time for each sample was 450 s, and the pixel size was 29.7 µm. Two images were taken per angular position and were averaged. After the reconstruction of the 3D structure part, CT analysis software was used to measure the total porosity of the internal section of the permeable part of the drip watering prototype. Furthermore, the isometric projections were obtained using DataViewer software.

2.7.2. Measurement and Analysis of the Water Flow in Real Conditions
Water Flow Measurement System

The measurement of the water flow rate of the drip watering prototype was carried out with the systems shown in Figure 3. The system ias characterized by the prototype of a watering cap placed on a PET bottle filled with 0.5 L of water and a beaker containing commercial soil. The adopted experimental method consisted of weighing the beaker/soil system at different times, following the passage of water from the bottle to the soil through the permeable volume of the prototype. The bottle/prototype system was maintained by a fixed support, while the beaker/soil system was placed on a mobile support so as to not manipulate the bottle and modify the internal pressure. Each measurement was made using a volume of soil equal to 90 mL, and this was repeated three times. The saturation point of the soil with a volume of 90 mL was equal to 30 ± 2 mL of water.

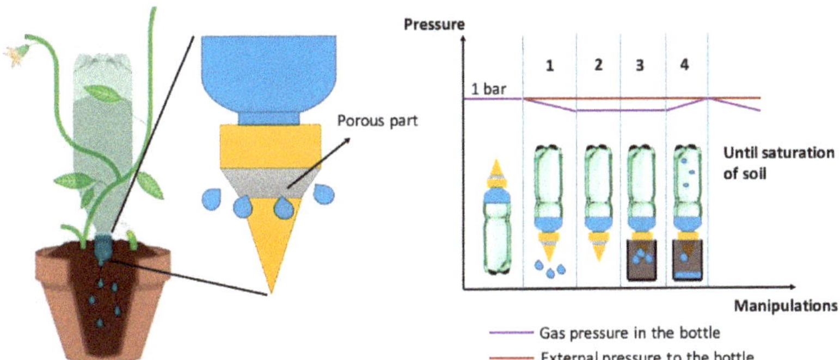

Figure 3. Simplified diagram of the operating principle of the drip watering system.

Analysis of the Water Flow through the Prototype System

A mathematical model was developed to quantitatively compare the results obtained by the water flow measurement. However, this paper goes beyond a detailed study of the fluid dynamics of the considered prototype. For this reason, simplifying hypotheses were considered, which will be explained in this paragraph.

The total flow of the water from the bottle to the soil was defined by considering a mass conservation balance.

The mass balance was simplified by taking into account only the resistance to water transport due to the permeable part of the prototype. The liquid water flow can be described as [50]

$$d(\rho\, V)/d(t) = \rho\, Q_D \tag{2}$$

where $d(\rho\, V)/d(t)$ is the transitory term related to the accumulation of water in the volume of the soil, ρ is the density of the water (kg/m^3), V is the volume of water absorbed by the soil (m^3), and Q_D is the volumetric flow rates of water through the permeable part of the prototype.

In particular, considering the laminar flow condition and the hypotheses on single-phase fluid flow, the term Q_D can be described by Darcy's law [51]:

$$Q_D = (\rho\, k\, A\, g\, \Delta h)/(\mu\, L) \tag{3}$$

where k is the permeability of the permeable part (m^2), A is the average cross-section area of the permeable part (m^2), g is the gravitational acceleration, Δh is the hydrostatic gradient of the water in the bottle which is considered constant over time (m), μ is the dynamic viscosity of the water (Pa s), and L is the length of the permeable part (m).

For verification of the laminarity conditions, the Reynolds number was calculated according to the following definition [52]:

$$Re = (\rho\, r\, v)/\mu \tag{4}$$

where r is the average radius of the permeable part (m) and v is the flow speed of the water through the permeable part (m/s). All the Reynolds number values confirmed a condition of laminarity (Table 5).

By substituting Equation (3) into Equation (2) and integrating the ordinary differential equation, we obtain

$$V = (\rho\, k\, A\, g\, \Delta h\, t)/(\mu\, L) \tag{5}$$

where t is the exposure time of the prototype in the soil. According to the simplifying conditions, this equation can fit the data in the initial conditions of the experiment, but it does not take into account the attenuation value due to soil saturation [53,54]. Equation

(6) reports Equation (5), normalized with respect to the volume of water in saturated soil conditions:

$$V/V_s = (\rho\, k\, A\, g\, \Delta h\, t)/(V_s\, \mu\, L).$$

(6)

Equation (6) could be used for a linear fitting of the data at the origin of the axes from which the hydraulic permeability k was extrapolated. The model parameters are reported in Table 3.

Table 3. Constant parameters used in the model.

Parameters	Values
ρ (kg/m^3)	1000
μ (Pa s)	0.0001
r (mm)	5
L (mm)	7
Δh (cm)	18
V_s (cm^3)	30

3. Results and Discussion

3.1. Thermal Properties

To investigate the effect of the pozzolan on the thermal properties of the studied composites, thermogravimetric analysis (TGA) and differential scanning calorimetric (DSC) analysis were performed. Figure 4a shows the TGA thermograms for all the composites as a function of the different percentages of pozzolan content. As can be observed, the decomposition kinetics were not perturbed significantly by the presence of fillers, maintaining the corresponding temperature to 50% of the mass loss around 500 °C. In other words, the thermal stability of the pure polymer used as a matrix was conserved. Figure 4b,d shows the second heating and cooling DSC thermograms, respectively. The curves depict insignificant differences in terms of the melting and cooling transitions having the same shape. However, the presence of the pozzolan reduced the melting enthalpy (ΔH_m) and crystallization enthalpy (ΔH_c) values. The numerical values are reported in Table 3. The possibility of melting the composites using less thermal energy can be an advantage for the extrusion of the material through the 3D printer nozzle, facilitating the phase transition of the polymeric filament in relation to the residence time of the material in the extruder [55]. Concerning crystallization kinetics, as is well known, the high-density polyethylene has fast crystallization kinetics [56]. Figure 4c shows the relative crystallization degree as a function of the time obtained at a constant temperature of 118 °C. Fast crystallization was observed for all the samples, and the half-crystallization time was in the range of 0.5–1.5 min. Meanwhile, there was a slight reduction in the degree of the crystallization after the pozzolan was added (Table 4), which was linked to the decrease in polymer chain mobility.

Table 4. Thermal properties obtained by differential scanning calorimetric analysis.

Pozzolan	ΔH_m (J/g)	T_m (°C)	ΔH_c (J/g)	T_c (°C)	X_c (%)
0%	112	135	127	114	39
20%	74	133	89	115	33
40%	58	133	64	116	34
60%	36	134	43	116	34

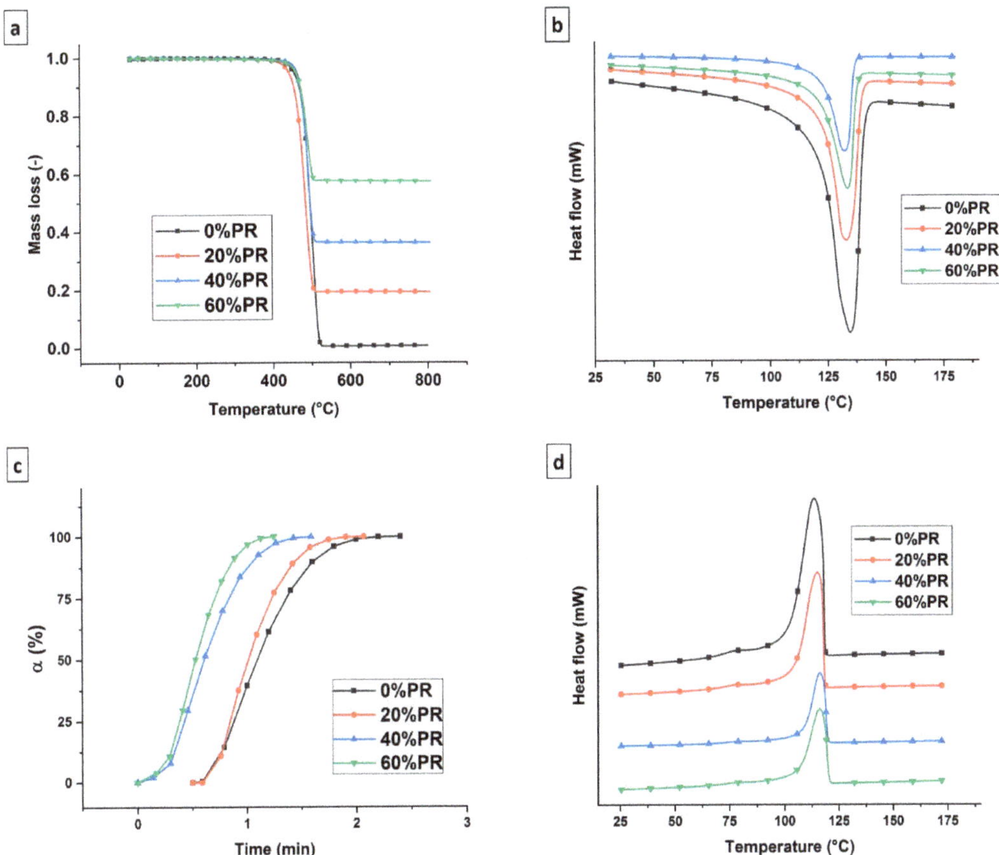

Figure 4. (**a**) Thermogravimetric analysis thermogram. (**b**) Differential scanning calorimetric second heating thermogram. (**c**) Relative crystallinity as a function of time. (**d**) Differential scanning calorimetric second cooling thermogram.

3.2. Rheological Properties

The rheological properties of the polymer composites were affected by the filler presence that interacted with the polymer matrix. In the case of highly filled composites for material extrusion process fabrication, the rheological behavior analysis had relevant importance to understanding the suitability of the material for the considered process. Figure 5a shows the evolution of the zero shear viscosity measured for samples at different temperatures (140 °C, 150 °C, 170 °C, and 180 °C). The results show that the zero shear viscosity increased by increasing the filler rate in the composites, leading to a reinforcing effect on the materials [57]. In particular, there was a slight increase in the viscosity for the composites at pozzolan rates of 20% and 40%, while a significant increase was observed for a pozzolan rate of 60%. This increase in viscosity that occurred in the case of the 60% pozzolan rate may indicate that the filler ratio was close to the higher maximum packing fraction [58]. In the meantime, the viscosity decreased linearly with the temperature (Figure 5a), except for the 60% composite, which had a faster reduction between 140 °C and 150 °C.

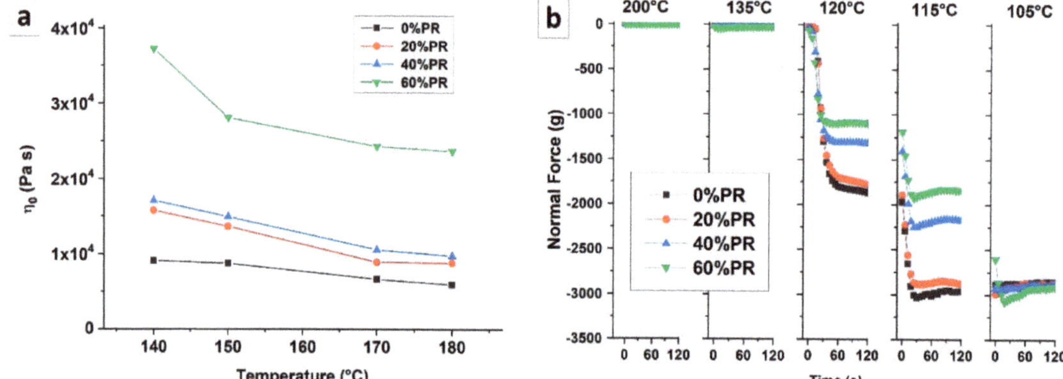

Figure 5. (**a**) Newtonian viscosity as a function of the rheological test temperature. (**b**) Normal force at a constant gap measured at different temperatures.

The gap test at a constant gap was used to analyze the effect of the pozzolan fillers on the volumetric contraction during the cooling of the composites. In particular, the test was performed by measuring the normal force evolution due to the temperature variation. The rheograms related to the normal force measurement are reported in Figure 5b, and the plotted curves are relative to the five temperatures chosen to focus on as the most important areas of the data results. The results show that the presence of the pozzolan slowed down the increase in normal force evolution for all the studied samples. In particular, at 120 °C and 115 °C, the normal force of the composites was lower compared with the pure polymer. A similar value of normal force was achieved for temperatures below 105 °C. Consequently, the presence of the pozzolan reduced the macromolecular mobility, limiting the phenomenon of volumetric shrinkage [59]. This behavior is interesting in light of the 3D printing process. In fact, having a slower volumetric reduction allows for greater cohesion between the deposited layers and greater geometric precision of the printed part, reducing the residual stress [60]. This analysis is in agreement with what was observed during the printing of the composite. Indeed, the composite materials showed a significant reduction in the warpage phenomenon and therefore better conservation of the geometry of the object compared with neat HDPE. Consequently, the detachment force from the printing plate decreased significantly by decreasing the normal force, resulting from the volumetric shrinkage, leading to the ease of printing pozzolan-based composites.

3.3. Mechanical Properties

Figure 6 shows the tensile properties and impact strength values measured for the samples printed with a neat matrix and with the composites at 20%, 40%, and 60% pozzolan content. As can be observed in Figure 6a, Young's modulus increased with the increase of the pozzolan content in the polymer matrix, highlighting a reinforcement of the matrix by the fillers. The ultimate strength and ultimate strain are reported in Figure 6b,c, respectively, and the results had the same value trends. In particular, the neat polymer samples and the composites samples at 20% pozzolan content did not present significant differences. However, by increasing the filler quantities, the elongation at break was reduced, and the maximum reduction was observed for the 60% pozzolan-filled composites (Figure 6b). This behavior is usually observed in polymeric composites, where the greater rigidity of the filler compared with the polymer matrix and the additional stress at the filler–polymer interface promote an increase in the elastic modulus of the composite. However, the presence of the filler generates phase heterogeneity and the discontinuity of the polymer matrix. Generally, this discontinuity translates into a lower ductility [61–63]. Concerning the stress at break (Figure 6c), an improvement of this property for all the composites was observed, showing a maximum value for the composites at 20% pozzolan content. The impact strength values measured for all the printed samples with different pozzolan content are shown in Figure 6d. As can be seen from the data, the trend was similar to the stress at break, where the maximum impact resistance value was read for the 20% pozzolan composite. Instead, the samples printed with 40% and 60% pozzolan-filled composites showed a reduction in impact resistance compared with the pure polymer. This unexpected result can be explained by considering that the composite at a filling ratio of 20% pozzolan mainly presented a reinforcing effect from the filler. However, with the increase in the filling ratio (40% and 60%), the discontinuity in the matrix generated by the pozzolan became important, reducing the stress at break and impact strength values [41,62]. Considering all the results, the optimal formulation was obtained for the composite with 20% pozzolan content.

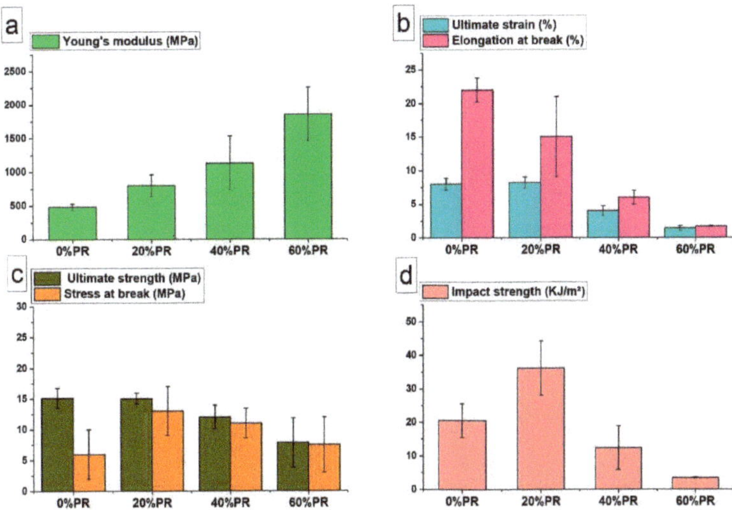

Figure 6. Tensile test results of 3D-printed specimens: (**a**) Young's modulus, (**b**) ultimate strain and elongation at break, and (**c**) ultimate strength and stress at break. (**d**) Impact strength results from the impact test for 3D-printed specimens.

These results are in accordance with the literature for the composites intended for 3D printing. For example, Kariz et al. (2018) [64] observed an improvement of 20% in the elastic modulus of polylactic-acid (PLA) by adding 50% (w/w) wood flour. Wu et al. (2017) [65] achieved, in a similar way, an improvement of 20% for polyhydroxyalcanoates (PHAs) with 40% (w/w) palm fiber in addition to a chemical compatibilizer. Concerning the literature studies on the BioPE-based composites, the maximum increase of the composites' stiffness obtained by Tarrés et al. [24] was two times higher compared with neat BioPE. In this work, the maximum increase of the elastic modulus for the composites was three times higher than pure BioPE.

Another advantage in this work was the ability to print BioPE correctly with filler mass ratios up to 60% without any chemical treatments performed on the filler. In fact, in the majority of the cases, the maximum mass ratio added to (Bio)PE for 3D printing was around 30% [24,26,66].

3.4. Drip Watering Prototype Characterization

In this section, a prototype for precision drip irrigation is studied. In particular, the effect of the print fill density on the void morphology and water permeability are analyzed.

For the printing of the prototypes, the optimal formulation obtained with the composite material, having a ratio of 20% pozzolan, is used.

3.4.1. Morphological Analysis

The morphological analysis of the permeable part of the drip watering prototype was carried out with X-ray tomography. Figure 7a shows the axonometric projection of the samples printed with the filament composites at 0% and 20% pozzolan and with different infill densities (50% and 70% ID). As can be seen from the images, the lattice structure had less microvoids between layers for the samples printed using a lower ID. However, at 70% ID, the 20% pozzolan composites demonstrated fewer microvoids between layers compared with the sample printed with a pure polymer. This behavior suggests that the pozzolan increases the adhesion between the layers of material deposited during printing. The total porosity values are reported in Figure 7b, quantifying this observation and showing the dependency of the porosity with respect to the height z of the permeable part. In particular, for each analyzed sample, the porosity was reduced with an increasing z value. This phenomenon was attributable to the greater cohesion of the material due to the reduction of the diameter of the piece as the height increased with the conical geometry, therefore leading to a smaller surface of the deposited layer. The smaller surface allowed a lower heat exchange time and therefore a lower cooling rate of the composite, permitting greater welding of the layers [47].

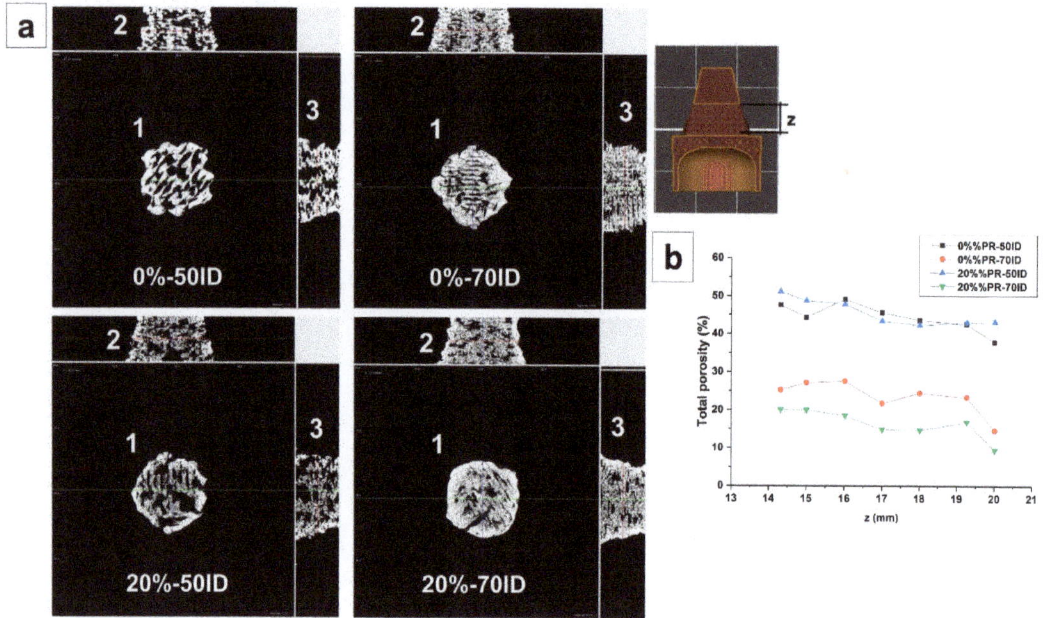

Figure 7. (**a**) Axonometric projection images analyzed with X-ray tomography. (**b**) Total porosity, measured along the permeable part of the prototype.

3.4.2. Water Flow Measuring

The physical (working) principle of the system is described by a simplified example in Figure 3. In particular, (1) when the bottle is rotated, the water flows out due to the hydrostatic gradient of the liquid volume in the bottle. (2) However, the flow stops due to the vacuum effect that occurs in the close volume of air present in the bottle. (3) At this point, the bottle is placed in the soil, which will tend to absorb the water present in the pores of the prototype. (4) When the volume of water in the pores is such that it allows the passage of air due to the pressure difference between the outside and the inside of the bottle, the initial pressure of the gas phase in the bottle will be restored. Therefore, in the absence of a plant, the cycle will continue until the soil is saturated. Furthermore, the flow of water is strictly dependent on the quantity of microvoids distributed on the surface of the permeable part, which is the contact surface between the soil and the water.

This operating system can allow for adjusting the supply of water according to the needs of the plant by controlling the morphology of the prototype voids through the printing parameters and the choice of the material (e.g., ID or fillers).

Figure 8 shows the normalized water volume absorbed by the soil in the beaker at different times. A significant decrease in the water volume absorbed by the soil for the samples printed at 70% ID could be observed. The reduction of the microvoids passing from the lower to higher ID allowed lower water flow, with an order of magnitude in the range from ml/h to ml/day. However, the composites filled with 20% pozzolan had a slower watering rate compared with the pure HDPE. This result highlights that the pozzolan improved the adhesion between the layers and consequently reduced the microvoids, as was observed by X-ray tomography for the samples printed at 70% ID [41,67]. This observation was confirmed by the hydraulic permeability values (Table 5) calculated starting from the linear fitting of the kinetic model reported in Figure 8. The values showed a reduction of the permeability as the 3D printing ID increased and for the samples printed with filament composites.

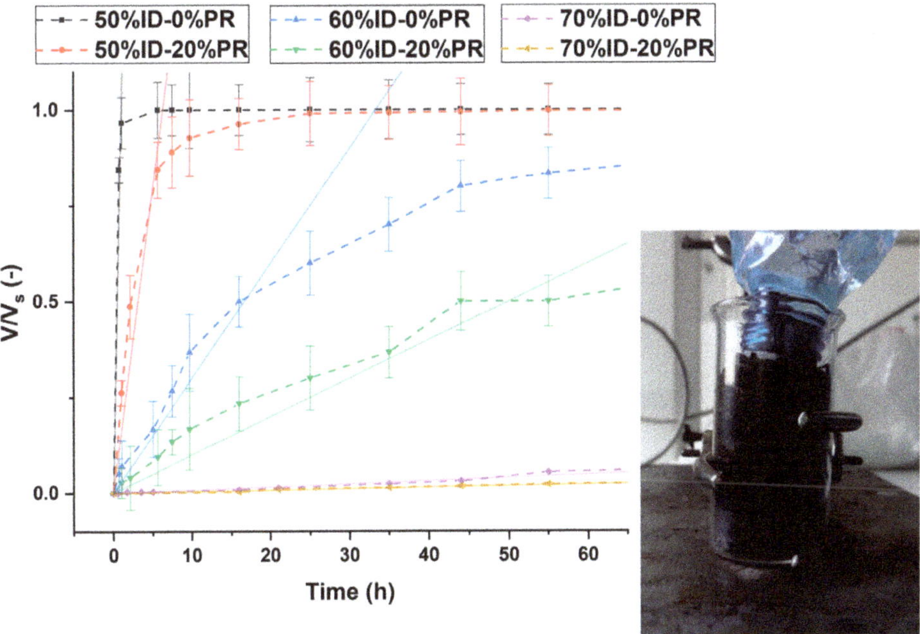

Figure 8. The volume of water absorbed and normalized on the sutured volume of the simulating soil as a function of time. The scatter curves show the data results, while the line curves show the model results.

Table 5. Results of the Reynolds number and hydraulic permeability of the permeable part of the prototype.

Samples	Reynolds Number	k (m^2)
50%ID–0%PR	1.2×10^0	3.2×10^{-18}
50%ID–20%PR	2.7×10^{-1}	5.3×10^{-19}
60%ID–0%PR	3.7×10^{-1}	9.9×10^{-20}
60%ID–20%PR	2.1×10^{-1}	3.3×10^{-20}
70%ID–0%PR	2.5×10^{-3}	2.6×10^{-21}
70%ID–20%PR	7.1×10^{-4}	1.2×10^{-21}

4. Conclusions

In this study, a new polymeric composite filament based on high-density biopolyethylene and pozzolan by-product was produced and used for the material extrusion process. In particular, four formulations at 0%, 20%, 40% and 60% pozzolan were obtained. The composites did not present a difference in terms of thermal decomposition compared with the neat matrix, showing good thermal stability. The presence of the pozzolan decreased the melting enthalpy and the crystallinity degree. In terms of viscoelastic behavior, the Newtonian viscosity increased significantly for the composite at a 60% pozzolan content, reaching the maximum packing fraction. Concerning the normal force evolution, it was observed that the presence of the pozzolan limited the volumetric shrinkage during the cooling of the composites.

The printing of objects with complex geometries was successfully achieved. The printed objects with the composite's filaments showed an increase in the elastic modulus, stress at break, and impact strength, but at the same time, the ultimate strain, elongation at break, and ultimate strength were reduced compared with the neat polymer. In other words, an improvement of the material rigidity with an optimal formulation at a 20% pozzolan ratio was observed. Finally, a drip watering prototype was conceived and printed with different infill densities to control the water flow of irrigation. As was shown by the X-ray tomography images, the pozzolan improved the cohesion between the layers of the final object, reducing the microvoids of the permeable part of the drip watering prototype.

From this study, it can be concluded that the huge amount of generated pozzolan by-products from quarries can be valorized as a filler for polymeric matrices, leading to the improvement of their properties. At the same time, it was demonstrated that the percentage of the pozzolan filler could reach up to 60% (w/w) for HDPE 3D printing filaments without any further treatment. Moreover, a decrease in the amount of the polymeric matrix used in the composites can lead to an important economic impact on the elaboration of new eco-friendly composites. Finally, some works are underway in our laboratory to develop pozzolan-based composites with a polymeric matrix that is fully biodegradable in the soil. An industrial application could be imagined as a seed pot directly implanted in the soil without further transplanting. After the matrix's biodegradation, the pozzolan remaining in the soil can play a fertilizing role. This work will be a subject for future publication.

Author Contributions: Conceptualization, N.S., V.V., and H.A.; methodology, N.S., V.V., and H.A.; software, N.S.; validation, V.V. and H.A.; formal analysis, N.S.; investigation, N.S.; data curation, N.S.; writing—original draft preparation, N.S.; writing—review and editing, V.V. and H.A.; visualization, N.S.; supervision, V.V. and H.A. All authors have read and agreed to the published version of the manuscript.

Funding: This research received no external funding.

Data Availability Statement: The data presented in this study are available on request from the corresponding author.

Conflicts of Interest: The authors declare no conflict of interest.

Polymers **2021**, *13*, 1908

References

1. Crump, S.S. Apparatus and Method for Creating Three-Dimensional Objects. U.S. Patent No. 5,121,329, 9 June 1992.
2. Braconnier, D.J.; Jensen, R.E.; Peterson, A.M. Processing Parameter Correlations in Material Extrusion Additive Manufacturing. *Addit. Manuf.* **2020**, *31*, 100924. [CrossRef]
3. Chard, J.; van Iersel, M.; Bugbee, B. *Mini-Lysimeters to Monitor Transpiration and Control Drought Stress: System Design and Unique Applications*; Merrill-Cazier Library: Logan, UT, USA, 2010.
4. Hess, L.; De Kroon, H. Effects of Rooting Volume and Nutrient Availability as an Alternative Explanation for Root Self/Non-Self Discrimination. *J. Ecol.* **2007**, *95*, 241–251. [CrossRef]
5. Schrader, J.A.; Kratsch, H.A.; Graves, W.R. *Bioplastic Container Cropping Systems: Green Technology for the Green Industry*; Sustainable Horticulture Research Consortium: Ames, IA, USA, 2016.
6. Poorter, H.; Bühler, J.; van Dusschoten, D.; Climent, J.; Postma, J.A. Pot Size Matters: A Meta-Analysis of the Effects of Rooting Volume on Plant Growth. *Funct. Plant Biol.* **2012**, *39*, 839–850. [CrossRef] [PubMed]
7. Ray, J.D.; Sinclair, T.R. The Effect of Pot Size on Growth and Transpiration of Maize and Soybean during Water Deficit Stress. *J. Exp. Bot.* **1998**, *49*, 1381–1386. [CrossRef]
8. Biran, I.; Eliassaf, A. The Effect of Container Size and Aeration Conditions on Growth of Roots and Canopy of Woody Plants. *Sci. Hortic.* **1980**, *12*, 385–394. [CrossRef]
9. Nikolaou, G.; Neocleous, D.; Katsoulas, N.; Kittas, C. Irrigation of Greenhouse Crops. *Horticulturae* **2019**, *5*, 7. [CrossRef]
10. Massetani, F.; Savini, G.; Neri, D. Effect of Rooting Time, Pot Size and Fertigation Technique on Strawberry Plant Architecture. *J. Berry Res.* **2014**, *4*, 217–224. [CrossRef]
11. Ujeniya, P.S.; Rachchh, N.V. A Review on Manufacturing, Machining, and Recycling of 3D Printed Composite Materials. In *IOP Conference Series: Materials Science and Engineering*; IOP Publishing: Bristol, UK, 2019; Volume 653, p. 012024.
12. Askanian, H.; Muranaka de Lima, D.; Commereuc, S.; Verney, V. Toward a Better Understanding of the Fused Deposition Modeling Process: Comparison with Injection Molding. *3D Print. Addit. Manuf.* **2018**, *5*, 319–327. [CrossRef]
13. Vidakis, N.; Petousis, M.; Tzounis, L.; Maniadi, A.; Velidakis, E.; Mountakis, N.; Kechagias, J. Sustainable Additive Manufacturing: Mechanical Response of Polyamide 12 over Multiple Recycling Processes. *Materials* **2021**, *14*, 466. [CrossRef]
14. Vidakis, N.; Vairis, A.; Petousis, M.; Savvakis, K.; Kechagias, J. Fused Deposition Modelling Parts Tensile Strength Characterisation. *Acad. J. Manuf. Eng.* **2016**, *14*, 87–94.
15. Kumar, M.; Ramachandran, R.; Omarbekova, A. 3D Printed Polycarbonate Reinforced Acrylonitrile–Butadiene–Styrene Composites: Composition Effects on Mechanical Properties, Micro-Structure and Void Formation Study. *J. Mech. Sci. Technol.* **2019**, *33*. [CrossRef]
16. Rodzeń, K.; Sharma, P.K.; McIlhagger, A.; Mokhtari, M.; Dave, F.; Tormey, D.; Sherlock, R.; Meenan, B.J.; Boyd, A. The Direct 3D Printing of Functional PEEK/Hydroxyapatite Composites via a Fused Filament Fabrication Approach. *Polymers* **2021**, *13*, 545. [CrossRef]
17. HDPE. Available online: https://polymerdatabase.com/Commercial%20Polymers/HDPE.html (accessed on 26 March 2021).
18. Zander, N.E.; Gillan, M.; Burckhard, Z.; Gardea, F. Recycled Polypropylene Blends as Novel 3D Printing Materials. *Addit. Manuf.* **2019**, *25*, 122–130. [CrossRef]
19. Sang, L.; Han, S.; Li, Z.; Yang, X.; Hou, W. Development of Short Basalt Fiber Reinforced Polylactide Composites and Their Feasible Evaluation for 3D Printing Applications. *Compos. Part B Eng.* **2019**, *164*, 629–639. [CrossRef]
20. Lee, J.-Y.; An, J.; Chua, C.K. Fundamentals and Applications of 3D Printing for Novel Materials. *Appl. Mater. Today* **2017**, *7*, 120–133. [CrossRef]
21. Wang, T.-M.; Xi, J.-T.; Jin, Y. A Model Research for Prototype Warp Deformation in the FDM Process. *Int. J. Adv. Manuf. Technol.* **2007**, *33*, 1087–1096. [CrossRef]
22. Plastics Market Size, Share & Trends Report, 2020–2027. Available online: https://www.grandviewresearch.com/industry-analysis/global-plastics-market (accessed on 26 March 2021).
23. Schirmeister, C.G.; Hees, T.; Licht, E.H.; Mülhaupt, R. 3D Printing of High Density Polyethylene by Fused Filament Fabrication. *Addit. Manuf.* **2019**, *28*, 152–159. [CrossRef]
24. Tarrés, Q.; Melbø, J.K.; Delgado-Aguilar, M.; Espinach, F.X.; Mutjé, P.; Chinga-Carrasco, G. Bio-Polyethylene Reinforced with Thermomechanical Pulp Fibers: Mechanical and Micromechanical Characterization and Its Application in 3D-Printing by Fused Deposition Modelling. *Compos. Part B Eng.* **2018**, *153*, 70–77. [CrossRef]
25. Balani, S.B.; Chabert, F.; Nassiet, V.; Cantarel, A. Influence of Printing Parameters on the Stability of Deposited Beads in Fused Filament Fabrication of Poly (Lactic) Acid. *Addit. Manuf.* **2019**, *25*, 112–121. [CrossRef]
26. Mazzanti, V.; Malagutti, L.; Mollica, F. FDM 3D Printing of Polymers Containing Natural Fillers: A Review of Their Mechanical Properties. *Polymers* **2019**, *11*, 1094. [CrossRef]
27. Stoof, D.; Pickering, K. Sustainable Composite Fused Deposition Modelling Filament Using Recycled Pre-Consumer Polypropylene. *Compos. Part B Eng.* **2018**, *135*, 110–118. [CrossRef]
28. Chong, S.; Pan, G.-T.; Khalid, M.; Yang, T.C.-K.; Hung, S.-T.; Huang, C.-M. Physical Characterization and Pre-Assessment of Recycled High-Density Polyethylene as 3D Printing Material. *J. Polym. Environ.* **2017**, *25*, 136–145. [CrossRef]
29. Peng, F.; Jiang, H.; Woods, A.; Joo, P.; Amis, E.J.; Zacharia, N.S.; Vogt, B.D. 3D Printing with Core–Shell Filaments Containing High or Low Density Polyethylene Shells. *ACS Appl. Polym. Mater.* **2019**, *1*, 275–285. [CrossRef]

30. Mohan, V.B.; Bhattacharyya, D. Mechanical, Electrical and Thermal Performance of Hybrid Polyethylene-Graphene Nanoplatelets-Polypyrrole Composites: A Comparative Analysis of 3D Printed and Compression Molded Samples. *Polym. Plast. Technol. Mater.* **2020**, *59*, 780–796. [CrossRef]

31. Zhang, F.; Ma, G.; Tan, Y. The Nozzle Structure Design and Analysis for Continuous Carbon Fiber Composite 3D Printing. In Proceedings of the 2017 7th International Conference on Advanced Design and Manufacturing Engineering (ICADME 2017), Shenzhen, China, 10–11 May 2017; Atlantis Press: Moscow, Russia, 2017; pp. 193–199.

32. Wang, X.; Jiang, M.; Zhou, Z.; Gou, J.; Hui, D. 3D Printing of Polymer Matrix Composites: A Review and Prospective. *Compos. Part B Eng.* **2017**, *110*, 442–458. [CrossRef]

33. Naigaga, E. An Examination of the Sustainability of Pozzolana Mining Processes in Uganda. *Int. J. Res. Chem. Metall. Civ. Eng.* **2014**, *1*, 25.

34. El Youbi, M.S.; Ahmed, E. Development and Study of Physical, Chemical and Mechanical Properties of a New Formulation of Cement of a Varying Percentage of Natural Pozzolan. *J. Chem. Technol. Met.* **2017**, *52*, 873–884.

35. Kim, H.; Choi, S.; Lee, B.; Kim, S.; Kim, H.; Cho, C.; Cho, D. Thermal Properties of Bio Flour-Filled Polypropylene Bio-Composites with Different Pozzolan Contents. *J. Therm. Anal. Calorim.* **2007**, *89*, 821–827. [CrossRef]

36. Sbaffoni, S.; Boni, M.R.; Vaccari, M. Potential of Compost Mixed with Tuff and Pozzolana in Site Restoration. *Waste Manag.* **2015**, *39*, 146–157. [CrossRef] [PubMed]

37. Seco, A.; Ramirez, F.; Miqueleiz, L.; Urmeneta, P.; García, B.; Prieto, E.; Oroz, V. *Types of Waste for the Production of Pozzolanic Materials—A Review*; Intech: Rabat, Morocco, 2012; p. 29.

38. Rocher, P. *Mémento Roches et Minéraux Industriels*; BRGM: Orléans, France, 1992; p. 47.

39. Les Services de l'État dans le Puy-de-Dôme. Schéma Départemental des Carriers. Available online: http://www.puy-de-dome.gouv.fr/schema-departemental-des-carrieres-r1292.html (accessed on 25 March 2021).

40. Pouzzolane. Available online: http://www.ciment.wikibis.com/pouzzolane.php (accessed on 25 March 2021).

41. Schiavone, N.; Verney, V.; Askanian, H. Pozzolan Based 3D Printing Composites: From the Formulation Till the Final Application in the Precision Irrigation Field. *Materials* **2021**, *14*, 43. [CrossRef]

42. Esposito Corcione, C.; Palumbo, E.; Masciullo, A.; Montagna, F.; Torricelli, M.C. Fused Deposition Modeling (FDM): An Innovative Technique Aimed at Reusing Lecce Stone Waste for Industrial Design and Building Applications. *Constr. Build. Mater.* **2018**, *158*, 276–284. [CrossRef]

43. Wasti, S.; Adhikari, S. Use of Biomaterials for 3D Printing by Fused Deposition Modeling Technique: A Review. *Front. Chem.* **2020**, *8*. [CrossRef] [PubMed]

44. Available online: https://www.hitachi-hightech.com/file/global/pdf/products/science/appli/ana/thermal/application_TA_026e.pdf (accessed on 25 March 2021).

45. Refaa, Z.; Boutaous, M.; Xin, S.; Siginer, D.A. Thermophysical Analysis and Modeling of the Crystallization and Melting Behavior of PLA with Talc. *J. Anal. Calorim.* **2017**, *128*, 687–698. [CrossRef]

46. Marek, A.A.; Verney, V. Rheological Behavior of Polyolefins during UV Irradiation at High Temperature as a Coupled Degradative Process. *Eur. Polym. J.* **2015**, *72*, 1–11. [CrossRef]

47. Schiavone, N.; Verney, V.; Askanian, H. Effect of 3D Printing Temperature Profile on Polymer Materials Behavior. *3D Print. Addit. Manuf.* **2020**, *7*, 311–325. [CrossRef]

48. Wypych, G. 2-Mechanisms of Adhesion. In *Handbook of Adhesion Promoters*; Wypych, G., Ed.; ChemTec Publishing: Scarborough, ON, Canada, 2018; pp. 5–44. ISBN 978-1-927885-29-1.

49. Graziano, A.; Jaffer, S.; Sain, M. Review on Modification Strategies of Polyethylene/Polypropylene Immiscible Thermoplastic Polymer Blends for Enhancing Their Mechanical Behavior. *J. Elastomers Plast.* **2019**, *51*, 291–336. [CrossRef]

50. *Perry's Chemical Engineers' Handbook*, 9th ed.; McGraw-Hill Education—Access Engineering: New York, NY, USA, 2019; Available online: https://www.accessengineeringlibrary.com/content/book/9780071834087 (accessed on 26 May 2021).

51. Whitaker, S. Flow in Porous Media I: A Theoretical Derivation of Darcy's Law. *Transp. Porous Med.* **1986**, *1*, 3–25. [CrossRef]

52. Bird, R.B.; Stewart, W.E.; Lightfoot, E.N. *Transport Phenomena*; John Wiley & Sons: Hoboken, NJ, USA, 2006; ISBN 978-0-470-11539-8.

53. Ouedraogo, F.; Cherblanc, F.; Naon, B.; Bénet, J.-C. Water Transfer in Soil at Low Water Content. Is the Local Equilibrium Assumption Still Appropriate? *J. Hydrol.* **2013**, *492*, 117–127. [CrossRef]

54. Fundamental Equations in Colloid and Surface Science. In *Introduction to Applied Colloid and Surface Chemistry*; John Wiley & Sons, Ltd.: Hoboken, NJ, USA, 2016; pp. 74–95. ISBN 978-1-118-88119-4.

55. Phan, D.D.; Horner, J.S.; Swain, Z.R.; Beris, A.N.; Mackay, M.E. Computational Fluid Dynamics Simulation of the Melting Process in the Fused Filament Fabrication Additive Manufacturing Technique. *Addit. Manuf.* **2020**, *33*, 101161. [CrossRef]

56. Patki, R.P.; Phillips, P.J. Crystallization Kinetics of Linear Polyethylene: The Maximum in Crystal Growth Rate–Temperature Dependence. *Eur. Polym. J.* **2008**, *44*, 534–541. [CrossRef]

57. Valentina, I.; Haroutioun, A.; Fabrice, L.; Vincent, V.; Roberto, P. Poly(Lactic Acid)-Based Nanobiocomposites with Modulated Degradation Rates. *Materials* **2018**, *11*, 1943. [CrossRef]

58. Escócio, V.A.; Pacheco, E.B.A.V.; da Silva, A.L.N.; de Paula Cavalcante, A.; Visconte, L.L.Y. Rheological Behavior of Renewable Polyethylene (HDPE) Composites and Sponge Gourd (*Luffa cylindrica*) Residue. *Int. J. Polym. Sci.* **2015**, *2015*, 714352. Available online: https://www.hindawi.com/journals/ijps/2015/714352/ (accessed on 2 February 2021). [CrossRef]

59. Shah, P.; Stansbury, J. Role of Filler and Functional Group Conversion in the Evolution of Properties in Polymeric Dental Restoratives. *Dent. Mater.* **2014**, *30*. [CrossRef]

60. Casavola, C.; Cazzato, A.; Moramarco, V.; Pappalettera, G. Residual Stress Measurement in Fused Deposition Modelling Parts. *Polym. Test.* **2017**, *58*, 249–255. [CrossRef]

61. Askanian, H.; Verney, V.; Commereuc, S.; Guyonnet, R.; Massardier, V. Wood Polypropylene Composites Prepared by Thermally Modified Fibers at Two Extrusion Speeds: Mechanical and Viscoelastic Properties. *Holzforschung* **2015**, *69*, 313–319. [CrossRef]

62. Kiran, M.D.; Govindaraju, H.K.; Jayaraju, T.; Kumar, N. Review-Effect of Fillers on Mechanical Properties of Polymer Matrix Composites. *Mater. Today: Proc.* **2018**, *5*, 22421–22424. [CrossRef]

63. Jancar, J. Chapter 31—Composite materiomics: Multi length scale hierarchical composites for structural and tissue engineering applications. In *Multifunctionality of Polymer Composites*; Friedrich, K., Breuer, U., Eds.; William Andrew Publishing: Oxford, UK, 2015; pp. 903–944. ISBN 978-0-323-26434-1.

64. Kariz, M.; Sernek, M.; Obućina, M.; Kuzman, M.K. Effect of Wood Content in FDM Filament on Properties of 3D Printed Parts. *Mater. Today Commun.* **2018**, *14*, 135–140. [CrossRef]

65. Wu, C.-S.; Liao, H.-T.; Cai, Y.-X. Characterisation, Biodegradability and Application of Palm Fibre-Reinforced Polyhydroxyalkanoate Composites. *Polym. Degrad. Stab.* **2017**, *140*, 55–63. [CrossRef]

66. Filgueira, D.; Holmen, S.; Melbø, J.K.; Moldes, D.; Echtermeyer, A.T.; Chinga-Carrasco, G. 3D Printable Filaments Made of Biobased Polyethylene Biocomposites. *Polymers* **2018**, *10*, 314. [CrossRef]

67. Liao, Y.; Liu, C.; Coppola, B.; Barra, G.; Di Maio, L.; Incarnato, L.; Lafdi, K. Effect of Porosity and Crystallinity on 3D Printed PLA Properties. *Polymers* **2019**, *11*, 1487. [CrossRef]

polymers

MDPI

Article

Effect of Silver Nanopowder on Mechanical, Thermal and Antimicrobial Properties of Kenaf/HDPE Composites

Vikneswari Sanmuham [1], Mohamed Thariq Hameed Sultan [1,2,3,*], A. M. Radzi [4,*], Ahmad Adlie Shamsuri [2], Ain Umaira Md Shah [2], Syafiqah Nur Azrie Safri [2] and Adi Azriff Basri [1]

1   Department of Aerospace Engineering, Faculty of Engineering, Universiti Putra Malaysia, UPM Serdang 43400, Malaysia; vikki.sanmuham@gmail.com (V.S.); adiazriff@upm.edu.my (A.A.B.)
2   Laboratory of Biocomposite Technology, Institute of Tropical Forestry and Forest Products (INTROP), Universiti Putra Malaysia, UPM Serdang 43400, Malaysia; adlie@upm.edu.my (A.A.S.); ainumaira91@gmail.com (A.U.M.S.); snasafri@gmail.com (S.N.A.S.)
3   Aerospace Malaysia Innovation Centre (944751-A), Prime Minister's Department, MIGHT Partnership Hub, Jalan Impact, Cyberjaya 63000, Malaysia
4   Malaysia-Japan International Institute of Technology, Universiti Teknologi Malaysia, Kuala Lumpur 54100, Malaysia
*   Correspondence: thariq@upm.edu.my (M.T.H.S.); mradzi9595@gmail.com (A.M.R.); Tel.: +603-9769-6396 (M.T.H.S.)

Abstract: This study aims to investigate the effect of AgNPs on the mechanical, thermal and antimicrobial activity of kenaf/HDPE composites. AgNP material was prepared at different contents, from 0, 2, 4, 6, 8 to 10 wt%, by an internal mixer and hot compression at a temperature of 150 °C. Mechanical (tensile, modulus and elongation at break), thermal (TGA and DSC) and antimicrobial tests were performed to analyze behavior and inhibitory effects. The obtained results indicate that the effect of AgNP content displays improved tensile and modulus properties, as well as thermal and antimicrobial properties. The highest tensile stress is 5.07 MPa and was obtained at 10wt, TGA showed 10 wt% and had improved thermal stability and DSC showed improved stability with increased AgNP content. The findings of this study show the potential of incorporating AgNP concentrations as a secondary substitute to improve the performance in terms of mechanical, thermal and antimicrobial properties without treatment. The addition of AgNP content in polymer composite can be used as a secondary filler to improve the properties.

Keywords: silver nanopowder; kenaf; high-density polyethylene; antimicrobial

Citation: Sanmuham, V.; Sultan, M.T.H.; Radzi, A.M.; Shamsuri, A.A.; Md Shah, A.U.; Safri, S.N.A.; Basri, A.A. Effect of Silver Nanopowder on Mechanical, Thermal and Antimicrobial Properties of Kenaf/HDPE Composites. *Polymers* **2021**, *13*, 3928. https://doi.org/10.3390/polym13223928

Academic Editor: Emin Bayraktar

Received: 21 June 2021
Accepted: 27 July 2021
Published: 14 November 2021

Publisher's Note: MDPI stays neutral with regard to jurisdictional claims in published maps and institutional affiliations.

1. Introduction

Nowadays, nanotechnology is growing rapidly in the food and manufacturing industries for its effectiveness in terms of heat resistance and antimicrobial agents on products. The use of nanoparticles on composites brings advantages, especially in the aspect of dispersion. In addition, an increase in the nanoparticle content will reduce the hydrophobic properties and increase the mechanical, physical and thermal properties [1]. Due to their small size, these materials can interact with cellulose easily and disperse in composite systems. As written by Theodore et al. [2], the supermolecular structure of biopolymers contributes to the moisture absorption into cellulose, hemicelluloses and lignin. Nanoparticle usage in order to modify the surface of the lignocellulose-based composite can contribute to physical or chemical barriers to control the accumulation of moisture.

One of the nanomaterials used is silver nanopowders (AgNPs). AgNPs are increasingly used to produce various products, such as aerospace parts, medical, packaging and automotive because AgNPs have their uniqueness [3–5]. Due to the uniqueness of AgNP properties, various products have been produced, such as antibacterial agents, furniture, optical sensor health products, food industry products and anticancer agents [3,4,6]. AgNPs have the nano-sized potential to transform various physical and chemical properties of a substance and be used for a variety of purposes depending on an application [7]. Nurani et al. [8] declared that silver nanopowder as a secondary filler will be very promising since it has a high aspect ratio, which can determine surface-related properties such as solubility and stability. The high aspect ratio of silver nanoparticles exhibits microbial resistance and improves resistant strains, which are reliable to the current research concerns about optical, catalytic and microbial properties [8,9].

Natural fibers are known by industry and experts as elongated objects obtained from animals or plants. Natural fibers are cost-effective, have low-density properties, have low carbon sequestration, have good mechanical properties (high toughness and high specific strength) and are highly biodegradable [10–13]. Furthermore, natural fibers possess little manufacturing vitality in comparison to synthetic fibers, which are commonly used [10,14,15]. This would lead to producing polymer composites with very few production costs. Natural fibers can also help to improve environmental sustainability since they are eco-friendly composites [16–18]. The roughness and toughness of the fiber bundle will be arise the limitations in processing of natural fibers. Therefore, suitable chemical treatments have been proposed to overcome this problem [19]. The use of chemicals for surface treatment on natural fibers has various limitations, such as an expensive price, the surface fiber will shrink, and it requires high energy and processing time [20].

Kenaf (Hibiscus cannabinus L.) is an annual plant, which has historically been used for rope, twine, sackcloth and canvas. Kenaf fibers have many beneficial aspects, such as acceptable specific strength, economic viability, low density, reduced tool wear, cost-effectiveness, renewability, good biodegradability, reduced dermal and respiratory irritations, and energy savings by reducing the manufacturing process compared to synthetic fibers [21,22]. Based on Nishino et al. [23], it has been reported that kenaf composites have higher mechanical strength and thermal properties compared to other kinds of natural fiber polymer composites [24]. Kenaf is considered to be environmentally friendly due to several reasons, such as kenaf absorbs carbon dioxide more than any crop, no chemical pest control is needed, it removes toxic elements such as phosphorous from soil and it contributes to soil remediation [25]. Azammi et al. [26] and Yahya et al. [27] claimed that kenaf natural fiber-reinforced plastics are low weight and easy to fabricate; additionally, they are cost-effective and completely recyclable.

However, the research regarding the relationship between kenaf and silver-powder-reinforced HDPE and how to reduce surface treatment on fiber was never looked at. Apart from that, information regarding the combination of both materials is less. In this study, filler materials, such as organic or inorganic, in polymer systems can support main fillers in the strengthening of composites [28,29]. Therefore, the selection of AgNPs as a secondary filler was used to reduce fiber surface treatment [3] and improve the tensile, thermal and antimicrobial properties of kenaf/HDPE composites. The addition of silver nanopowder was suggested to improve the thermal stability of kenaf/HDPE composites. The antibacterial mechanism and silver nanopowder anti-biofilm may withstand the development of microorganisms and are expected to improve the sustainability of the kenaf/HDPE composites.

2. Materials and Methods

2.1. Materials

The main filler of the composites is kenaf with a size of 420 μm, which was obtained from the National Kenaf and Tobacco Board, Malaysia (NKTB). The secondary filler used in this research is a silver nanopowder (AgNP) with a size <100 nm. The matrix polymer that was used is High-Density Polyethylene (HDPE) with a density of 0.95×10^3 kg/m^3, which was purchased from Lotte Chemical Titan Sdn. Bhd. (Pasir Gudang, Malaysia). All materials in this study were used directly without any treatment or further refinement.

2.2. Sample Preparation

The kenaf and AgNPs were prepared through a Brabender Plastograph internal mixer as shown in Figure 1 to mix homogeneously at 150 °C for 50 min, and the speed was fixed at 60 rpm. HDPE with a mass of 24 g was added into the chamber, and then kenaf with a mass of 16 g [28] and AgNPs was varied to 0.82 g (2 wt%), 1.67 g (4 wt%), 2.55 g (6 wt%), 3.48 g (8 wt%), and 4.44 g (10 wt%). The composites were produced in the form of a sheet (1 mm) using a hot-press machine (Figure 1) at 150 °C and cooled for 5 min. All composites were dried in an oven for 24 h at a temperature of 70 °C. Figure 2 shows the samples before and after the compression process. All tests were carried out at Universiti Putra Malaysia (UPM), Malaysia.

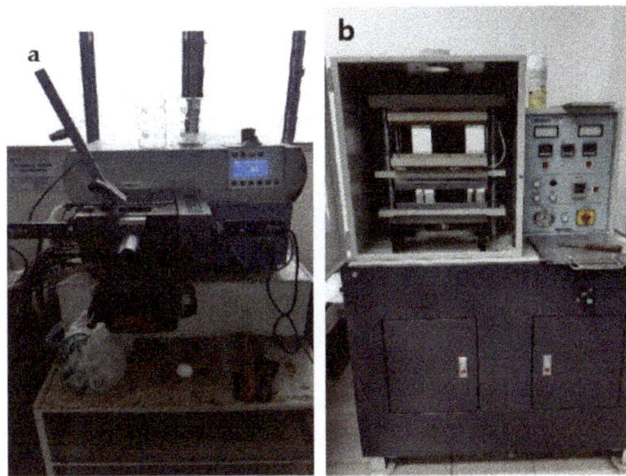

Figure 1. (**a**) Brabender Plastograph internal and (**b**) hot press machine.

Figure 2. Sample before and after the compression process.

2.3. Thermal Analysis

Thermal analysis is a technique for identifying the thermophilic and kinetic properties of materials. The technique used in this research is TGA (thermogravimetric analysis) and DSC (Differential Scanning Calorimetry). All tests were carried out at Universiti Putra Malaysia (UPM), Malaysia.

2.3.1. Thermogravimetric Analysis (TGA)

The thermal properties in composites are practically significant for studies related to temperature, and it is important to determine the degradation of kenaf/HDPE composites. Thermogravimetric analysis (TGA) was conducted with the TA Instrument TGA (Q500 model) (TA Instruments, New Castle, DE, USA) according to ASTM E1131 [30], which is displayed in Figure 3 for the ultimate decomposition temperatures of the prepared composite materials. All tests were carried out at Universiti Putra Malaysia (UPM), Malaysia.

Figure 3. Thermogravimetric analyzer.

The heating levels of 10 °C min^{-1} for the experiment were determined, with a temperature range of between 30 and 600 °C. Approximately 10 to 11 mg of sample was heated in a platinum sample pan of the prepared sample in a 50 mL min^{-1} flow rate of nitrogen gas atmosphere. All tests were carried out at Universiti Putra Malaysia (UPM), Malaysia.

2.3.2. Differential Scanning Calorimetry (DSC)

A Differential Scanning Calorimeter (DSC) (TA Instruments, model DSC Q20, New Castle, DE, USA), as shown in Figure 4, was used to calculate the melting point temperatures of the blends that had been prepared. A DSC test was examined in a differential scanning calorimeter after heating the sample in the temperature range 35–200 °C at a heating rate of 10 °C min^{-1}. All tests were carried out at Universiti Putra Malaysia (UPM), Malaysia.

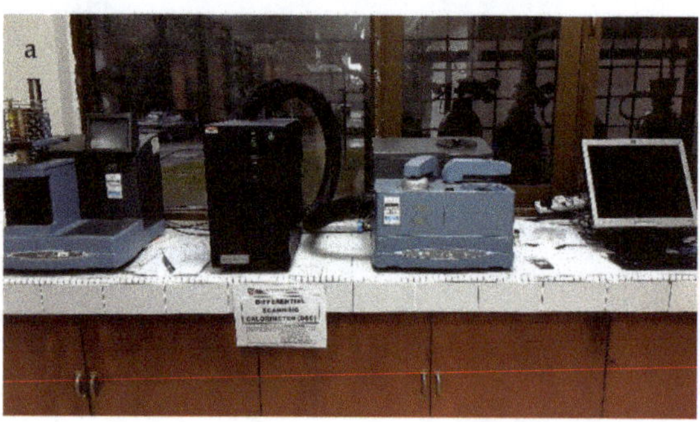
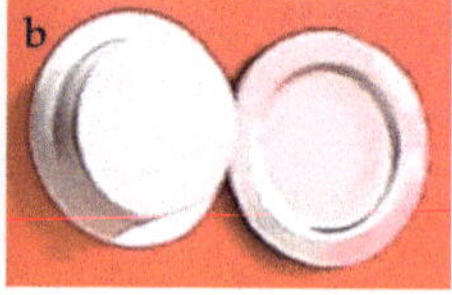

Figure 4. (**a**) Differential Scanning Calorimeter (DSC) and (**b**) aluminum sample pan (5 mm).

2.4. Antimicrobial Analysis

Two kinds of mediums were utilized: bacteria-growing sterile Mueller–Hinton agar (MHA) and yeast-growing sterile polydopamine (PDA). All tests were carried out at Universiti Putra Malaysia (UPM), Malaysia.

By incubating pure micro-organism culture at 37 °C, for 18 h, an inoculum suspension was created. The implanted microbes were shifted into the liquid media to 0.5 McFarland turbidity standard for adjusting their turbidity. With a sterile cotton swab, a sample of Escherichia coli (E. coli), UPMC 25921, Pseudomonas aeruginosa (P. aeruginosa) ATCC 15442, and Candida albicans (*C. albicans*) ATCC 90029 was dispersed over the entire agar surface and then swabbed over the sampling area. The kenaf/HDPE composites with and without AgNP samples were cut into 6 mm discs in diameter prior to testing and sterilized for 30 min using UV radiation. Then, put it into the Petri platter. A Petri dish was incubated for 18 to 24 h at 37 °C for bacteria, but for fungi it was incubated for 72 h at 30 °C. A clear area has been observed and the inhibition area size has been registered. Every sample was performed three times.

3. Results and Discussion

3.1. Tensile Properties

The tensile stress properties of kenaf/HDPE biocomposites with five different concentrations of AgNPs, which are 0%, 2%, 4%, 6%, 8% and 10%, were demonstrated. The changes in the failure of each composite tested can be seen in its properties. The results of the test determine whether the composite is acceptable or not in terms of function, structure and visuals. Figures 5 and 6 show the tensile stress and modulus properties of kenaf/HDPE composites with the different AgNP contents. A summary of the analysis of mechanical properties is also presented. From the result, the trends of tensile stress and modulus show an increase with increasing AgNP contents compare to without AgNPs. This indicates exceptional compatibility between kenaf/HDPE composites shows a major increase in their tensile stress at the values of 3.78, 4.03, 4.15, 4.67 and 5.07 MPa with the addition of AgNPs of 0 to 10%, respectively. The main factor of this increase is due to the physical interaction between the components. It has been proven by Rhim [4] that the interaction and combination of materials need to have a relationship between all the components to ensure increased tensile properties. Besides, the performance increase is also due to the nature of AgNPs having a nano-chain structure and high nano-mechanical performance [31].

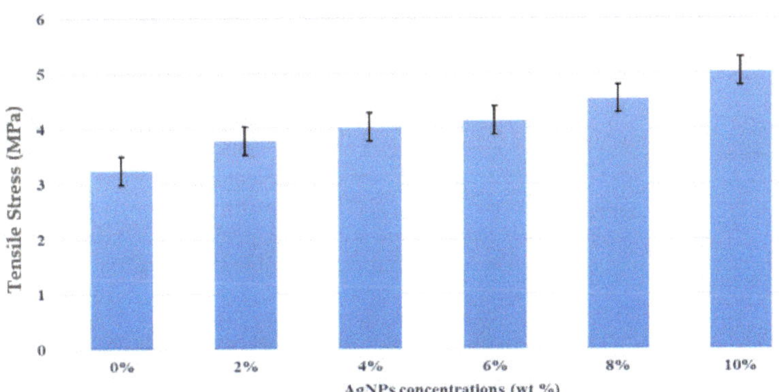

Figure 5. Effects of AgNP content on tensile stress of kenaf/HDPE composites.

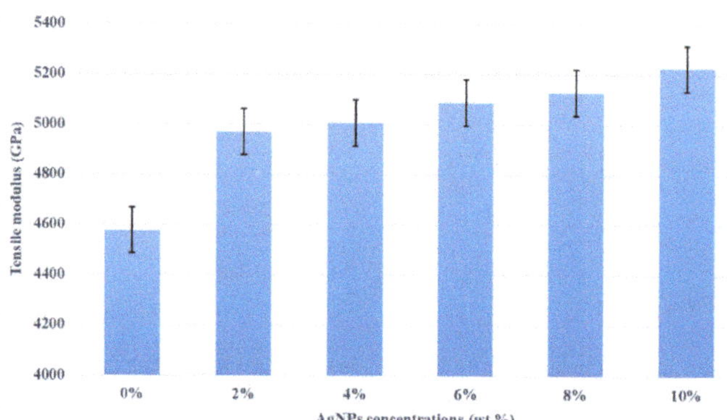

Figure 6. Effects of AgNP content on tensile modulus of kenaf/HDPE composites.

Due to the very limited inclusion of AgNPs in the kenaf/HDPE composites, slight increases in the tensile stress and tensile modulus were predicted. Figure 6 shows the tensile modulus of the kenaf/HDPE composites with different AgNP contents. The figure shows a similar trend with tensile stress increased with increasing the AgNP content. The increase in tensile modulus is due to several factors, such as good adhesion, abrasive surface, less void and microcracking [32] between kenaf and AgNP fillers. Moreover, AgNPs could act as a secondary filler in kenaf/HDPE composite to enhance the tensile and stiffness characteristics. N. Sapiai et al. [33] reported that the efficiency of a filler also depends on various factors, such as filler dispersion, filler volume, the type of matrix used and matrix bonding [34]. Such a relationship was also reported for composites such as kenaf and dolomite-reinforced Low-Density Polyethylene (LDPE) composites [28]. Therefore, dolomite in the composite material could act as a secondary filler to enhance their stiffness characteristics.

Figure 7 shows the elongation at the break of the kenaf/HDPE composites with different concentrations of AgNPs. From the observations, the elongation of kenaf/HDPE composites shows decreased with increasing the AgNP content compared to composites without AgNPs. This is due to the higher ductile behavior of kenaf/HDPE composites with increased AgNP content compared to those without AgNPs. The results for the elongation at break for various percentages of AgNP addition in kenaf/HDPE composites are illustrated in Figure 7. Such results show the decrease in elongation as a larger percentage of AgNPs is added to the composites. The kenaf/HDPE composites with a maximum addition of 10% AgNPs were able to display the lowest elongation, among other samples. The elongation at break follows a reverse trend with that of tensile strength and modulus. As the concentration of AgNPs increases, the amount of elongation at break decreased. This phenomenon may be due to excessive polymer and fiber interactions, hard domain rotation, interface slippages and fibrillation on the matrix [35,36]. In addition, it might occur because AgNP content is a large percentage of crystallinity, slowly losing its ductile properties, with additional rigid silver nanoparticles. The impact of AgNP addition to kenaf/HDPE composites has, therefore, been shown to be more essential for tensile stress and tensile modulus, whereas the elongation is dropped significantly due to the addition of AgNPs that restrict the molecule chain to move, which makes the chain lose its ductility and high in stiffness.

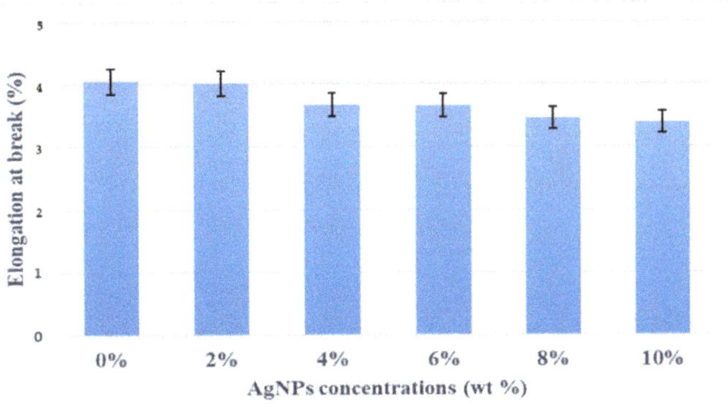

Figure 7. Effects of AgNP content on elongation at break of kenaf/HDPE composites.

3.2. Thermal Properties

3.2.1. Thermogravimetric Analysis (TGA)

TGA measured the thermal stability of the pure kenaf/HDPE composites with different concentrations of silver nanopowder. The degradation temperatures of kenaf/HDPE composites with different contents of AgNPs (0 %,2 %,4 %,6 %,8 % and 10 %) are exhibited in the TGA graph shown in Figure 8. The curves indicate that thermal degradation only occurred after some concentrations of heat energy had been incorporated into the components. Additionally, it is clear that all the kenaf/HDPE composites with different AgNP contents showed that there was no significant difference in the value of thermal stability from each other. The onset degradation temperature was observed.

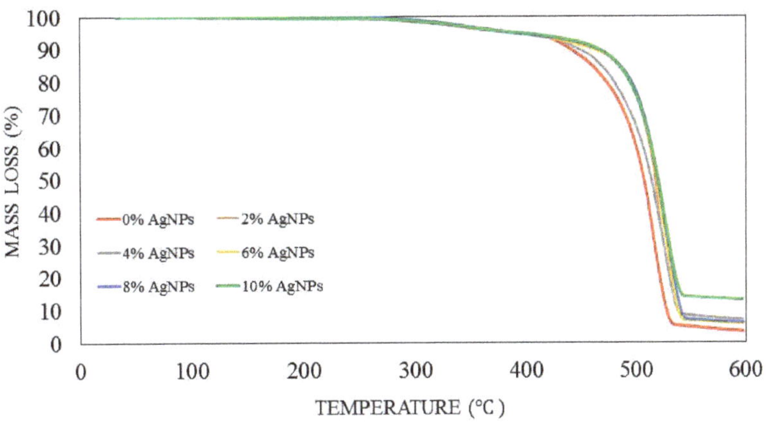

Figure 8. TGA of kenaf/HDPE composites at different AgNP contents.

The experimental result shows the same thermal degradation trend for all composites. The trend graph shows a decrease for all composites with and without AgNP content. It is also observed from the diagram that kenaf/HDPE composites with AgNP content are more stable compared to those without AgNPs at a temperature of 400–550 °C. In thermal experiments, there are four visible phases. The first phase is at a temperature of 30–100 °C, where, at this temperature, the mass of the sample will decrease because the evaporation moisture process takes place on all composite samples. Due to naturally hydrophilic characteristics, kenaf has moisture content at room temperature and humidity before the composite fabrication process [32,37,38]. The second phase is the decomposition of the element lignocellulosic at a temperature range of 200–320 °C [13,39], followed by the third phase, which is the decomposition of hemicellulose at a temperature range of 350–450 °C [40,41], and finally, the last phase is the decomposition of cellulose lignin and ash [16]. Table 1 shows the result degradation of different AgNP contents on kenaf/HDPE composites. From the result, AgNP wt10% shows that the highest percentage char residue was 12.96%.

Table 1. The initial and maximum degradation temperature values of the kenaf/HDPE composites from TGA thermograms.

AgNPs (%)	Maximum Degradation Temperature (°C)	Weight Loss (%)	Char at 600 °C (wt%)
0	472	80.45	3.27
2	479	80.42	6.02
4	493	80.40	6.97
6	494	80.43	5.47
8	495	80.50	6.77
10	496	80.41	12.96

The result shows that the addition of AgNP content of kenaf/HDPE composites showed an increase in thermal stability compared to without AgNPs. It has been known that AgNPs have excellent thermal stability and volume ratio properties [3,42]. With the increase in the AgNP content, the thermal decomposition properties of kenaf/HDPE composites were further increased. This occurs due to the increment of the molecular weight of the kenaf/HDPE composites. Besides, the addition of AgNP content of kenaf/HDPE composites has improved its thermal properties and also because the AgNP content has high decomposition properties. The same finding has been investigated by [28], where the addition of other materials, such as dolomite, as well as increasing its content into the composite system can improve the thermal stability properties. The TGA thermograms are of the LDPE/kenaf composites with different contents of dolomite, and the dolomite possesses the highest thermal decomposition characteristics. With the increase in the content of dolomite in the composites, the thermal stability properties also increase, and this is due to the properties of dolomite possessing the highest thermal decomposition characteristics. Overall, the addition of AgNP content as a secondary filler provides benefits to improve the thermal stability properties of kenaf/HDPE composites.

3.2.2. Differential Scanning Calorimetry (DSC)

The DSC thermograms, melting point (T_m) and glass transition temperature (T_g) values of the kenaf/HDPE composites with and without AgNPs are shown in Figure 9 and Table 2.

DSC study on materials is important because it can analyze the properties of composite materials [43]. From Figure 9, the glass transition temperatures, T_m (midpoint values), of kenaf/HDPE composites with AgNPs are greater than those of kenaf/HDPE composites without AgNPs. The results indicate that the addition of AgNPs tends to increase the prohibition of the movement of the polymer matrix chain. The T_m (onset value) of the kenaf/HDPE composites also significantly increased with increasing the AgNPs contents from 2 wt% to 10 wt%, and it indicates that the addition of AgNPs into Kenaf/HDPE composites is more stable than without AgNP content. Kenaf/HDPE composites with more AgNPs have a T_m (onset value) slightly higher than those with less concentration of AgNPs, and therefore the thermal properties of kenaf/HDPE composites have also been impacted by AgNPs. The T_m (onset value) for kenaf/HDPE with 10 wt% is 128.53 °C when compared to those with a lower concentration of AgNPs. The T_m (onset value) increases due to AgNP content, which can make the structure between kenaf and HDPE change and increase the structure by the content of AgNPs. Combining AgNPs into composites enhanced the thermal stability and was similarly reported by another author [44–46]. Variations in the T_m (midpoint value) might also adversely determine the compatibility of the samples. From Table 2, Tm (midpoint value) levels of kenaf/HDPE composites with 0 wt% of AgNPs is lesser (129.97 °C) than kenaf/HDPE composites with 10 wt% of AgNPs (133.83 °C). The intermolecular forces between kenaf/HDPE composites matrixes are due to AgNPs in mixes, which can strengthen the interfacial bond between mixing components [37,45]. The combination of kenaf/HDPE composites with AgNPs provided the advantages to produce quality multifunctional materials and it is also suitable for various manufacturing applications.

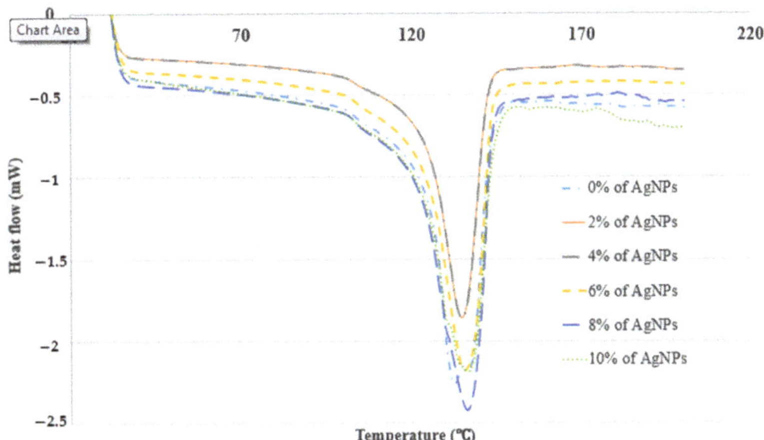

Figure 9. DSC thermograms of the kenaf/HDPE composites containing AgNPs.

Table 2. Thermal properties of kenaf/HDPE composites without and with AgNPs, as measured from DSC thermograms.

AgNPs (wt %)	T_m (Onset Value) (°C)	T_m (Midpoint Value) (°C)
0	120.76	129.97
2	120.90	129.98
4	121.17	132.20
6	121.90	132.75
8	122.78	133.45
10	128.53	133.83

3.3. Antimicrobial Testing

The test is usually conducted for the antibiotic sample that is most successful in treating bacteria or fungal infections. Antimicrobial/Antifungal Test Microbial reactions to antimicrobial agents distinguish from each other (sensitivities/resistances). Two widespread pathogenic bacterial, P. aeruginosa and E. coli, were chosen to assess the antibacterial activity of the kenaf/HDPE with Silver Nanopowder (AgNP), while C. *albicans* was analyzed for the fungal inhibition in Table 3. The bacteria that is commonly used in research can be sorted into two groups, i.e., Gram-negative bacteria and Gram-positive bacteria. In this research, foodborne pathogenic bacteria, which are E. *coli* (ATCC 25922) and P. *aeruginosa* (ATCC 15442) [47], were used to carry out the antibacterial testing.

Table 3. Antimicrobial properties of kenaf/HDPE composite materials.

Concentration of AgNPs (wt %)	Target Microbes		
	Escherichia Coli UPMC 25922	Pseudomonas Aeruginosa ATCC 15442	Candida Albicans ATCC 90028
0	−	−	−
2	−	−	−
4	−	−	−
6	−	−	−
8	−	−	+
10	+	+	+

+ve: the presence of inhibition zone against microbes. −ve: no indication of inhibition zone against microbes.

Based on Tables 3 and 4, the kenaf/HDPE composites with AgNPs revealed a stronger activity against the gram-negative bacteria compared without AgNP composition, which is significantly lower than the activity seen in the earlier literature [5,48]. Antimicrobial variation may be attributable to their different configurations of cell walls [42,49]. According to Priyadarshini et.al. [50], Gram-negative bacteria consist of thin peptidoglycan (7–8 nm). Consequently, the thin layer of peptidoglycan allows the penetration of silver nanoparticles in microbial cells to destroy. Additional microorganisms, specifically the pathogenic fungus *C. albicans*, were also tested, and the processed kenaf/HDPE composites filled with AgNPs exhibited greater areas of inhibition compared to Gram-negative bacteria, which followed the work of Veranitisagul et al. [49]. Once again, the process is specifically connected to the microorganisms' cell wall structure.

Table 4. Antibacterial study of kenaf/HDPE composites.

Microbes Concentration of AgNPs (wt%)	Inhibition Zone (mm)	
	Escherichia Coli UPMC 25922	Pseudomonas Aeruginosa ATCC 15442
10	12.2	14.3

Kenaf/HDPE composites with 10wt% of AgNPs demonstrated the strongest inhibition zone against E. coli (ATCC 25922) above all the samples. *C. albicans*, underlining the significance of Ag particles in 8wt% and 10wt% of kenaf/HDPE composites, filled AgNP contents. The findings show that silver nanoparticles exhibit greater sensitivity towards Pseudomonas aeruginosa than Escherichia coli in extracts for bacterial activity. Such bacteria displayed an inhibition zone ranging from 10–15 to 10–13 mm, respectively.

Table 5 shows the inhibition zone measurement of bacteria, whereas Table 6 shows the inhibition zone measurement of yeast that had been used for the antimicrobial testing.

Table 5. Antifungal study of kenaf/HDPE composites.

Microbes Concentration of AgNPs (wt%)	Inhibition Zone (mm)
	Candida Albicans ATCC 90028
8	6
10	17.5

Table 6. Positive tests for antimicrobial properties on E. coli, P. aeruginosa and *C. albicans* using the agar disk diffusion method.

Microbes	Images
Escherichia coli UPMC 25922	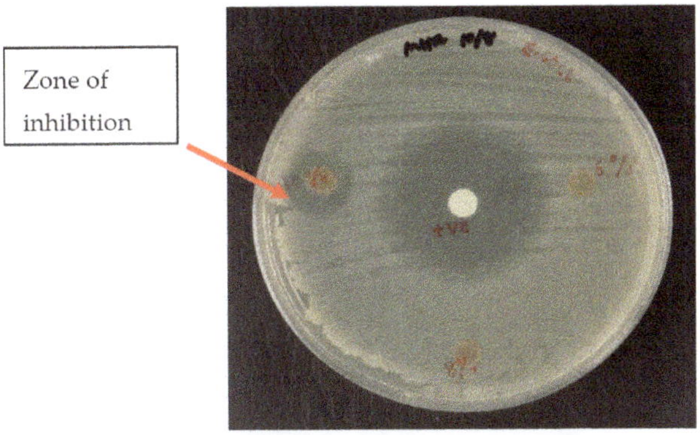
Pseudomonas aeruginosa ATCC 15442	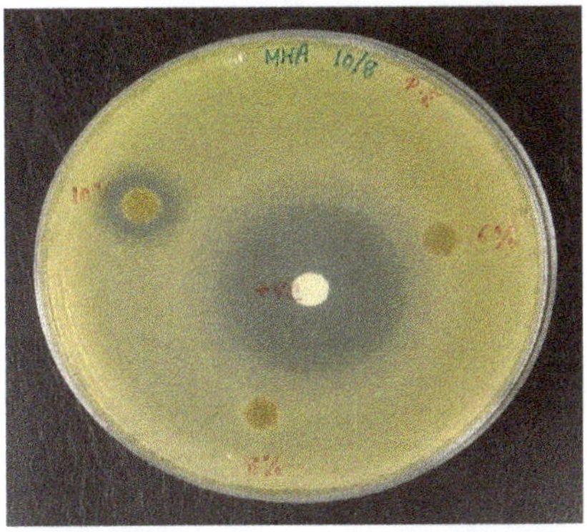
Candida albicans ATCC 90028	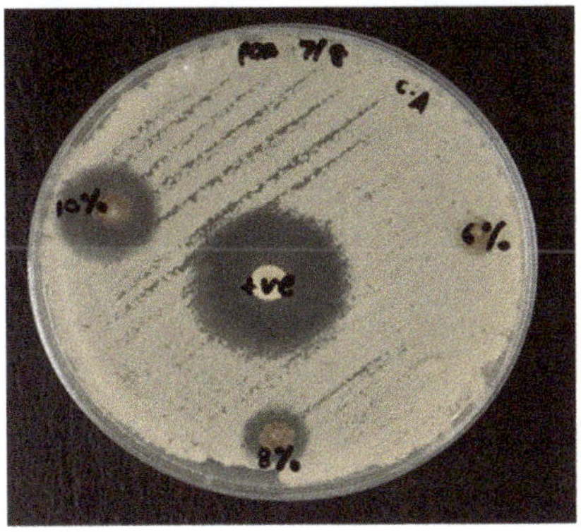

Table 6 displays improved results for fungal infection relative to the Candida albicans test where positive results are acquired for samples of kenaf/HDPE composites with 8 wt% and 10 wt % of AgNPs. Silver nanopowder and the microbial reaction were specifically on the surface when the materials were placed for a long time in the surrounding environment. This stresses the fact that although kenaf/HDPE composites filled AgNPs, they are antimicrobial and biodegradable materials. Those kenaf/HDPE-composite-filled

AgNPs may also be one of the potential technologies for forthcoming plastics that are environmentally sustainable.

4. Conclusions

The effect of AgNPs contents on tensile, thermal and antimicrobial properties of kenaf/HDPE composites was investigated in this study. The tensile strength and modulus of the kenaf/HDPE composites show improvement with the presence of AgNPs, and elongation at break was reduced at the highest content, as shown by the mechanical testing result. Kenaf/HDPE composites revealed the highest tensile stress and tensile modulus at 10 wt% AgNPs, which are 5.05 MPa and 5.23 GPa, respectively. Meanwhile, with regard to thermal properties, there was a large increase in the alternative values of T_g and T_m due to the participation of AgNP content. The temperature of T_g has changed positively. Relative to kenaf/HDPE composites, HDPE with 10 wt% AgNP content displayed the maximum degradation temperature of 496 °C. The main factors in this research are kenaf/HDPE composites with 10 wt% of AgNPs that exhibit antimicrobial activity against E. coli, P. aeruginosa and *C. albicans* inhibition using an agar disk diffusion test as an output. Kenaf/HDPE composites with AgNP content are expected to be one of the reasonable manufacturing materials, as they showed strong antimicrobial properties that may minimize the requirements for the cleaning process. The findings obtained include the fact that AgNPs can inhibit the growth of microbes. This shows that kenaf/HDPE composites are biodegradable and exhibit antimicrobial properties that can be implemented in the interior part of aircraft and food packaging.

Author Contributions: The effect of silver nanopowder presented here was conducted in cooperation between all authors. V.S. wrote the manuscript, M.T.H.S. supervised and conceived the research project, A.M.R. edited the manuscript, A.A.S. constructed the research methods, A.U.M.S. undertook project management, S.N.A.S. analyzed data and A.A.B. gathered resources. All authors have read and agreed to the published version of the manuscript.

Funding: Please add: This research was funded by Universiti Putra Malaysia (UPM), grant number Geran Putra Berimpak 9668200 and the APC was funded by Universiti Putra Malaysia.

Institutional Review Board Statement: Not applicable.

Informed Consent Statement: Not applicable.

Data Availability Statement: The data presented in this study are available on request from the corresponding author.

Acknowledgments: The authors would like to thank Universiti Putra Malaysia for the financial support through Geran Putra Berimpak 9668200. The authors would like to thank the Department of Aerospace Engineering, Faculty of Engineering, Universiti Putra Malaysia, and Laboratory of Biocomposite Technology, Institute of Tropical Forestry and Forest Product (INTROP), Universiti Putra Malaysia (HICOE) and Malaysia-Japan International Institute of Technology, Universiti Teknologi Malaysia for the close collaboration in this research.

Conflicts of Interest: The authors declare no conflict of interest.

References

1. Mosiewicki, M.A.; Aranguren, M.I. A short review on novel biocomposites based on plant oil precursors. *Eur. Polym. J.* **2013**, *49*, 1243–1256. [CrossRef]
2. Wegner, T.H.; Jones, E.P. A fundamental review of the relationships between nanotechnology and lignocellulosic biomass. *Nanosci. Technol. Renew. Biomater.* **2009**, *1*, 1–41.
3. Zhang, X.F.; Liu, Z.G.; Shen, W.; Gurunathan, S. Silver nanoparticles: Synthesis, characterization, properties, applications, and therapeutic approaches. *Int. J. Mol. Sci.* **2016**, *17*, 1534. [CrossRef] [PubMed]
4. Rhim, J.W.; Wang, L.F.; Hong, S.I. Preparation and characterization of agar/silver nanoparticles composite films with antimicrobial activity. *Food Hydrocoll.* **2013**, *33*, 327–335. [CrossRef]
5. Shankar, S.; Jaiswal, L.; Aparna, R.S.L.; Prasad, R.G.S.V. Synthesis, characterization, in vitro biocompatibility, and antimicrobial activity of gold, silver and gold silver alloy nanoparticles prepared from Lansium domesticum fruit peel extract. *Mater. Lett.* **2014**, *137*, 73–75. [CrossRef]

6.   Chernousova, S.; Epple, M. Silver as Antibacterial Agent: Ion, Nanoparticle, and Metal. *Angew. Chem. Int. Ed.* **2013**, *52*, 1636–1653. [CrossRef]

7.   Saba, N.; Tahir, P.M.; Jawaid, M. A review on potentiality of nano filler/natural fiber filled polymer hybrid composites. *Polymers* **2014**, *6*, 2247–2273. [CrossRef]

8.   Nurani, S.J.; Saha, C.K. Silver Nanoparticles Synthesis, Properties, Applications and Future Perspectives: A Short Review Silver Nanoparticles Synthesis, Properties, Applications and Future Perspectives: A Short Review Sharif Masnad Hossain Sunny. *IOSR J. Electr. Electron. Eng.* **2015**, *10*, 117–126.

9.   Elnour, K.M.M.A.; Alwarthan, A.; Ammar, R.A.A. Synthesis and applications of silver nanoparticles. *Arab. J. Chem.* **2010**, *3*, 135–140.

10.   Aji, I.S.; Sapuan, S.M.; Zainudin, E.S.; Abdan, K. Kenaf fibres as reinforcement for polymeric composites: A review. *Int. J. Mech. Mater. Eng.* **2009**, *4*, 239–1248.

11.   Anuar, H.; Zuraida, A. Improvement in mechanical properties of reinforced thermoplastic elastomer composite with kenaf bast fibre. *Compos. Part B Eng.* **2011**, *42*, 462–465. [CrossRef]

12.   Ali, M.R.; Salit, M.S.; Jawaid, M.; Mansur, M.R.; Manap, M.F.A. *Polyurethane Polymers Composites and Nanocomposites*; Elsevier: Amsterdam, The Netherlands, 2018; pp. 521–546.

13.   Radzi, A.M.; Sapuan, S.M.; Jawaid, M.; Mansor, M.R. Influence of fibre contents on mechanical and thermal properties of roselle fibre reinforced polyurethane composites. *Fibers Polym.* **2017**, *18*, 1353–1358. [CrossRef]

14.   Kumar, V.; Kumari, M. Processing and characterization of natural cellulose fibers/thermoset polymer composites. *Carbohydr. Polym.* **2014**, *109*, 102–117.

15.   Prithivirajan, R.; Jayabal, S.; Bharathiraja, G. Bio-based composites from waste agricultural residues: Mechanical and morphological properties. *Cellul. Chem. Technol.* **2015**, *49*, 65–68.

16.   Radzi, A.M.; Sapuan, S.M.; Jawaid, M.; Mansor, M.R. Water absorption, thickness swelling and thermal properties of roselle/sugar palm fibre reinforced thermoplastic polyurethane hybrid composites. *J. Mater. Res. Technol.* **2019**, *8*, 3988–3994. [CrossRef]

17.   Radzi, A.M.; Sapuan, S.M.; Jawaid, M.; Mansor, M.R. Effect of Alkaline Treatment on Mechanical, Physical and Thermal Properties of Roselle/Sugar Palm Fiber Reinforced Thermoplastic Polyurethane Hybrid Composites. *Fibers Polym.* **2019**, *20*, 847–855. [CrossRef]

18.   Venkateshwaran, N.; ElayaPerumal, A. Mechanical and water absorption properties of woven jute/banana hybrid composites. *Fibers Polym.* **2012**, *13*, 907–914. [CrossRef]

19.   Abdallah, F.B.; Cheikh, R.B.; Baklouti, M.; Denchev, Z.; Cunha, A.M. Effect of surface treatment in cork reinforced composites. *J. Polym. Res.* **2010**, *17*, 519–528. [CrossRef]

20.   Shamsuri, A.A.; Azid, M.K.A.; Ariff, A.H.M.; Sudari, A.K. Influence of surface treatment on tensile properties of low-density polyethylene/cellulose woven biocomposites: A preliminary study. *Polymers.* **2014**, *6*, 2345–2356. [CrossRef]

21.   Joshi, S.V.; Drzal, L.T.; Mohanty, A.K.; Arora, S. Are natural fiber composites environmentally superior to glass fiber reinforced composites? *Compos. Part A Appl. Sci. Manuf.* **2004**, *35*, 371–376. [CrossRef]

22.   Yahaya, R.; Sapuan, S.M.; Jawaid, M.; Leman, Z.; Zainudin, E.S. Effect of layering sequence and chemical treatment on the mechanical properties of woven kenaf-aramid hybrid laminated composites. *Mater. Des.* **2015**, *67*, 173–179. [CrossRef]

23.   Nishino, T.; Hirao, K.; Kotera, M.; Nakamae, K.; Inagaki, H. Kenaf reinforced biodegradable composite. *Compos. Sci. Technol.* **2003**, *63*, 1281–1286. [CrossRef]

24.   El-shekeil, Y.A.; Sapuan, S.M.; Abdan, K.; Zainudin, E.S. Influence of fiber content on the mechanical and thermal properties of Kenaf fiber reinforced thermoplastic polyurethane composites. *Mater. Des.* **2012**, *40*, 299–303. [CrossRef]

25.   Zampaloni, M.; Pourboghrat, F. Kenaf natural fiber reinforced polypropylene composites: A discussion on manufacturing problems and solutions. *Compos. Part A Appl. Sci. Manuf.* **2007**, *38*, 1569–1580. [CrossRef]

26.   Azammi, A.M.N.; Sapuan, S.M.; Ishak, M.R.; Sultan, M.T.H. Mechanical and Thermal Properties of Kenaf Reinforced Thermoplastic Polyurethane (TPU)-Natural Rubber (NR) Composites. *Fibers Polym.* **2018**, *19*, 446–451. [CrossRef]

27.   Yahaya, R.; Sapuan, S.M.; Jawaid, M.; Leman, Z.; Zainudin, E.S. Mechanical performance of woven kenaf-Kevlar hybrid composites. *J. Reinf. Plast. Compos.* **2014**, *33*, 2242–2254. [CrossRef]

28.   Shamsuri, A.A.; Zolkepli, M.N.M.; Ariff, A.H.M.; Sudari, A.K.; Zarin, M.A. A Preliminary Investigation on Processing, Mechanical and Thermal Properties of Polyethylene/Kenaf Biocomposites with Dolomite Added as Secondary Filler. *J. Compos.* **2015**, *2015*, 1–7. [CrossRef]

29.   Dhawan, V.; Singh, S.; Singh, I. Effect of Natural Fillers on Mechanical Properties of GFRP Composites. *J. Compos.* **2013**, 1–8. [CrossRef]

30.   ASTM. Standard Test Method for Compositional Analysis by Thermogravimetry 1. *ASTM Int.* **2015**, *8*, 6.

31.   Barud, H.S.; Regiani, T.; Marques, R.F.C.; Lustri, W.R.; Messaddeq, Y.; Ribeiro, S.J.L. Antimicrobial bacterial cellulose-silver nanoparticles composite membranes. *J. Nanomater.* **2011**, *2011*, 721631. [CrossRef]

32.   Radzi, A.M.; Sapuan, S.M.; Jawaid, M.; Mansor, M.R. Mechanical and Thermal Performances of Roselle Fiber-Reinforced Thermoplastic Polyurethane Composites. *Polym. Plast. Technol. Eng.* **2018**, *57*, 601–608. [CrossRef]

33.   Sapiai, N.; Jumahat, A.; Mahmud, J. Flexural and Tensile Properties of Kenaf/Glass Fibres Hybrid Composites. *J. Teknol.* **2015**, *3*, 115–120.

34. Zakaria, M.R.; Akil, H.M.; Kudus, M.H.A.; Kadarman, A.H. Improving flexural and dielectric properties of MWCNT/epoxy nanocomposites by introducing advanced hybrid filler system. *Compos. Struct.* **2015**, *132*, 50–64. [CrossRef]
35. Finnigan, B.; Martin, D.; Halley, P.; Truss, R.; Campbell, K. Morphology and properties of thermoplastic polyurethane nanocomposites incorporating hydrophilic layered silicates. *Polymer* **2004**, *45*, 2249–2260. [CrossRef]
36. Xie, W.; Gao, Z.; Pan, W.P.; Hunter, D.; Singh, A.; Vaia, R. Thermal degradation chemistry of alkyl quaternary ammonium Montmorillonite. *Chem. Mater.* **2001**, *13*, 2979–2990. [CrossRef]
37. Kian, L.K.; Jawaid, M.; Ariffin, H.; Karim, Z.; Sultan, M.T.H. Morphological, physico-chemical, and thermal properties of cellulose nanowhiskers from roselle fibers. *Cellulose* **2019**, *26*, 6599–6613. [CrossRef]
38. Chee, S.S.; Jawaid, M.; Sultan, M.T.H. Thermal Stability and Dynamic Mechanical Properties of Kenaf/Bamboo Fibre Reinforced Epoxy Composites. *BioResources* **2017**, *12*, 7118–7132.
39. Asim, M.; Paridah, M.T.; Saba, N. Thermal, physical properties and flammability of silane treated kenaf/pineapple leaf fibres phenolic hybrid composites. *Compos. Struct.* **2018**, *202*, 1330–1338. [CrossRef]
40. Khalil, H.A.; Suraya, N.; Atiqah, N.; Jawaid, M.; Hassan, A. Mechanical and thermal properties of chemical treated kenaf fibres reinforced polyester composites. *J. Compos. Mater.* **2013**, *47*, 3343–3350. [CrossRef]
41. El-Shekeil, Y.A.; Sapuan, S.M.; Abdan, K.; Zainudin, E.S.; Al-SHUJA'A, O.M. Effect of pMDI isocyanate additive on mechanical and thermal properties of Kenaf fibre reinforced thermoplastic polyurethane composites Y. *Bull. Mater. Sci.* **2012**, *35*, 1151–1155.
42. Shankar, S.; Wang, L.F.; Rhim, J.W. Preparations and characterization of alginate/silver composite films: Effect of types of silver particles. *Carbohydr. Polym.* **2016**, *146*, 208–216. [CrossRef]
43. Sahari, J.; Sapuan, S.M.; Zainudin, E.S.; Maleque, M.A. Physico-chemical and thermal properties of starch derived from sugar palm tree (Arenga pinnata). *Asian J. Chem.* **2014**, *26*, 955–959. [CrossRef]
44. Volova, T.G.; Shumilova, A.A.; Shidlovskiy, I.P. Antibacterial properties of films of cellulose composites with silver nanoparticles and antibiotics. *Polym. Test.* **2018**, *65*, 54–68. [CrossRef]
45. Yu, H.; Sun, B.; Zhang, D.; Chen, G.; Yang, X.; Yao, J. Reinforcement of biodegradable poly(3-hydroxybutyrate-co-3-hydroxyvalerate) with cellulose nanocrystal/silver nanohybrids as bifunctional nanofillers. *J. Mater. Chem. B.* **2014**, *2*, 8479–8489. [CrossRef] [PubMed]
46. Yu, H.Y.; Yang, X.Y.; Lu, F.F.; Chen, G.Y.; Yao, J.M. Fabrication of multifunctional cellulose nanocrystals/poly(lactic acid) nanocomposites with silver nanoparticles by spraying method. *Carbohydr. Polym.* **2016**, *140*, 209–219. [CrossRef] [PubMed]
47. Jumaidin, R.; Sapuan, S.M.; Jawaid, M.; Ishak, M.R.; Sahari, J. Characteristics of Thermoplastic Sugar Palm Starch/Agar Blend: Thermal, Tensile, and Physical Properties. *Int. J. Biol. Macromol.* **2016**, *89*, 575–581. [CrossRef] [PubMed]
48. Syafiq, R.; Sapuan, S.M.; Zuhri, M.R.M. Antimicrobial activities of starch-based biopolymers and biocomposites incorporated with plant essential oils: A review. *Polymers* **2020**, *12*, 2403. [CrossRef] [PubMed]
49. Veranitisagul, C.; Wattanathana, W.; Wannapaiboon, S.; Hanlumyuang, Y.; Sukthavorn, K.; Nootsuwan, N.; Chotiwan, S.; Phuthong, W.; Jongrungruangchok, S.; Laobuthee, A. Antimicrobial, conductive, and mechanical properties of AgCB/PBS composite system. *J. Chem.* **2019**. [CrossRef]
50. Priyadarshini, S.; Gopinath, V.; Priyadharsshini, N.M.; MubarakAli, D.; Velusamy, P. Synthesis of anisotropic silver nanoparticles using novel strain, Bacillus flexus and its biomedical application. *Colloids Surf. B* **2013**, *102*, 232–237. [CrossRef] [PubMed]

Article

Non Edible Oil-Based Epoxy Resins from Jatropha Oil and Their Shape Memory Behaviors

Lu Lu Taung Mai [1,2], Min Min Aung [1,3,4,*], Sarah Anis Muhamad Saidi [3], Paik San H'ng [1,5,*], Marwah Rayung [1] and Adila Mohamad Jaafar [3,4]

1 Higher Education Centre of Excellence (HiCoE), Institute of Tropical Forestry and Forest Products, University Putra Malaysia, Serdang 43400, Malaysia; lulutawngmai@gmail.com (L.L.T.M.); marwahrayung@yahoo.com (M.R.)
2 Department of Chemistry, Universiry of Myitkyina, Kachine State 01011, Myanmar
3 Department of Chemistry, Faculty of Science, Universiti Putra Malaysia, Serdang 43400, Malaysia; sarah.anisss@gmail.com (S.A.M.S.); adilamj@upm.edu.my (A.M.J.)
4 Chemistry Division, Centre of Foundation Studies for Agricultural Science, University Putra Malaysia, Serdang 43400, Malaysia
5 Department of Forestry and Environment, Faculty of Forestry, Universiti Putra Malaysia, Serdang 43400, Malaysia
* Correspondence: minmin_aung@upm.edu.my (M.M.A.); ngpaiksan@upm.edu.my (P.S.H.)

Abstract: The use of bio-based polymers in place of conventional polymers gives positives effects in the sense of reduction of environmental impacts and the offsetting of petroleum consumption. As such, in this study, jatropha oil was used to prepare epoxidized jatropha oil (EJO) by the epoxidation method. The EJO was used to prepare a shape memory polymer (SMP) by mixing it with the curing agent 4-methylhexahydrophthalic anhydride (MHPA) and a tetraethylammonium bromide (TEAB) catalyst. The resulting bio-based polymer is slightly transparent and brown in color. It has soft and flexible properties resulting from the aliphatic chain in jatropha oil. The functionality of SMP was analyzed by Fourier transform infrared (FTIR) spectroscopy analysis. The thermal behavior of the SMP was measured by thermogravimetric analysis (TGA), and it showed that the samples were thermally stable up to 150 °C. Moreover, the glass transition temperature characteristic was obtained using differential scanning calorimetry (DSC) analysis. The shape memory recovery behavior was investigated. Overall, EJO/MHPA was prepared by a relatively simple method and showed good shape recovery properties.

Keywords: epoxidized jatropha oil; shape memory polymer; bio-based polymer; jatropha oil

check for **updates**

Citation: Taung Mai, L.L.; Aung, M.M.; Muhamad Saidi, S.A.; H'ng, P.S.; Rayung, M.; Jaafar, A.M. Non Edible Oil-Based Epoxy Resins from Jatropha Oil and Their Shape Memory Behaviors. *Polymers* **2021**, *13*, 2177. https://doi.org/10.3390/polym13132177

Academic Editors: Emin Bayraktar, S. M. Sapuan and R. A. Ilyas

Received: 27 April 2021
Accepted: 19 May 2021
Published: 30 June 2021

Publisher's Note: MDPI stays neutral with regard to jurisdictional claims in published maps and institutional affiliations.

1. Introduction

Shape memory polymers (SMPs) are a class of emerging smart materials that can change shape and "remember" their original shape. They have a sensitive response to external stimuli such as pH, humidity, light, electricity, temperature, and so on [1–3]. To date, most of the studies reported on SMPs are stimulated by temperature. In general, SMPs contain two phases: the permanent shape and the temporary shape. The permanent shape can be either covalent or physical cross-links. The temporary shape is set by deformation above a certain transition temperature [4]. They typically rely on vitrification, crystallization, or some other physical interaction. When the polymers are reheated above the transition temperature, the oriented polymer chains are released, resulting in recovery of the permanent shape. SMPs have attracted significant attention due to their inherent advantages like good processability, low cost, and high recovery ability [5]. At present, shape memory polymers have been applied in many fields such as aerospace [6], medicine [7,8], self-finishing smart textiles [9], electronic devices [10], and self-assembling structures [11]. Various types of polymers have been found to have shape memory effects including polyurethane [12,13], trans-polyisoprene [14], styrene-butadiene copolymer [15], and epoxy material [16].

Recently, considerable efforts have been expended to develop bio-based SMPs. Bio-based polymers are gaining attention as they have renewability features couples with their economic and environmental benefits. In essence, there have been substantial studies reported on bio-based epoxy materials for shape memory polymers [17–19]. Among all renewable resources, vegetable oils are expected to be an ideal alternative to chemical feedstocks in preparing epoxy materials. Vegetable oil is one of the cheapest and most abundant types of biomass, which is available in large quantities [20]. Vegetable oil is predominantly made up of triglyceride molecules with varying compositions of fatty acids. Some examples of vegetable oils that can be used include palm oil, soybean oil, linseed oil, sunflower oil, castor oil, jatropha oil, and many others [19].

This study focuses on the utilization of jatropha oil as the raw material to prepare epoxy resins for SMP production. Jatropha oil is obtained from the seeds of jatropha curcas tree. Jatropha curcas is a succulent plant that belongs to the Euphorbiaceae family, widely grown in South America, South-West Asia, India, and Africa. The oil content of jatropha seeds was reported at 63%, which is higher than that of palm kernel, linseed, and soybean oil content [21]. Jatropha oil is classified as a non-edible oil due to the presence of toxic components in the structure, such as phorbol ester, curcains, and jatropherol [22], and thus it cannot be used for nutritional purposes without detoxification. By being a non-edible oil, there will be no unnecessary competition with the food-related industries. Other than that, jatropha oil contains high amounts of unsaturation, which is made up mostly by oleic and linoleic acid. The chemical compositions of the oil differ depending on the climate and locality of the jatropha tree. The high degree of unsaturation of the oil offers a wide alternative for chemical modification. In this study, we describe the preparation and characterization of bio-based shape memory materials from epoxidized jatropha oil. Further, the thermal and shape memory properties of the resulting materials are investigated.

2. Materials and Methods

2.1. Materials

Crude jatropha oil was supplied by Biofuel Bionas Sdn Bhd. Formic acid (98%), hydrogen peroxide (30%), sulphuric acid (95%), methanol (99%), and crystal violet (96%) were supplied by R&M Chemicals, Dundee, UK. Magnesium sulphate (97%), hydrogen bromide (33%), 4-methylhexhydrophthalic anhydride (MHPA) (98%), and tetraethyl ammonium bromide (TEAB) (98%) were obtained from ACROS Organic, Carlsbad, CA, USA. Potassium hydrogen phthalate (99%) was obtained from AJAX Chemicals, Taren Point, Australia, and chlorobenzene (99%) was supplied by Merck, Kenilworth, NJ, USA. All chemicals were used as received.

2.2. Preparation of Epoxidized Jatropha Oil

The epoxidized jatropha oil (EJO) was prepared in bulk according to the procedures reported in the literature [23,24]. A mixture of jatropha oil and formic acid was heated to 40 °C under continuous stirring in a water bath. Once the temperature reached 40 °C, hydrogen peroxide was added in a dropwise manner to avoid overheating. Then, the temperature was increased and maintained at 60 °C for 5 h. Oxirane oxygen content (OOC) analysis was conducted every 1 h. After 5 h of stirring, the mixture was cooled down to room temperature. After that, the aqueous layer was discarded, and the produced epoxy was washed with an excess amount of distilled water until neutral pH was obtained. Lastly, a rotary evaporator was used to remove the remaining trace of water in the epoxy to obtain a pure EJO. The sample was kept with $MgSO_4$ drying agent and stored in a desiccator.

2.3. Preparation of EJO/MHPA

The shape memory polymer was prepared by mixing EJO, MHPA, and TEAB, as shown in Figure 1. TEAB was added as the catalyst in the reaction. Various molar ratios of anhydride to epoxy were prepared in this study, which were 0.6, 0.8, 1.0, 1.2, and 1.4. For the molar ratio of 1.0, 8.0 g of EJO, 5.6 g of MHPA, and 0.07 g of TEAB were mixed together [5]. The mixture was degassed under vacuum conditions for 10 min at 80 °C. Then, the mixture was poured into a Teflon mold and was kept in the oven for 22 h at 150 °C. The produced EJO/MPHA polymer was cooled down to room temperature. The same procedure was repeated to prepare SMP by using another molar ratio.

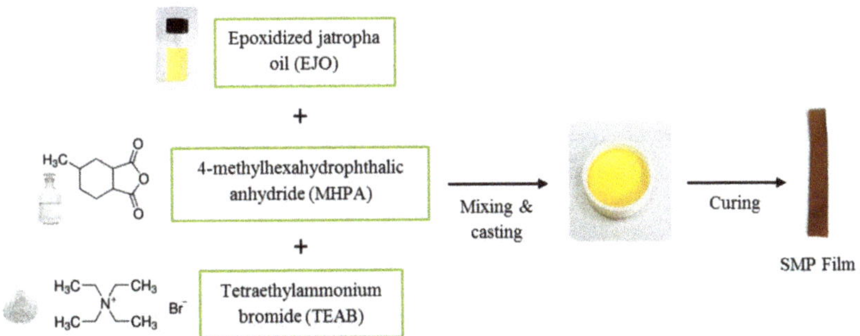

Figure 1. Schematic view of the preparation of jatropha oil-based shape memory polymer (SPM).

2.4. Characterization

2.4.1. Oxirane Oxygen Content

The measurement of oxirane oxygen (OOC) content during the preparation of EJO was carried out according to the ASTM D1652–97 Test Method A Standard [25]. This measurement was used to calculate the conversion of the double bond of the oil to oxirane rings. A total of 0.4 g of EJO was dissolved in 10 mL of chlorobenzene and 3–4 drops of crystal violet were used as the indicator. The mixture was titrated with standardized hydrogen bromide until it reached the endpoint, which was observed by color changes to green. The OOC was calculated based on Equation (1).

$$OOC\ (\%) = \frac{1.6N(V - B)}{W} \tag{1}$$

where N is the normality of standardized hydrogen bromide solution (N), V is the volume of standardized hydrogen bromide used for the EJO sample (mL), B is the volume of standardized hydrogen bromide used for the blank sample (mL), and W is the weight of the sample (g).

2.4.2. Fourier Transform Infrared Spectroscopy

Fourier transform infrared (FTIR) spectroscopy was used to analyze the presence of different functional groups in the samples. The analysis was conducted by using a SHIDMADZU IR Tracer-100 series spectrophotometer. The IR spectra of samples recorded were in the wavenumber range of 4000 cm^{-1} to 400 cm^{-1} at a scanning rate of 4 cm^{-1}.

2.4.3. Nuclear Magnetic Resonance Spectroscopy

The ^{1}H-NMR (proton NMR) and ^{13}C-NMR (carbon NMR) spectra were recorded using a 500 MHz JEOL Nuclear Magnetic Resonance (Peabody, MA, USA) at room temperature. The samples were dissolved in deuterated chloroform (CDCl$_3$). The ^{1}H and ^{13}C chemical shifts were recorded in ppm and referenced to tetramethylsilane (TMS) at 0.0 ppm.

2.4.4. Thermal Analysis

The thermal properties of the EJO/MHPA polymers with various ratios of anhydride to oxirane were evaluated by using a Perkin Elmer TGA7, Cleveland, OH, USA, thermogravimetric analyzer. This analysis was performed at temperatures ranging from 25 °C to 600 °C at a constant heating rate of 10 °C min^{-1} under a nitrogen atmosphere (flow rate: 50 mL min^{-1}). The changes in the weight of the sample as a function of temperature and/or time were measured. Additionally, the glass transition behavior of the SMP was investigated by using a Mettler Toledo, Grelfensee, Switzerland, differential scanning calorimetry (DSC) instrument. The scan was conducted at the rate of 10 °C min^{-1} from −20 °C to 120 °C with a nitrogen gas flow rate of 50° mL min^{-1}.

2.4.5. Investigation of the Shape Memory Recovery Behavior of EJO/MHPA Polymer

The shape memory behavior of the EJO/MPHA polymer was investigated by using the cured EJO/MPHA polymer. The samples were cut to produce a linear rectangular shape, and this was taken as their original shape. Then, the sample was heated at 150 °C and cooled down to room temperature, forming a temporary shape. When the temporary shape was heated again at 150 °C, the temporary shape returned back to its original shape.

3. Results

3.1. Oxirane Oxygen Content

During the epoxidation process, the double bonds of the unsaturated oil were converted into epoxy rings by using in situ generated performic acid. It was prepared by mixing formic acid and hydrogen peroxide. In this case, formic acid functioned as the oxygen carrier, and hydrogen peroxide acted as the oxygen donor. The reaction was carried out in situ, as the reaction is highly exothermic [26]. The conversion double bonds to epoxy groups were monitored using oxirane oxygen content (OOC) analysis. Figure 2 shows the OOC value against the reaction time pattern. As can be seen in Figure 2, the OOC value gradually increased as the reaction time increased and reached a maximum at 5 h. The produced epoxy jatropha oil had a maximum oxirane oxygen content of 3.63%. Prolonged reaction time will cause the OOC value to drop due to the possibility of the ring-opening reaction to occur. During the epoxidation process, the temperature must be controlled, as a high reaction temperature may intensify the rate of hydrogen peroxide decomposition and may exceed the rate of epoxy formation. As oxygen supplied by the hydrogen peroxide depleted, the epoxy formation ended earlier than anticipated. These unfavorable reactions resulted in lower oxirane oxygen content than the theoretical value. A similar observation of the OOC value trend of EJO was reported by a previous study [27]. Subsequently, the produced EJO in this study was used as the epoxide monomers to prepare EJO/MHPA polymers.

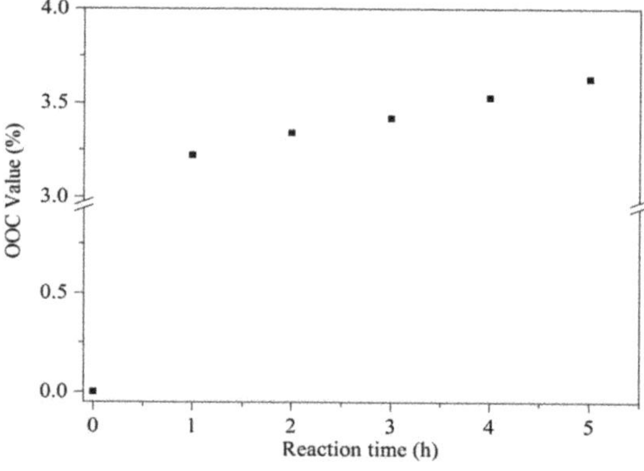

Figure 2. Oxirane oxygen content (OOC) value with time for the epoxidized jatropha oil process.

3.2. FTIR Analysis

FTIR analysis was conducted to identify chemical changes that occurred during the preparation process to the end product. The epoxidation of double bonds in jatropha oil forming an oxirane group and the ring opening of the oxirane group in the polymerization were observed. Figure 3 shows the FTIR spectra of jatropha oil, EJO, MHPA, and EJO/MHPA polymer. Based on Figure 3a for jatropha oil, the absorption bands at 2853 and 2922 cm^{-1} were attributed to the –CH symmetric and asymmetric stretching, respectively, of –CH$_2$ groups [28]. Moreover, the peaks at 721 and 1464 cm^{-1} were ascribed to –CH$_2$ rocking and bending vibrations. The intense peak at 1744 cm^{-1} indicated the C=O stretching vibration. The C–O–C stretching vibration of the ester group was located at 1163 cm^{-1}. The band at 3009 cm^{-1} was attributed to the C=C double bonds in the oil, which were oleic and linoleic acid [29]. In Figure 3b, the disappearance of the double bond functional group peak at 3009 cm^{-1}, which was initially present in the JO spectra, indicates the conversion of the double bond to the epoxy group. The new peaks formed at 823 and 841 cm^{-1} corresponded to the conversion of double bonds into epoxy groups. This is in agreement with similar findings reported by other researchers previously [27,30,31].

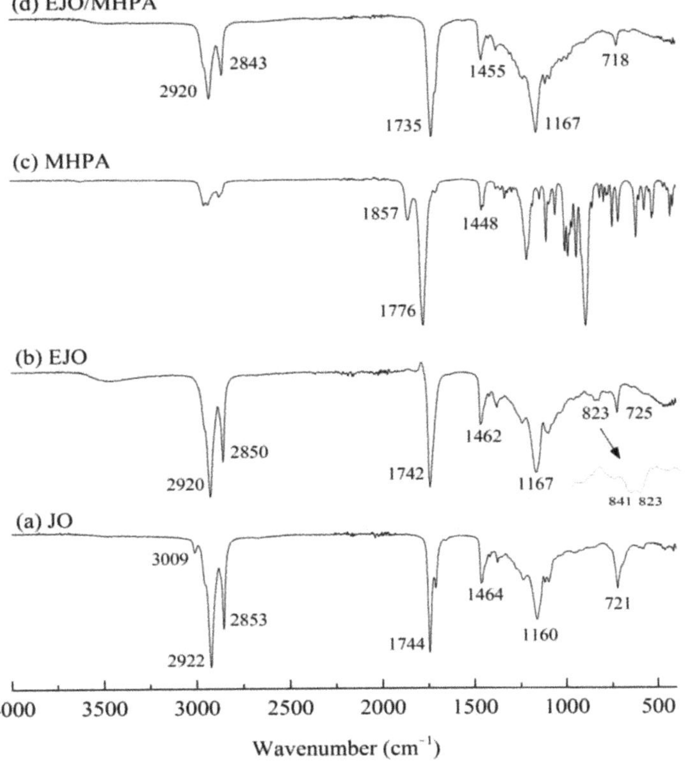

Figure 3. FTIR spectra of (a) JO, (b) EJO, (c) MHPA, and (d) EJO/MHPA SMP.

For the MHPA spectrum presented in Figure 3c, peaks of symmetric and asymmetric C=O stretching could be observed at the 1776 and 1857 cm^{-1} regions. Moreover, the spectrum for EJO/MHPA polymer, displayed in Figure 3d, showed a combination of both EJO and MHPA characteristics. A strong adsorption peak of C=O stretching of a carbonyl group present at 1735 cm^{-1} showed the formation of ester bonds between the MHPA and EJO [32]. Furthermore, the epoxy peaks also disappeared for EJO/MHPA polymer, confirming the formation of the bond. It can be deduced that there was a hydrolysis reaction of MHPA [5] and a ring opening of the oxirane group forming the EJO/MHPA polymer. The cross-linked ester was formed, as shown in Figure 4.

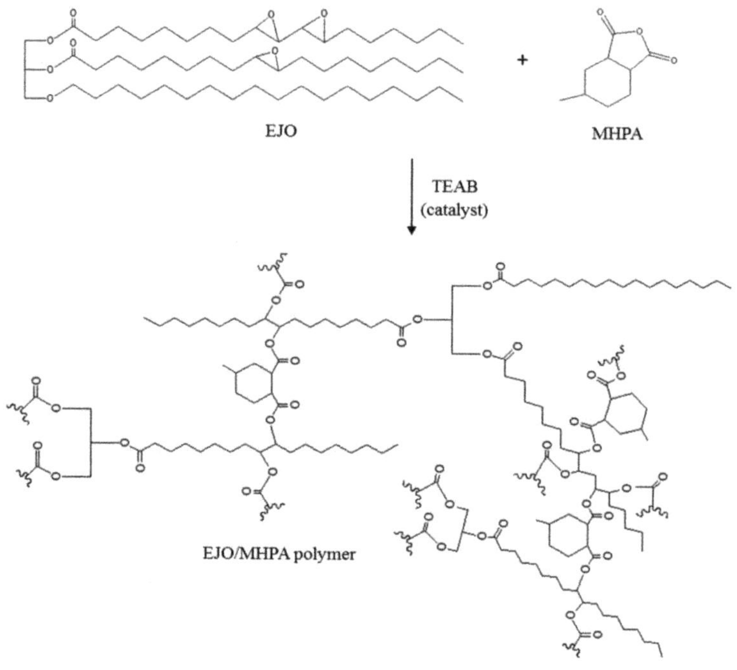

EJO

MHPA

+

TEAB
(catalyst)

EJO/MHPA polymer

Figure 4. Schematic reaction of EJO and MHPA.

The ^{1}H NMR and ^{13}C NMR spectra of epoxidized jatropha oil are depicted in Figure 5. Upon epoxidation, there was a notable chemical shift in the spectra that could be observed. The unsaturation peak in the jatropha oil almost completely disappeared in the ^{1}H NMR spectrum for epoxidized jatropha oil. Moreover, the formation of new peaks in the region of 2.8–3.1 ppm indicated the presence of epoxy group protons [31]. Further, the characteristic unsaturation peaks at 120–130 ppm for ^{13}C NMR was absent and accompanied by the appearance of carbon signals at 54–58 ppm showed the presence of epoxy groups. It can be deduced that double bonds in the jatropha oil were successfully converted to epoxy groups. This finding was in agreement with the previous study reported by Sammaiah and co-workers [28].

3.3. Thermal Analysis

Thermogravimetric analysis (TGA) is a commonly used technique to study the thermal properties of a polymer based on the degradation temperature, residuals, and other information obtained from the analysis [33]. The TGA curves of the EJO/MHPA polymers after curing at various ratios are shown in Figure 6, whereas their first-derivative weight curves (DTG) are depicted in Figure 7.

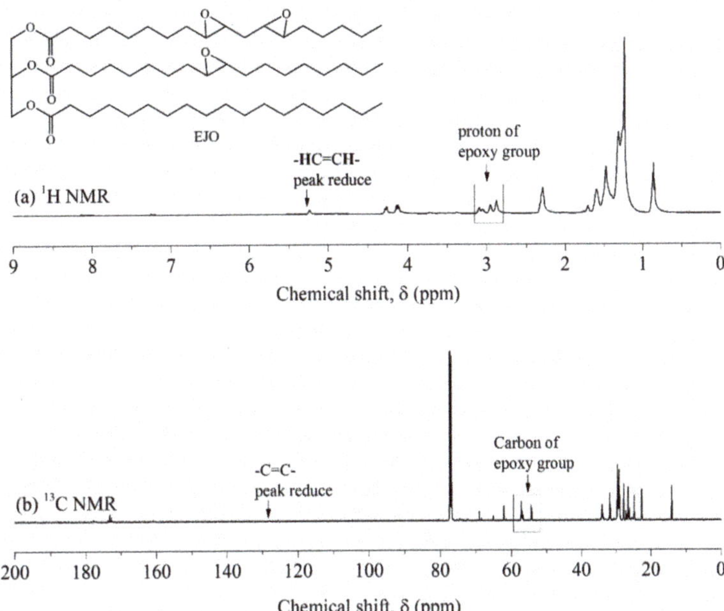

Figure 5. (a) ^{1}H NMR and (b) ^{13}C NMR spectra of epoxidized jatropha oil.

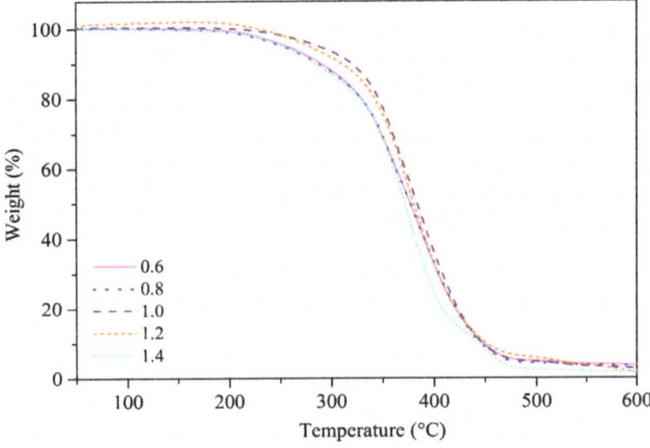

Figure 6. Thermogravimetric analysis (TGA) curves of EJO/MHPA with various ratios of anhydride to oxirane.

TGA data were analyzed based on its initial onset temperature (T_i), decomposition temperature (T_d) at different weight losses (wt), which were 5, 25, and 50% ($T_{5\%}$, $T_{25\%}$, and $T_{50\%}$), and weight residue % at 600 °C. All of the EJO/MHPA polymers had a single-step degradation pattern, and there was no evidence of unreacted chemicals that would volatilize before reaching the temperature of polymer degradation [34]. The temperatures corresponding to the $T_{5\%}$ for all samples were recorded above 240 °C, indicating good thermal stability. Overall, the analysis of various molar ratios showed that a ratio of 1.0 had the best thermal stability. This can be explained by the optimum crosslinking of the polymer network for the EJO/MHPA polymer with a molar ratio of 1.0 compared to other ratios, whereby it required more heat to break the bond. The peak temperature of mass losses significantly increased from a ratio of 0.6 to 1.0 and decreased for ratios of 1.2 and 1.4, which could be observed from the DTG thermograms. Details on the thermal properties of the EJO/MHPA polymers are shown in Table 1. From this, it can be concluded that the molar ratio of 1.0 of the EJO/MHPA polymer has better thermal stability than other ratios.

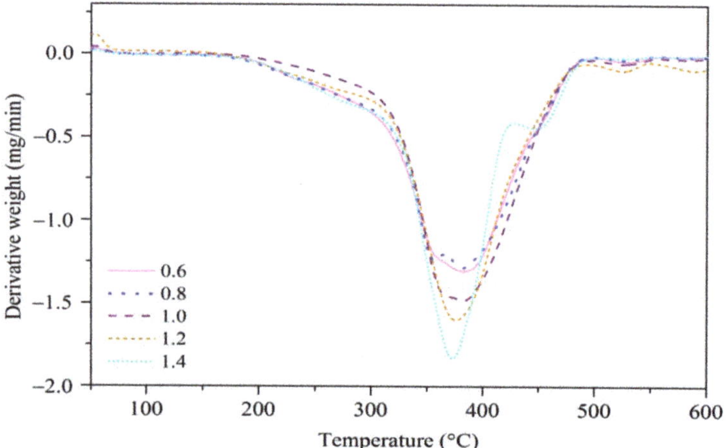

Figure 7. First-derivative weight curves (DTG) of EJO/MHPA polymers with various molar ratios of anhydride to oxirane.

Table 1. Thermal properties of EJO/MHPA polymers from TGA and DTG.

EJO/MHPA Ratio	T_i (°C)	T_d at Different wt. Losses (°C)			$T_{d\max}$ (°C)	Wt. Residue at 600 °C (%)
		$T_{5\%}$	$T_{25\%}$	$T_{50\%}$		
0.6	163	252	334	369	382	3.69
0.8	147	249	335	371	381	3.44
1.0	168	283	347	375	383	2.64
1.2	150	273	343	372	376	1.00
1.4	151	254	338	368	373	1.75

The glass transition temperature (T_g) is considered as a fundamental polymer characteristic related to polymer properties and processing. In general, polymers with high crosslinking density have higher T_g; however, the composition in the polymer within the cross-linked structure also plays an important role in the T_g behavior. As is well-known, DSC is a widely used instrument to characterize T_g of polymer materials [35]. Figure 8 and Table 2 summarize the thermal properties of EJO/MHPA polymers found in DSC curves, such as glass transition (T_g) and enthalpy of transition (ΔH). All synthesized EJO/MHPA polymers showed only glass transition temperature values, but there was no presence of melting or crystallization peaks in DSC curves, which suggests that these EJO/MHPA polymers were amorphous. The curves also did not show an exothermic transition above 100 °C, which confirm the complete cross-linking between the epoxy group and anhydride during the curing cycle. The various molar ratios of EJO/MHPA polymers showed different values possibly due to the chain flexibility. For the amorphous polymer, T_g values can serve as the shape transition temperature (T_g). DSC data suggested that the T_{g1} of the EJO/MPHA polymers were T_{g1} between 7.8 and 11.3 °C. The second glass transition was above 100 °C, where they could recover back to their original shape. However, the peak of T_{g2} was too small to detect, and the ratios of 1.2, 1.0, and 0.8 had higher temperatures of T_g peak, as can be seen in Figure 8. SMPs are divided based on their switch type into either T_g-type SMPs with an amorphous phase or T_m-type SMPs with a crystalline phase.

As for the T_g- type SMPs, a large rubber modulus can usually be maintained above the T_g. T_g-type SMPs exhibit relatively slow shape recovery compared with T_m- or T_g-type SMPs due to their broader glass transition interval, which hinders their application where immediate shape recovery is required [36].

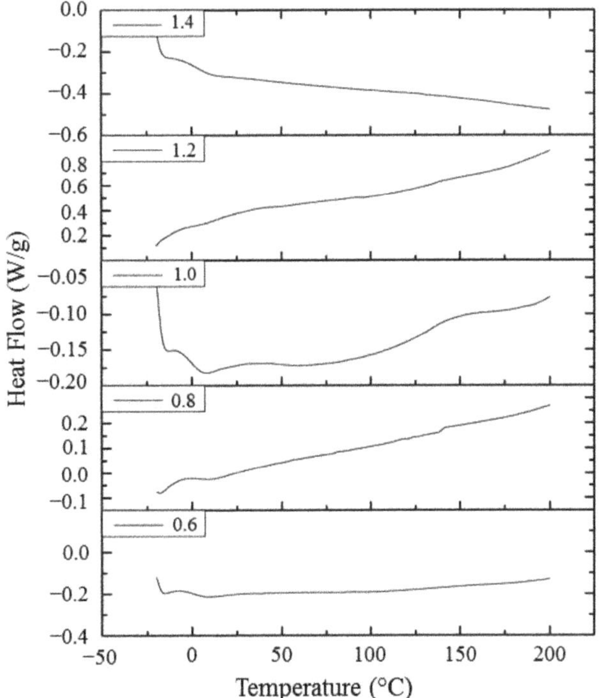

Figure 8. DSC curves of EJO/MHPA polymers with various ratios of anhydride to oxirane.

Table 2. Thermal properties of EJO/MHPA polymers found in DSC analysis.

EJO/MHPA Ratio	$T_{g\,1}$ (°C)	$T_{g\,2}$ (°C)	ΔH (J/g)
0.6	9.8	135	−3.76
0.8	10.8	140	−1.07
1.0	9.6	144	−2.86
1.2	7.8		−0.76
1.4	11.3		−2.00

3.4. Shape Memory-Recovery Behaviours

A study on the shape memory recovery behaviors of the EJO/MHPA polymers was performed. The primary shape of EJO/MHPA was in a linear rectangular shape. Then, the polymer was heated at 150 °C and cooled at room temperature. All of the EJO/MHPA polymers with various molar ratios showed good recovery ability and reached their original shape quickly within 30 s when they were heated above their glass transition temperature. The fundamental mechanism behind the shape memory behavior is the activation and freezing of the motion of the SMP chains above and below the T_g [36], and the mechanism can be explained as follows: Firstly, the macroscopic deformation is translated to conformational change of the polymer molecular segments. Above T_g, the polymer is in a rubbery state and can deform easily. Then, the temporary shape is fixed with the internal stress of the polymer network by cooling. When the EJO/MHPA is reheated above the T_g, the heating causes rearrangement of molecular segments of the polymer network. Then the microscopic deformation is released, and the original shape of the EJO/MHPA polymers is recovered [5]. Shape-memory behavior can be exploited and demonstrated in various polymer systems significantly different in molecular structure and morphology. The common conventional SMP systems include cross-linked ethylene vinyl acetate copolymer, styrene-based polymers, acrylate-based polymers, polynorbornene, cross-linked polycyclooctene, epoxy-based polymers, thio-ene-based polymers, segmented polyurethane, and segmented PU ionomers. The intrinsic mechanism for shape memory behavior in thermal responsive SMPs is the reversible freezing and activation of polymeric chain motion in the switching segments below and above the transition temperature (T_{trans}). The net-points of an SMP network, which maintain its dimensional stability, could be either covalent or physically crosslinked [36,37].

The shape memory effect (SME) is an extrinsic property of smart materials that allows them to restore or maintain their original shape even after being severely and quasiplastically distorted in the presence of the right stimuli [38]. According to previous research, one-way SME and two-way SME are two types of shape memory polymers with non-reversible and reversible shape-shifting properties, respectively [37]. One-way SME can retain a temporary shape once the stimulus is terminated. For two-way SME, the temporary shapes can be recovered to the initial shape when the stimulus is terminated. Reversible behavior occurs during heating and cooling in the presence or absence of external stress. In this study, EJO/MHPA polymer agreed with the two-way SME theoretically reported [39]. Figure 9 shows the demonstration of the shape-memory behaviors of EJO/MHPA polymers at molar ratio of 1.0. Moreover, the composition of epoxidized oil and anhydride curing agent play important roles in shape recovery due to crosslink density and chain flexibility.

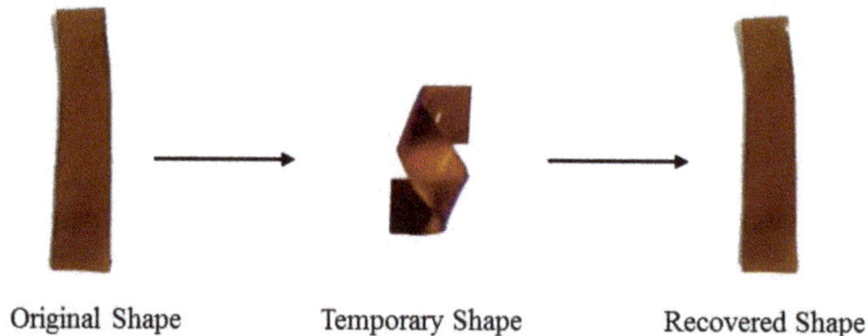

Figure 9. A demonstration of the shape memory recovery behaviors of EJO/MHPA polymers reported in this paper.

4. Conclusions

We presented the preparation and characterization of bio-based epoxy material from jatropha oil for use in a shape memory polymer application. FTIR analysis showed the transition from jatropha oil to epoxidized oil through the disappearance of unsaturation peaks and formation of new epoxy peaks. The EJO was used to prepare bio-based polymers by curing this functionalized oil with MHPA acid anhydride and catalyzed with TEAB. The reaction produced slightly transparent and brown color polymers. The thermal properties of the EJO/MHPA polymers of different feed molar ratios of anhydride to oxirane were significantly different, perhaps due to the cross-link density and chain flexibility. The glass transition temperature of the EJO/MHPA polymers was between 7.8 and 11.3 °C. The shape memory polymer showed that the ratio of the anhydride to oxirane of 1.0 had better thermal stability compared to other ratios. Conclusively, the EJO/MHPA polymers showed good shape memory recovery behaviors. It is expected that this work could provide insight and an alternative approach for the production of shape memory polymers from current technology.

Author Contributions: Conceptualization, M.M.A.; methodology, M.M.A.; formal analysis, L.L.T.M. and S.A.M.S.; investigation, L.L.T.M. and S.A.M.S.; resources, P.S.H.; writing—original draft preparation, L.L.T.M. and S.A.M.S.; writing—review and editing, M.R. and A.M.J.; supervision, M.M.A. and P.S.H.; funding acquisition, M.M.A. All authors have read and agreed to the published version of the manuscript.

Funding: This research was funded by Universiti Putra Malaysia, grant IPB–9532000.

Institutional Review Board Statement: Not applicable.

Informed Consent Statement: Not applicable.

Data Availability Statement: The data presented in this study are available on request from the corresponding author.

Acknowledgments: The authors would like to acknowledge INTROP and the Faculty of Science (UPM) for their technical support.

Conflicts of Interest: The authors declare no conflict of interest.

References

1. Chan, B.Q.Y.; Low, Z.W.K.; Heng, S.J.W.; Chan, S.Y.; Owh, C.; Loh, X.J. Recent Advances in Shape Memory Soft Materials for Biomedical Applications. *ACS Appl. Mater. Interfaces* **2016**, *8*, 10070–10087. [CrossRef] [PubMed]
2. Zhang, X.; Tan, B.H.; Li, Z. Biodegradable polyester shape memory polymers: Recent advances in design, material properties and applications. *Mater. Sci. Eng. C* **2018**, *92*, 1061–1074. [CrossRef]
3. Liu, T.; Zhou, T.; Yao, Y.; Zhang, F.; Liu, L.; Liu, Y.; Leng, J. Stimulus methods of multi-functional shape memory polymer nanocomposites: A review. *Compos. Part A* **2017**, *100*, 20–30. [CrossRef]
4. Guo, B.; Chen, Y.; Lei, Y.; Zhang, L.; Zhou, W.Y.; Rabie, A.B.M.; Zhao, J. Biobased poly(propylene sebacate) as shape memory polymer with tunable switching temperature for potential biomedical applications. *Biomacromolecules* **2011**, *12*, 1312–1321. [CrossRef] [PubMed]
5. Tsujimoto, T.; Takeshita, K.; Uyama, H. Bio-based epoxy resins from epoxidized plant oils and their shape memory behaviors. *J. Am. Oil Chem. Soc.* **2016**, *93*, 1663–1669. [CrossRef]
6. Liu, Y.; Du, H.; Liu, L.; Leng, J. Shape memory polymers and their composites in aerospace applications: A review. *Smart Mater. Struct.* **2014**, *23*, 023001. [CrossRef]
7. Zhao, W.; Liu, L.; Zhang, F.; Leng, J.; Liu, Y. Shape memory polymers and their composites in biomedical applications. *Mater. Sci. Eng. C* **2019**, *97*, 864–883. [CrossRef]
8. Lendlein, A.; Behl, M.; Hiebl, B.; Wischke, C. Shape-memory polymers as a technology platform for biomedical applications. *Expert Rev. Med. Devices* **2010**, *7*, 357–379. [CrossRef]
9. Purwar, R.; Sachan, R. Thermoresponsive shape memory polymers for smart textiles. In *Advances in Functional and Protective Textiles*; Elsevier BV: Amsterdam, The Netherlands, 2020; pp. 37–62.
10. Kong, D.; Li, J.; Guo, A.; Xiao, X. High temperature electromagnetic shielding shape memory polymer composite. *Chem. Eng. J.* **2021**, *408*, 1–7. [CrossRef]
11. Zhou, H.; Luo, H.; Yao, Y.; Wang, H.; Yi, G.; Lin, W.; Gu, Y. Low-voltage-triggered rapid shape memory actuation with interfacial self-assembled silver nanowires. *Mater. Lett.* **2019**, *252*, 76–79. [CrossRef]
12. Kumar, B.; Noor, N.; Thakur, S.; Pan, N.; Narayana, H.; Yan, S.C.; Wang, F.; Shah, P. Shape memory polyurethane-based smart polymer substrates for physiologically responsive, dynamic pressure (re)distribution. *ACS Omega.* **2019**, *4*, 15348–15358. [CrossRef] [PubMed]
13. Petrović, Z.S.; Milić, J.; Zhang, F.; Ilavsky, J. Fast-responding bio-based shape memory thermoplastic polyurethanes. *Polymer* **2017**, *121*, 26–37. [CrossRef] [PubMed]
14. Tsukada, G.; Tokuda, M.; Torii, M. Temperature triggered shape memory effect of transpolyisoprene-based polymer. *J. Endod.* **2014**, *40*, 1658–1662. [CrossRef] [PubMed]
15. Wang, Y.; Ye, J.; Tian, W. Shape memory polymer composites of poly(styrene-b-butadiene-b-styrene) copolymer/liner low density polyethylene/Fe3O4 nanoparticles for remote activation. *Appl. Sci.* **2016**, *6*, 333. [CrossRef]
16. Zheng, N.; Fang, G.; Cao, Z.; Zhao, Q.; Xie, T. High strain epoxy shape memory polymer. *Polym. Chem.* **2015**, *6*, 3046–3053. [CrossRef]
17. Liu, T.; Hao, C.; Wang, L.; Li, Y.; Liu, W.; Xin, J.; Zhang, J. Eugenol-derived biobased epoxy: Shape memory, repairing, and recyclability. *Macromolecules* **2017**, *50*, 8588–8597. [CrossRef]
18. Tsujimoto, T.; Takayama, T.; Uyama, H. Biodegradable shape memory polymeric material from epoxidized soybean oil and polycaprolactone. *Polymers* **2015**, *7*, 2165–2174. [CrossRef]
19. Tsujimoto, T.; Uyama, H. Full biobased polymeric material from plant oil and poly(lactic acid) with a shape memory property. *ACS Sustain. Chem. Eng.* **2014**, *2*, 2057–2062. [CrossRef]

20. Ca, V.; Lligadas, G.; Ronda, J.C.; Galia, M. Renewable polymeric materials from vegetable oils: A perspective. *Mater. Today* **2013**, *16*, 337–343.
21. Akbar, E.; Yaakob, Z.; Kamarudin, S.K.; Ismail, M.; Salimon, J. Characteristic and composition of jatropha curcas oil seed from Malaysia and its potential as biodiesel feedstock. *Eur. J. Sci. Res.* **2009**, *29*, 396–403.
22. Rayung, M.; Aung, M.M.; Ahmad, A.; Su'ait, M.S.; Abdullah, L.C.; Jamil, S.N.A. Characteristics of ionically conducting jatropha oil-based polyurethane acrylate gel electrolyte doped with potassium iodide. *Mater. Chem. Phys.* **2019**, *222*, 110–117. [CrossRef]
23. Aung, M.M.; Yaakob, Z.; Abdullah, L.C.; Rayung, M.; Li, W.J. A comparative study of acrylate oligomer on jatropha and palm oil-based UV-curable surface coating. *Ind. Crops Prod.* **2015**, *77*, 1047–1052. [CrossRef]
24. Tajau, R.; Mahmood, M.H.; Salleh, M.Z.; Dahlan, K.Z.M.; Ismail, R.C.; Faisal, S.M.; Rahman, S.M.Z.S. Production of uv-curable palm oil resins/oligomers using laboratory scale and pilot scale systems. *Sains Malays.* **2013**, *42*, 459–467.
25. *Standard Test Method for Epoxy Content of Epoxy Resins*; ASTM International: Conshohocken, PA, USA, 2019.
26. Derawi, D.; Salimon, J.; Ahmed, W.A. Preparation of epoxidized palm olein as renewable material by using peroxy acids. *Malays. J. Anal. Sci.* **2014**, *18*, 584–591.
27. Hazmi, A.S.A.; Aung, M.M.; Abdullah, L.C.; Salleh, M.Z.; Mahmood, M.H. Producing Jatropha oil-based polyol via epoxidation and ring opening. *Ind. Crops Prod.* **2013**, *50*, 563–567. [CrossRef]
28. Sammaiah, A.; Padmaja, K.V.; Badari, R.; Prasad, N. Synthesis of epoxy jatropha oil and its evaluation for lubricant properties. *J. Oleo Sci.* **2014**, *63*, 637–643. [CrossRef]
29. Taib, E.R.J.; Abdullah, L.C.; Aung, M.M.; Basri, M.; Salleh, M.Z.; Saalah, S.; Mamat, S.; Chee, C.Y.; Wong, J.L. Physico-chemical characterisation of epoxy acrylate resin from jatropha seed oil. *Pigment Resin Technol.* **2017**, *46*, 485–495. [CrossRef]
30. Venkatesh, D.; Jaisankar, V. Synthesis and characterization of bio-polyurethanes prepared using certain bio-based polyols. *Mater. Today Proc.* **2019**, *14*, 482–491. [CrossRef]
31. Zhang, J.; Tang, J.J.; Zhang, J.X. Polyols prepared from ring-opening epoxidized soybean oil by a castor oil-based fatty diol. *Int. J. Polym. Sci.* **2015**, *2015*, 529235. [CrossRef]
32. Venkatakoteswararao, R.; Vinayagamoorthy, G. Review on thermal analysis of polymer matrix composites. *IOP Conf. Ser. Mater. Sci. Eng.* **2020**, *954*, 1–7. [CrossRef]
33. Gogoi, P.; Boruah, M.; Sharma, S.; Dolui, S.K. Blends of epoxidized alkyd resins based on jatropha oil and the epoxidized oil cured with aqueous citric acid solution: A green technology approach. *ACS Sustain. Chem. Eng.* **2015**, *3*, 261–268. [CrossRef]
34. Wang, R.; Schuman, T.P. Vegetable oil-derived epoxy monomers and polymer blends: A comparative study with review. *Express Polym. Lett.* **2012**, *7*, 272–292. [CrossRef]
35. Lu, H.; Wang, X.; Yao, Y.; Fu, Y.Q. A Frozen volume transition model and working mechanism for the shape memory effect in amorphous polymers. *IOP Publ.* **2018**, *27*, 1–35. [CrossRef]
36. Hu, J.; Zhu, Y.; Huang, H.; Lu, J. Recent advances in shape–memory polymers: Structure, mechanism, functionality, modeling and applications. *Prog. Polym. Sci.* **2012**, *37*, 1720–1763. [CrossRef]
37. Xu, J.; Song, J. Thermal Responsive Shape Memory Polymers for Biomedical Applications. In *Biomedical Engineering—Frontiers and Challenges*; IntechOpen: London, UK, 2011; pp. 126–142.
38. Thomas, N. *A Review on Properties and Applications of Shape Memory Polymers*; ResearchGate: Berlin, Germany, 2020; pp. 1–13.
39. Hu, J. *Shape Memory Polymers: Fundamentals, Advances and Applications*; Smithers Rapra: Akron, OH, USA, 2014.

Article

The Influence of Reaction Time on Non-Covalent Functionalisation of P3HT/MWCNT Nanocomposites

N.M. Nurazzi [1], N. Abdullah [1,*], S.Z.N. Demon [1], N.A. Halim [1] and I.S. Mohamad [2]

[1] Centre for Defence Foundation Studies, National Defence University of Malaysia, Kem Sungai Besi, Kuala Lumpur 57000, Malaysia; mohd.nurazzi@gmail.com (N.M.N.); zulaikha@upnm.edu.my (S.Z.N.D.); norhana@upnm.edu.my (N.H.A.)

[2] Department of Diploma Studies, Faculty of Mechanical Engineering, University Teknikal Malaysia Melaka, Hang Tuah Jaya, Durian Tunggal, Melaka 76100, Malaysia; imran@utem.edu.my

* Correspondence: norli.abdullah@upnm.edu.my

Abstract: Non-covalent functionalisation of the carbon nanotube (CNT) sidewall through polymer wrapping is the key strategy for improving well-dispersed CNTs without persistent alteration of their electronic properties. In this work, the effect of reaction time on regioregular poly (3-hexylthiophene-2,5-diyl) (P3HT)-wrapped hydroxylated multi-walled CNT (MWCNT-OH) nanocomposites was investigated. Five different reaction times (24, 48, 72, 96, and 120 h) were conducted at room temperature in order to clearly determine the factors that influenced the quality of wrapped MWCNT-OH. Morphological analysis using Field Emission Scanning Electron Microscopic (FESEM) and High-Resolution Transmission Electron Microscope (HRTEM) analysis showed that P3HT successfully wrapped the MWCNT-OH sidewall, evidenced by the changes in the mean diameter size of the nanocomposites. Results obtained from Raman spectroscopy, X-ray Photoelectron Spectroscopy (XPS) as well as Thermogravimetric Analysis (TGA) showed a significant effect of the wrapped polymer on the CNT sidewall as the reaction time increased. Overall, the method used during the preparation of P3HT-wrapped MWCNT-OH and the presented results significantly provided a bottom-up approach to determine the effect of different reaction times on polymer wrapping to further expand this material for novel applications, especially chemical sensors.

Keywords: CNT; MWCNT; non-covalent functionalisation; polythiophene; P3HT; reaction time

Citation: Nurazzi, N.M.; Abdullah, N.; Demon, S.Z.N.; Halim, N.A.; Mohamad, I.S. The Influence of Reaction Time on Non-Covalent Functionalisation of P3HT/MWCNT Nanocomposites. *Polymers* **2021**, *13*, 1916. https://doi.org/10.3390/polym13121916

Academic Editors: Emin Bayraktar, S. M. Sapuan and R. A. Ilyas

Received: 8 April 2021
Accepted: 26 May 2021
Published: 9 June 2021

Publisher's Note: MDPI stays neutral with regard to jurisdictional claims in published maps and institutional affiliations.

1. Introduction

Since the discovery of polyacetylene in 1977 by Hideki Shirakawa et al., research in the field of conductive polymers has heavily impacted the discovery and development of π-conjugated conductive polymers [1]. Since then, many studies have been conducted by various researchers from different fields using leading conductive polymers, such as polypyrrole (PPy), polyaniline (PANI), polythiophene (PTh), poly (3,4-ethylenedioxythiophene) (PEDOT), and poly (p-phenylene vinylene) (PPV). These conductive polymers demonstrated great potential for integration into future optical and electronic devices due to their capacity to transition between semiconducting and conducting states, as well as the ability to alter mechanical properties by controlled doping, chemical modification, and stacking or creating composites with other materials [2]. In general, conductive polymers possess alternating single (σ) and double (π) bonds, and these π-conjugated systems provide the conductive polymers with intrinsic optical, electrochemical, and electrical and electronic properties [3]. However, the development of the properties of conductive polymers has not been completely proportionate with those of their metallic and inorganic semiconductor counterparts. Therefore, conductive polymers have been modified or hybridised with other heterogeneous material or other carbonaceous material components to overcome and improve their inherent boundaries in terms of solubility, conductivity, and long-term stability [4]. Amongst the conductive polymers and the π-conjugated polymers,

PTh and its derivatives have shown promising characteristics, which were comparable to those of both PANI and PPy, for wide applications such as in chemical and biosensors [5–8], organic photovoltaics (OPVs) [9–11], electronic magnetic shielding (EMI) [12,13], battery [14,15], microwave absorption [16,17], water purification devices [18,19], and hydrogen storage [20,21].

PTh-based nanocomposites containing nanocarbon species, such as graphene, carbon nanofibers (CNFs), and CNTs, were developed and show promising results in the targeted applications [22–25]. These nanocarbon species improved the structural ordering of the nanocomposite chains and facilitated delocalisation of the charge carriers, resulting in enhanced conductivity. Some conductive polymers can behave like semiconductors due to their heterocyclic compounds, which display physicochemical characteristics such as solubility, hydrogen bonding, surface activity, thermal expansion, and electrical conductivity. As a result, reversible changes in the sensing layer's conductivity could be detected upon polar chemical adsorption on the sidewall at room temperature [26]. This effect is believed to be caused by the charge transfer between gas molecules and the polymer or swelling of the polymer film's [27]. Amongst the carbonaceous materials considered to date, CNT has been widely investigated because of its excellent electrical, mechanical, and thermal properties [28]. Since then, various techniques to incorporate CNT in polymer matrices were designed with a desire to fabricate new advanced materials with multifunctional properties. The exceptional mechanical properties associated with CNT vary in the literature regarding the exact properties of CNT. Theoretical and experimental results have shown an extraordinarily high elastic modulus, greater than 1 TPa (the elastic modulus of diamond is 1.2 TPa) and reported strengths 10 to 100 times higher than those of the strongest steel at a fraction of the weight [29]. CNT also has superior thermal and electric properties: thermally stable up to 2800 °C in a vacuum, electrical conductivity is about 10^3 S/cm (the electric current-carrying capacity is 1000 times higher than those of metals like copper), thermal conductivity is about 1,900 W m^{-1} K^{-1}, (the thermal conductivity is about twice as high as that of diamond [30–32]), and the current density of individual metallic single-walled carbon nanotubes (SWCNT) is about 4×10^9 A/cm^2, which is 1000 times higher than that of copper wires [33].

Besides the exceptional characteristics mentioned above, low compatibility and limitation of dispersion of CNTs in polymeric matrices occurred, but the interaction between CNTs and the polymer remained weak. CNTs were in the form of long bundles due to stabilisation by π–π electron interactions and high surface energy. CNTs incorporated into the polymeric matrices was an attractive method to combine and complement the optical, electrical, and mechanical properties of individual CNTs with the unique properties of the desired polymer. These unique properties make CNTs an ideal reinforcing agent in a number of applications [34]. In previous research on homogeneously dispersed CNTs as a conductive filler in polymer matrices, researchers covalently functionalised the CNT sidewall with a monomer such as thiophene [35,36] and specific functional groups or active elements such as carboxylation, amides, etc. [37–43]. Based on the aforementioned studies, the covalent modification towards CNT composites produced high stability functionalisation. In particular, this mechanism would lead to the efficient load transfer from the polymer/CNT matrix through the covalent bonding [44]. However, in most cases, covalent functionalisation changes the structure of the CNTs, thereby degrading their unique electrical and mechanical properties. Furthermore, functionalisation causes shortening of the CNTs, which reduces the advantage of CNTs in regard to the ratio aspect [29].

The optimisation strategy for the improvement of CNT dispersion in polymers matrices was done by Liu and Choi (2012). The dispersion is a spatial property, whereby the individual CNTs are spread with a roughly uniform number density throughout the continuous polymer matrix. The challenge is to separate the nanotubes from their initial aggregated nature, which is usually achieved by local shear forces. Direct manual mixing of CNTs with polymer resin, though the simplest approach, did not create sufficient local shear force and therefore led to poor dispersion of CNTs inside the polymer matrix. A more effective separation and dispersion of CNT bundles requires the overcoming of the inter-tube van der Waals attraction forces. Physical approaches such as shear mixing, mechanical stirring, further sonication, ball milling, and micro-bead milling processes have been employed for this purpose. Although these techniques might appear very different, they are all governed by the transfer of physical shear stress onto nanotubes, which break down the bundles. Usually, dispersion via shear mixing is only achievable for specific types of CNTs, with a high shear rate in a rather viscous medium. Huang et al. (2006) demonstrated that nanocomposites containing high loading concentrations of CNTs (up to 7 wt.%) could be dispersed via this technique [45]. However, the processing time significantly increases as the loading concentration rises. More importantly, shear mixing tends to sectionalise CNTs into shorter length, thereby reducing their conductivity significantly, an undesired attribute for nanocomposite, which was intended for use as a sensor material.

Another concern of the optimisation is the sonication frequency. A higher power of sonication, greater than 500 W, is desired as it provides a higher shearing force to break down the CNT bundles. Nevertheless, prolonged exposure could possibly damage or shorten the nanotubes that finally leads to decreased conductivity of the CNTs. For the dispersion related to optimisation of the mechanical stirring, a specific speed (rpm) could shorten the dissolution time. Prolonging the additional time of mechanical stirring led the dispersion to become less stable, and CNT agglomerations were visibly seen. Therefore, the use of non-covalent functionalisation, which maintained the CNT structure, was proposed as a key strategy to achieve well-dispersed CNTs in PTh matrices and better PTh wrapping towards the CNT sidewall. Referring to Bose et al. (2010), non-covalent functionalisation was considered an efficient alternative strategy to tailor the CNT and polymer interface and preserve the reliability of the nanotubes [46]. This route is particularly attractive because of the possibility of adsorbing various groups of ordered architectures on the CNT sidewall without disturbing the extended p-conjugation of the nanotubes. Besides the effect of functionalisation during the preparation of polymer wrapping towards the CNT sidewall, the effect of reaction time is also one of the crucial criteria. Hence, in this paper, a composition of 1:1 of P3HT and MWCNT-OH was used, and the physical properties of nanocomposites under the different reaction times were investigated. FESEM and HRTEM were performed to study the structural and the wrapping dispersion of nanotubes in a polymer matrix, whereas, the structural, morphological, and thermal stability properties were evaluated by Raman, XPS, and TGA, respectively.

2. Experimental Section

2.1. Materials

Commercially available MWCNT-OH (purity, >95%) was purchased from Nanostructured & Amorphous Materials, Inc., Houstan, Texas, USA. Regioregular P3HT (molecular weight, Mw 50,000 to 100,000; purity, ≥90%) and organic solvent tetrahydrofuran (THF) were purchased from Sigma–Aldrich, Selangor, Malaysia. Methanol (CH_3OH), a solvent used for washing, was supplied by R&M Chemicals, Selangor, Malaysia. The materials and solvent were used as received without further purification.

2.2. Preparation of P3HT/MWCNT-OH Nanocomposites

The preparation of P3HT/MWCNT nanocomposites was adopted from a previous methodology [47], with a revised version. The schematic procedure followed for preparation of the nanocomposites is presented in Figure 1. Approximately 5 mg of MWCNT and 5 mg of P3HT were added to a 25 mL volumetric flask consisting of a magnetic bar (0.5 (L) cm × 0.2 (D) cm). Then, 5 mL of THF was poured into the volumetric flask. The volumetric flask was placed in a ceramic heating plate, and the suspension was stirred consistently at a speed of 650 rpm for 24 h at 50 °C (sample labelled as MWCNT-24). The reaction mixture was then immersed in a water bath and sonicated at a frequency of 50 Hz for 2 h at room temperature. The resultant precipitate of the P3HT/MWCNT nanocomposite was then carefully washed several times with methanol and further filtered using a vacuum

Buchner funnel. The obtained black powder was dried at 25 °C for 24 h (MWCN-24). With the same amount of materials and procedure, the preparation of nanocomposites was repeated at different reaction times for 48, 72, 96, and 120 h and labelled as MWCNT-48, MWCNT-72, MWCNT-96, and MWCNT-120, respectively. The yield of the nanocomposite from each set of experiments was around 80 to 90%.

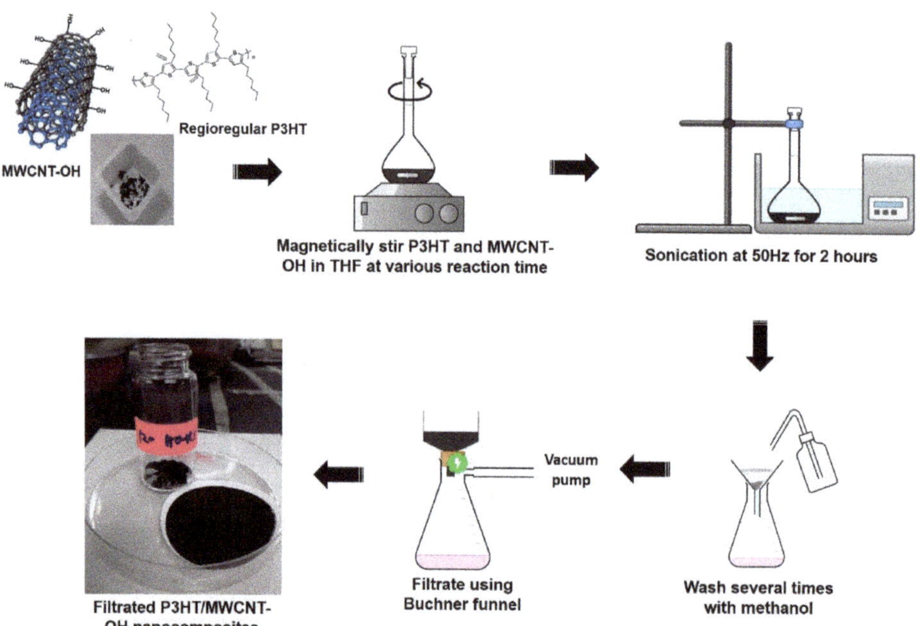

Figure 1. Schematic procedure for preparation of P3HT-wrapped MWCNT-OH nanocomposites.

2.3. Characterisation Methods

The structural analysis through the Raman spectroscopic measurement was carried out using Renishaw inVia Reflex Confocal Micro Raman System (Renishaw plc, Wotton-under-Edge, UK). The HPNIR laser for sample excitation was set at a wavelength of 787 nm, 1200 mm^{-1} gratings, and the magnification was a 100× objective lens. The chemical state of the element compositions of P3HT/MWCNT was analysed using an auger electron spectroscope with an x-ray photoelectron spectrometer (XPS), Kratos/Shimadzu (Shimadzu Corporation, Kyoto, Japan). The binding energy values of XPS lines were calibrated using Al Kα radiation at 1000 eV (hv = 1000 eV) and a spot size of 100 μm. The surface morphological analysis and the diameter size distribution of the nanocomposites were characterised using a Field Emission Scanning Electron Microscopic (FESEM), Carl Zeiss Gemini FESEM 500 (Carl Zeiss AG, Jena, Germany). Prior to the analysis, samples were initially coated with gold for about 1 min in order to avoid charging. The effectiveness of polymer wrapping onto MWCNT was also overserved under a High-Resolution Transmission Electron Microscope (HRTEM), JEOL JEM 2,100F HRTEM (Tokyo, Japan) at an acceleration voltage of 200 kV. For HRTEM analysis, the sample was initially dispersed in acetone by ultra-sonication for 60 s. Thereafter, a drop of the suspension was transferred onto a carbon-coated copper grid and mounted on the microscope, and the images were recorded. The TGA was conducted using a TGA-DSC HT 3 analyser (Mettler Toledo, Selangor, Malaysia) from a temperature of 25 to 900 °C at a heating rate of 10 °C/minute under a nitrogen atmosphere to study the stability of the prepared nanocomposites.

3. Results and Discussion

3.1. Raman Spectroscopy Analysis

Raman spectroscopy was used to study the possible interactions between pristine MWCNT-OH and the P3HT. The pristine MWCNT-OH spectrum was compared with the P3HT-wrapped MWCNT-OH nanocomposites that were prepared at different reaction times, as can be seen in Figure 2. The summarised intensity ratio between the D band and G band is presented in Table 1. From the Raman spectra it could be observed that pristine P3HT had two distinct peaks at wavelengths of 1384 cm^{-1} and 1449 cm^{-1}, which corresponded to the C-C skeletal stretching vibrations and C=C skeletal stretching of the thiophene ring deformation of the side chain of P3HT, respectively. The Raman spectra of MWCNT-OH demonstrated two prominent peaks characteristic at 1300 cm^{-1} (D band) attributed to the disorder and imperfection of the carbon crystallites and sp^3 vibration present in the MWCNT-OH structures [48]. Peaks at 1571 cm^{-1} (G band) corresponded to one of the two E$_{2g}$ modes corresponding to stretching vibrations in the basal plane (sp^2 domains) of single-crystal graphene [49]. After the non-covalent functionalisation of MWCNT-OH with P3HT, both the D band and G band were shifted to a higher wavelength and an additional peak was observed at 1445 cm^{-1}, which represented the characteristic of protonated P3HT. A relative increase and shift in the D band led to the increased intensity ratio, which indicated the slightly defected structures of the MWCNT-OH embedded by the P3HT [6,50].

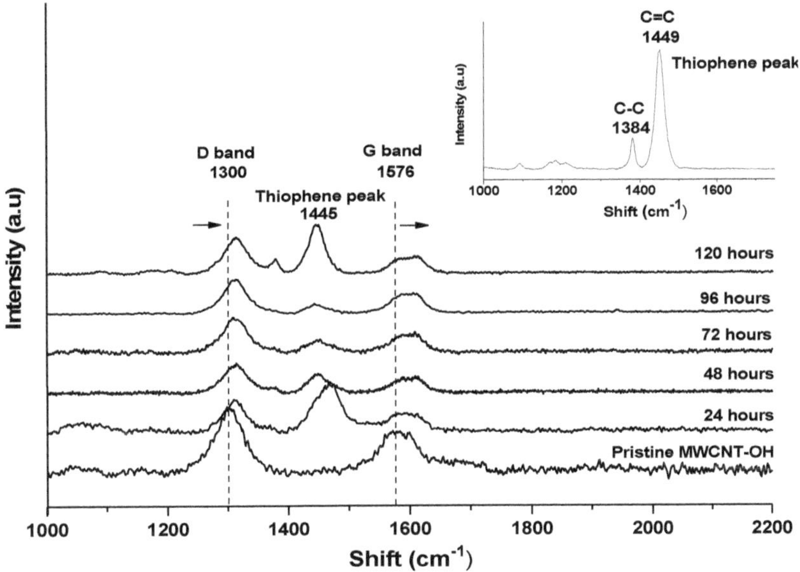

Figure 2. Raman spectra of (**a**) pristine P3HT, (**b**) pristine MWCNT-OH and P3HT/MWCNT-OH nanocomposites prepared at different reaction times.

The intensity ratio data presented in Table 1 showed that increasing the reaction time of the suspension from 24 h to 120 h resulted in a linear increase in the intensity ratio of I$_D$/I$_G$ Raman spectra from 0.83 (pristine MWCNT-OH) to 0.98 (MWCNT-120). This trend indicated that pristine MWCNT displayed a slightly less disordered structure compared to nanocomposite samples. Gomez et al. (2016) increased the intensity ratio of the Raman spectra corresponding to a higher number of sp^3 carbon, which was generally attributed to the presence of more structural defects [51]. The more structural defects in this study referred to the successful functionalisation of the P3HT towards the wall of the MWCNT-OH as a non-covalent interaction. This interaction led to less penetration of the electron for the detection of the graphitic surface (G band) rather than the penetration of the electron on the wrapped area classified as a defective and imperfect surface.

Table 1. Intensity ratio of the I_D/I_G band of pristine MWCNT-OH and P3HT/MWCNT-OH nanocomposites.

Sample	I_D	I_G	I_D/I_G
Pristine MWCNT-OH	1297	1571	0.83
MWCNT-24	1312	1458	0.90
MWCNT-48	1313	1414	0.93
MWCNT-72	1372	1456	0.94
MWCNT-96	1390	1431	0.97
MWCNT-120	1398	1427	0.98

Additionally, since the glass transition temperature (T_g) of P3HT was low at 12 °C [52], the longer contact time between the nanotubes and P3HT in suspension was expected to increase the heat generation upon the stirring process over time. This synergised the wrapping process of the P3HT onto the MWCNT wall and tips, whereas the additional new peak at 1449 cm^{-1} for all the nanocomposite samples was due to C=C stretching vibrations from polymer P3HT. The intensity of this peak was prominent for sample MWCNT-24 and showed a decreasing of trend as the reaction time increased. This indicated the functionalisation of P3HT towards the MWCNTs-OH sidewall was possibly well formed towards the success of P3HT wrapping (Figure 2). However, for sample MWCNT-120, a well-defined peak could be seen at 1445 cm^{-1}, which indicated that more a concentrated thiophene group was present on the nanotube wall. Furthermore, interactions between MWCNT-OH and P3HT blended for nanocomposites at 24-hour reaction time caused a shift in the D band of about 14 cm^{-1}, from 1297 to 1312 cm^{-1}, and in the G band of about 80 cm^{-1} from 1571 to 1492 cm^{-1}. For the longer reaction time at 120 h, the nanocomposites caused a shift in the D band of about 8 cm^{-1}, from 1297 to 1398 cm^{-1}, and in the G band of about 26 cm^{-1} from 1571 to 1427 cm^{-1}. The shifting of both the D band and G band in the Raman spectra was due to the wrapped polymer on the nanotube walls and this was in line with the observation from FESEM images and was supported by the diameter size distribution histogram. A clear enhancement in the intensity ratio observed on P3HT wrapping was also reported by other researchers, which correlated with the effect of non-covalent polymer wrapping on the CNT sidewall [53–56].

3.2. XPS Analysis

The presence of the functional group on the sidewall of P3HT/MWCNT-OH nanocomposites, prepared at different reaction times, was further investigated using XPS analysis (Figure 3 and Table 2). It could be seen that the intensity peak for pristine MWCNT-OH exhibited an intense peak at 283.45 corresponding to C 1s (97.25%) and a relatively intense peak at 531.6 eV corresponding to O 1s (2.75%) [57]. There was no S 2s or S 2p spectrum in the case of the MWCNT-OH. The S 2p core-level spectrum of P3HT and P3HT/MWCNT-OH could be deconvoluted into at least two spin-orbit-split doublet (S-2p$_{3/2}$ and S-2p$_{1/2}$) peaks at approximately 165.1 eV and 165.8 eV, respectively, which were attributed to the neutral sulfur atoms [58]. In a study by Karim et al. (2006), the successful wrapping of PTh towards the CNT sidewall could be observed by the two peaks of S 2p, which appeared at 163.7 and around 230 eV for S 2s [50]. This means that the chemical environment of the S element, in pure P3HT and P3HT/MWCNT-OH, was almost identical. The MWCNT-OH nanocomposites for all samples wrapped with P3HT showed a decrease in carbon and oxygen contents, whilst the peak for the sulfur content started to appear for all nanocomposites samples. An enhancement in the intensity for all nanocomposite samples, such as P3HT/MWCNT-OH at 72 h, exhibited characteristic peaks of sulfur at 227 eV (S 2s) and 163 eV (S 2p) in addition to C 1s and O 1s peaks, which confirmed the presence of thiophene moieties from P3HT on the MWCNT-OH sidewall [59–62].

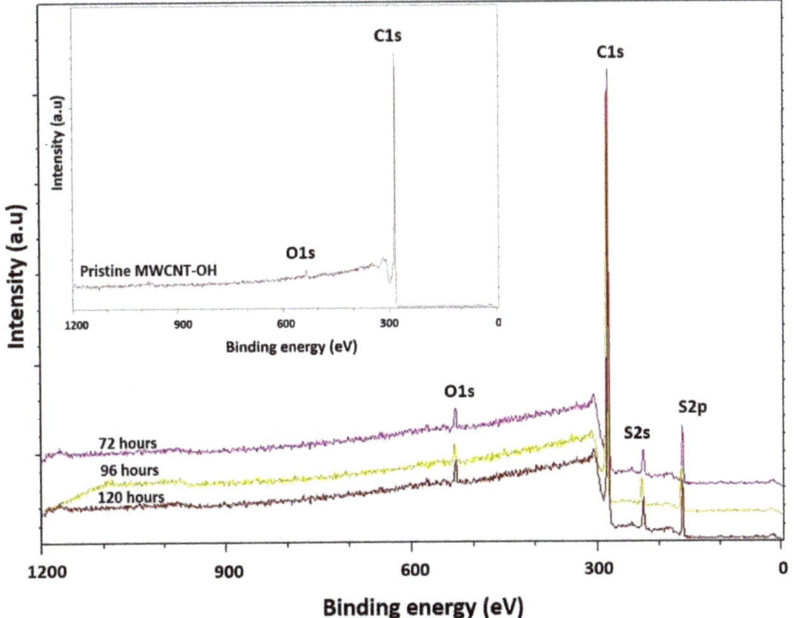

Figure 3. XPS spectra of pristine MWCNT-OH and P3HT/MWCNT-OH nanocomposites prepared at different reaction times.

Table 2. XPS data for pristine P3HT, pristine MWCNT-OH, and P3HT/MWCNT-OH nanocomposites.

Sample	Peak	Binding Energy (eV)	Atomic (%)	Ref.
Pristine P3HT	C 1s	284.6	-	[59]
	O 1s	~530.0	-	
	S 2s	229.1	-	
	S 2p	165.1, 165.8	-	
Pristine MWCNT-OH	C 1s	284.0	97.25	
	O 1s	531.0	2.75	
	S 2s	-	-	
	S 2p	-	-	
MWCNT-72	C 1s	284.0	80.5	Current work
	O 1s	531.0	2.7	
	S 2s	227.0	8.5	
	S 2p	163.0	8.3	
MWCNT-96	C 1s	281.0	78.2	
	O 1s	529.0	2.70	
	S 2s	225.0	6.6	
	S 2p	161.0	12.5	
MWCNT-120	C 1s	281.0	77.07	
	O 1s	529.0	2.63	
	S 2s	225.0	6.5	
	S 2p	161.0	13.8	

The P3HT/MWCNT-OH with 120 h of reaction time contained the higher atomic composition of P3HT at S 2s and S 2p, which was evident for the higher P3HT wrapped on the MWCNT-OH sidewall and increased the diameter size distribution (Table 3). The intensity of the carbon atomic percentage was found to decrease as the reaction time increased up to 120 h of reaction time, compared with the pristine MWCNT-OH carbon atomic percentage because the graphite surface of the MWCNT-OH was covered by the P3HT. This complemented the Raman results (Table 1), whereby prolonging the reaction time led to an increase in the defective band (D band), which was related to the Raman intensity ratio. The increased defective band indicated the graphitic or the carbon atoms had been covered or wrapped with the P3HT on the MWCNT-OH sidewall.

Table 3. Mean diameter of P3HT/MWCNT-OH nanocomposites at different reaction times.

Sample	Diameter (nm)	Standard Deviation
Pristine MWCNT-OH	27	8.10
MWCNT-24	33	12.48
MWCNT-48	35	14.80
MWCNT-72	42	18.30
MWCNT-96	45	12.58
MWCNT-120	50	19.24

3.3. FESEM and Mean Diameter Size Distribution Analysis

Figure 4 depicts the texture surface morphology of pristine P3HT, pristine MWCNT-OH, and the nanocomposite images that were prepared at different reaction times under FESEM. As shown in Figure 4a, the pristine P3HT showed a smooth surface texture without any notable morphology, and it was clearly visible, whereas, the morphology of pristine MWCNT-OH in Figure 4b showed that the nanotubes were not well aligned and were randomly entangled. In certain areas, the nanotubes were highly interconnected with each other, and agglomeration of MWCNT-OH was highly visible. This behaviour was explained by the fact that due to strong intrinsic van der Waals forces, CNTs tend to hold together as bundles, which eventually leads to low solubility in many solvents and therefore poor dispersion when mixed into various polymers.

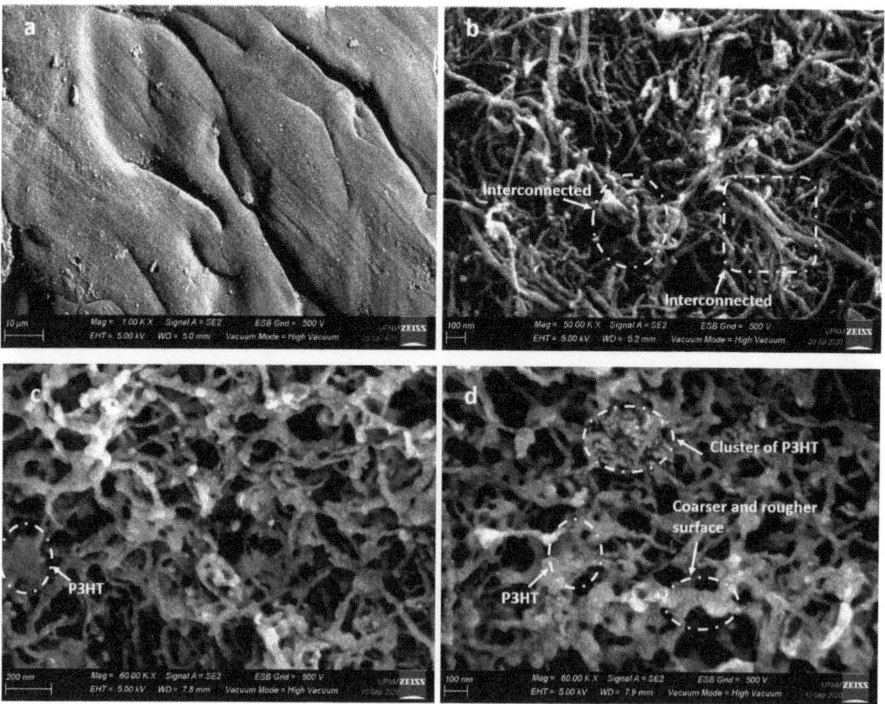

Figure 4. *Cont.*

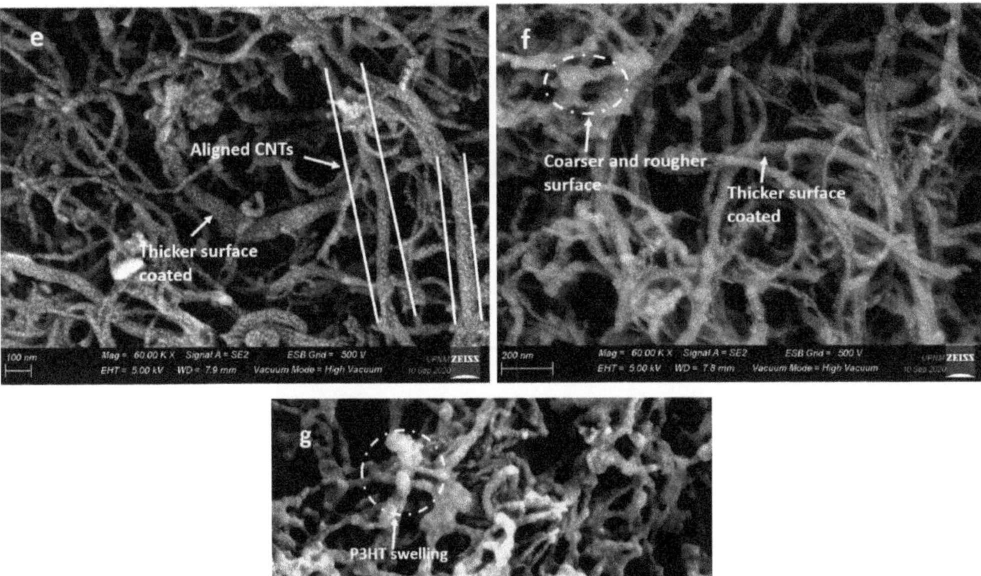

Figure 4. FESEM image of (**a**) pristine P3HT, (**b**) pristine MWCNT-OH, P3HT/MWCNT-OH nanocomposites at different reaction times (**c**) 24 h, (**d**) 48 h, (**e**) 72 h, (**f**) 96 h, and (**g**) 120 h.

For the P3HT/MWCNT-OH nanocomposites, the surface texture became coarser than the pristine MWCNT-OH, indicating that they were successfully wrapped and covered by the P3HT matrix. From the optimised reaction time for polymer wrapping, the non-covalent π–π interaction and CH–π interaction took place on wrapping of P3HT over MWCNT-OH walls. From the images in Figure 4c,d for MWCNT-24 and MWCNT-48, it was observed that there was not much difference in the images. The FESEM image showed P3HT were agglomerates and were inhomogeneously distributed on the nanotube surface. Furthermore, highly saturated, thicker and unsmooth wrapping spots were observed in certain areas. The entangled nanotubes were still visible in certain areas. This might be due to insufficient contact time for the P3HT to overcome the van der Walls forces towards the nanotube surface that led to better dispersion and wrapping. For MWCNT-72 and MWCNT-96 nanocomposite images in Figure 4e,f, less tangling and better alignment of MWCNT-OH were observed. This was due to the successful covering and wrapping of the P3HT, which made the nanotube sidewall thicker. In some areas, a highly uniform coating of P3HT along the nanotube surface could be seen. However, bigger clusters and a coarser surface were clearly shown in the images observed for MWCNT-120 in Figure 4g. This might be attributed to the swelling of P3HT due to a longer reaction time that made it saturated in a certain area along the nanotube wall.

The mean diameter size of nanocomposites was obtained using ImageJ software based on 100 points plotted in FESEM images. The histogram average of the variation of the diameter size distribution and standard deviation data observed for pristine MWCNT-OH and the P3HT/MWCNT-OH is shown in Figure 5 and Table 3, respectively. An average diameter size (together with standard deviation (stdev)) of the nanocomposite continued to increase from 27 nm (stdev = 8.10), 33 nm (stdev = 12.48), 35 nm (stdev = 14.80), 42 nm (stdev = 18.30), 45 nm (stdev = 12.58), and 50 nm (stdev = 19.24). This showed that the increase in the reaction time of the suspension increased the effectiveness of P3HT wrapping towards the MWCNT-OH sidewall by the increased diameter of MWCNT-OH. This also correlated with the intensity ratio of I_D/I_G Raman spectra, whereby the intensity ratio increased as the reaction time of the polymer wrapping increased.

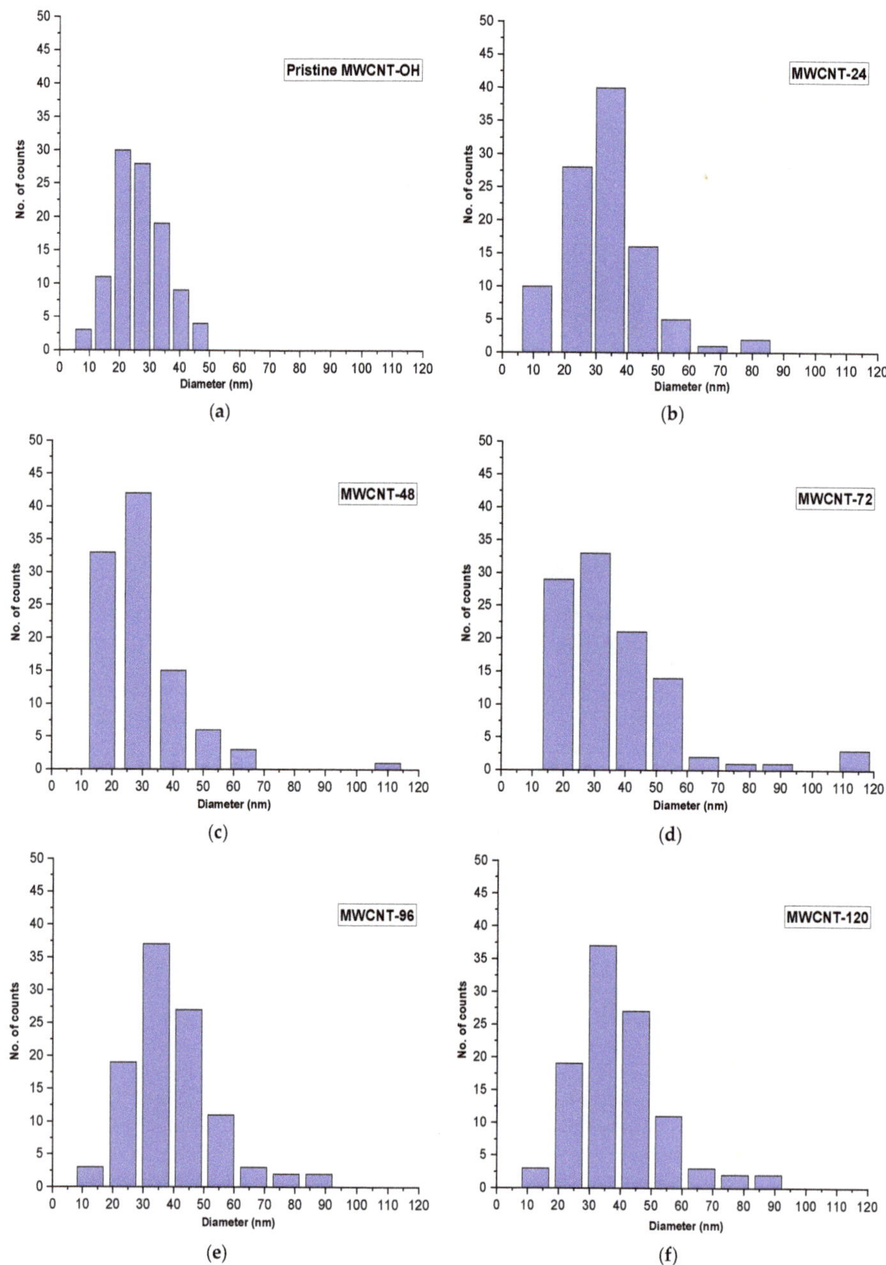

Figure 5. Histogram of diameter size distribution of (**a**) pristine MWCNT-OH and (**b–f**) P3HT/MWCNT-OH nanocomposites at different reaction times.

3.4. High-Resolution Transmission Electron Microscopy Analysis

In order to have a better resolution of the behaviour of individual nanotubes wrapped/unwrapped with P3HT, the nanocomposite was imaged using HRTEM. A close analysis of isolated nanocarbon confirmed that MWCNT-OH dispersed well in the polymer matrix. From the images in Figure 6, at lower reaction times of 24 h and 48 h during the preparation of nanocomposites, it could be seen clearly that there was a heterogeneous distribution of MWCNT-OH in the polymer matrix. There were unwrapped and non-uniformly wrapped P3HTs on nanotubes that adhered to the MWCNT-OH sidewalls with a diameter of about 30 nm. This slightly expanded the diameter of nanotubes, which were close to the diameter of pristine MWCNT-OH, and this might explain that a longer reaction time was required to enhance the interaction between P3HT and MWCNT-OH. Such ununiform P3HT wrapped over the MWCNT-OH sidewall could be attributed to a low electronic interaction between the lone pair of sulphur ions in P3HT and the π-bond from MWCNT-OH [63], whereas nanocomposites that were prepared at 72- and 96-hour reaction times showed that the nanotubes were dispersed homogenously with less entanglement and more aligned tubes. This might be due to sufficient time for the P3HT to swell, make an interaction and wrap the MWCNT-OH sidewalls. The formation of highly uniform and thick coatings of the P3HT layer was found on the sidewall of MWCNT-OH, which accounted for the stability of the P3HT/MWCNT-OH nanocomposites. This might be explained by the non-covalent of π–π interaction and CH-π interaction, which occurred on the wrapping of P3HT over the MWCNT-OH that resulted in uniform dispersion by overcoming the van der Waals forces between the MWCNT-OH [48].

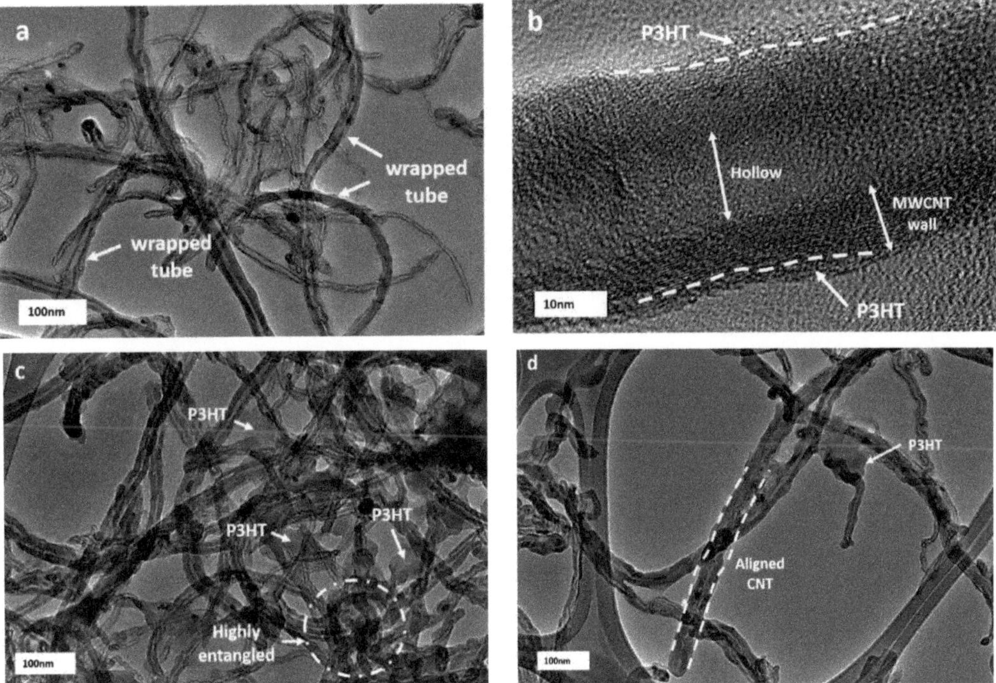

Figure 6. *Cont.*

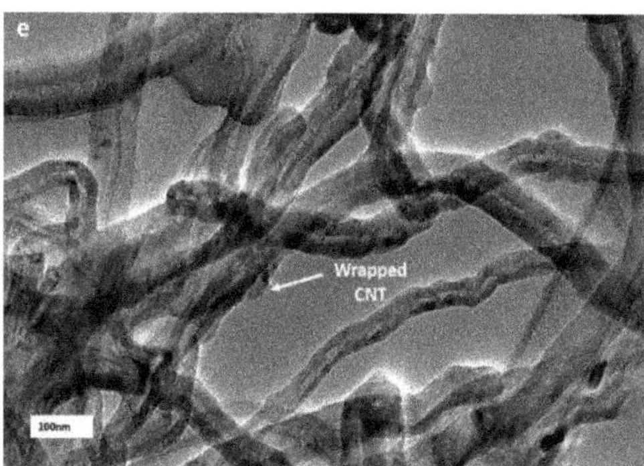

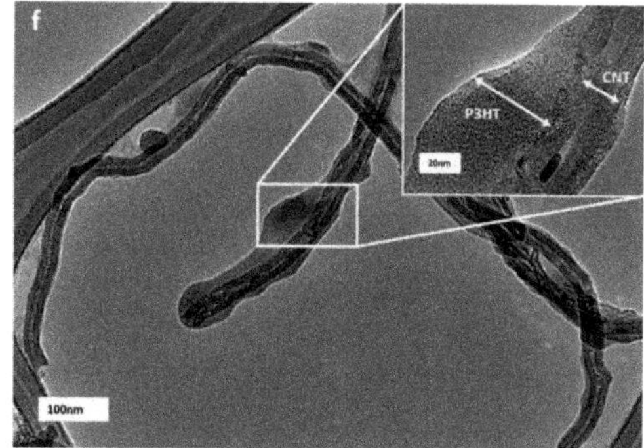

Figure 6. HRTEM image of P3HT/MWCNT-OH nanocomposites at different reaction times (**a**) 24 h, (**b**) 24 h at 10 nm view (**c**) 48 h, (**d**) 72 h, (**e**) 96 h, and (**f**) 120 h.

However, the maximum reaction time at 120 h might experience a different behaviour of the nanocomposite texture. The apparent formation of excess P3HT on the surface that resulted in thickening of the diameter of nanotubes with ununiform wrapping occurred at a longer reaction time. This was in agreement with the Raman spectra analysis and FESEM images. From Figure 6, it was observed that at a lower reaction time of 24-hour preparation of nanocomposites, there were unwrapped parts of MCNT-OH by the P3HT and non-uniform wrapping, which adhered to the MWCNT-OH sidewalls with a diameter of 30 nm, which was around the diameter of pristine MWCNT-OH. Such ununiform P3HT that wrapped over the MWCNT-OH sidewall could be accredited to less electronic interaction between the lone pair of sulphur ions in P3HT and the π-bond from MWCNT-OH [63]. This showed that a longer reaction time of the suspension was needed for better interaction between P3HT and MWCNT-OH. A longer reaction time obviously showed a reduction in the parts of unwrapped MCNT-OH. This might be due to sufficient time for the P3HT to swell, form an interaction and then wrap the MWCNT-OH sidewalls. Hussein et al. (2020) suggested that highly efficient π-conjugated systems might facilitate extra movement of charge carriers, as the nanocomposites would show a significant improvement in the direct current (DC) electrical conductivity, electrochemical behaviour, sensing ability, and the limit of detection (LOD) [64].

3.5. Thermogravimetric Analysis

The thermogravimetric analysis (TGA) and the differential thermogravimetry (DTG) curves for pristine P3HT, pristine MWCNT-OH, and P3HT/MWCNT-OH nanocomposites at different reaction times are presented in Figure 7a,b. A single degradation step was observed for pristine P3HT, which corresponded to the degradation of the alkyl side chain and main chain backbone at a temperature of 487.2 °C accompanied by a symmetric exothermic peak. Hacaloglu et al. (1997) stated that the degradation referred to the unsaturation along the chain, and the breaking of the thiophene ring was assigned to be the most possible decomposition pathway [65]. The MWCNT-OH was comparatively very stable in the range of 0 to 900 °C, with only 1.3% weight loss recorded at a temperature above 600 °C (2nd stage). This corresponded to the degradation temperature of impurities, C=C and C–OH [36]. Peng et al. (2008) indicated that the degradation of MWCNT significantly commenced at a temperature above 599.85 °C [66]. The 1st stage of decomposition of MWCNT-OH was observed at a temperature of 310.5 °C, which corresponded to the removal of hydroxyl groups on MWCNT [67,68]. Overall, based on the T_{onset} (Table 4), the thermal stability of the P3HT/MWCNT-OH was improved when the reaction time increased. This might have been influenced by the sufficient time for the P3HT to swell, form an interaction and then wrap the MWCNT-OH sidewalls that finally led to the improvement of thermal stability. This stage might be due to the chain scission of the sulfur atoms from the P3HT. However, the onset temperature for the P3HT/MWCNT-OH for all samples was observed from 342.7 to 355.5 °C.

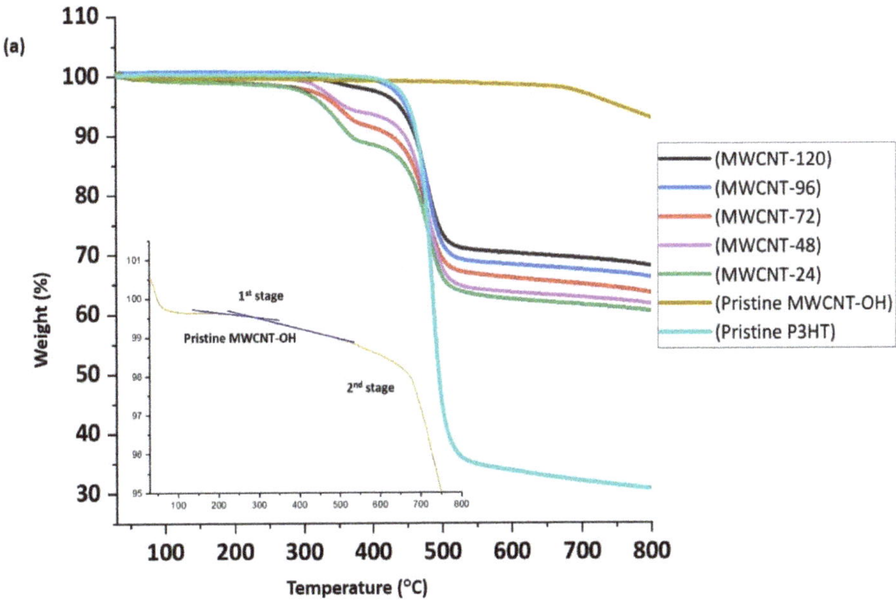

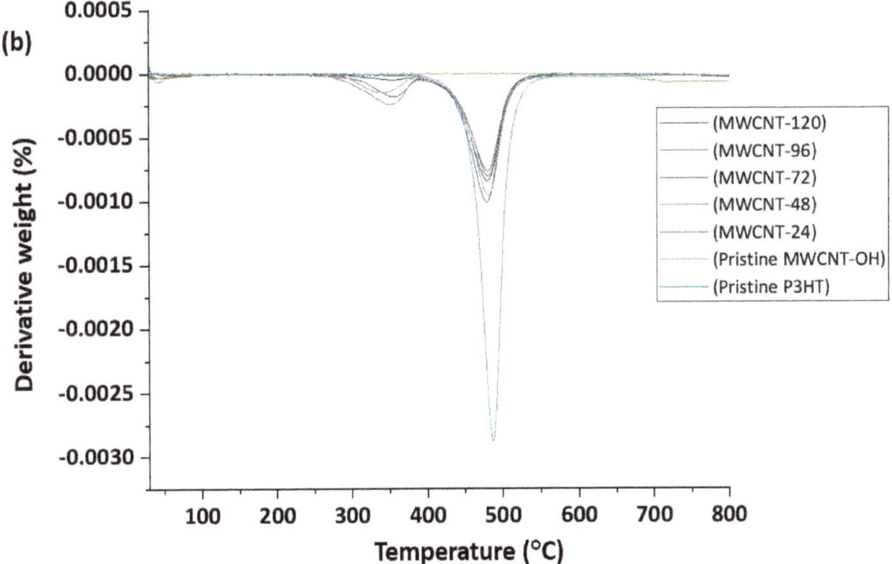

Figure 7. (**a**) TGA (**b**) DTG curves for pristine P3HT, pristine MWCNT-OH, and P3HT/MWCNT-OH nanocomposites at different reaction times.

Table 4. Characteristic temperatures of nanocomposites at elevated weight loss.

Sample	T_{onset} (°C)	T_{max} (°C)	$T_{10\%}$ (°C)	$T_{25\%}$ (°C)	Weight Loss (%) at 500 °C
Pristine P3HT	-	487.2	462.9	479.6	55.0
Pristine MWCNT-OH	310.5	-	-	-	1.3
MWCNT-24	342.7	481.0	364.8	478.4	33.8
MWCNT-48	351.3	480.7	446.1	483.5	31.7
MWCNT-72	351.5	481.0	450.2	488.9	31.0
MWCNT-96	354.5	480.2	455.4	490.0	30.5
MWCNT-120	355.5	479.5	461.5	493.8	26.1

It was observed that there was no significant trend in the degradation temperature for the T_{max} from the P3HT/MWCNT-OH nanocomposites. At this stage, the degradation, which occurred for P3HT/MWCNT-OH nanocomposites, might correspond to the side chain and the polymer backbone degradation. The thermal stability of P3HT/MWCNT-OH upon increased reaction time also supported the observation for $T_{10\%}$ and $T_{25\%}$, which referred to the degradation temperature at 10% of weight loss and 25% of weight loss, respectively. While at a temperature of 500 °C, the weight loss percentage was reduced as the reaction time increased from 24 h to 120 h. Compared to pristine P3HT, a reduction of weight loss was observed, from 55 to 26.1%. In addition, the phenyl ring in the P3HT backbone could increase thermal stability due to its role in the conjugation length. This proved that the effect of reaction time might influence the thermal stability and the weight loss of the nanocomposites. As a result, the greater thermal stability of the nanocomposites could result in the excellent stability of the conductivity of P3HT/MWCNT-OH nanocomposites at higher temperatures. The results in Table 4 show that the thermoelectric performance of the nanocomposites could be investigated under 300 °C, without destroying the structure of P3HT.

4. Conclusions

In this study, non-covalent functionalisation of a P3HT-wrapped MWCNT-OH nanocomposite was successfully prepared. It was found that the textural, structural, morphological, and thermal stability of the resultant nanocomposite properties changed significantly as the reaction time during the stirring process increased from 24 h to 120 h. The HRTEM and FESEM images revealed the incorporation of P3HT wrapped homogenously on the MWCNT wall at a lower reaction time with thickness in the range of 6 to 8 nm, whereas as the reaction time increased, the distribution of P3HT wrapped on the CNT wall became less homogeneous with the distribution of a thick P3HT layer of 20 to 23 nm wrapped onto the outer walls in certain areas. The intensity ratio from the Raman spectra increased up to 19% at a longer reaction time of 120 h compared with pristine MWCNT-OH, which caused the D band to increase and lowered the G band, which provides evidence of successful dispersion and wrapping of MWCNT-OH into the P3HT matrix. The stability of the nanocomposites increased as the reaction time increased, which influenced the thermal stability and the weight loss of the nanocomposites. In addition, this result would be expected for an effective electronic as well as π–π interaction between the P3HT and MWCNT-OH system that prolonged the degradation of the composite system at a certain point. Therefore, this study suggests that greater thermal stability and electronic interaction of the nanocomposites could drive the excellent potential application in the fabrication of polymer-based devices.

Author Contributions: Conceptualization N.M.N. and N.A.; validation, N.A.; writing—original draft preparation, N.M.N. and N.A.; writing—review and editing, N.M.N. and N.A.; supervision, N.A., S.Z.N.D., N.A.H. and I.S.M.; project administration, N.A.; funding acquisition, N.A., S.Z.N.D. and N.A.H. All authors have read and agreed to the published version of the manuscript.

Funding: We gratefully acknowledge the financial support from the Department of Higher Education (JPT), Ministry of Higher Education Malaysia and the Science and Technology Facilities Council, United Kingdom (STFC) under Newton fund's Programme and Malaysia Partnership and Alliances in Research (MyPAiR) for research grant ISIS-NEWTON/2019/SG/01 and the Chemical

Defence Research Centre (CHEMDEF), National Defence University of Malaysia, for the research grant UPNM/2018/CHEMDEFF/ST/3.

Institutional Review Board Statement: Not applicable.

Informed Consent Statement: Not applicable.

Data Availability Statement: Not applicable.

Conflicts of Interest: The authors declare no conflict of interest.

References

1. Shirakawa, H.; Louis, E.J.; MacDiarmid, A.G.; Chiang, C.K.; Heeger, A.J. Synthesis of electrically conducting organic polymers: Halogen derivatives of polyacetylene,(CH) x. *J. Chem. Soc. Chem. Commun.* **1977**, 578–580. [CrossRef]
2. Kaloni, T.P.; Giesbrecht, P.K.; Schreckenbach, G.; Freund, M.S. Polythiophene: From fundamental perspectives to applications. *Chem. Mater.* **2017**, *29*, 10248–10283. [CrossRef]
3. Mohd Nurazzi, N.; Muhammad Asyraf, M.R.; Khalina, A.; Abdullah, N.; Sabaruddin, F.A.; Kamarudin, S.H.; Ahmad, S.; Mahat, A.M.; Lee, C.L.; Aisyah, H.A. Fabrication, Functionalization, and Application of Carbon Nanotube-Reinforced Polymer Composite: An Overview. *Polymers* **2021**, *13*, 1047. [CrossRef]
4. Le, T.H.; Kim, Y.; Yoon, H. Electrical and electrochemical properties of conducting polymers. *Polymers* **2017**, *9*, 150. [CrossRef]
5. Zhang, Y.; Bunes, B.R.; Wu, N.; Ansari, A.; Rajabali, S.; Zang, L. Sensing methamphetamine with chemiresistive sensors based on polythiophene-blended single-walled carbon nanotubes. *Sens. Actuators B Chem.* **2018**, *255*, 1814–1818. [CrossRef]
6. Husain, A.; Ahmad, S.; Mohammad, F. Electrical conductivity and ammonia sensing studies on polythiophene/MWCNTs nanocomposites. *Materialia* **2020**, *14*, 100868. [CrossRef]
7. Nurazzi, N.M.; Harussani, M.M.; Siti Zulaikha, N.D.; Norhana, A.H.; Syakir, M.I.; Norli, A. Composites based on conductive polymer with carbon nanotubes in DMMP gas sensors–an overview. *Polimery* **2021**, *66*, 85–97. [CrossRef]
8. Norizan, M.N.; Zulaikha, N.D.S.; Norhana, A.B.; Syakir, M.I.; Norli, A. Carbon nanotubes-based sensor for ammonia gas detection–an overview. *Polimery* **2021**, *66*, 175–186. [CrossRef]
9. Liang, Z.; Li, M.; Wang, Q.; Qin, Y.; Stuard, S.J.; Peng, Z.; Deng, Y.; Ade, H.; Ye, L.; Geng, Y. Optimization Requirements of Efficient Polythiophene: Nonfullerene Organic Solar Cells. *Joule* **2020**, *4*, 1278–1295. [CrossRef]
10. Liang, Q.; Jiao, X.; Yan, Y.; Xie, Z.; Lu, G.; Liu, J.; Han, Y. Separating Crystallization Process of P3HT and O-IDTBR to Construct Highly Crystalline Interpenetrating Network with Optimized Vertical Phase Separation. *Adv. Funct. Mater.* **2019**, *29*, 1807591. [CrossRef]
11. Wang, Q.; Qin, Y.; Li, M.; Ye, L.; Geng, Y. Molecular Engineering and Morphology Control of Polythiophene: Nonfullerene Acceptor Blends for High-Performance Solar Cells. *Adv. Energy Mater.* **2020**, *10*, 2002572. [CrossRef]
12. Iqbal, S.; Shah, J.; Kotnala, R.K.; Ahmad, S. Highly efficient low cost EMI shielding by barium ferrite encapsulated polythiophene nanocomposite. *J. Alloy. Compd.* **2019**, *779*, 487–496. [CrossRef]
13. Dar, M.A.; Majid, K.; Najar, M.H.; Kotnala, R.K.; Shah, J.; Dhawan, S.K.; Farukh, M. Surfactant-assisted synthesis of polythiophene/Ni 0.5 Zn 0.5 Fe 2− x Ce x O 4 ferrite composites: Study of structural, dielectric and magnetic properties for EMI-shielding applications. *Phys. Chem. Chem. Phys.* **2017**, *19*, 10629–10643. [CrossRef] [PubMed]
14. Kwon, Y.H.; Park, J.J.; Housel, L.M.; Minnici, K.; Zhang, G.; Lee, S.R.; Lee, S.W.; Chen, Z.; Noda, S.; Takeuchi, E.S. Carbon Nanotube Web with Carboxylated Polythiophene "Assist" for High-Performance Battery Electrodes. *ACS Nano* **2018**, *12*, 3126–3139. [CrossRef] [PubMed]
15. Xu, D.; Wang, P.; Yang, R. Enhanced electrochemical performance of core-shell Li4Ti5O12/PTh as advanced anode for rechargeable lithium-ion batteries. *Ceram. Int.* **2017**, *43*, 7600–7606. [CrossRef]
16. Peymanfar, R.; Mohammadi, A.; Javanshir, S. Preparation of graphite-like carbon nitride/polythiophene nanocomposite and investigation of its optical and microwave absorbing characteristics. *Compos. Commun.* **2020**, *21*, 100421. [CrossRef]
17. Li, L.; Liu, S.; Lu, L. Synthesis and significantly enhanced microwave absorption properties of cobalt ferrite hollow microspheres with protrusions/polythiophene composites. *J. Alloys Compd.* **2017**, *722*, 158–165. [CrossRef]
18. Dehghani, Z.; Dadfarnia, S.; Haji Shabani, A.M.; Ehrampoush, M.H. Magnetic Multi-Walled Carbon Nanotubes Modified with Polythiophene as a Sorbent for Simultaneous Solid Phase Microextraction of Lead and Cadmium from Water and Food Samples. *Anal. Bioanal. Chem. Res.* **2020**, *7*, 509–523.
19. Dutta, K.; Rana, D. Polythiophenes: An emerging class of promising water purifying materials. *Eur. Polym. J.* **2019**, *116*, 370–385. [CrossRef]
20. Kao, E.; Liang, Q.; Bertholet, G.R.K.; Zang, X.; Park, H.S.; Bae, J.; Lu, J.; Lin, L. Electropolymerized polythiophene photoelectrodes for photocatalytic water splitting and hydrogen production. *Sens. Actuators A Phys.* **2018**, *277*, 18–25. [CrossRef]
21. Ng, C.H.; Winther-Jensen, O.; Ohlin, C.A.; Winther-Jensen, B. Enhanced catalytic activity towards hydrogen evolution on polythiophene via microstructural changes. *Int. J. Hydrogen Energy* **2017**, *42*, 886–894. [CrossRef]
22. Husain, A.; Ahmad, S.; Mohammad, F. Synthesis, characterisation and ethanol sensing application of polythiophene/graphene nanocomposite. *Mater. Chem. Phys.* **2020**, *239*, 122324. [CrossRef]

23. Bhardwaj, P.; Grace, A.N. Antistatic and microwave shielding performance of polythiophene-graphene grafted 3-dimensional carbon fibre composite. *Diam. Relat. Mater.* **2020**, *106*, 107871. [CrossRef]
24. Liu, W.; Zhu, L. Nanofibers for Gas Sensing. *Adv. Fiber Sens. Technol.* **2020**, 133–153.
25. Demon, S.Z.N.; Kamisan, A.I.; Abdullah, N.; Noor, S.A.M.; Khim, O.K.; Kasim, N.A.M.; Yahya, M.Z.A.; Manaf, N.A.A.; Azmi, A.F.M.; Halim, N.A. Graphene-based Materials in Gas Sensor Applications: A Review. *Sens. Mater.* **2020**, *32*, 759–777. [CrossRef]
26. Shah, A.H. Applications of carbon nanotubes and their polymer nanocomposites for gas sensors. In *Carbon Nanotubes-Current Progress of their Polymer Composites*; IntechOpen (United Kingdom): London, UK, 2016; pp. 459–494.
27. Hunter, G.W.; Akbar, S.; Bhansali, S.; Daniele, M.; Erb, P.D.; Johnson, K.; Liu, C.C.; Miller, D.; Oralkan, O.; Hesketh, P.J. Editors' choice—Critical review—A critical review of solid state gas sensors. *J. Electrochem. Soc.* **2020**, *167*, 1–30. [CrossRef]
28. Norizan, M.N.; Moklis, M.H.; Demon, S.Z.N.; Halim, N.A.; Samsuri, A.; Mohamad, I.S.; Knight, V.F.; Abdullah, N. Carbon nanotubes: Functionalisation and their application in chemical sensors. *RSC Adv.* **2020**, *10*, 43704–43732. [CrossRef]
29. Ahmadi, M.; Zabihi, O.; Masoomi, M.; Naebe, M. Synergistic effect of MWCNTs functionalization on interfacial and mechanical properties of multi-scale UHMWPE fibre reinforced epoxy composites. *Compos. Sci. Technol.* **2016**, *134*, 1–11. [CrossRef]
30. Maruyama, B.; Alam, K. Carbon nanotubes and nanofibers in composite materials. *Sampe J.* **2002**, *38*, 59–70.
31. Collins, P.G.; Avouris, P. Nanotubes for Electronics–Scientific American. *San Fr. Nat. Publ. Group* **2000**, *283*, 62–69.
32. Abdullah, A.; Mohamad, I.S.; Bani Hashim, A.Y.; Abdullah, N.; Poh, B.W.; Md Isa, M.H.; Zainal Abidin, S. Thermal conductivity and viscosity of deionised water and ethylene glycol-based nanofluids. *J. Mech. Eng. Sci.* **2016**, *10*, 2249–2261. [CrossRef]
33. Kou, J.; Qian, H.; Lu, H.; Liu, Y.; Xu, Y.; Wu, F.; Fan, J. Optimizing the design of nanostructures for improved thermal conduction within confined spaces. *Nanoscale Res. Lett.* **2011**, *6*, 1–8. [CrossRef]
34. Elashmawi, I.S.; Gaabour, L.H. Raman, morphology and electrical behavior of nanocomposites based on PEO/PVDF with multi-walled carbon nanotubes. *Results Phys.* **2015**, *5*, 105–110. [CrossRef]
35. Mescoloto, A.D.F.; Pulcinelli, S.H.; Santilli, C.V.; Gonçalves, V.C. Structural and thermal properties of carboxylic acid functionalized polythiophenes. *Polímeros* **2014**, *24*, 25–30. [CrossRef]
36. Bachhav, S.; Patil, D. Preparation and Characterization of Multiwalled Carbon Nanotubes-Polythiophene Nanocomposites and its Gas Sensitivity Study at Room Temperature. *J. Nanostructures* **2017**, *7*, 247–257.
37. Ribeiro, B.; Botelho, E.C.; Costa, M.L.; Bandeira, C.F. Carbon nanotube buckypaper reinforced polymer composites: A review. *Polímeros* **2017**, *27*, 247–255. [CrossRef]
38. Mallakpour, S.; Soltanian, S. Surface functionalization of carbon nanotubes: Fabrication and applications. *Rsc Adv.* **2016**, *6*, 109916–109935. [CrossRef]
39. Karim, M.R. Synthesis and characterizations of poly (3-hexylthiophene) and modified carbon nanotube composites. *J. Nanomater.* **2012**, *2012*, 174353. [CrossRef]
40. Janudin, N.; Abdullah, L.C.; Abdullah, N.; Md Yasin, F.; Saidi, N.M.; Kasim, N.A.M. Comparison and characterization of acid functionalization of multi walled carbon nanotubes using various methods. *Solid State Phenom.* **2017**, *264*, 83–86. [CrossRef]
41. Mohamad Saidi, N.; Kasim, N.A.M.; Osman, M.J.; Janudin, N.; Mohamad, I.S.; Abdullah, N. The Influences of Chemical and Mechanical Treatment on the Morphology of Carbon Nanofibers. *Solid State Phenom.* **2017**, *264*, 107–111. [CrossRef]
42. Janudin, N.; Abdullah, L.C.; Abdullah, N.; Yasin, F.M.; Saidi, N.M.; Kasim, N.A.M. Characterization of Amide and Ester Functionalized Multiwalled Carbon Nanotubes. *Asian J. Chem.* **2018**, *30*, 1613–1616. [CrossRef]
43. Janudin, N.; Abdullah, N.; Wan Yunus, W.M.Z.; Yasin, F.M.; Yaacob, M.H.; Mohamad Saidi, N.; Mohd Kasim, N.A. Effect of functionalized carbon nanotubes in the detection of benzene at room temperature. *J. Nanotechnol.* **2018**, *2018*, 2107898. [CrossRef]
44. Ma, P.C.; Siddiqui, N.A.; Marom, G.; Kim, J.K. Dispersion and functionalization of carbon nanotubes for polymer-based nanocomposites: A review. *Compos. Part A Appl. Sci. Manuf.* **2010**, *41*, 1345–1367. [CrossRef]
45. Huang, Y.Y.; Ahir, S.V.; Terentjev, E.M. Dispersion rheology of carbon nanotubes in a polymer matrix. *Phys. Rev. B* **2006**, *73*, 125422. [CrossRef]
46. Bose, S.; Khare, R.A.; Moldenaers, P. Assessing the strengths and weaknesses of various types of pre-treatments of carbon nanotubes on the properties of polymer/carbon nanotubes composites: A critical review. *Polymer* **2010**, *51*, 975–993. [CrossRef]
47. Sa'aya, N.S.N.; Demon, S.Z.N.; Abdullah, N.; Abd Shatar, V.F.K.E.; Halim, N.A. Optical and Morphological Studies of Multiwalled Carbon Nanotube-incorporated Poly (3-hexylthiophene-2, 5-diyl) Nanocomposites. *Sens. Mater.* **2019**, *31*, 2997–3006.
48. Rathore, P.; Negi, C.M.S.; Yadav, A.; Verma, A.S.; Gupta, S.K. Influence of MWCNT doping on performance of polymer bulk heterojunction based devices. *Optik* **2018**, *160*, 131–137. [CrossRef]
49. Abdulla, S.; Mathew, T.L.; Pullithadathil, B. Highly sensitive, room temperature gas sensor based on polyaniline-multiwalled carbon nanotubes (PANI/MWCNTs) nanocomposite for trace-level ammonia detection. *Sens. Actuators B Chem.* **2015**, *221*, 1523–1534. [CrossRef]
50. Karim, M.R.; Lee, C.J.; Lee, M.S. Synthesis and characterization of conducting polythiophene/carbon nanotubes composites. *J. Polym. Sci. Part A Polym. Chem.* **2006**, *44*, 5283–5290. [CrossRef]
51. Gómez, S.; Rendtorff, N.M.; Aglietti, E.F.; Sakka, Y.; Suárez, G. Surface modification of multiwall carbon nanotubes by sulfonitric treatment. *Appl. Surf. Sci.* **2016**, *379*, 264–269. [CrossRef]
52. Zhao, Y.; Yuan, G.; Roche, P.; Leclerc, M. A calorimetric study of the phase transitions in poly (3-hexylthiophene). *Polymer* **1995**, *36*, 2211–2214. [CrossRef]

53. Li, C.; Chen, Y.; Wang, Y.; Iqbal, Z.; Chhowalla, M.; Mitra, S. A fullerene–single wall carbon nanotube complex for polymer bulk heterojunction photovoltaic cells. *J. Mater. Chem.* **2007**, *17*, 2406–2411. [CrossRef]
54. Giulianini, M.; Waclawik, E.R.; Bell, J.M.; Scarselli, M.; Castrucci, P.; De Crescenzi, M.; Motta, N. Microscopic and spectroscopic investigation of poly (3-hexylthiophene) interaction with carbon nanotubes. *Polymers* **2011**, *3*, 1433–1446. [CrossRef]
55. Saini, V.; Li, Z.; Bourdo, S.; Dervishi, E.; Xu, Y.; Ma, X.; Kunets, V.P.; Salamo, G.J.; Viswanathan, T.; Biris, A.R. Electrical, optical, and morphological properties of P3HT-MWNT nanocomposites prepared by in situ polymerization. *J. Phys. Chem. C* **2009**, *113*, 8023–8029. [CrossRef]
56. Fujigaya, T.; Nakashima, N. Non-covalent polymer wrapping of carbon nanotubes and the role of wrapped polymers as functional dispersants. *Sci. Technol. Adv. Mater.* **2015**, *16*, 024802. [CrossRef] [PubMed]
57. Datsyuk, V.; Kalyva, M.; Papagelis, K.; Parthenios, J.; Tasis, D.; Siokou, A.; Kallitsis, I.; Galiotis, C. Chemical oxidation of multiwalled carbon nanotubes. *Carbon* **2008**, *46*, 833–840. [CrossRef]
58. Kang, E.T.; Neoh, K.G.; Tan, K.L. X-ray photoelectron spectroscopic studies of poly (2, 2'-bithiophene) and its complexes. *Phys. Rev. B* **1991**, *44*, 10461. [CrossRef]
59. Karim, M.R.; Yeum, J.H.; Lee, M.S.; Lim, K.T. Synthesis of conducting polythiophene composites with multi-walled carbon nanotube by the γ-radiolysis polymerization method. *Mater. Chem. Phys.* **2008**, *112*, 779–782. [CrossRef]
60. Swathy, T.S.; Antony, M.J. Tangled silver nanoparticles embedded polythiophene-functionalized multiwalled carbon nanotube nanocomposites with remarkable electrical and thermal properties. *Polymer* **2020**, *189*, 122171. [CrossRef]
61. Dumitru, A.; Vulpe, S.; Radu, A.; Antohe, S. Influence of nitrogen environment on the performance of conducting polymers/CNTs nanocomposites modified anodes for microbial fuel cells (MFCs). *Rom. J. Phys* **2018**, *63*, 605–625.
62. Erten-Ela, S.; Cogal, S.; Cogal, G.C.; Oksuz, A.U. Highly conductive polymer materials based multi-walled carbon nanotubes as counter electrodes for dye-sensitized solar cells. *Fuller. Nanotub. Carbon Nanostruct.* **2016**, *24*, 380–384. [CrossRef]
63. Husain, A.; Ahmad, S.; Mohammad, F. Polythiophene/graphene/zinc tungstate nanocomposite: Synthesis, characterization, DC electrical conductivity and cigarette smoke sensing application. *Polym. Polym. Compos.* **2020**. [CrossRef]
64. Husain, A.; Shariq, M.U.; Mohammad, F. DC electrical conductivity and liquefied petroleum gas sensing application of polythiophene/zinc oxide nanocomposite. *Materialia* **2020**, *9*, 100599. [CrossRef]
65. Hacaloglu, J.; Yigit, S.; Akbulut, U.; Toppare, L. Thermal degradation of polythiophene natural rubber and polythiophene-synthetic rubber conducting polymer composites. *Polymer* **1997**, *38*, 5119–5124. [CrossRef]
66. Peng, Z.; Holm, A.H.; Nielsen, L.T.; Pedersen, S.U.; Daasbjerg, K. Covalent sidewall functionalization of carbon nanotubes by a "formation– degradation" approach. *Chem. Mater.* **2008**, *20*, 6068–6075. [CrossRef]
67. Li, S.; Wang, Z.; Jia, J.; Hou, C.; Hao, X.; Zhang, H. Preparation of hydroxyl and (3-aminopropyl) triethoxysilane functionalized multiwall carbon nanotubes for use as conductive fillers in the polyurethane composite. *Polym. Compos.* **2018**, *39*, 1212–1222. [CrossRef]
68. Liang, S.; Li, G.; Tian, R. Multi-walled carbon nanotubes functionalized with a ultrahigh fraction of carboxyl and hydroxyl groups by ultrasound-assisted oxidation. *J. Mater. Sci.* **2016**, *51*, 3513–3524. [CrossRef]

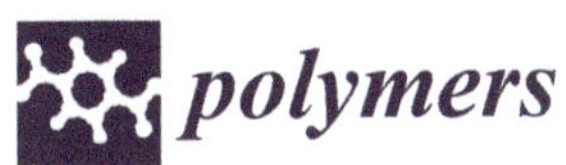

Review

Natural Fiber Reinforced Composite Material for Product Design: A Short Review

M. A. Azman [1], M. R. M. Asyraf [2], A. Khalina [3], Michal Petrů [4], C. M. Ruzaidi [1,5], S. M. Sapuan [3,6], W. B. Wan Nik [1,5], M. R. Ishak [2], R. A. Ilyas [7,8,*] and M. J. Suriani [1,5,*]

1   Faculty of Ocean Engineering Technology and Informatics, Universiti Malaysia Terengganu, Kuala Nerus 21030, Terengganu, Malaysia; asyrafazman23@yahoo.com (M.A.A.); ruzaidi@umt.edu.my (C.M.R.); niksani@umt.edu.my (W.B.W.N.)
2   Department of Aerospace Engineering, Faculty of Engineering, Universiti Putra Malaysia, Serdang 43400, Selangor, Malaysia; asyrafriz96@gmail.com (M.R.M.A.); mohdridzwan@upm.edu.my (M.R.I.)
3   Laboratory of Biocomposite Technology, Institute of Tropical Forestry and Forest Products (INTROP), Universiti Putra Malaysia, Serdang 43400, Selangor, Malaysia; khalina@upm.edu.my (A.K.); sapuan@upm.edu.my (S.M.S.)
4   Faculty of Mechanical Engineering, Technical University of Liberec, Studentská 2, 461 17 Liberec, Czech Republic; michal.petru@tul.cz
5   Marine Materials Research Group, Faculty of Ocean Engineering Technology and Informatics, Universiti Malaysia Terengganu, Kuala Nerus 21030, Terengganu, Malaysia
6   Advanced Engineering Materials and Composites Research Centre (AEMC), Department of Mechanical and Manufacturing Engineering, Faculty of Engineering, Universiti Putra Malaysia, Serdang 43400, Selangor, Malaysia
7   School of Chemical and Energy Engineering, Faculty of Engineering, Universiti Teknologi Malaysia, Johor Bahru 81310, Johor, Malaysia
8   Centre for Advanced Composite Materials (CACM), Universiti Teknologi Malaysia, Johor Bahru 81310, Johor, Malaysia
*   Correspondence: ahmadilyas@utm.edu.my (R.A.I.); surianimatjusoh@umt.edu.my (M.J.S.)

Citation: Azman, M.A.; Asyraf, M.R.M.; Khalina, A.; Petrů, M.; Ruzaidi, C.M.; Sapuan, S.M.; Wan Nik, W.B.; Ishak, M.R.; Ilyas, R.A.; Suriani, M.J. Natural Fiber Reinforced Composite Material for Product Design: A Short Review. *Polymers* **2021**, *13*, 1917. https://doi.org/10.3390/polym13121917

Academic Editor: Francisco Javier Espinach Orús

Received: 30 April 2021
Accepted: 2 June 2021
Published: 9 June 2021

Publisher's Note: MDPI stays neutral with regard to jurisdictional claims in published maps and institutional affiliations.

Abstract: Natural fibers have attracted great attention from industrial players and researchers for the exploitation of polymer composites because of their "greener" nature and contribution to sustainable practice. Various industries have shifted toward sustainable technology in order to improve the balance between the environment and social and economic concerns. This manuscript aims to provide a brief review of the development of the foremost natural fiber-reinforced polymer composite (NFRPC) product designs and their applications. The first part of the manuscript presents a summary of the background of various natural fibers and their composites in the context of engineering applications. The behaviors of NFPCs vary with fiber type, source, and structure. Several drawbacks of NFPCs, e.g., higher water absorption rate, inferior fire resistance, and lower mechanical properties, have limited their applications. This has necessitated the development of good practice in systematic engineering design in order to attain optimized NRPC products. Product design and manufacturing engineering need to move in a mutually considerate manner in order to produce successful natural fiber-based composite material products. The design process involves concept design, material selection, and finally, the manufacturing of the design. Numerous products have been commercialized using natural fibers, e.g., sports equipment, musical instruments, and electronic products. In the end, this review provides a guideline for the product design process based on natural fibers, which subsequently leads to a sustainable design.

Keywords: natural fiber composite; product design; sustainability design; design process

1. Introduction

A new product begins with an idea and ends with the physical production of the product. According to Milton and Rodgers [1], to minimize or reduce the impact of a product on the environment, it is necessary to reconsider its impact throughout its life cycle, such as how it is produced, the development process, usage, packaging, preservation, and recycling or disposal. When a product ignores environmental factors in its design process, designers are likely to face backlash from their customers. The competition of the product market nowadays is increasing; therefore, designers should consider the selection of environmentally friendly materials as the main criteria in their design.

Natural fiber composites are environmentally friendly materials that have attracted attention in the field of product manufacturing engineering [2,3]. Starting 3000 years ago, straw-reinforced clay was the first composite material to be used by the ancient Egyptians in their building construction. Research and development have proven that natural fibers have been successfully applied as reinforcements in the composites industry, such as for transportation, interior components, building, aircraft, and construction [4–8]. Furthermore, natural fiber composites have the advantages of being cheaper than synthetic composites, bio-degradable, abundantly available, renewable, and lightweight [9–16]. Natural fibers originate from three sources, namely, plants, animals, and minerals. There are more than 2000 types of fiber plants in the world, and these are mostly composed of cellulose, e.g., kenaf, sugar palm, bamboo, corn, cotton, flax, hay (from grass cutting), hemp, henequen, jute, pineapple leaf, banana, ramie, and sisal [17,18]. The use of natural fibers in composites can also solve some other problems, such as moderate energy consumption during production, leaving almost no carbon footprint, and reducing disposal problems [19–22].

Design is the first step in the manufacturing process; at this stage, many important decisions need to be made that will affect the result of a product. Therefore, several things need to be considered, such as manufacturing, assembly, cost, sales, maintenance, disposal, and recycling, early on in the design process. In addition, 70% of product manufacturing costs are determined at an early stage of the design process [23]. Product design using natural fiber composite uses the same method as other product design processes [24–26]. Product designers will determine the formal qualities of products manufactured by industry by focusing on aptness in function, use, ease of production, materials, cost, and the number of constituent parts. Additionally, they also concentrate on user experience—the interaction between users and products, types of meanings products evoke, and what sorts of emotions the products elicit [17]. Marzuki [27] also reported that designers need three things, namely, material, machinery, and method of manufacturing. This means that designers do not only have to produce quality designs, but they are also responsible for proposing and determining the appropriate materials so that the product can be produced at an affordable cost. This shows that the use of materials, human factors, and design are interrelated and can serve as a guideline to designers.

Fundamentally, design and manufacturing need to move in an integrated manner to complete the design process, including design concepts, material selection, and manufacturing process selection. Each problem needs to be addressed according to its respective expertise. Some experts will focus on the materials to be used, while others will focus on design concepts [24,28]. Design concepts and materials can be combined in a computer system, namely, computer-aided drawing (CAD) and finite element analysis (FEA).

2. Natural Fiber Reinforced Composite Material

Natural Fiber

Natural fiber materials have become increasingly popular in the manufacturing industry and have been studied by many researchers. Natural fibers are divided into three categories, namely, cellulose-based, protein-based, and mineral-based, as shown in Table 1. Natural fibers are sustainable materials that are available in nature and have advantages as listed in Table 2. The compositions of natural fibers can be divided into three main components, which are cellulose, hemicellulose, and lignin. Table 3 displays the chemical composition of the natural fibers, in which the chemical composition and cell structures are quite complex and differ between plant parts and origins. Depending on the cellulose crystallinity, the physical, chemical, and mechanical behaviors of the lignocellulosic fibers vary from one to another [29,30]. Generally, natural fiber's main constituent is cellulose, at 30–80%, followed by hemicellulose at 7–40%, and 3–33% lignin, as shown in Table 3.

Table 1. Types of natural fibers [31–34].

Natural Fiber	Cellulose/Lignocellulose	Grass/Reed	Bamboo, corn
		Stalk	Wheat, maize, oat, rice
		Wood	Hardwood, softwood
		Fruit	Coir
		Seed	Cotton
		Leaf	Abaca, banana, pineapple, sisal
		Bast	Flax, hemp, jute, kenaf, ramie
	Animal	Wool/hair	Cashmere, goat hair, horse hair, lamb wool
		silk	Mulberry
	Mineral	-	Asbestos, ceramic, metal

Table 2. The advantages of natural fibers.

Author (Year)	Advantages of Natural Fibers
Bakar et al. [10]	Low cost, low elongation, low density, non-conductivity, corrosion resistance, absorb significant amounts and able to solve environmental pollution.
Corona et al. [35]	Renewable, moderate energy consumption for production and disposal can reduce environmental problems.
Hanan et al. [4]	Has certain strength properties, non-rough surface, lightweight, renewable, has specific modulus properties, can reduce pollution, biodegradable, require less energy to produce, and inexpensive.
Aji et al. [36]	Low density, cost-saving during manufacturing, less rough surface, harmless biodegradation, renewable, comparable mechanical properties with inorganic fiber, recyclable in most countries, and the surface is easily modified.
Amir et al. [37]	Substitute for synthetic fibers and as a reinforcing material in composites.
Nordin et al. [11]	In terms of mechanical properties, natural fibers are a good substitute for polymer composites because of their renewable material source, light weight, inexpensiveness, low density, and the materials are readily available.
Maleque et al. [38]	Ease of use in chemical and mechanical modifications.
Rognoli et al. [17]	Environmentally friendly materials.
Taekema and Karana, [39]	Low density, high specific strength, renewable, recyclable according to the mixture of materials used, high thermal and acoustic insulation, energy consumption savings of up to 60% in the production process (average for automotive component manufacturing), can be produced with low technology and investment and highly recommended for developing countries.
Sapuan and Maleque [40]	Mechanical properties are comparable to existing conventional materials that include low production costs, renewability, and environmentally friendly materials.
Shekar and Ramachandra [41]	Good mechanical properties, renewable, non-abrasive to process equipment, and can be burned at the end of its life cycle for energy recovery, and also abundantly available.
Elanchezhian et al. [9]	Renewable, inexpensive, completely or partially recyclable material, and biodegradable. In addition, this material has low density, low cost, and has environmentally friendly mechanical properties. It is also an alternative material for fiberglass, carbon, and human-made fibers for composite manufacturing.

Table 2. *Cont.*

Author (Year)	Advantages of Natural Fibers
Ilyas et al. [42–44]	Cost-effective, biodegradable, and renewable materials.
Peças et al. [31]	Renewable, low production costs, low density, acceptable modulus–weight ratio, low manufacturing energy consumption, low carbon, and biodegradable.
Huda et al. [45]	Cheaper, less energy required in the production of fiber reinforcement compared to conventional fibers such as glass and carbon.
Thyavihalli Girijappa et al. [46]	Abundantly available and cost-effective production.
Arpitha et al. [47]	Good mechanical properties, light weight, low cost, high specific strength, less rough surface, environmentally friendly, and good biodegradation characteristics.
Madhu et al. [48]	Creates huge employment opportunities in the rural plantation sector, available in large quantities, biodegradable, recyclable, better energy recovery, low production costs, lightweight materials, high strength and specific modulus, lower health risks, low density, low cost, less skin irritation, less abrasion of equipment, reduced tool wear, improved energy recovery, and reduced skin irritation and respiration

Table 3. Chemical composition of selected common natural fibers.

Fibers	Holocellulose (wt. %)		Lignin (wt. %)	Ash (wt. %)	Extractives (wt. %)	Crystallinity (%)	Ref.
	Cellulose (wt. %)	Hemicellulose (wt. %)					
Arecanut husk	34.18	20.83	31.60	2.34	-	37	[49]
Banana	7.5	74.9	7.9	0.01	9.6	15.0	[50]
Curauna	70.2 ± 0.7	18.3 ± 0.8	9.3 ± 0.9	-	-	64	[51]
Helicteres isora plant	71 ± 2.6	3.1 ± 0.5	21 ± 0.9	-	-	38	[52]
Kenaf bast	63.5 ± 0.5	17.6 ±1.4	12.7 ± 1.5	2.2 ± 0.8	4.0 ± 1.0	48.2	[53]
Kenaf core powder	80.26		23.58	-	-	48.1	[54]
Mengkuang leaves	37.3 ± 0.6	34.4 ± 0.2	24 ± 0.8		2.5 ± 0.02	55.1	[55]
Oil palm empty fruit bunch (OPEFB)	37.1 ± 4.4	39.9 ± 0.75	18.6 ± 1.3	-	3.1 ± 3.4	45.0	[56]
Oil palm empty fruit bunch (OPEFB)	40 ±2	23 ±2	21 ± 1	-	2.0 ± 0.2	40	[57]
Oil palm frond (OPF)	45.0 ± 0.6	32.0 ± 1.4	16.9 ± 0.4	-	2.3 ± 1.0	54.5	[56]
Oil palm mesocarp fiber (OPMF)	28.2 ± 0.8	32.7 ± 4.8	32.4 ± 4.0	-	6.5 ± 0.1	34.3	[56]
Phoenix dactylifera palm leaflet	33.5	26.0	27.0	6.5	-	50	[58]
Phoenix dactylifera palm rachis	44.0	28.0	14.0	2.5	-	55	[58]
Pineapple leaf	81.27 ± 2.45	12.31 ± 1.35	3.46 ± 0.58	-	-	35.97	[59]
Ramie	69.83	9.63	3.98	-	-	55.48	[60]
Rubber wood	45 ±3	20 ± 2	29 ± 2	-	2.5 ± 0.5	46	[57]
Soy hull	56.4 ± 0.92	12.5 ± 0.72	18.0 ± 2.5	-	-	59.8	[61]
Sugar beet	44.95 ± 0.09	25.40 ± 2.06	11.23 ± 1.66	17.67 ± 1.54	-	35.67	[62]
Sugar palm	43.88	7.24	33.24	1.01	2.73	55.8	[63]
Sugarcane bagasse	43.6	27.7	27.7	-	-	76	[64]
Water hyacinth	42.8	20.6	4.1	-	-	59.56	[65]
Wheat straw	43.2 ± 0.15	34.1 ± 1.2	22.0 ± 3.1	-	-	57.5	[66]

3. Composites

In 1980, a fiberglass-reinforced plastic composite (GFRP) known as fiberglass and carbon fiber-reinforced polymer composite (CFRP) was designed. Composites are formed from a combination of two or more materials of physical and chemical difference [41].

This combination consists of the reinforcement phase, in the form of fibers, pieces, or particles [67,68], being embedded in other materials, referred to as the matrix phase [69–71]. Reinforcement is load-bearing, while the matrix phase serves as a binder of the reinforcing material and distributes the load between the fibers. According to Elanchezhian et al. [9], the matrix also acts as a material to protect the fiber material from damage, before, during, and after composite processing. This combination is able to produce new materials with better properties than the individual material [15,72].

On the other hand, composite hybrids involve a combination of two or more fibers in a matrix [10]. Composite hybrids also have broad prospects in product design as manufacturing materials. Hybrid composites can overcome several natural fiber deficiencies, e.g., low mechanical properties, high absorption properties, poor adhesion, and poor thermal stability during the process [4]. The results of the studies by Rashid et al. [73] on Kevlar reinforced with woven coir found that the impact strength showed minimal enhancement while the breakable properties of pure epoxy composites were decreased. According to Jawaid et al. [74], the addition of fibers and coupling agents significantly improved the thermal stability (e.g., decomposition and residue content) of the hybrids. In addition, a study conducted by Masoodi and Pillai [75] found that hybrid jute composites possessed high resistance to water absorption. However, the strength of hybrid jute composites was decreased as the humidity was increased. To reduce this effect, more jute fiber fractions are needed. Past studies have also found that the combination of natural fiber with synthetic fiber should be preferably recommended. For instance, a study on long kenaf fibers with Kevlar highlighted the effectiveness of materials, as well as cost material savings. The result of the study showed that reinforcing 20% of Kevlar's weight within the composite kenaf enabled it to absorb a maximum energy of 12.76 J [10]. This has proven that the combination of Kevlar fiber with a kenaf composite is capable of improving energy absorption and imparting higher strength properties.

4. Product Design for Natural Fiber Composite (NFC)

Industrial design (ID) is the professional practice of designing products, devices, objects, and services used by millions of people around the world every day [76]. Product design is one of the sub-areas in industrial design that include medical and safety equipment and home appliances. A good product design needs to go through a long process, namely, the design process, before entering the manufacturing stage. According to Abidin et al. [77], furniture design has a very wide scope and comprises furniture in houses, offices, and public places. For transportation design, it consists of land, sea, and air vehicles, e.g., cars, motorcycles, buses, sea trucks, ships, jet skis, helicopters, and airplanes. According to Ramani et al. [78], in 2007, the industrial sector in the United States of America had produced over 1235×106 metric tons of carbon dioxide gas that would further complicate the restoration of the greenhouse gases.

4.1. Selection Material in Product Design

Product designers can use natural fiber composite materials in design proposals in order for the design to be promoted as an eco-design. Eco-design is also known as design for the environment, and is defined as the process of "integrating a systematic environmental system into product design and development" [79]. Designers need to be more careful in choosing the right natural fiber for a product. According to research conducted by Karana [80], the choice of material often depends on the material that has been used before, to ensure that the material to be used is safe. However, this method causes the selection of materials to be limited. The selection of material plays an important role in the production of an innovative product. According to a study by Taekema and Karana [39], materials can be distinguished according to their sensory properties, e.g., blurred texture and transparency, or mechanical properties such as tensile strength, thermal conductivity, and the ability of materials to be processed and shaped, for example, they can be painted or injected. This process needs to be completed in the conceptual design phase. Figure 1 shows the design process starting from the initial stage of idea sketching to 2D rendering [81]. Therefore, as an introduction, designers should form an overview of these key properties, e.g., sensorial properties (such as its velvet-like texture and its transparency), technical properties (such as it specific tensile strength), and formability properties (such as its ability to undergo injection molding or being paintable) to inspire and stimulate them to decide on a particular material. Unless technical requirements are defined at the outset of the project, product designers consider the technical properties at an overview level, and not in detail at the conceptual design stage [39]. According to Elvin Karana et al. [82], designers can use the Meanings of Material (MoM) model as a guideline for material selection. These guidelines can also be applied by product designers to select natural fibers that are appropriate for the designed product. In addition, natural fibers also have intangible properties, such as their relation to trends and value to the culture, and emotions evoked by a material. These circumstances play important roles in helping product designers to make decisions in material selection.

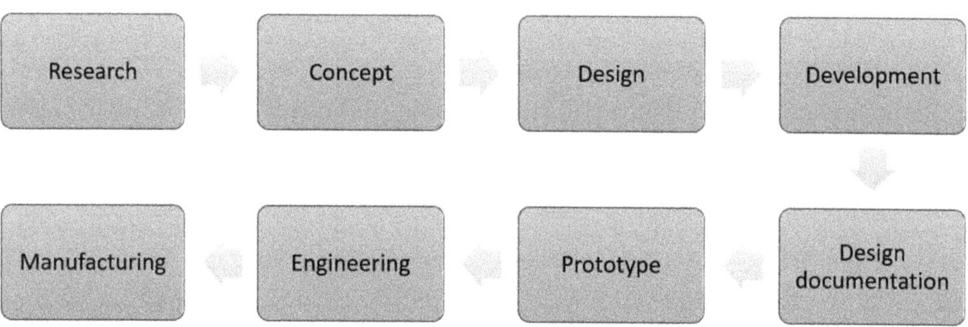

Figure 1. Design process flow.

4.2. Evaluation Concept for Product Design

Product designers need reliable, rigorous, and robust methods for evaluating and selecting their design proposals. Choosing the right method is very important, and choosing the wrong product design proposal to be developed can be very costly to the manufacturer in terms of money, time, and other valuable resources. Designers need to constantly evaluate the direction of their design concept while, at the same time, creating many concepts to choose from. When considering the selection of proposals, product design specifications (PDS) are very useful because they serve as evaluation parameters in the process [1]. Poor PDS is also one of the reasons for low-quality or unsuccessful products. Specifications are desired as measurable parameters of features that facilitate the realization of a function [83]. Table 4 shows some of the key points that should appear in a PDS. These are taken from a pilot study conducted by Azman et al. [84] into redeveloping a face mask design for hajj people.

Matrix evaluation, or the Paugh method, is a quantitative technique used by designers to evaluate their proposed design concepts by ranking them against the set criteria stated in the PDS [1]. This method was invented by a British engineering design professor Stuart Paugh, who was considered a pioneer in product design development, and this method has been used worldwide in the field of design for manufacturing [23]. The selection of concept design proposals is one of the processes involved in narrowing down a number of alternative proposals and aims to select one for further development and refinement. Below is the decision matrix model (Paugh's method). As stated by Mahmud et al. [83], a significant increase in the information available on product design specifications (PDS) during the design process leads to a lower rate of product desertion. The process of selecting the most satisfactory design proposal according to the PDS is very important in ensuring that the proposed design concept does not deviate from the guidelines as stated in the PDS. Matrix evaluation can help designers, engineers, manufacturers, marketing staff, users, clients, and buyers to reduce ambiguity and confusion in the evaluation and selection process, resulting in clearer communication and the delivery of successful new products to market more frequently.

Table 4. Product design specification (PDS).

Design Specifications	Explanation
Universal design	Usable by both genders; availability of different sizes; usable by pilgrims with a beard, other facial hair, or other conditions that prevent a good seal between the face and the sealing surface of the face mask
Comfortable	Ergonomic; large breathing space (or dead space) for relaxed breathing; reduced facial covering without compromising the face mask's efficiency
Effectiveness	Therapeutic effectiveness of the face masks against airborne infectious diseases is highly critical
Low cost	The low cost can allow face masks to be given for free by Tabung Haji, as preferred by the pilgrims

4.3. Development of Product Design by Integrating Design for Sustainability with Other Concurrent Engineering Techniques

During the process of the development of NFCs products, product engineers have to implement the concept of design for sustainability (DfS) in order to promote sustainable products. Design for sustainability (DfS) could play a vital role in directing us towards sustainable consumption and production, which is defined based on the four pillars of sustainability ((1) ecological, (2) social, (3) economic, and (4) institutional, as shown in Figure 2) that are essential to achieving sustainable life quality [85]. For designers to practice sustainability, they should include and assess these four pillars, from obtaining the resources to producing final products [86]. It is fundamental to incorporate constituents that adhere to consumption and production standards, including the use of the most appropriate technology, materials, and production processes to achieve zero carbon emissions and minimal non-renewable resource use, whilst paying attention to the impacts on human well-being [87].

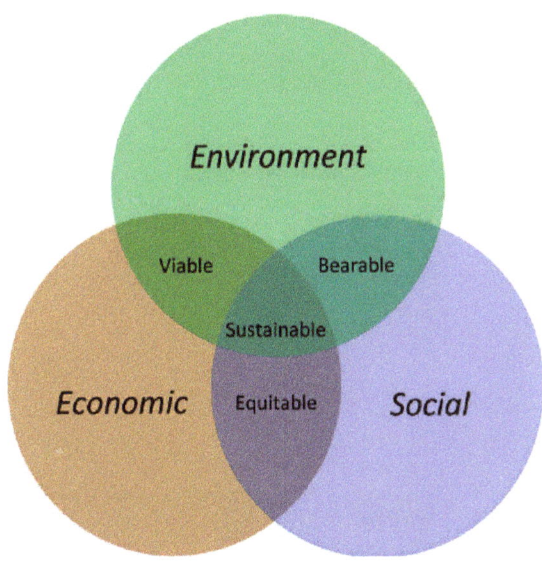

Figure 2. The pillars of sustainability [88].

Commonly, the DfS approach exploits the design for excellent (DfX) to produce a sustainable product. This process involves the analysis of the environmental impacts of specific design attributes, comprising safety and biodegradability prospects, in the development of sustainable components/products. Jawahir et al. [89] proposed a conceptual framework for DfS based on the DfX principles as displayed in Figure 3. The implementation of concurrent engineering in product development is essential to satisfying human needs and developing sustainable products before the manufacturing process [90]. In particular, a biocomposite product has to meet the requirements of life cycle analysis and sales trends, e.g., raw material and production costs, product's performance, and consumer's demands [91,92].

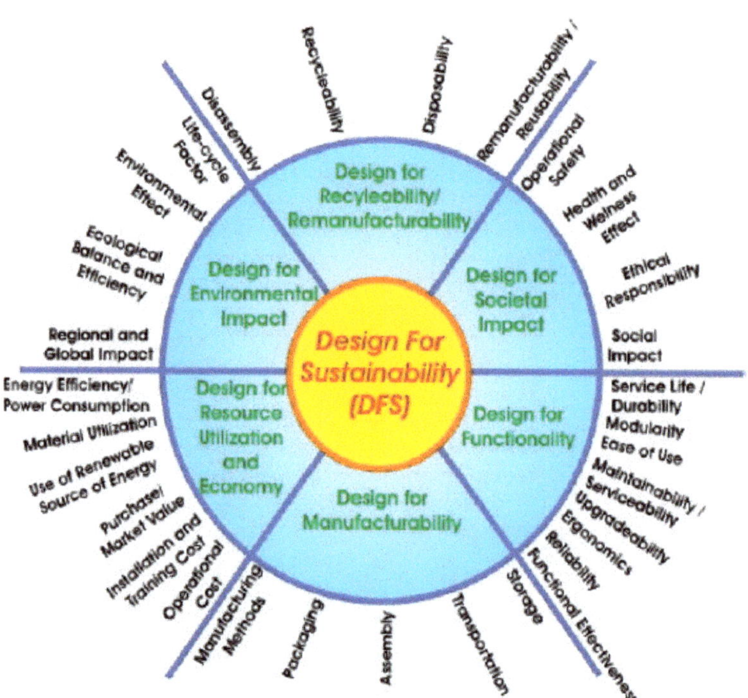

Figure 3. Elements within design for sustainability (DfS) [89].

4.3.1. Theory of Inventive Problem Solving (TRIZ)

TRIZ, or the theory of inventive problem solving, is a tool used by concurrent engineers to develop various solutions using inventive principles to cater to the problems that arise [93]. The tool also eliminates any negative drawbacks that may arise during the development of the solution, as it is focused on the root cause of the problem [94–96]. Initially, the tool is applied to determine the design intentions (purpose subject of the design) before the development of inventive solutions, specifically for the design. The TRIZ tool can be categorized into four main techniques: (1) Su-field modeling, (2) algorithms of inventive problem solving (ARIZ), (3) prediction of technology trends, and (4) contradiction engineering with 40 inventive principles [97]. The application of the techniques depends on the complexity level of the targeted problems when attempting to systematically solve a problem by identifying opportunity and innovation techniques [98]. In this case, Cascini et al. [99] developed a new concept of sheet metal snips based on TRIZ contradiction methods, which compare the improving and worsening parameters to select suitable inventive principles. At the end of the product development process, they refined the design concepts via a CAD optimization tool. Figure 4 displays the conceptual design process conducted by Asyraf et al. [100].

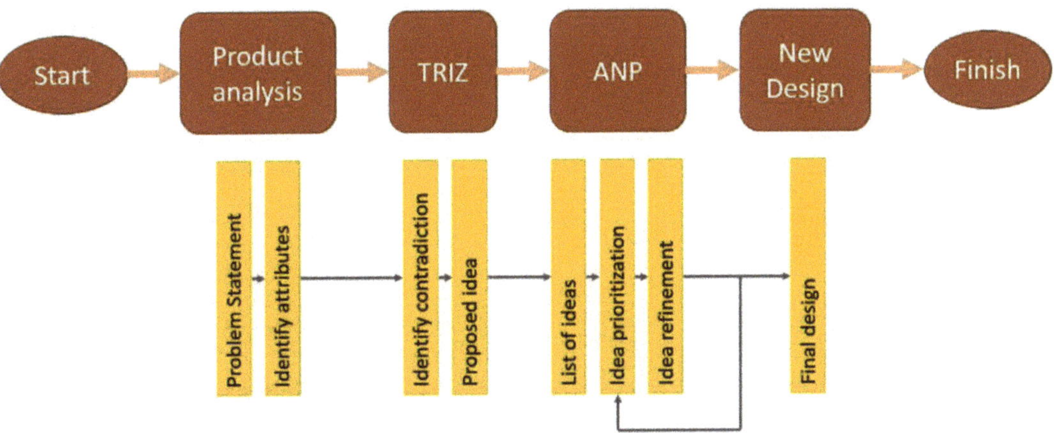

Figure 4. The application of TRIZ in the concurrent engineering conceptual design framework to develop the product [100].

4.3.2. Voice of Customer

Voice of customer (VOC) is one of the main approaches in concurrent engineering to generate ideas for design intends. The voice of the customer is obtained via many techniques, including direct customer specifications, observation, surveys, discussion or interviews, focus groups, warranty data, and field reports. From these VOC data, this information is then incorporated in a product planning matrix or quality function deployment (QFD) [101]. The QFD is used to define customer requests and turn this information into systematic plans to produce products to meet those desires [101].

4.3.3. Morphological Chart

The morphological chart is a concurrent engineering technique that implements a chart with various arrangements to aid designers in selecting new combinations of attributes/elements. The "morphology" term refers to the study of the form or shape of the material, whereas "morphological chart" is defined as a systematic approach to generating and analyzing the form or characteristics of a product that might be selected [102,103]. The chart functions to offer a series of choices for each element and component that can be combined to become a solution idea. The combination of elements and components would create multiple design features beneficial for the product's functions. Figure 5 displays an example of a morphological chart used to develop and model conceptual designs for a natural fiber composite fire extinguisher.

TRIZ solution principles	Design features	Solution			
		1	2	3	4
#40 Parameter Changes - Apply a hybrid fibres for whole fire extinguisher shell to withstand high internal pressure with lesser raw material cost	Natural fibre	Sisal	Jute	Flax	Hemp
	Man-made fibre	Aramid	E-glass	Carbon	
	Material	Single natural fibre	Hybrid natural fibre and conventional fibre		
	Position and pattern of hybridization fibres	Interlayer	Intralayer (yarn by yarn)	Intralayer (fibre by fibre)	
- Change the internal profile of tank into two sphere profile to have consistency of distribution of internal pressure.	Internal profile	Sphere	Cylindrical	Cuboid	
	Compartment	Single	Double		
- Fabricate the exterior tank into cylindrical shape by combing all winding pattern to make it resist from cracking when accidentally impact with hard object.	Manufacturing process	Hand-layup	Pultrusion	Filament winding	Vacuum bagging
	Winding pattern	Hoop	Helical	Polar	Combined
	External profile	Cylindrical	Square	Triangular	Hexagonal

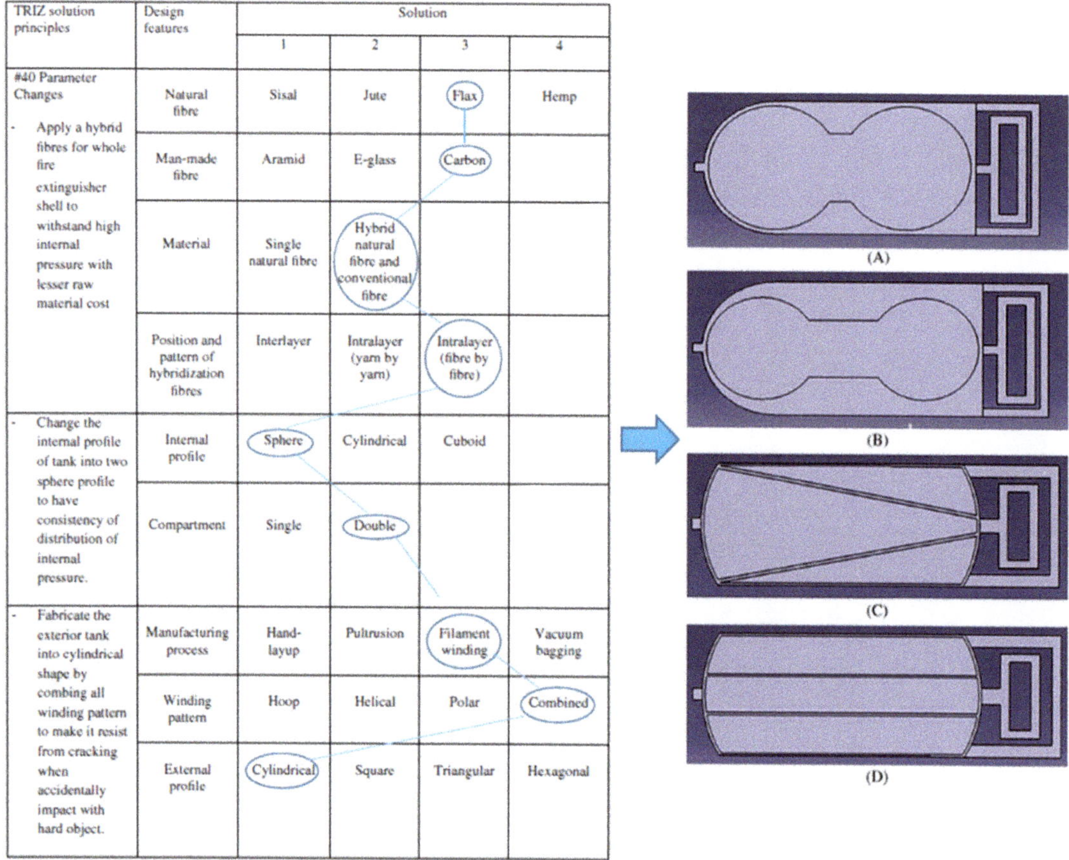

Figure 5. Morphological chart used to elaborate the design characteristics of a fire extinguisher, by Asyraf et al. [104]. (**A**)—concept design A, (**B**)—concept design B, (**C**)—concept design C, (**D**)—concept design D.

4.3.4. Extending the Search Space

This concurrent engineering approach is also called the "Why? Why? Why?" technique, which is used to elaborate the search option by questioning the root cause of the problem [105]. For this case, the question would be "Why do we need safety in composite products?" After getting the answer to the question, it would be followed with another "why" question, until a conclusive solution is reached. For this case, the method is highly dependent on luck, thus brainstorming is suggested to solve the problems of the product's development.

4.3.5. Gallery Method

This technique is used by designers to produce ideas by displaying many generated concepts simultaneously with a conducted discussion. Typically, these concepts are visualized by sketching and taping them on the wall of the designer's design room to review each aspect of the idea. Designers may consequentially be able to suggest improvements for the concept, or they might even suddenly generate related ideas via this process [106].

4.3.6. Brainstorming

Analysis of the current systems or products is one of the methods used by many designers and researchers to initiate new models or prototypes with better solutions [107]. This type of method is also called brainstorming, which involves discussions covering the physical analysis of current products. The discussion might produce a clearer picture by mind mapping problems, generating ideas, producing concept designs, and determining fabrication processes and finalized prototypes [91]. Usually, the analysis of the existing product would be in relation to competitors' products, older products of one's own company, and similar products that have several sub-functions of function structures.

According to Sapuan et al. [108], several generation techniques have been used by designers to develop conceptual designs for products, e.g., theory of inventive problem solving (TRIZ), brainstorming, strengths, weaknesses, opportunities and threats (SWOT) analysis, gallery method, and systematic exploitation of proven ideas of experience [109,110]. For this project, the simplest way to generate the idea for conceptualizing the design concepts was using the brainstorming approach. A design focus group was formed for discussion among the members of Advanced Engineering Materials and Composites Research Centre (AEMC), Department Mechanical and Manufacturing Engineering, Universiti Putra Malaysia. Every concept design was outlined and listed based on discussion outputs and PDS documents of the previous design stage. In the end, around five conceptual designs of a roselle fiber-reinforced polymer composite smartphone holder were developed. The details of each concept design are laid out in Table 5.

Table 5. New conceptual designs of a roselle fiber-reinforced polymer composite smartphone holder and their descriptions [108].

	Conceptual Design	Description
1.		• Concept inspired by high heel shoes. • Lifts the smartphone higher than regular holder. • Good artistic structure.
2.		• Easy to manufacture and assemble. • Simple and minimalist design. • High stability due to wider baseline.
3.		• Simple and minimalist design. • Easy for storage. • Easy to fabricate and manufacture. • High stability due to wider baseline.
4.		• A hollow triangle feature can be used to hold stationary (i.e., pen, pencil, key, etc.). • Three strip lines for aesthetics purposes.

Table 5. *Cont.*

	Conceptual Design	Description
5.		• A hollow triangle feature can be used to hold stationery (i.e., pen, pencil, key, etc.). • Simple and minimal design. • Isometric design, hence easy to fabricate.

In another study conducted by Ilyas et al. [111], a focus group was formed to comprehensively discuss and produce ideas on the conceptual design of a biocomposite mug pad among members of the Advanced Engineering Materials and Biocomposites Research Centre (AEMC), Department of Mechanical and Manufacturing Engineering, Universiti Putra Malaysia. After the brainstorming output, every concept design was listed based on the previous PDS document. Specifically, for this research activity, five design concepts of a roselle fiber biocomposite mug pad with details were produced and are tabulated in Table 6. Creative and innovative variations were developed to add value to the ideas, in addition to using roselle biocomposites.

Table 6. Proposed conceptual designs of a roselle fiber biocomposite mug pad [111].

	Conceptual Design	Description
1.		• Concept inspired by the bottle cap. • Composes of a main pad and two stand legs. • Has a good artistic design.
2.		• Comprises a bottom rectangle plate and a circle shape pad on the top. • Simple and minimalist design. • High stability due to a wider baseline.
3.		• Concept inspired by a square coffee table. • Composed of four legs, one ring to hold the mug, and one square pad. • Complex shape with artistic value.
4.		• Simple and minimalist design. • Easy for storage. • Easy to fabricate and manufacture. • High stability. • High strength due to wider baseline. • Concept inspired by a leaf.

Table 6. *Cont.*

Conceptual Design	Description
5.	• Simple and minimalist design. • Circular and isometric design, hence easy to fabricate. • Easy for storage. • Easy to fabricate and manufacture. • High stability.

5. Natural Fibers Composite Applications

The applications for NFCs are growing rapidly in numerous engineering fields. Various types of natural fibers have been used as reinforcements in polymer composites, including corn [112], water hyacinth [113], coir [114], ginger [115,116], cotton [117,118], kenaf [7,91,119–122], sugarcane [123–125], flax [126], ramie [60], hemp [127], kapok [128], sisal [129], wood [22], oil palm [130,131], banana [132], as well as sugar palm [43,63,133–141]. Along with biodegradability, natural fibers come with many other advantages, e.g., substituting timber for wood plastic composite, being less costly, availability, and reducing deforestation [21]. Natural fibers have huge potential to be converted into useful products [20], as it was revealed by Ilyas et al. [142] that natural fibers are the right material for the replacement of glass and carbon fibers. Different natural fibers, e.g., jute, hemp, kenaf, oil palm, and bamboo-reinforced polymer composites, have become of great importance in different automotive applications, structural components, packaging applications, furniture, and constructions [46,143]. NFCs are used in electrical and electronic industries, aerospace, sports, recreation equipment, boats, machinery, office products, and so forth. A roselle fiber-reinforced polymer composite smartphone holder developed using the design for sustainability (DfS) approach was achieved by Sapuan et al. [108], as shown in Figure 6. The concept development of the environmentally friendly smartphone holder product was carried out using concept generation and concept evaluation techniques. The roselle composite smartphone holder development process involved market analysis, product design specification (PDS) document generation, conceptual design creation, and detailed design of the finished product. The mold of the product was fabricated using a 3D printing method. Then, the roselle fiber composite smartphone holder was fabricated via a hand lay-up process.

Another study was conducted by Ilyas et al. [111] on a roselle fiber-reinforced polymer composite mug pad's product development process using the sustainability (DfS) approach, as shown in Figure 7. The concept development of the environmentally friendly mug pad product was performed using concept generation and concept evaluation techniques. The processes involved in their study were similar to those in the study of Sapuan et al. [108], in which the final design of the molded product was fabricated using a 3D printer, and the roselle fiber composite mug pad was fabricated using a hand lay-up process. The final product was completed and demonstrated easy fabrication, light weight, low overall cost, and an appropriate balance between functionality and aesthetics. Table 7 shows the example applications of the natural fiber composite.

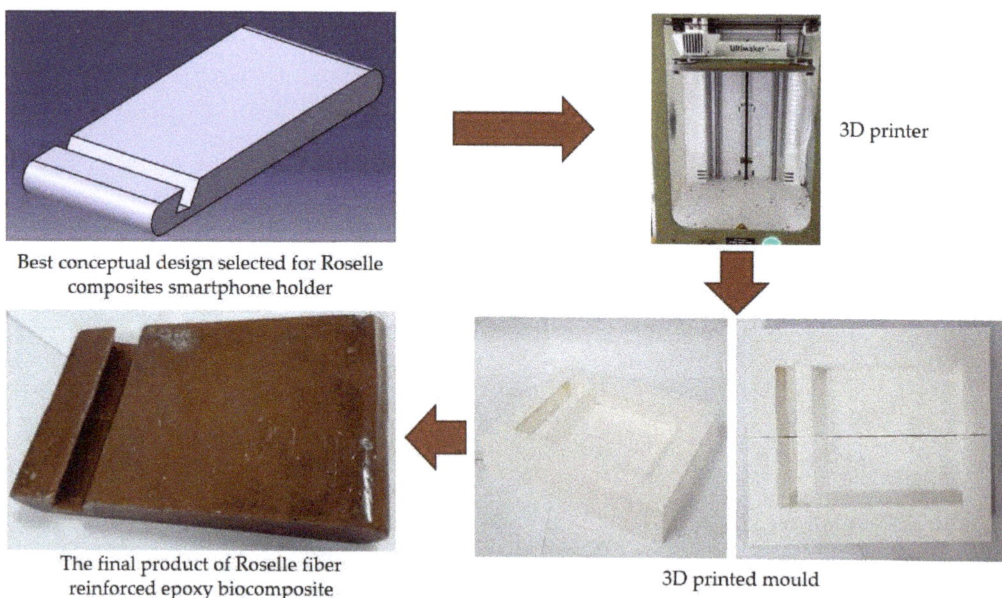

Figure 6. Smartphone holder using roselle fiber-reinforced polymer composites.

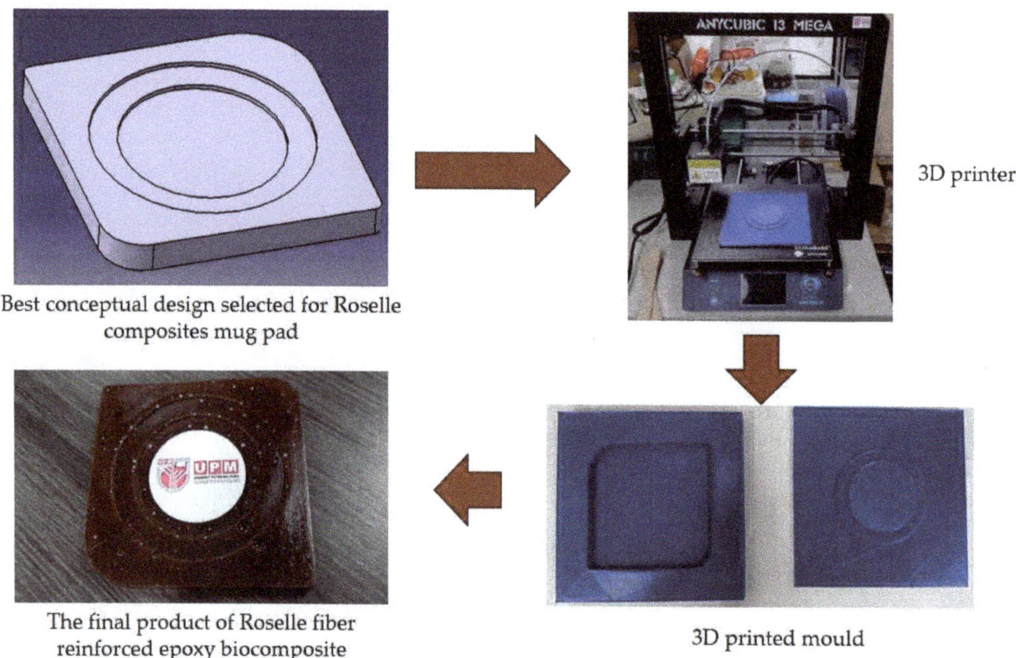

Figure 7. Mug pad holder using roselle fiber-reinforced polymer composites.

Table 7. Applications of natural fibers composites [143].

Natural Fiber Composite	Applications
Bamboo	Application in building, construction, and others
Roselle	Mug pad, smartphone holder, furniture, automotive applications
Hemp	Construction products, textile, cordage, geotextile, paper and packaging, furniture, electrical, banknote, and pipe
Oil palm	Building materials such as window, door frame, structural insulated panel building system, siding, fencing, roofing, decking, and others
Wood	Window frame, panel, door shutter, decking, railing system, and fencing
Flax	Window frame, panel, decking, railing system, fencing, tennis racket, bicycle frame, fork, seat post, snowboarding, and laptop case
Rice husk	Building materials such as building panel, brick, window frame, panel, decking, railing system, and fencing
Bagasse	Window frame, panel, decking, railing systems, and fencing
Sisal	Used in the construction industry such as in panels, doors, shutting plates, and roofing sheet; also, in the manufacturing of paper and pulp
Stalk	Building panel, furniture panel, brick, drain, and pipeline
Kenaf	Packing material, mobile case, bag, insulation, clothing-grade cloth, soilless potting mix, animal bedding, and material that absorbs oil and liquids
Cotton	Furniture industry, textile and yarn, food packaging, and cordage
Coir	Building panel, flush door shutter, roofing sheet, storage tank, packing material, helmet and postbox, mirror casing, paperweights, projector cover, voltage stabilizer cover, filling material for seat upholstery, brush and broom, rope and yarn for net, bag, and mat, as well as padding for mattress and seat cushion
Ramie	Industrial sewing thread, packing material, fishing net, and filter cloth. It is also made into fabrics for household furnishings (upholstery, canvas) and clothing, as well as paper manufacture
Jute	Building panel, roofing sheet, door frame, door shutter, transport, packaging, geotextiles, and chipboard

5.1. Natural Fibers Composites' Applications in Electrical and Electronic Components

Currently, the increased importance of raw materials derived from renewable resources, as well as the recyclability or biodegradability of products, are causing a transformation from petroleum-based synthetics to natural fibers in electrical and electronic applications [31]. The broad advantages of natural fiber-reinforced composites, such as high stiffness to weight ratio, light weight, and biodegradability, make them suitable for different applications in electrical and electronic industries. "FOMA(R) N701iECO" utilize kenaf fibers in their eco-mobile phone casing, as shown in Figure 8.

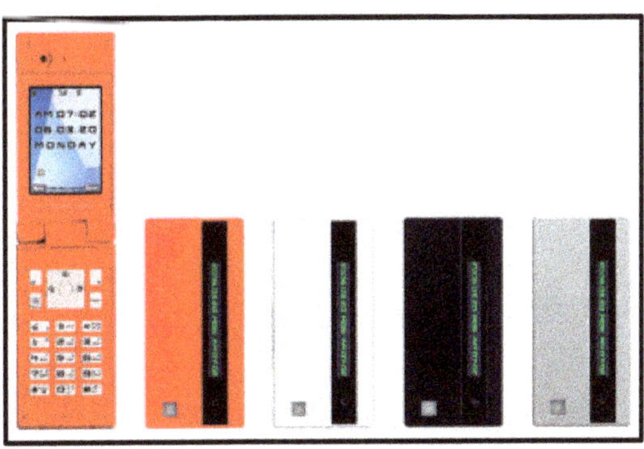

Figure 8. First mobile phone using a casing made of natural fiber material.

5.2. Natural Fibers Composites' Applications in Packaging

More recently, natural fiber composites have provided an alternative solution for better packaging. Previously, most of the petroleum-based plastics being used for food packaging have been non-degradable, causing many environmental problems associated with their disposal, including damage to the environment and eco-systems, water supplies, sewer systems, rivers and streams [144]. Moreover, they are non-renewable, and their prices are rising and unstable, given the impending depletion of petroleum resources. According to Ngo [145], the utilization of coir (coconut) fiber reinforced with natural latex, in place of synthetic materials, is of great interest for reducing the utilization of non-renewable and petroleum-based resources. Coconut fiber is a very tough yet also elastic material that hardly deteriorates at all over time. It is a durable material that can be re-used many times. After it is used, it can be recycled or disposed of without problem. After molding the material into the right shape, the material is heated to vulcanize the natural latex. The result is a very open structure that is strong and resilient. Figure 9 shows the packaging products produced by Enkev Manufacturer out of coconut fiber.

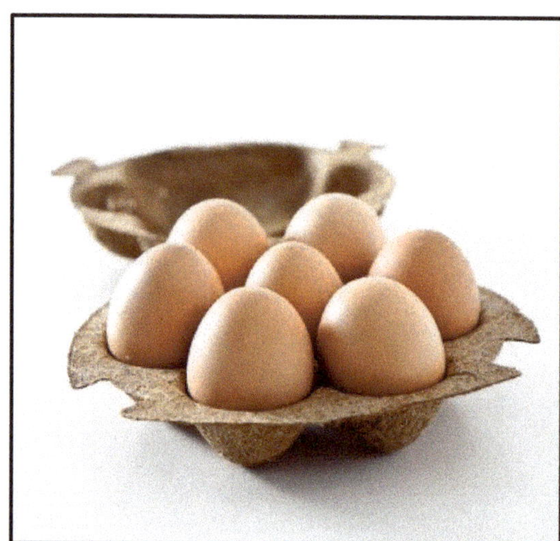

Figure 9. Packagings produced by Enkev Manufacturer from coconut fiber.

5.3. Natural Fibers Composites' Applications in Sports Equipment

Despite the most prominent applications of natural fiber composites being in the automotive industry, there are additional fields of application for natural fiber composites, such as in sports products. Before the advent of fiber-reinforced composites, sports equipment materials were made of wood, steel, stainless steel, aluminum, and alloy. In comparison with these materials, fiber reinforced composite materials have obvious advantages in the following aspects. The lower weight and relatively lower cost of natural fibers are the main aspects referred to as the reasons for the use of natural fiber composites in sports equipment. Most sports equipment relies on humans to move; therefore, lightweight equipment is desirable [146]. According to a study conducted by Yusup et al. [147], an oil palm empty fruit bunch fiber/epoxy composite that went through 24 h fiber treatment showed high potential to be used as a reinforcement to epoxy as a suitable material for sports equipment. Based on the results obtained related to the mechanical and physical properties, the composite of OPEFB fiber/epoxy had a flexural strength between 67.90 MPa and 83.63 MPa, which falls in the range of field hockey sticks' strength requirements. The longboard shown in Figure 10 is one of the products made from AmpliTex®, bCores®, powerRibs, and natural fiber (flax, balsa wood) composite reinforcement materials [148].

Figure 10. Bcomp Manufacturer's longboards made from natural fiber.

6. Conclusions

This review article provides a compact and informative summary of natural fiber-reinforced polymer matrices from the perspective of product design development. Among the three main sources of natural fibers are plants, animals, and minerals, and these plant fibers or cellulosic fibers are in high demand, having developed since the resources they require are widely available, consume less energy, and are non-toxic to nature and humans. In general, natural fibers are made up of several main constituents, such as cellulose, hemicellulose, lignin, and pectin. Many researchers have discovered the good mechanical performance of these natural fibers due to the cellulose, which provides the good shape and structural integrity of the fibers. Thus, the integration of natural fibers with a polymer matrix in composites benefits various industries, as they exhibit low density, lower solidity, biodegradability, and cheapness compared to synthetic composites. Natural fiber composites are an effective way of improving the quality of products developed from them, in terms of environmental suitability, and economic and technical feasibility. The most common natural fibers used in composite products are flax, coir, hemp, and jute, while roselle, sugar palm, and kenaf are examples of emerging fibers due to their high mechanical strength and stiffness that are suitable for many engineering applications. It can be concluded that appropriate product design and manufacturing processes of NF-PCs are required to enhance the properties of the products and their materials toward optimized strength and functionality. To ensure the optimization of the strength and functionality of natural fiber composite products, engineering design processes and techniques such as TRIZ, brainstorming, the voice of customers (VOCs), and morphological charts are essential. These techniques could define the problems of users and refine them in terms of the product's functionality. In the end, an appropriate manufacturing process incorporates the product's design and its applications. In the future, further research will be required to develop optimized engineering design techniques that complement the strength of the natural fiber composites, manufacturing processes, and functionality for heavy industry applications. Even now, natural fiber composites have the potential to be used in many applications that do not require very high load-bearing or high-temperature working capabilities.

Author Contributions: Conceptualization, M.A.A. and M.J.S.; validation, M.A.A. and M.J.S.; investigation, M.A.A. and M.J.S.; writing—original draft preparation, M.A.A. and M.J.S.; writing—review and editing, C.M.R., S.M.S., W.B.W.N., R.A.I., A.K., M.P., M.R.M.A. and M.R.I.; supervision, M.J.S.; funding acquisition, M.P. All authors have read and agreed to the published version of the manuscript.

Funding: The result was obtained through the financial support of the Ministry of Education, Youth and Sports of the Czech Republic and the European Union (European Structural and Investment Funds—Operational Programme Research, Development and Education) in the context of the project "Modular platform for autonomous chassis of specialized electric vehicles for freight and equipment transportation," Reg. No. CZ.02.1.01/0.0/0.0/16_025/0007293. This research was also funded by the Fundamental Research Scheme (FRGS) entitled "Correlation of Manufacturing Defects, Interfacial Adhesion, Physical and Mechanical Properties of Plant Fiber Reinforced Hybrid Composite Material Towards Compatibility Behaviors", grant number 59624.

Institutional Review Board Statement: Not applicable.

Informed Consent Statement: Not applicable.

Data Availability Statement: Not applicable.

Acknowledgments: The authors would like to thank the production team from MSET Inflatable. Composite Corporation Sdn. Bhd. located in Gong Badak Industrial Zone, Kuala Nerus, Terengganu, Malaysia for their support during completing this article.

Conflicts of Interest: The authors declare no conflict of interest.

References

1. Milton, A.; Rodgers, P. Product Design Process and Methods. In *Research Methods for Product Design*; Laurence King Publishing: London, UK, 2013; ISBN 9781780673028.
2. Bocci, E.; Prosperi, E.; Mair, V.; Bocci, M. Ageing and Cooling of Hot-Mix-Asphalt During Hauling and Paving—A Laboratory and Site Study. *Sustainabilty* **2020**, *12*, 8612. [CrossRef]
3. Corrado, A.; Polini, W. Measurement of high flexibility components in composite material by touch probe and force sensing resistors. *J. Manuf. Process.* **2019**, *45*, 520–531. [CrossRef]
4. Hanan, F.; Jawaid, M.; Tahir, P.M. Mechanical performance of oil palm/kenaf fiber-reinforced epoxy-based bilayer hybrid composites. *J. Nat. Fibers* **2018**, *17*, 155–167. [CrossRef]
5. Asyraf, M.R.M.; Ishak, M.R.; Sapuan, S.M.; Yidris, N.; Ilyas, R.A.; Rafidah, M.; Razman, M.R. Potential Application of Green Composites for Cross Arm Component in Transmission Tower: A Brief Review. *Int. J. Polym. Sci.* **2020**, *2020*, 1–15. [CrossRef]
6. Alsubari, S.; Zuhri, M.Y.M.; Sapuan, S.M.; Ishak, M.R.; Ilyas, R.A.; Asyraf, M.R.M. Potential of Natural Fiber Reinforced Polymer Composites in Sandwich Structures: A Review on Its Mechanical Properties. *Polymers* **2021**, *13*, 423. [CrossRef]
7. Sapuan, S.; Hemapriya, G.; Ilyas, R.; Atikah, M.; Asyraf, M.; Mansor, M.R. Implementation of design for sustainability in developing trophy plaque using green kenaf polymer composites. In *Design for Sustainability*; Elsevier: Amsterdam, The Netherlands, 2021; pp. 85–103.
8. Amir, A.; Ishak, M.; Yidris, N.; Zuhri, M.; Asyraf, M. Potential of Honeycomb-Filled Composite Structure in Composite Cross-Arm Component: A Review on Recent Progress and Its Mechanical Properties. *Polymers* **2021**, *13*, 1341. [CrossRef] [PubMed]
9. Elanchezhian, C.; Ramnath, B.; Ramakrishnan, G.; Rajendrakumar, M.; Naveenkumar, V.; Saravanakumar, M. Review on mechanical properties of natural fiber composites. *Mater. Today Proc.* **2018**, *5*, 1785–1790. [CrossRef]
10. Bakar, N.H.; Hyie, K.M.; Ramlan, A.S.; Hassan, M.K.; Jumahat, A. Mechanical Properties of Kevlar Reinforcement in Kenaf Composites. *Appl. Mech. Mater.* **2013**, *465-466*, 847–851. [CrossRef]
11. Nordin, N.A.; Yussof, F.M.; Kasolang, S.; Salleh, Z.; Ahmad, M.A. Wear Rate of Natural Fibre: Long Kenaf Composite. *Procedia Eng.* **2013**, *68*, 145–151. [CrossRef]
12. Asyraf, M.R.M.; Rafidah, M.; Azrina, A.; Razman, M.R. Dynamic mechanical behaviour of kenaf cellulosic fibre biocomposites: A comprehensive review on chemical treatments. *Cellulose* **2021**, *28*, 2675–2695. [CrossRef]
13. Johari, A.; Ishak, M.; Leman, Z.; Yusoff, M.; Asyraf, M.R.M. Influence of CaCO3 in pultruded glass fiber/unsaturated polyester resin composite on flexural creep behavior using conventional and time-temperature superposition principle methods. *Polimery* **2020**, *65*, 792–800. [CrossRef]

14. Asyraf, M.; Ishak, M.; Sapuan, S.; Yidris, N. Utilization of Bracing Arms as Additional Reinforcement in Pultruded Glass Fiber-Reinforced Polymer Composite Cross-Arms: Creep Experimental and Numerical Analyses. *Polymers* **2021**, *13*, 620. [CrossRef]
15. Nurazzi, N.M.; Asyraf, M.; Khalina, A.; Abdullah, N.; Sabaruddin, F.; Kamarudin, S.; Ahmad, S.; Mahat, A.; Lee, C.; Aisyah, H.; et al. Fabrication, Functionalization, and Application of Carbon Nanotube-Reinforced Polymer Composite: An Overview. *Polymers* **2021**, *13*, 1047. [CrossRef] [PubMed]
16. Ilyas, R.; Sapuan, S.; Asyraf, M.; Dayana, D.; Amelia, J.; Rani, M.; Norrrahim, M.; Nurazzi, N.; Aisyah, H.; Sharma, S.; et al. Polymer Composites Filled with Metal Derivatives: A Review of Flame Retardants. *Polymery* **2021**, *13*, 1701. [CrossRef]
17. Rognoli, V.; Karana, E.; Pedgley, O. Natural fibre composites in product design: An investigation into material perception and acceptance. In Proceedings of the 2011 Conference on Designing Pleasurable Products and Interfaces, Milan, Italy, 22–25 June 2011; pp. 1–4.
18. Ilyas, R.A.; Sapuan, S.M.; Asyraf, M.R.M.; Atikah, M.S.N.; Ibrahim, R.; Norrrahim, M.N.F.; Yasim-Anuar, T.A.T.; Megashah, L.N. Mechanical and Dynamic Mechanical Analysis of Bio-based Composites. In *Mechanical and Dynamic Properties of Biocomposites*; Krishnasamy, S., Nagarajan, R., Thiagamani, S.M.K., Siengchin, S., Eds.; WILEY-VCH GmbH: Weinheim, Germany, 2021.
19. Dicker, M.P.; Duckworth, P.F.; Baker, A.B.; Francois, G.; Hazzard, M.K.; Weaver, P.M. Green composites: A review of material attributes and complementary applications. *Compos. Part A Appl. Sci. Manuf.* **2014**, *56*, 280–289. [CrossRef]
20. Omran, A.A.B.; Mohammed, A.A.B.A.; Sapuan, S.M.; Ilyas, R.A.; Asyraf, M.R.M.; Koloor, S.S.R.; Petrů, M. Micro- and Nanocellulose in Polymer Composite Materials: A Review. *Polymers* **2021**, *13*, 231. [CrossRef] [PubMed]
21. Nurazzi, N.; Asyraf, M.; Khalina, A.; Abdullah, N.; Aisyah, H.; Rafiqah, S.; Sabaruddin, F.; Kamarudin, S.; Norrrahim, M.; Ilyas, R.; et al. A Review on Natural Fiber Reinforced Polymer Composite for Bullet Proof and Ballistic Applications. *Polymers* **2021**, *13*, 646. [CrossRef]
22. Asyraf, M.; Ishak, M.; Sapuan, S.; Yidris, N.; Ilyas, R. Woods and composites cantilever beam: A comprehensive review of experimental and numerical creep methodologies. *J. Mater. Res. Technol.* **2020**, *9*, 6759–6776. [CrossRef]
23. Sapuan, S.M.; Nukman, Y. The Relationship Between Manufacturing and Design for Manufacturing in Product Development of Natural Fibre Composites. In *Manufacturing of Coir Fiber Reinforced Polymer Composites Using Hot Compression Technique*; Springer: Berlin/Heidelberg, Germany, 2014; p. 2.
24. Mansor, M.; Sapuan, S.; Zainudin, E.S.; Nuraini, A.; Hambali, A. Conceptual design of kenaf fiber polymer composite automotive parking brake lever using integrated TRIZ–Morphological Chart–Analytic Hierarchy Process method. *Mater. Des.* **2014**, *54*, 473–482. [CrossRef]
25. Asyraf, M.R.M.; Ishak, M.R.; Sapuan, S.M.; Yidris, N. Comparison of Static and Long-term Creep Behaviors between Balau Wood and Glass Fiber Reinforced Polymer Composite for Cross-arm Application. *Fibers Polym.* **2021**, *22*, 793–803. [CrossRef]
26. Asyraf, M.; Ishak, M.; Sapuan, S.; Yidris, N. Influence of Additional Bracing Arms as Reinforcement Members in Wooden Timber Cross-Arms on Their Long-Term Creep Responses and Properties. *Appl. Sci.* **2021**, *11*, 2061. [CrossRef]
27. Marzuki, I. Jurureka Perindustrian. In *Reka Bentuk Produk*; Dewan Bahasa dan Pustaka: Kuala Lumpur, Malaysia, 2013; pp. 4–5, ISBN 978-983-46-1439-3.
28. Asyraf, M.R.M.; Ishak, M.R.; Sapuan, S.M.; Yidris, N.; Ilyas, R.A.; Rafidah, M.; Razman, M.R. Evaluation of design and simulation of creep test rig for full-scale cross arm structure. *Adv. Civ. Eng.* **2020**, 6980918. [CrossRef]
29. Ilyas, R.; Sapuan, S.; Harussani, M.; Hakimi, M.; Haziq, M.; Atikah, M.; Asyraf, M.; Ishak, M.; Razman, M.; Nurazzi, N.; et al. Polylactic Acid (PLA) Biocomposite: Processing, Additive Manufacturing and Advanced Applications. *Polymers* **2021**, *13*, 1326. [CrossRef] [PubMed]
30. Suriani, M.; Rapi, H.; Ilyas, R.; Petrů, M.; Sapuan, S. Delamination and Manufacturing Defects in Natural Fiber-Reinforced Hybrid Composite: A Review. *Polymers* **2021**, *13*, 1323. [CrossRef] [PubMed]
31. Peças, P.; Carvalho, H.; Salman, H.; Leite, M. Natural Fibre Composites and Their Applications: A Review. *J. Compos. Sci.* **2018**, *2*, 66. [CrossRef]
32. Sanjay, M.; Arpitha, G., Yogesha, B Study on Mechanical Properties of Natural—Glass Fibre Reinforced Polymer Hybrid Composites: A Review. *Mater. Today Proc.* **2015**, *2*, 2959–2967. [CrossRef]
33. Bharath, K.N.; Basavarajappa, S. Applications of biocomposite materials based on natural fibers from renewable resources: A review. *Sci. Eng. Compos. Mater.* **2016**, *23*, 123–133. [CrossRef]
34. Aditya, P.H.; Kishore, K.S.; Prasad, D.V.V.K. Characterization of Natural Fiber Reinforced Composites. *Int. J. Eng. Appl. Sci.* **2017**, *4*, 1–10.
35. Corona, A.; Madsen, B.; Hauschild, M.Z.; Birkved, M. Natural fibre selection for composite eco-design. *CIRP Ann.* **2016**, *65*, 13–16. [CrossRef]
36. Padmavathi, T.; Naidu, S.V.; Rao, R. Studies on Mechanical Behavior of Surface Modified Sisal Fibre–Epoxy Composites. *J. Reinf. Plast. Compos.* **2012**, *31*, 519–532. [CrossRef]
37. Amir, N.; Abidin, K.A.Z.; Shiri, F.B.M. Effects of Fibre Configuration on Mechanical Properties of Banana Fibre/PP/MAPP Natural Fibre Reinforced Polymer Composite. *Procedia Eng.* **2017**, *184*, 573–580. [CrossRef]
38. Maleque, M.A.; Belal, F.Y.; Sapuan, S.M. Mechanical properties study of pseudo-stem banana fiber reinforced epoxy composite. *Arab. J. Sci. Eng.* **2007**, *32*, 359–364.
39. Taekema, J.; Karana, E. Creating awareness on natural fibre composites in design. In Proceedings of the DS 70: Proceedings of DESIGN 2012, the 12th International Design Conference, Dubrovnik, Croatia, 21–24 May 2012; pp. 1141–1150.

40. Sapuan, S.; Maleque, A. Design and fabrication of natural woven fabric reinforced epoxy composite for household telephone stand. *Mater. Des.* **2005**, *26*, 65–71. [CrossRef]

41. Shekar, H.S.; Ramachandra, M. Green Composites: A Review. *Mater. Today: Proc.* **2018**, *5*, 2518–2526. [CrossRef]

42. Ilyas, R.A.; Sapuan, S.M.; Norizan, M.N.; Atikah, M.S.N.; Huzaifah, M.R.M.; Radzi, A.M.; Ishak, M.R.; Zainudin, E.S.; Izwan, S.; Azammi, A.M.N.; et al. Potential of natural fibre composites for transport industry: A review. In *Prosiding Seminar Enau Kebangsaan 2019*; Institute of Tropical Forest and Forest Products (INTROP); Universiti Putra Malaysia: Bahau, Malaysia, 2019; pp. 2–11.

43. Ilyas, R.A.; Sapuan, S.M.; Atiqah, A.; Ibrahim, R.; Abral, H.; Ishak, M.R.; Zainudin, E.S.; Nurazzi, N.M.; Atikah, M.S.N.; Ansari, M.N.M.; et al. Sugar palm (*Arenga pinnata* [*Wurmb.*] *Merr*) starch films containing sugar palm nanofibrillated cellulose as reinforcement: Water barrier properties. *Polym. Compos.* **2019**, *41*, 459–467. [CrossRef]

44. Ilyas, R.; Sapuan, S.; Atikah, M.; Asyraf, M.; Rafiqah, S.A.; Aisyah, H.; Nurazzi, N.M.; Norrrahim, M. Effect of hydrolysis time on the morphological, physical, chemical, and thermal behavior of sugar palm nanocrystalline cellulose (*Arenga pinnata* (*Wurmb.*) *Merr*). *Text. Res. J.* **2021**, *91*, 152–167. [CrossRef]

45. Huda, M.S.; Drzal, L.T.; Ray, D.; Mohanty, A.K.; Mishra, M. Natural-fiber composites in the automotive sector. In *Properties and Performance of Natural-Fibre Composites*; Elsevier: Amsterdam, The Netherlands, 2008; pp. 221–268.

46. Girijappa, Y.G.T.; Rangappa, S.M.; Parameswaranpillai, J.; Siengchin, S. Natural Fibers as Sustainable and Renewable Resource for Development of Eco-Friendly Composites: A Comprehensive Review. *Front. Mater.* **2019**, *6*, 1–14. [CrossRef]

47. Arpitha, G.; Sanjay, M.; Senthamaraikannan, P.; Barile, C.; Yogesha, B. Hybridization Effect of Sisal/Glass/Epoxy/Filler Based Woven Fabric Reinforced Composites. *Exp. Tech.* **2017**, *41*, 577–584. [CrossRef]

48. Madhu, P.; Sanjay, M.R.; Senthamaraikannan, P.; Pradeep, S.; Saravanakumar, S.S.; Yogesha, B. A review on synthesis and characterization of commercially available natural fibers: Part-I. *J. Nat. Fibers* **2018**, *11*, 25–36.

49. Chandra, C.S.J.; George, N.; Narayanankutty, S.K. Isolation and characterization of cellulose nanofibrils from arecanut husk fibre. *Carbohydr. Polym.* **2016**, *142*, 158–166.

50. Tibolla, H.; Pelissari, F.M.; Menegalli, F.C. Cellulose nanofibers produced from banana peel by chemical and enzymatic treatment. *LWT* **2014**, *59*, 1311–1318. [CrossRef]

51. Corrêa, A.C.; Teixeira, E.D.M.; Pessan, L.; Mattoso, L.H.C. Cellulose nanofibers from curaua fibers. *Cellulose* **2010**, *17*, 1183–1192. [CrossRef]

52. Chirayil, C.J.; Joy, J.; Mathew, L.; Mozetic, M.; Koetz, J.; Thomas, S. Isolation and characterization of cellulose nanofibrils from *Helicteres isora* plant. *Ind. Crop. Prod.* **2014**, *59*, 27–34. [CrossRef]

53. Jonoobi, M.; Harun, J.; Shakeri, A.; Misra, M.; Oksmand, K. Chemical composition, crystallinity, and thermal degradation of bleached and unbleached kenaf bast (*Hibiscus cannabinus*) pulp and nanofibers. *BioResources* **2009**, *4*, 626–639. [CrossRef]

54. Chan, H.C.; Chia, C.H.; Zakaria, S.; Ahmad, I.; Dufresne, A. Production and Characterisation of Cellulose and Nano-Crystalline Cellulose from Kenaf Core Wood. *Bioresources* **2012**, *8*, 785–794. [CrossRef]

55. Sheltami, R.M.; Abdullah, I.; Ahmad, I.; Dufresne, A.; Kargarzadeh, H. Extraction of cellulose nanocrystals from mengkuang leaves (*Pandanus tectorius*). *Carbohydr. Polym.* **2012**, *88*, 772–779. [CrossRef]

56. Megashah, L.N.; Ariffin, H.; Zakaria, M.R.; Hassan, M.A. Properties of Cellulose Extract from Different Types of Oil Palm Biomass. *IOP Conf. Series: Mater. Sci. Eng.* **2018**, *368*, 012049. [CrossRef]

57. Jonoobi, M.; Khazaeian, A.; Tahir, P.M.; Azry, S.S.; Oksman, K. Characteristics of cellulose nanofibers isolated from rubberwood and empty fruit bunches of oil palm using chemo-mechanical process. *Cellulose* **2011**, *18*, 1085–1095. [CrossRef]

58. Bendahou, A.; Habibi, Y.; Kaddami, H.; Dufresne, A. Physico-Chemical Characterization of Palm from Phoenix Dactylifera–L, Preparation of Cellulose Whiskers and Natural Rubber–Based Nanocomposites. *J. Biobased Mater. Bioenergy* **2009**, *3*, 81–90. [CrossRef]

59. Cherian, B.M.; Leão, A.L.; de Souza, S.F.; Thomas, S.; Pothan, L.A.; Kottaisamy, M. Isolation of nanocellulose from pineapple leaf fibres by steam explosion. *Carbohydr. Polym.* **2010**, *81*, 720–725. [CrossRef]

60. Syafri, E.; Kasim, A.; Abral, H.; Asben, A. Cellulose nanofibers isolation and characterization from ramie using a chemical-ultrasonic treatment. *J. Nat. Fibers* **2018**, *16*, 1145–1155. [CrossRef]

61. Alemdar, A.; Sain, M. Isolation and characterization of nanofibers from agricultural residues—Wheat straw and soy hulls. *Bioresour. Technol.* **2008**, *99*, 1664–1671. [CrossRef]

62. Li, M.; Wang, L.-J.; Li, D.; Cheng, Y.-L.; Adhikari, B. Preparation and characterization of cellulose nanofibers from de-pectinated sugar beet pulp. *Carbohydr. Polym.* **2014**, *102*, 136–143. [CrossRef]

63. Ilyas, R.; Sapuan, S.; Ishak, M. Isolation and characterization of nanocrystalline cellulose from sugar palm fibres (*Arenga pinnata*). *Carbohydr. Polym.* **2018**, *181*, 1038–1051. [CrossRef] [PubMed]

64. Teixeira, E.D.M.; Bondancia, T.; Teodoro, K.B.R.; Corrêa, A.C.; Marconcini, J.M.; Mattoso, L.H.C. Sugarcane bagasse whiskers: Extraction and characterizations. *Ind. Crop. Prod.* **2011**, *33*, 63–66. [CrossRef]

65. Abral, H.; Dalimunthe, M.H.; Hartono, J.; Efendi, R.P.; Asrofi, M.; Sugiarti, E.; Sapuan, S.M.; Park, J.-W.; Kim, H.-J. Characterization of Tapioca Starch Biopolymer Composites Reinforced with Micro Scale Water Hyacinth Fibers. *Starch Stärke* **2018**, *70*, 1–8. [CrossRef]

66. Alemdar, A.; Sain, M. Biocomposites from wheat straw nanofibers: Morphology, thermal and mechanical properties. *Compos. Sci. Technol.* **2008**, *68*, 557–565. [CrossRef]

67. Wang, F.; Xie, Z.; Liang, J.; Fang, B.; Piao, Y.; Hao, M.; Wang, Z. Tourmaline-Modified FeMnTiOx Catalysts for Improved Low-Temperature NH3-SCR Performance. *Environ. Sci. Technol.* **2019**, *53*, 6989–6996. [CrossRef]
68. Piao, Y.; Jiang, Q.; Li, H.; Matsumoto, H.; Liang, J.; Liu, W.; Pham-Huu, C.; Liu, Y.; Wang, F. Identify Zr Promotion Effects in Atomic Scale for Co-Based Catalysts in Fischer–Tropsch Synthesis. *ACS Catal.* **2020**, *10*, 7894–7906. [CrossRef]
69. Syafiq, R.; Sapuan, S.M.; Zuhri, M.Y.M.; Ilyas, R.A.; Nazrin, A.; Sherwani, S.F.K.; Khalina, A. Antimicrobial Activities of Starch-Based Biopolymers and Biocomposites Incorporated with Plant Essential Oils: A Review. *Polymers* **2020**, *12*, 2403. [CrossRef]
70. Nazrin, A.; Sapuan, S.M.; Zuhri, M.Y.M.; Ilyas, R.A.; Syafiq, R.; Sherwani, S.F.K. Nanocellulose Reinforced Thermoplastic Starch (TPS), Polylactic Acid (PLA), and Polybutylene Succinate (PBS) for Food Packaging Applications. *Front. Chem.* **2020**, *8*, 1–12. [CrossRef]
71. Diyana, Z.; Jumaidin, R.; Selamat, M.; Ghazali, I.; Julmohammad, N.; Huda, N.; Ilyas, R. Physical Properties of Thermoplastic Starch Derived from Natural Resources and Its Blends: A Review. *Polymers* **2021**, *13*, 1396. [CrossRef]
72. Sapuan, S.; Aulia, H.; Ilyas, R.; Atiqah, A.; Dele-Afolabi, T.; Nurazzi, M.; Supian, A.; Atikah, M. Mechanical Properties of Longitudinal Basalt/Woven-Glass-Fiber-reinforced Unsaturated Polyester-Resin Hybrid Composites. *Polymers* **2020**, *12*, 2211. [CrossRef]
73. Abdul, R.A.H.; Roslan, A.F.; Jaafar, M.; Roslan, M.N.; Ariffin, S. Mechanical Properties Evaluation of Woven Coir and Kevlar Reinforced Epoxy Composites. *Adv. Mater. Res.* **2011**, *277*, 36–42. [CrossRef]
74. Jawaid, M.; Khalil, H.A.; Bhat, A.; Abu Baker, A. Impact Properties of Natural Fiber Hybrid Reinforced Epoxy Composites. *Adv. Mater. Res.* **2011**, *264-265*, 688–693. [CrossRef]
75. Masoodi, R.; Pillai, K.M. A study on moisture absorption and swelling in bio-based jute-epoxy composites. *J. Reinf. Plast. Compos.* **2012**, *31*, 285–294. [CrossRef]
76. What Is Industrial Design? Available online: https://www.idsa.org/what-industrial-design (accessed on 18 March 2021).
77. Abidin, S.Z.; Abdullah, M.H.; Yusoff, Z. *Seni reka perindustrian: Daripada Idea Kepada Lakaran Dewan Bahasa dan Pustaka*; Seni Reka Perindustrian: Kuala Lumpur, Malaysia, 2013; ISBN 9834615337.
78. Ramani, K.; Ramanujan, D.; Bernstein, W.Z.; Zhao, F.; Sutherland, J.; Handwerker, C.; Choi, J.-K.; Kim, H.; Thurston, D. Integrated sustainable life cycle design: A review. *J. Mech. Des.* **2010**, *132*, 1–15. [CrossRef]
79. Yung, W.K.; Chan, H.K.; So, J.H.; Wong, D.W.; Choi, A.C.; Yue, T.M. A life-cycle assessment for eco-redesign of a consumer electronic product. *J. Eng. Des.* **2011**, *22*, 69–85. [CrossRef]
80. Karana, E. Materials selection in design: From research to education. In Proceedings of the the 1st International Symposium for Design Education Researchers, Paris, France, 18–19 May 2011.
81. Marzuki, I. Proses Reka Bentuk Produk. In *Reka Bentuk Produk*; Dewan Bahasa dan Pustaka: Kuala Lumpur, Malaysia, 2013; p. 16, ISBN 978-983-46-1439-3.
82. Karana, E.; Hekkert, P.; Kandachar, P. A tool for meaning driven materials selection. *Mater. Des.* **2010**, *31*, 2932–2941. [CrossRef]
83. Mahmud, J.; Khor, S.; Ismail, M.M.; Taib, J.M.; Ramlan, N.; Ling, K. Design for paraplegia: Preparing product design specifications for a wheelchair. *Technol. Disabil.* **2015**, *27*, 79–89. [CrossRef]
84. Azman, M.A.; Yusof, S.A.M.; Abdullah, I.; Mohamad, I.; Mohammed, J.S. Factors influencing face mask selection and design specifications: Results from pilot study amongst malaysian umrah pilgrims. *J. Teknol.* **2017**, *79*, 79. [CrossRef]
85. Spangenberg, J.H.; Fuad-Luke, A.; Blincoe, K. Design for Sustainability (DfS): The interface of sustainable production and consumption. *J. Clean. Prod.* **2010**, *18*, 1485–1493. [CrossRef]
86. Ali, S.; Razman, M.; Awang, A.; Asyraf, M.; Ishak, M.; Ilyas, R.; Lawrence, R. Critical Determinants of Household Electricity Consumption in a Rapidly Growing City. *Sustainabilty* **2021**, *13*, 4441. [CrossRef]
87. Spangenberg, J.H. Sustainable development indicators: Towards integrated systems as a tool for managing and monitoring a complex transition. *Int. J. Glob. Environ. Issues* **2009**, *9*, 318. [CrossRef]
88. Von Keyserlingk, M.A.G.; Martin, N.P.; Kebreab, E.; Knowlton, K.F.; Grant, R.J.; Stephenson, M.; Sniffen, C.J.; Harner, J.R., III; Wright, A.D.; Smith, S.I. Invited review: Sustainability of the US dairy industry. *J. Dairy Sci.* **2013**, *96*, 5405–5425. [CrossRef] [PubMed]
89. Jawahir, I.S.; Rouch, K.E.; Dillon, O.W.; Holloway, L.; Hall, A.; Knuf, J. Design for sustainability (DFS): New challenges in developing and implementing a curriculum for next generation design and manufacturing engineers. *Int. J. Eng. Educ.* **2007**, *23*, 1053–1064.
90. Hambali, A.; Sapuan, S.M.; Ismail, N.; Nukman, Y. Application of analytical hierarchy process in the design concept selection of automotive composite bumper beam during the conceptual design stage. *Sci. Res. Essays* **2009**, *4*, 198–211.
91. Mazani, N.; Sapuan, S.; Sanyang, M.; Atiqah, A.; Ilyas, R. Design and Fabrication of a Shoe Shelf from Kenaf Fiber Reinforced Unsaturated Polyester Composites. In *Lignocellulose for Future Bioeconomy*; Elsevier: Amsterdam, The Netherlands, 2019; pp. 315–332.
92. Pahl, G.; Beitz, W. *Engineering design: A Systematic Approach*; Springer: London, UK, 1996; ISBN 1846283183.
93. Asyraf, M.; Ishak, M.; Sapuan, S.; Yidris, N. Conceptual design of multi-operation outdoor flexural creep test rig using hybrid concurrent engineering approach. *J. Mater. Res. Technol.* **2020**, *9*, 2357–2368. [CrossRef]
94. Rosli, M.U.; Ariffin, M.K.A.; Sapuan, S.M.; Sulaiman, S. Integrated AHP-TRIZ innovation method for automotive door panel design. *Int. J. Eng. Technol.* **2013**, *5*, 3158–3167.

95. Li, M.; Ming, X.; He, L.; Zheng, M.; Xu, Z. A TRIZ-based Trimming method for Patent design around. *Comput. Des.* **2015**, *62*, 20–30. [CrossRef]

96. Ahmad, S.A.; Ang, M.C.; Ng, K.W.; Abdul Wahab, A.N. Reducing home energy usage based on TRIZ concept. *Adv. Environ. Biol.* **2015**, *9*, 6–11.

97. San, Y.T.; Jin, Y.T.; Li, S.C. *TRIZ: Systematic Innovation in Manufacturing*; Firstfruit Sdn. Bhd.: Selangor, Malaysia, 2011; ISBN 9838040266.

98. Li, T. Retracted article: Applying TRIZ and AHP to develop innovative design for automated assembly systems. *Int. J. Adv. Manuf. Technol.* **2009**, *46*, 301–313. [CrossRef]

99. Cascini, G.; Rissone, P.; Rotini, F.; Russo, D. Systematic design through the integration of TRIZ and optimization tools. *Procedia Eng.* **2011**, *9*, 674–679. [CrossRef]

100. Asyraf, M.; Ishak, M.; Sapuan, S.; Yidris, N. Conceptual design of creep testing rig for full-scale cross arm using TRIZ-Morphological chart-analytic network process technique. *J. Mater. Res. Technol.* **2019**, *8*, 5647–5658. [CrossRef]

101. Yeh, C.H.; Huang, J.C.Y.; Yu, C.K. Integration of four-phase QFD and TRIZ in product R&D: A notebook case study. *Res. Eng. Des.* **2010**, *22*, 125–141. [CrossRef]

102. Ullman, D.G. *The Mechanical Design Process*; McGraw-Hill: Maidenhead, UK, 1992; ISBN 0-07-065739-4. [CrossRef]

103. Ulrich, K.; Eppinger, S. *Product Design and Development*; McGraw Hill: New York, NY, USA, 1995.

104. Asyraf, M.R.M.; Rafidah, M.; Ishak, M.R.; Sapuan, S.M.; Yidris, N.; Ilyas, R.A.; Razman, M.R. Integration of TRIZ, morphological chart and ANP method for development of FRP composite portable fire extinguisher. *Polym. Compos.* **2020**, *41*, 2917–2932. [CrossRef]

105. Eder, W. Engineering design methods. *Des. Stud.* **1990**, *11*, 54. [CrossRef]

106. McKoy, F.L.; Vargas-Hernández, N.; Summers, J.D.; Shah, J.J. Influence of Design Representation on Effectiveness of Idea Generation. In Proceedings of the Volume 4: 13th International Conference on Design Theory and Methodology; ASME International, 2001, Pittsburgh, PA, USA, 9–12 September 2001; Volume 4, pp. 39–48.

107. Pahl, G.; Beitz, W. *Engineering Design*; Design Council: London, UK, 1984; ISBN 9781405146012.

108. Sapuan, S.M.; Ilyas, R.A.; Asyraf, M.R.M.; Suhrisman, A.; Afiq, T.M.N.; Atikah, M.S.N.; Ibrahim, R. Application of Design for Sustainability to Develop Smartphone Holder using Roselle Fiber-Reinforced Polymer Composites. In *Roselle: Production, Processing, Products and Biocomposites*; Sapuan, S.M., Razali, N., Radzi, A.M., Ilyas, R.A., Eds.; Elsevier Academic Press: Cambridge, MA, USA, 2021; pp. 1–300, ISBN 978-0323852135.

109. Sapuan, S. Concurrent Engineering in Natural Fibre Composite Product Development. *Appl. Mech. Mater.* **2015**, *761*, 59–62. [CrossRef]

110. Sapuan, S.M. A Conceptual Design of the Concurrent Engineering Design System for Polymeric-Based Composite Automotive Pedals. *Am. J. Appl. Sci.* **2005**, *2*, 514–525. [CrossRef]

111. Ilyas, R.A.; Asyraf, M.R.M.; Sapuan, S.M.; Afiq, T.M.N.; Suhrisman, A.; Atikah, M.S.N.; Ibrahim, R. Development of Roselle Fiber Reinforced Polymer Biocomposites Mug Pad using Hybrid Design for Sustainability and Pugh Method. In *Roselle: Production, Processing, Products and Biocomposites*; Sapuan, S.M., Razali, N., Radzi, A.M., Ilyas, R.A., Eds.; Elsevier Academic Press: Cambridge, MA, USA, 2021; pp. 1–300, ISBN 978-0323852135.

112. Sari, N.H.; Pruncu, C.I.; Sapuan, S.M.; Ilyas, R.A.; Catur, A.D.; Suteja, S.; Sutaryono, Y.A.; Pullen, G. The effect of water immersion and fibre content on properties of corn husk fibres reinforced thermoset polyester composite. *Polym. Test.* **2020**, *91*, 106751. [CrossRef]

113. Syafri, E.; Sudirman; Mashadi; Yulianti, E.; Deswita; Asrofi, M.; Abral, H.; Sapuan, S.; Ilyas, R.; Fudholi, A. Effect of sonication time on the thermal stability, moisture absorption, and biodegradation of water hyacinth (*Eichhornia crassipes*) nanocellulose-filled bengkuang (*Pachyrhizus erosus*) starch biocomposites. *J. Mater. Res. Technol.* **2019**, *8*, 6223–6231. [CrossRef]

114. Siakeng, R.; Jawaid, M.; Asim, M.; Saba, N.; Sanjay, M.R.; Siengchin, S.; Fouad, H. Alkali treated coir/pineapple leaf fibres reinforced PLA hybrid composites: Evaluation of mechanical, morphological, thermal and physical properties. *Polym. Lett.* **2020**, *14*, 717–730. [CrossRef]

115. Abral, H.; Ariksa, J.; Mahardika, M.; Handayani, D.; Aminah, I.; Sandrawati, N.; Sapuan, S.; Ilyas, R. Highly transparent and antimicrobial PVA based bionanocomposites reinforced by ginger nanofiber. *Polym. Test.* **2020**, *81*, 106186. [CrossRef]

116. Abral, H.; Ariksa, J.; Mahardika, M.; Handayani, D.; Aminah, I.; Sandrawati, N.; Pratama, A.B.; Fajri, N.; Sapuan, S.; Ilyas, R. Transparent and antimicrobial cellulose film from ginger nanofiber. *Food Hydrocoll.* **2020**, *98*, 105266. [CrossRef]

117. Prachayawarakorn, J.; Limsiriwong, N.; Kongjindamunee, R.; Surakit, S. Effect of Agar and Cotton Fiber on Properties of Thermoplastic Waxy Rice Starch Composites. *J. Polym. Environ.* **2011**, *20*, 88–95. [CrossRef]

118. Kumar, T.S.M.; Chandrasekar, M.; Senthilkumar, K.; Ilyas, R.A.; Sapuan, S.M.; Hariram, N.; Rajulu, A.V.; Rajini, N.; Siengchin, S. Characterization, Thermal and Antimicrobial Properties of Hybrid Cellulose Nanocomposite Films with in-Situ Generated Copper Nanoparticles in *Tamarindus indica* Nut Powder. *J. Polym. Environ.* **2021**, *29*, 1134–1142. [CrossRef]

119. Aisyah, H.A.; Paridah, M.T.; Sapuan, S.M.; Khalina, A.; Berkalp, O.B.; Lee, S.H.; Lee, C.H.; Nurazzi, N.M.; Ramli, N.; Wahab, M.S.; et al. Thermal Properties of Woven Kenaf/Carbon Fibre-Reinforced Epoxy Hybrid Composite Panels. *Int. J. Polym. Sci.* **2019**, *2019*, 1–8. [CrossRef]

120. Jaafar, C.A.; Zainol, I.; Ishak, N.; Ilyas, R.; Sapuan, S. Effects of the liquid natural rubber (LNR) on mechanical properties and microstructure of epoxy/silica/kenaf hybrid composite for potential automotive applications. *J. Mater. Res. Technol.* **2021**, *12*, 1026–1038. [CrossRef]
121. Sabaruddin, F.A.; Tahir, P.M.; Sapuan, S.M.; Ilyas, R.A.; Lee, S.H.; Abdan, K.; Mazlan, N.; Roseley, A.S.M.; Hps, A.K. The Effects of Unbleached and Bleached Nanocellulose on the Thermal and Flammability of Polypropylene-Reinforced Kenaf Core Hybrid Polymer Bionanocomposites. *Polymers* **2020**, *13*, 116. [CrossRef]
122. Suriani, M.; Zainudin, H.; Ilyas, R.; Petrů, M.; Sapuan, S.; Ruzaidi, C.; Mustapha, R. Kenaf Fiber/Pet Yarn Reinforced Epoxy Hybrid Polymer Composites: Morphological, Tensile, and Flammability Properties. *Polymers* **2021**, *13*, 1532. [CrossRef]
123. Jumaidin, R.; Ilyas, R.A.; Saiful, M.; Hussin, F.; Mastura, M.T. Water Transport and Physical Properties of Sugarcane Bagasse Fibre Reinforced Thermoplastic Potato Starch Biocomposite. *J. Adv. Res. Fluid Mech. Therm. Sci.* **2019**, *61*, 273–281.
124. Asrofi, M.; Sujito, S.; Syafri, E.; Sapuan, S.; Ilyas, R. Improvement of Biocomposite Properties Based Tapioca Starch and Sugarcane Bagasse Cellulose Nanofibers. *Key Eng. Mater.* **2020**, *849*, 96–101. [CrossRef]
125. Asrofi, M.; Sapuan, S.; Ilyas, R.; Ramesh, M. Characteristic of composite bioplastics from tapioca starch and sugarcane bagasse fiber: Effect of time duration of ultrasonication (Bath-Type). *Mater. Today Proc.* **2020**. [CrossRef]
126. Nassiopoulos, E.; Njuguna, J. Thermo-mechanical performance of poly(lactic acid)/flax fibre-reinforced biocomposites. *Mater. Des.* **2015**, *66*, 473–485. [CrossRef]
127. Battegazzore, D.; Noori, A.; Frache, A. Hemp hurd and alfalfa as particle filler to improve the thermo-mechanical and fire retardant properties of poly (3-hydroxybutyrate-co-3-hydroxyhexanoate). *Polym. Compos.* **2019**, *40*, 3429–3437. [CrossRef]
128. Prachayawarakorn, J.; Chaiwatyothin, S.; Mueangta, S.; Hanchana, A. Effect of jute and kapok fibers on properties of thermoplastic cassava starch composites. *Mater. Des.* **2013**, *47*, 309–315. [CrossRef]
129. Gupta, M.; Singh, R. PLA-coated sisal fibre-reinforced polyester composite: Water absorption, static and dynamic mechanical properties. *J. Compos. Mater.* **2019**, *53*, 65–72. [CrossRef]
130. Ayu, R.S.; Khalina, A.; Harmaen, A.S.; Zaman, K.; Isma, T.; Liu, Q.; Ilyas, R.A.; Lee, C.H. Characterization Study of Empty Fruit Bunch (EFB) Fibers Reinforcement in Poly(Butylene) Succinate (PBS)/Starch/Glycerol Composite Sheet. *Polymers* **2020**, *12*, 1571. [CrossRef]
131. Suriani, M.; Radzi, F.; Ilyas, R.; Petrů, M.; Sapuan, S.; Ruzaidi, C. Flammability, Tensile, and Morphological Properties of Oil Palm Empty Fruit Bunches Fiber/Pet Yarn-Reinforced Epoxy Fire Retardant Hybrid Polymer Composites. *Polymers* **2021**, *13*, 1282. [CrossRef] [PubMed]
132. Jumaidin, R.; Diah, N.; Ilyas, R.; Alamjuri, R.; Yusof, F. Processing and Characterisation of Banana Leaf Fibre Reinforced Thermoplastic Cassava Starch Composites. *Polymers* **2021**, *13*, 1420. [CrossRef]
133. Rozilah, A.; Jaafar, C.N.A.; Sapuan, S.M.; Zainol, I.; Ilyas, R.A. The Effects of Silver Nanoparticles Compositions on the Mechanical, Physiochemical, Antibacterial, and Morphology Properties of Sugar Palm Starch Biocomposites for Antibacterial Coating. *Polymers* **2020**, *12*, 2605. [CrossRef]
134. Atiqah, A.; Jawaid, M.; Sapuan, S.; Ishak, M.; Ansari, M.; Ilyas, R. Physical and thermal properties of treated sugar palm/glass fibre reinforced thermoplastic polyurethane hybrid composites. *J. Mater. Res. Technol.* **2019**, *8*, 3726–3732. [CrossRef]
135. Atikah, M.; Ilyas, R.; Sapuan, S.; Ishak, M.; Zainudin, E.; Ibrahim, R.; Atiqah, A.; Ansari, M.; Jumaidin, R. Degradation and physical properties of sugar palm starch/sugar palm nanofibrillated cellulose bionanocomposite. *Polimery* **2019**, *64*, 680–689. [CrossRef]
136. Ilyas, R.; Sapuan, S.; Ibrahim, R.; Abral, H.; Ishak, M.; Zainudin, E.; Atikah, M.; Nurazzi, N.M.; Atiqah, A.; Ansari, M.; et al. Effect of sugar palm nanofibrillated cellulose concentrations on morphological, mechanical and physical properties of biodegradable films based on agro-waste sugar palm (*Arenga pinnata* (*Wurmb.*) Merr) starch. *J. Mater. Res. Technol.* **2019**, *8*, 4819–4830. [CrossRef]
137. Kedzierski, M.; Wiejak, A.; Janiszewska, J.; Wiśniewska, A.; Grzywa-Niksinska, I.; Kurzepa, K. Efficiency of selected biocide compounds in the protection of building coatings against colonization by mold fungi, cyanobacteria and algae. *Polimery* **2020**, *65*, 371–379. [CrossRef]
138. Suriani, M.; Sapuan, S.; Ruzaidi, C.; Nair, D.; Ilyas, R. Flammability, morphological and mechanical properties of sugar palm fiber/polyester yarn-reinforced epoxy hybrid biocomposites with magnesium hydroxide flame retardant filler. *Text. Res. J.* **2021**, 1–12. [CrossRef]
139. Ilyas, R.; Sapuan, S.; Ishak, M.; Zainudin, E.S. Development and characterization of sugar palm nanocrystalline cellulose reinforced sugar palm starch bionanocomposites. *Carbohydr. Polym.* **2018**, *202*, 186–202. [CrossRef]
140. Sanyang, M.L.; Sapuan, S.M.; Jawaid, M.; Ishak, M.R.; Sahari, J. Recent developments in sugar palm (*Arenga pinnata*) based biocomposites and their potential industrial applications: A review. *Renew. Sustain. Energy Rev.* **2016**, *54*, 533–549. [CrossRef]
141. Ilyas, R.A.; Sapuan, S.M.; Ibrahim, R.; Abral, H.; Ishak, M.R.; Zainudin, E.S.; Atiqah, A.; Atikah, M.S.N.; Syafri, E.; Asrofi, M.; et al. Thermal, Biodegradability and Water Barrier Properties of Bio-Nanocomposites Based on Plasticised Sugar Palm Starch and Nanofibrillated Celluloses from Sugar Palm Fibres. *J. Biobased Mater. Bioenergy* **2020**, *14*, 234–248. [CrossRef]
142. Ilyas, R.A.; Sapuan, S.M.; Sanyang, M.L.; Ishak, M.R.; Zainudin, E.S. Nanocrystalline Cellulose as Reinforcement for Polymeric Matrix Nanocomposites and its Potential Applications: A Review. *Curr. Anal. Chem.* **2018**, *14*, 203–225. [CrossRef]
143. Mohammed, L.; Ansari, M.N.M.; Pua, G.; Jawaid, M.; Islam, M.S. A Review on Natural Fiber Reinforced Polymer Composite and Its Applications. *Int. J. Polym. Sci.* **2015**, *2015*, 1–15. [CrossRef]

144. Majeed, K.; Jawaid, M.; Hassan, A.; Abu Bakar, A.; Khalil, H.A.; Salema, A.; Inuwa, I. Potential materials for food packaging from nanoclay/natural fibres filled hybrid composites. *Mater. Des.* **2013**, *46*, 391–410. [CrossRef]
145. Ngo, T.-D. Natural Fibers for Sustainable Bio-Composites. In *Natural and Artificial Fiber-Reinforced Composites as Renewable Sources*; Günay, E., Ed.; InTech: London, UK, 2018; pp. 107–126.
146. Zhang, L. The Application of Composite Fiber Materials in Sports Equipment. In *Proceedings of the 2015 International Conference on Education, Management, Information and Medicine*; Atlantis Press: Washington, DC, USA, 2015; pp. 450–453.
147. Yusup, E.; Mahzan, S.; Kamaruddin, M. Natural Fiber Reinforced Polymer for the Application of Sports Equipment using Mold Casting Method. In *Proceedings of the IOP Conference Series: Materials Science and Engineering*; IOP Publishing: Bristol, UK, 2019; Volume 494, p. 012040.
148. Floating on Flax—CAPiTA Snowboards with ampliTexTM Fusion Tape. Available online: https://www.bcomp.ch/news/capita-snowboards-with-amplitex-fusion-tape/ (accessed on 18 March 2021).

Article

Manufacturing Pitch and Polyethylene Blends-Based Fibres as Potential Carbon Fibre Precursors

Salem Mohammed Aldosari [1,2,*]**, Muhammad A. Khan** [3] **and Sameer Rahatekar** [1,*]

[1] Enhanced Composite and Structures Centre School of Aerospace, Transport, and Manufacturing, Cranfield University, Cranfield MK43 0AL, UK
[2] National Centre for Aviation Technology, King Abdulaziz City for Science and Technology (Kacst), Riyadh 11442, Saudi Arabia
[3] Centre of Life-Cycle Engineering and Management School of Aerospace, Transport, and Manufacturing, Cranfield University, Cranfield MK43 0AL, UK; Muhammad.A.Khan@cranfield.ac.uk
[*] Correspondence: s.m.aldosari@cranfield.ac.uk (S.M.A.); S.S.Rahatekar@cranfield.ac.uk (S.R.)

Abstract: The advantage of mesophase pitch-based carbon fibres is their high modulus, but pitch-based carbon fibres and precursors are very brittle. This paper reports the development of a unique manufacturing method using a blend of pitch and linear low-density polyethylene (LLDPE) from which it is possible to obtain precursors that are less brittle than neat pitch fibres. This study reports on the structure and properties of pitch and LLDPE blend precursors with LLDPE content ranging from 5 wt% to 20 wt%. Fibre microstructure was determined using scanning electron microscopy (SEM), which showed a two-phase region having distinct pitch fibre and LLDPE regions. Tensile testing of neat pitch fibres showed low strain to failure (brittle), but as the percentage of LLDPE was increased, the strain to failure and tensile strength both increased by a factor of more than 7. DSC characterisation of the melting/crystallization behaviour of LLDPE showed melting occurred around 120 °C to 124 °C, with crystallization between 99 °C and 103 °C. TGA measurements showed that for 5 wt%, 10 wt% LLDPE thermal stability was excellent to 800 °C. Blend pitch/LLDPE carbon fibres showed reduced brittleness combined with excellent thermal stability, and thus are a candidate as a potential precursor for pitch-based carbon fibre manufacturing.

Keywords: pitch; polyethylene; carbon fibres; extrusion; blend

Citation: Aldosari, S.M.; Khan, M.A.; Rahatekar, S. Manufacturing Pitch and Polyethylene Blends-Based Fibres as Potential Carbon Fibre Precursors. *Polymers* **2021**, *13*, 1445. https://doi.org/10.3390/polym13091445

Academic Editor: Emin Bayraktar

Received: 7 April 2021
Accepted: 27 April 2021
Published: 29 April 2021

Publisher's Note: MDPI stays neutral with regard to jurisdictional claims in published maps and institutional affiliations.

1. Introduction

Carbon fibres are widely used in many industries, including aerospace, defence, construction, and healthcare, because of their high mechanical, thermal and electrical properties. However, fast expansion of the application of carbon fibres to industry will only continue if they can be produced at low cost without compromising physical properties [1]. Carbon fibres using synthetic rayon as the precursor have been developed since 1960 [2,3], and carbon fibres from pitch have been produced with a high elastic modulus since 1963 [4]. Carbon fibres derived from pitch precursors are primarily categorised into two kinds on the basis of their properties and type of pitch precursor. They are either mesophase pitch-based carbon fibres with high modulus and tensile strength and, in this study, with a molecular weight around 2600 g/mol [5], or isotropic pitch-derived carbon fibres with desirable mechanical properties. Isotropic pitch manufacture is not easy owing to the sudden appearance of mesophase spheres above a certain temperature [6]. The ease with which the mesophase appears is one of the reasons that makes mesophase pitch-based carbon fibre (MPCF) a very promising candidate and a target material for extensive research [7,8]. Other reasons include the occurrence of a high degree of anisotropy with greater alignment within the fibres and, hence, higher values of mechanical properties than those of polyacrylonitrile (PAN) or isotropic pitch-derived carbon fibres [3]. However, despite being a cheap precursor, purification of the pitch is quite expensive, and its final performance depends strongly on the extent of the defects introduced during processing. Manufacturing carbon fibres using pitch, which, along with synthetic polymers, is abundant, should reduce material costs, allowing for the wider use of carbon fibres. When compared to the high cost of PAN-based fibres, the mechanical performance of such cheap and readily available materials, such as lignin, has been explored but has been found to be inadequate [9], while carbon fibres are lightweight with excellent mechanical properties [10]. Textile-grade polyethylene is chemically simple and attractive as a precursor for carbon fibre production. It has the potential to reduce manufacturing costs by 50% relative to the PAN [11–13]. The PE is less expensive than PAN, because 50–65% of production costs are attributed to the synthesis of the PAN precursor, and that is not required in the case of PE. PE production via fusion spinning is cheaper, faster, more energy-saving and more environmentally friendly. PE has a higher carbon content and carbonation rate in addition to its congruence nature [11,13,14]. Its inherent defects make MPCF brittle, and that makes the spinning of these fibres difficult [3,4]. The stabilisation of MPCF is the most important stage in production, but it is so slow that it is by far the most inefficient stage [15].

Huang refers to a patent that describes a fresh technique of how melt-spun, sulphonated and carbonised polyethylene (PE) precursor could be used to manufacture carbon fibres with a yield of 75% [16]. Subsequently, there has been much research to develop low-cost PE-based carbon fibres with good mechanical properties, good compatibility, and high carbon content and carbonisation ratio precursor [11,13,17–24].

LLDPE has a high molecular weight, in the range of 50,000 to 200,000, which is higher than the molecular weight of both high-density and low-density polyethylene (HDPE and LDPE), and LLDPE also has a higher carbon content [25]. Our final aim is to develop carbon fibre-based PE and mesosphase pitch, where the pitch's lower molecular weight of 400 to 800 may help to improve the amount of carbon in the carbon fibre [19,20,26,27].

To overcome the limitations imposed by poor spinnability due to the brittle nature of mesophase pitch, we use mixtures of mesophase pitch and LLDPE to reduce the brittleness of the pitch precursor fibres and to improve the fibre's spinnability. LLDPE offers high ductility and can be formed into carbon fibres, hence it could be an excellent blending material with which to prepare mesophase pitch carbon fibre precursors with reduced brittleness. The characteristic behaviour of both MPCF and PE carbon can be improved by innovatively producing MP/PE-derived carbon fibres with the benefits of low cost, improved ductility, and good mechanical properties with minimum imperfections. In a review paper by Salem et al., it was shown that the aforementioned objectives could be realised using a Pitch/PE precursor blend to manufacture carbon fibre precursors with lower brittleness [1]. Research on PE/pitch carbon fibres is still in an early stage. This paper aims to contribute to that research by reporting an investigation of the properties of carbon fibre precursors manufactured from blends of mesophase pitch and PE with the goal of reduced brittleness. We can show that the brittle nature of the mesophase pitch can be reduced by adding a small fraction of LLDPE. Such fibres with lower brittleness will offer potentially better precursors for future carbon fibre manufacturing.

2. Methodology

2.1. Material

Precursors used in this study were LLDPE with density 0.918 g/cm^3, melting point 121 °C, and softening point 99 °C, which was purchased from Sabic, mesophase pitch having density 1.425 g/cm^3, melting point 298 °C, and softening point 268 °C, and mesophase pitch in which the content was mesophase-92%, which was received from Bonding Chemical.

2.2. Materials Processing

Fibre melt spinning was carried out using a single screw extruder (Noztek Pro Filament Extruder), with a bespoke screw design, swappable barrel, and filament tolerances of 1.75 ± 0.04 mm. The nozzle diameter was 0.5 mm, with a length of 1.5 mm. Pitch/PE pellets were fed into the machine in different ratios to obtain varying proportions of pitch blends. The extruder was operated at 315 °C, at an extrusion speed of 2.5 m/min, and a rotating winder was used for cold stretching as shown in Figure 1. When collecting the fibres, the stretching speed was fixed at 2.5 m/min [28]. The processing steps are shown in Figure 2. The designation of samples with respect to varying proportions of pitch and LLDPE is shown in Table 1.

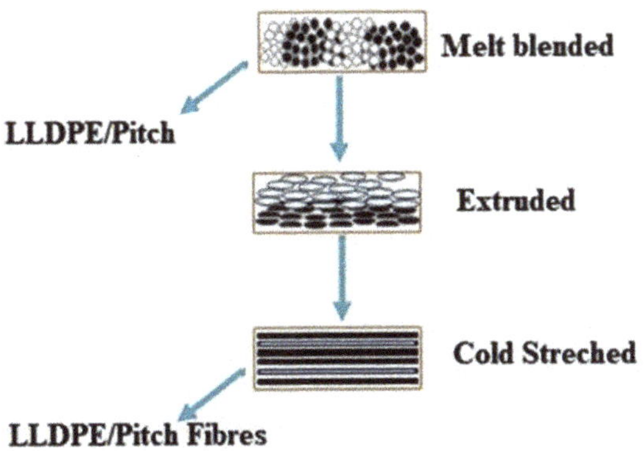

Figure 1. Production of fibres from melt blends.

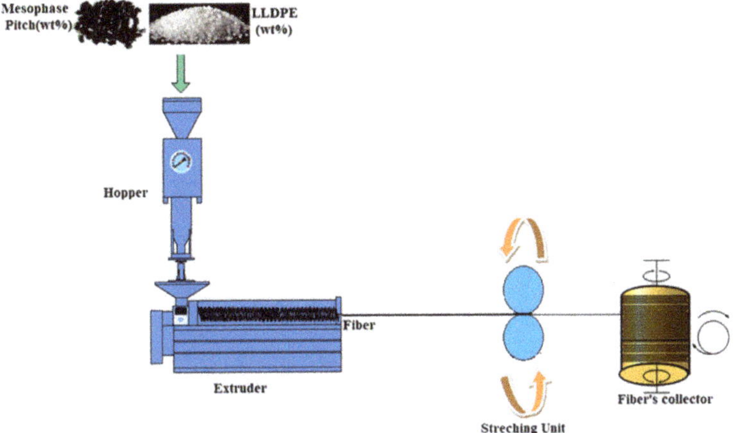

Figure 2. Schematic diagram of material processing.

Table 1. Blends in terms of wt% mesophase pitch and LLDPE.

Blend Designation	Mesophase Pitch	LLDPE
Mesophase Pitch	100 wt%	0 wt%
LLDPE (5 wt%)/Mesophase Pitch	95 wt%	5 wt%
LLDPE (10 wt%)/Mesophase Pitch	90 wt%	10 wt%
LLDPE (15 wt%)/Mesophase Pitch	85 wt%	15 wt%
LLDPE (20 wt)/Mesophase Pitch	80 wt%	20 wt%
LLDPE (100 wt%)	0 wt%	100 wt%

3. Characterisation Methods

3.1. Microscopy

3.1.1. Optical

A Nikon ECLIPSE ME600 optical microscope was used to determine the fibre diameter; the optical images were recorded at 20× magnification. The microscope can be used in differential interference contrast (DIC), which is a method of imaging based on contrast difference of samples.

3.1.2. Scanning Electron Microscopy (SEM)

The microstructure of prepared samples was analysed using a scanning electron microscope, a Tescan VEGA 3 with Oxford Instruments EDS software (Aztec). Prior to analysis, the Quorum 150T ES sputter coater was used to coat a gold layer onto the sample. The samples were prepared by cutting lengths with a clean blade and mounting both vertically (to image end-on) and horizontally (to image surface features) onto aluminium SEM stubs with double-sided carbon tabs. The gold-coated samples were then loaded into the SEM chamber separately due to the height difference. Images were taken at the magnifications noted on the data bar; typically, $500\times$, $1000\times$, and $4000\times$. EDS analysis was conducted to distinguish the pitch from the polyethylene. Measurements were taken of the diameter on the horizontally mounted fibres.

3.2. Static Mechanical Analysis (Tensile Test)

Tensile testing of melt-spun MP/PE fibres was performed using a DEBEN Microtest fibre tensile tester coupled with Leica EC4 Microscope; the test was conducted according to Standard ISO 11,566, 1996 Figure A1: Carbon fibre—Determination of the tensile properties of single-filament specimens. Six tests were carried out for each sample, and the repeatability of the tests was confirmed. Stress and strain were calculated using Equations (1) and (2).

$$Stress\ (\sigma) = \frac{F}{A_c} = \frac{F}{\pi \frac{d^2}{4}} \tag{1}$$

$$Strain\ (\epsilon) = \frac{Change\ in\ length}{Orignal\ length} = \frac{Elongation}{Guage\ Length} = \frac{\Delta L}{L_0} = \frac{L - L_0}{L_0} \tag{2}$$

d = diameter of fibres (mm), F = Axial load (N).

3.3. Differential Scanning Calorimetery (DSC)

A Mettler Toledo Differential Scanning Calorimeter, DSC Q 2000, was used to perform the DSC analysis. This equipment was used to measure the properties of prepared samples. The samples were heated from $-50\ °C$ to $200\ °C$ at a rate of $20\ °C/min$ under a nitrogen atmosphere, and the samples were held at $200\ °C$ for 5 min to eradicate their previous thermal history. Nonisothermal behaviour and kinetics were investigated by cooling these samples at a rate of $20\ °C/min$. The results were obtained after the heating and cooling cycles. The crystallinity of the samples was measured using the following equation:

$$\chi(\%) = \frac{Enthalpy\ of\ fusion\ (\Delta H)}{Enthalphy\ of\ fusion\ for\ Matrix(\Delta H^0)} \times 100 \tag{3}$$

$\chi = Crystallinity\ of\ the\ system.$
$\Delta H = Enthalpy\ of\ fusion\ of\ LLDPE\ and\ Pitch\ fibres.$
$\Delta H^0 = Enthalpy\ of\ fusion\ of\ neat\ Pitch.$

3.4. Thermogravimetric Analysis (TGA)

A Mettler Toledo, Thermogravimetric Analysis TGA Q 500, was used to perform the TGA analysis; to measure the properties of the prepared samples. The samples were heated from 50 to $800\ °C$ at a heating rate of $20\ °C/min$ under a nitrogen atmosphere, and the samples were held at $800\ °C$ for 5 min to eradicate their previous thermal history.

4. Results and Discussion

4.1. Microscopy

4.1.1. Optical Microscopy of Fibre

It was observed that the different samples produced different fibre diameters, which depended upon the LLDPE content. The fibre diameter decreased with an increase in LLDPE content. This is shown in Figure 3 where fibre diameter is seen with LLDPE content in LLDPE/mesophase pitch blends as shown in Table 2.

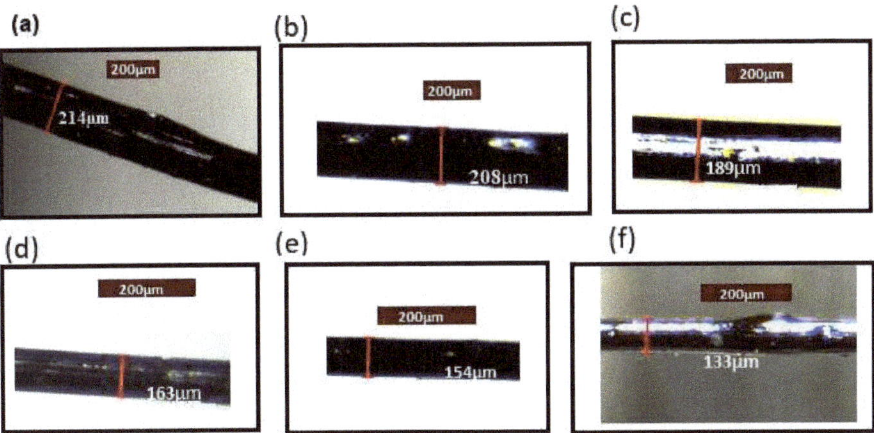

Figure 3. Optical microscope image of fibres with different LLDPE content. (**a**) Mesophase pitch, (**b**) 5 wt% LLDPE/mesophase pitch (**c**) 10 wt% LLDPE/mesophase pitch, (**d**) 15 wt% LLDPE/mesophase pitch, (**e**) 20 wt% LLDPE/mesophase pitch and (**f**) LLDPE.

Table 2. Blends in terms of wt% mesophase pitch and LLDPE fibres diameter.

Blend Designation	Fibre Diameter, µm
Mesophase Pitch	214 (±0.43)
LLDPE (5 wt%)/Mesophase Pitch	208 (±0.54)
LLDPE (10 wt%)/Mesophase Pitch	189 (±0.24)
LLDPE (15 wt%)/Mesophase Pitch	163 (±0.36)
LLDPE (20 wt%)/Mesophase Pitch	154 (±0.38)
LLDPE (100 wt%)	133 (±0.53)

We see from Figures 3 and 4 and Table 2 that the fibre diameter decreases with increasing LLPDE content. The diameter of the fibre varies due to the change in the applied axial elongation force required during the drawing process due to the change in LLDPE content. It has previously been reported [28] that the morphology of the blends is changed by varying the content of the blend's partner.

The diameter of fibres in the extrusion process is greatly affected by many factors, such as extrusion speed, the viscosity of the polymers, the diameter of orifice, stretching or drawing ratio, and speed of extrusion. The optical microscope was used to determine the diameter of the cold-drawn samples, and optical microscope images of these with respect to LLDPE content are presented in Figure 3, which shows that different samples produced different fibre diameters, which decreased with an increase in LLDPE content.

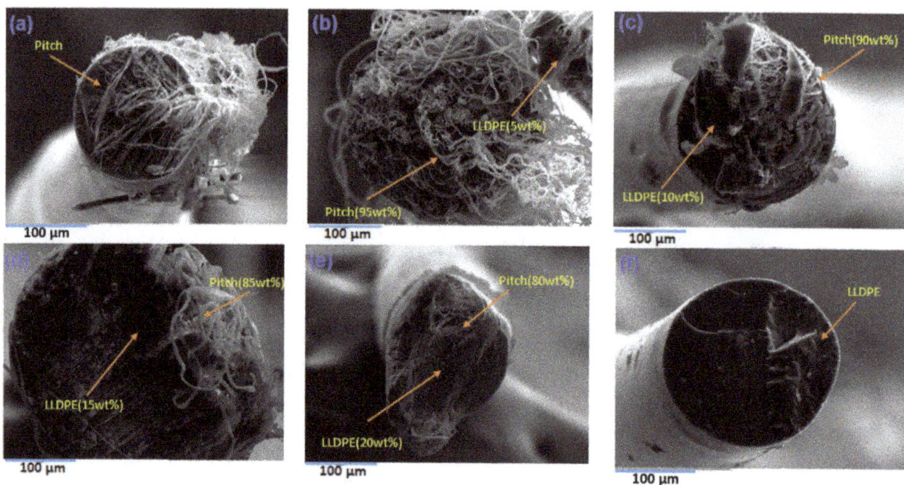

Figure 4. Morphology of fibres with LLDPE content in LLDPE/mesophase pitch blend. (**a**) Pure mesophase pitch, (**b**) 5 wt% LLDPE/mesophase, (**c**) 10 wt% LLDPE/mesophase pitch, (**d**) 15 wt% LLDPE/mesophase pitch, (**e**) 20 wt% LLDPE/mesophase pitch and (**f**) 100% LLDPE.

4.1.2. Scanning Electron Microscopy of Fibres

Figure 4 shows the morphology of a range of LLDPE/mesophase pitch blends. The SEM image shows that pitch fibres have a tendency to form microfibrils inside the fibres; Figure 4a [29]. As the LLDPE content is increased, there is a tendency for microfibre formation inside the blend fibres to reduce; see Figure 4b–e. Post adding of LLDPE with pitch, a decrease in the cluster formation and low alignment is observed in the mesophase pitch. The pure/neat LLDPE fibre does not show any microfibrous pitch (silver), only LLDPE (black solid). Therefore, we can conclude that the relative content of LLDPE has a very significant effect on fibre morphology and diameter. In mesophase pitch fibre fabrication, the molten liquid crystalline mesophase can be oriented within the die by shear treatment during a soften-spinning process. Overall, the shear forces involved lead to a strongly oriented structure, with the mesophase domains elongated in cylinders parallel to the axis of the fibres (circumferential fibres). The microstructural study suggests that the radial orientation of the fibres obtained from the mesophase pitch originates from the flow to throw across the die [30]. The spinning stage is the most important stage in the manufacture of mesophase pitch blend LLDPE, to control the shape, transverse texture and orientation of the fibre, which results in higher performance.

The mesophase pitch shows the liquid crystalline phase and preferential orientation of the pitch molecules, giving appropriate reference of the liquid crystal and flow orientation behaviour of pitch. The flow orientation behaviour may be causing microfibre formation. The polyethylene does not show the liquid crystalline behaviour, hence, when you add polyethylene in pitch, it decreases the microfibre content in pitch, and when you extrude the neat polyethylene, there are no microfibres due to lack of liquid crystal behaviour.

Pitch microfibers formation occurs because of the liquid crystalline mesophase nature of the pitch. It allows alignment and therefore microfiber formation, LLDPE not liquid crystalline, therefore does not make microfiber.

4.2. Tensile Tests of LLDPE/Mesophase Pitch Blends

LLDPE is inexpensive, flexible, and is extensively employed in several different forms. One study has shown that the tensile strength of LLDPE is typically in the region of 6.1 MPa and that the tensile strength can be increased by the addition of straw fibres [31]. Another study has shown the tensile strength of LLDPE can reach 9.9 MPa [32]. Others have reported that the tensile strength of LLDPE can be extended to 22 MPa, by the inclusion of multiwalled carbon nanotubes [33]. These improvements in the properties of LLDPE suggest extending its possible use in combination with new applications.

To understand the effect of extrusion on static mechanical properties, two types of extruded strands were used for tensile testing. These strands were related to the number of the extrusion cycle, in which sample-one is the designation for a single extruded strand, and sample-two is the designation for double extruded strands. Tensile tests were performed for both sample-one and sample-two strands. It was found that the tensile strength and tensile modulus were different for sample-one and sample-two. For sample one, the extruded tensile strength and modulus were 3.30 MPa and 151 MPa, respectively, and for the double extruded strands, the respective values were 3.06 MPa and 21.42 MPa. It was clearly seen that both tensile modulus and fracture elongation were higher for sample-one in comparison with sample-two. However, tensile strength was very slightly higher for sample-two in comparison with sample one, and from this we conclude that sample-two has a lower modulus and elongation at failure in comparison with sample-one due to degradation and damage that occurred in the LLDPE chains during the second extrusion. This damage may be due to the higher temperature and shearing during mixing in the extrusion process [34–36].

Figure 5a shows the stress vs. strain curve of pure extruded LLDPE. It can be seen that the curve is smooth, and tensile strength is typically in the range of 40 MPa, which indicates that the LLDPE plays a key role in load-bearing as well as toughening materials. The effect of increasing the proportion of LLDPE on the brittle mesophase pitch can be seen in Figure 5b, where the mechanical strength of LLDPE/mesophase pitch blends comes from both LLDPE and mesophase pitch.

Figure 5b shows the stress vs. strain curve of LLDPE/mesophase pitch blends for neat/pure pitch (0 wt%), 5 wt%, 10 wt%, 15 wt% and 20 wt% of LLDPE. It can be seen that the samples show different stress vs. strain behaviour depending on the LLDPE content. The lowest tensile strength (1.0 MPa) and lowest strain (5%) were found in the case of neat mesophase pitch. The tensile strength and strain increased steadily with LLDPE content up to the maximum of 20 wt% (10.0 MPa) and maximum strain (25%). We define brittleness as low strain to failure, therefore the pitch fibre sample shows brittle behaviour. When the LLDPE content increased, the strain to failure of these samples increased; thus, adding LLDPE to the fibres reduces brittleness. Table 3 shows a summary of the tensile strength, tensile modulus, and strain to failure of the neat pitch fibres and blends of pitch and LLDPE-based fibres along with the standard deviation for each type of fibre. It can be seen that, for the prepared samples, the higher the LLDPE content, the greater the tensile strength, tensile modulus, and strain at failure.

Table 3. Tensile strength, modulus, and strain at failure of the prepared samples.

Samples	Tensile Strength (MPa)	Tensile Modulus (MPa)	Strain at Failure
Mesophase Pitch	1.38 (±0.26)	428 (±4.3)	0.03 (±0.021)
LLDPE (5 wt%)/Mesophase Pitch	2.23 (±0.34)	477 (±5.7)	0.15 (±0.023)
LLDPE (10 wt%)/Mesophase Pitch	4.01 (±0.43)	628 (±5.4)	0.19 (±0.022)
LLDPE (15 wt%)/Mesophase Pitch	5.90 (±0.58)	682 (±4.5)	0.21 (±0.024)
LLDPE (20 wt%)/Mesophase Pitch	10.3 (±0.87)	763 (±5.3)	0.23 (±0.025)
LLDPE (100 wt%)	40.0 (±0.98)	994 (±6.4)	0.80 (±0.022)

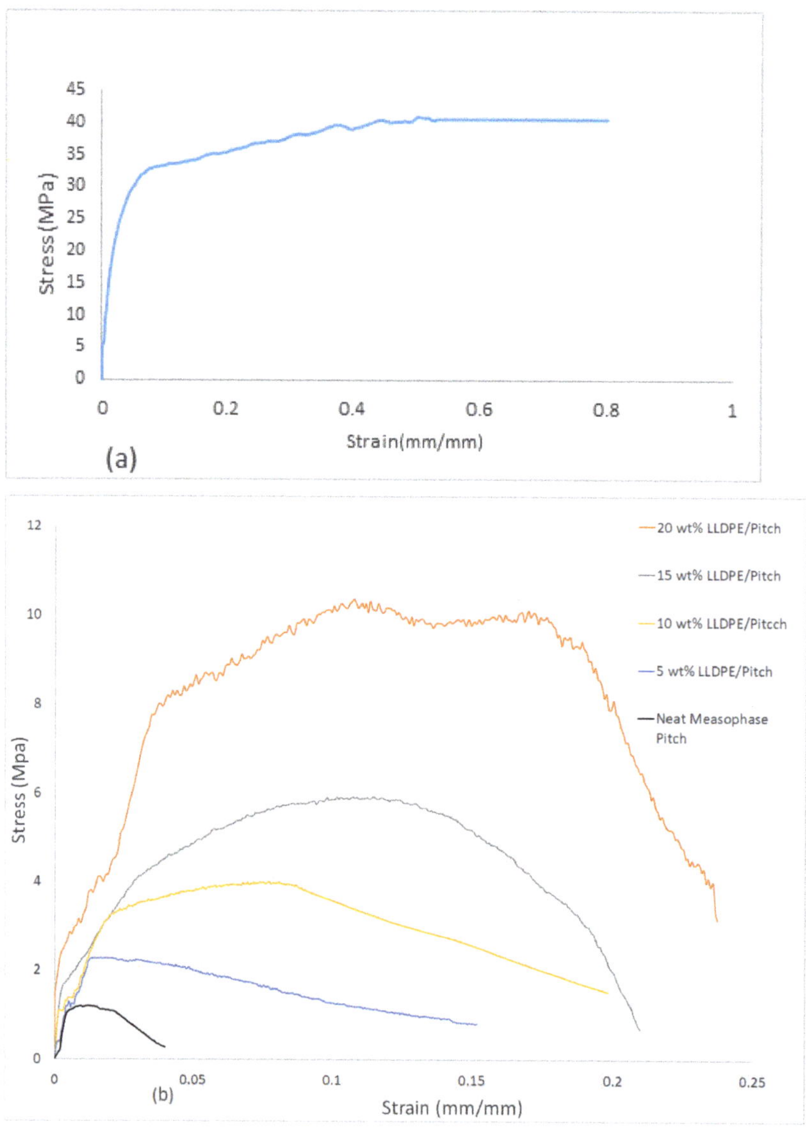

Figure 5. (a) Tensile strength of neat LLDPE; (b) Tensile strength of neat mesophase pitch fibres and LLDPE/mesophase pitch blend-based fibres with varying content of LLDPE.

4.3. Differential Scanning Calorimetery for LLDPE/Mesophase Pitch Blends

The crystallization of several LDPE and LLDPE blends has been evaluated using differential scanning calorimetry (DSC). It was observed that in the region between 110 and 120 °C, where pure LDPE does not melt, there was an increase in the population of crystallites melting. A reduction in the crystallite population across the range where LDPE shows its primary melting peaks (90–110 °C) has been observed, suggesting that a proportion of the lamella had shifted towards a higher melt temperature [37]. DSC has shown that blends of LDPE and LLDPE show different single melting peak endotherms for different compositions in the range 10 to 90 wt% LLDPE [38].

Kinetic data were acquired by least-squares analyses of experimental points obtained by differentiating primary thermograms. Degradation was found to be higher for LDPE than for LLDPE. Degradation has been seen as a primary order for LDPE and second order for LLDPE [38].

The thermal behaviour of LLDPE/mesophase pitch, over the range 0 wt% LLDPE/mesophase pitch to 20 wt% LLDPE/mesophase pitch, was investigated using DSC. The DSC experiments were carried out over a range of melting and crystallization temperatures of LLDPE to avoid any degradation of the LLDPE.

Figure 6a shows the crystallization behaviour of prepared samples, and it is noticed that neat mesophase pitch does not show any marked variation in crystallization behaviour, with no peaks or troughs occurring on the plot [39].

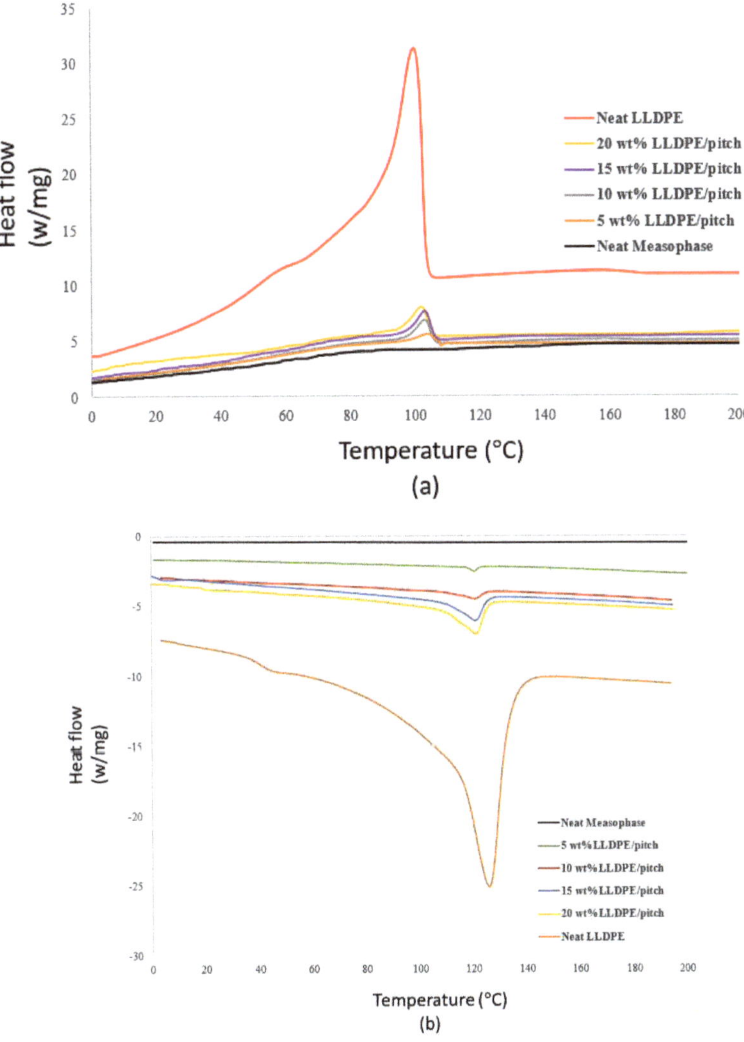

Figure 6. (a) Crystallisation, and (b) melting behaviour of LLDPE and LLDPE/mesophase pitch blend.

In Figure 6a, we see that neat LLDPE shows a well-organised crystallization peak and temperature. We also see that adding different proportions of LLDPE to the LLDPE/mesophase pitch blend clearly affects the crystallization behaviour of the blend, both the crystallization temperature and the enthalpy of crystallization (ΔH_c, area under the exotherm curve). The enthalpy of crystallization for LLDPE/mesophase pitch blend increases with the proportion of LLDPE. Figure 6 shows that the crystallization temperature of LLDPE/mesophase pitch blend decreases with an increase in the proportion of LLDPE, which also indicates that the crystallization temperature of blends is dependent upon the LLDPE content.

Similarly, the melting temperature and enthalpy of fusion of the samples change with the proportion of LLDPE in the LLDPE/mesophase pitch blend, see Figure 6b, and Table 4. The melting temperature increases with the proportion of LLDPE in the blend, so that the maximum melting temperature is for pure LLDPE. The enthalpy of fusion (ΔH_m, the area under the endotherm) is highest for pure LLDPE; this is due to the absence of crystalline domains in pure LLDPE compared to LLDPE/mesophase pitch blends. We see that the presence of LLDPE in LLDPE/mesophase pitch blends affects the melting and crystallization behaviour of the blends.

Table 4. Melting and crystallisation temperatures of LLDPE and LLDPE/mesophase pitch blend.

Samples	Melting Temperature (°C)	Crystallization Temperature (°C)	Enthalpy of Fusion (J/g) Sample	Enthalpy of Fusion (J/g) LLDPE
LLDPE 5 wt%/Mesophase Pitch	120	99	4	80
LLDPE 10 wt%/Mesophase Pitch	120.7	100	25	250
LLDPE 15 wt%/Mesophase Pitch	121	101	27	180
LLDPE 20 wt%/Mesophase Pitch	123	102	31	155
LLDPE 100 wt%	124	103	241	241

4.4. Thermogravimetric Analysis of LLDPE/Mesophase Pitch Blends

The objectives of the thermogravimetric analysis (TGA) tests were to measure the thermal decomposition of several blends of LLDPE/mesophase pitch as they burned in an air environment. Analysis of the thermal stability of a material is generally done by TGA and derivative TGA. TGA was also used for the thermal degradation of the composites. A higher decomposition temperature indicates greater thermal stability of the material [33,40].

TGA is used to measure the rate and amount of weight transformation in the material, either as a function of temperature or isothermally as a function of time, in a controlled atmosphere. It can be used to characterise any material that exhibits a weight change during combustion and to discover phase changes due to decomposition, oxidation or dryness. This data helps classify the percentage weight transformation and chemical structure, handling, plus end-use performance.

The TGA was carried out in the temperature range of 100 to 800 °C. The thermal degradation of neat LLDPE and different wt% LLDPE/mesophase pitch fibres was studied, and typical thermal stability curves are shown in Figure 7 for a range of mixes from neat mesophase pitch to neat LLDPE.

Figure 7 shows the weight loss vs. temperature of prepared samples, and it is seen that TGA shows the complete decomposition of the LLDPE at 530 °C. However, the 20 wt% LLDPE/mesophase pitch showed only 20% decomposition at 530 °C, confirming that mesophase pitch has greater stability than neat LLDPE, though the proportion of LLDPE will affect the thermal stability of an LLDPE/mesophase blend. By blending different proportions of LLDPE with mesophase pitch, the decomposition temperature of different LLDPE/mesophase pitch blends was found; see Figure 7. The fibres containing over 5 wt% and 10 wt% LLDPE could retain over 70 wt% of the fibre mass even at 800 °C. These two samples with reduced brittleness and high thermal stability can potentially offer excellent precursors for manufacturing carbon fibres.

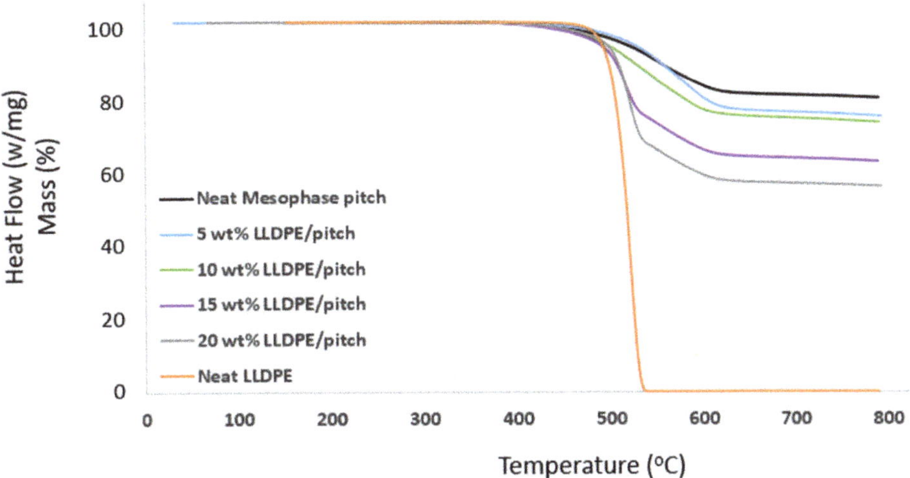

Figure 7. TGA curves of various proportions of LLDPE in LLDPE/mesophase pitch blends.

5. Conclusions

The aim of this study was to reduce the brittleness of pitch precursor fibres. These fibres are commonly used for manufacturing carbon fibres blended with a small fraction of LLDPE. We have demonstrated a simple but effective method of blending mesophase pitch with 5% to 20% (i.e., weight percentage) of LLDPE. The addition of LLDPE showed an increase of strain to failure value up to 25%, hence increases ductility. Thus, these blends reduce brittleness in all the LLDPE/pitch fibres. The LLDPE/pitch blend fibres also showed improved tensile strength and Young modulus from 1.3 to 10.3 MPa and from 428 to 763 MPa, respectively. The precursors used in this study along with blend can be used to produce potential carbon fibres with superior mechanical properties.

Morphology observed using SEM analysis revealed that neat LLDPE fibres did not show any microfibrils, whereas the neat pitch fibres showed microfibrillated structure. As the LLDPE content was increased, the microfibrous formation was seen to decrease, until with 100% LLDPE no microfibrous areas were observed and the appearance was close to solid black.

DSC analysis showed crystallisation temperature of LLDPE in the pitch/LLDPE blend ranged from 99 to 103 °C, and the melting temperature was in the range 120 to 124 °C.

TGA analysis showed high thermal stability, making the obtained pitch/LLDPE fibre precursor suitable to develop less brittle carbon fibres for high-performance composites, particularly for use in the aerospace sector.

A comprehensive investigation of the PE/pitch morphology is still required. It is key to understand how morphology can be controlled during fabrication as it affects the crack resistance and ductility of the obtained fibre. The stabilisation, carbonisation and graphitisation of PE/pitch precursors need to be explored to increase the possibility of producing more ductile carbon fibres. The carbonisation step, which deals with the presence of polyethylene chain, needs to be explored further. It can potentially increase the bent and loop regions of the graphitic planes and hence reduce the crack formation in the fibre.

Author Contributions: S.M.A. and S.R. conceptualized the theory and the method. S.M.A. performed the experimentation and the analysis. M.A.K. enhanced the data analysis. S.M.A. and M.A.K. worked on the original draft preparation. All authors have read and agreed to the published version of the manuscript.

Funding: This research received no external funding.

Institutional Review Board Statement: Not Applicable.

Informed Consent Statement: Not Applicable.

Data Availability Statement: Data can be available on a request from the corresponding author via email.

Conflicts of Interest: The authors declare no conflict of interest.

Appendix A

The tensile test procedure was conducted according to the standard ISO 11,566, 1996 (Figure A1).

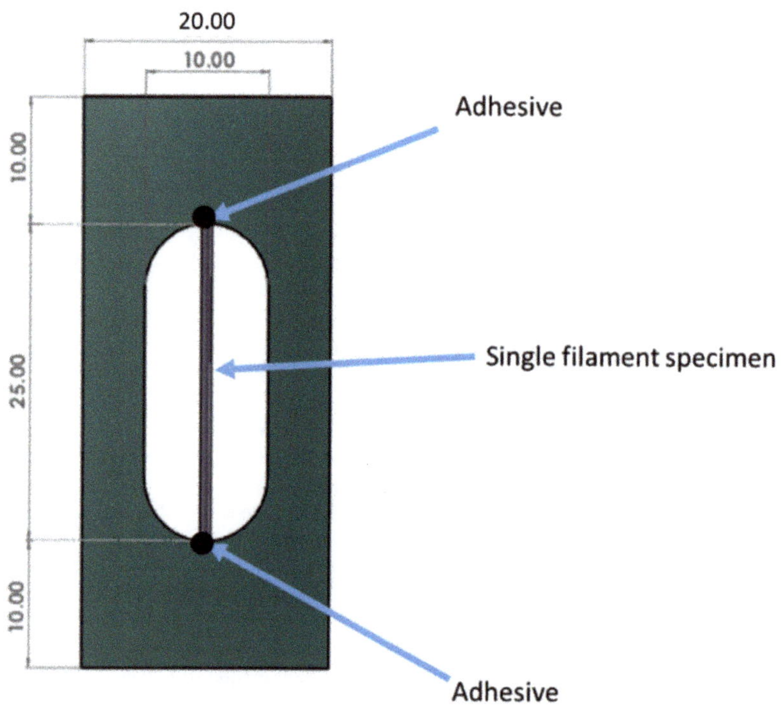

Figure A1. Tensile test setup for melt spun MP/PE fibre.

References

1. Aldosari, S.M.; Khan, M.; Rahatekar, S. Manufacturing carbon fibres from pitch and polyethylene blend precursors: A review. *J. Mater. Res. Technol.* **2020**, *9*, 7786–7806. [CrossRef]
2. Fitzer, E. Carbon Fibres Present State and Future Expectations. In *Carbon Fibers Filaments and Composites*; Figueired, J.L., Bernardo, C.A., Baker, R.T., Huttinger, K.J., Eds.; Kluwer Academic: Karlsruhe, Germany, 1990; pp. 3–41.
3. Liu, Y.; Kumar, S. Recent progress in fabrication, structure, and properties of carbon fibers. *Polym. Rev.* **2012**, *52*, 234–258. [CrossRef]
4. Park, S.J.; Lee, S.Y. History and Structure of Carbon Fibers. In *Carbon Fibers*; Springer: Incheon, Korea, 2015; pp. 1–30. [CrossRef]
5. Yuan Guanming, C.Z. Preparation, Characterization, and Applications of Carbonaceous Mesophase: A Review. In *Nematic Liquid Crystals*, 1st ed.; Carlescu, I., Ed.; IntechOpen: London, UK, 2019; pp. 1–20. [CrossRef]

6. Zeng, S.M.; Maeda, T.; Tokumitsu, K.; Mondori, J.; Mochida, I. Preparation of isotropic pitch precursors for general purpose carbon fibers (GPCF) by air blowing-II. Air blowing of coal tar, hydrogenated coal tar, and petroleum pitches. *Carbon* **1993**, *31*, 413–419. [CrossRef]

7. Fortin, F.; Yoon, S.H.; Korai, Y.; Mochida, I. Structure of round-shaped methylnaphthalene-derived mesophase pitch-based carbon fibres prepared by spinning through a Y-shaped die hole. *J. Mater. Sci.* **1995**, *30*, 4567–4583. [CrossRef]

8. Xiao, B.; Huang, Q.; Chen, H.; Chen, X.; Long, G. A fractal model for capillary flow through a single tortuous capillary with roughened surfaces in fibrous porous media. *Fractals* **2021**, *29*, 2150017. [CrossRef]

9. Gao, Z.; Zhu, J.; Rajabpour, S.; Joshi, K.; Kowalik, M.; Croom, B.; Schwab, Y.; Zhang, L.; Bumgardner, C.; Brown, K.R.; et al. Graphene reinforced carbon fibers. *Sci. Adv.* **2020**, *6*, 1–11. [CrossRef] [PubMed]

10. Tran, T.Q.; Lee, J.K.Y.; Chinnappan, A.; Loc, N.H.; Tran, L.T.; Ji, D.; Jayathilaka, W.; Kumar, V.V.; Ramakrishna, S. High-performance carbon fiber/gold/copper composite wires for lightweight electrical cables. *J. Mater. Sci. Technol.* **2020**, *42*, 46–53. [CrossRef]

11. De Palmenaer, A.; Wortberg, G.; Drissen, F.; Seide, G. Production of Polyethylene Based Carbon Fibres. *Chem. Eng. Trans.* **2015**, *43*, 1699–1704. [CrossRef]

12. Wortberg, G.; De Palmenaer, A.; Beckers, M.; Seide, G.; Gries, T. Polyethylene-Based Carbon Fibers by the Use of Sulphonation for Stabilization. *Fibers* **2015**, *3*, 373–379. [CrossRef]

13. Kim, K.-W.; Lee, H.-M.; Kim, B.S.; Hwang, S.-H.; Kwac, L.-K.; An, K.-H.; Kim, B.-J. Preparation and thermal properties of polyethylene-based carbonized fibers. *Carbon Lett.* **2015**, *16*, 62–66. [CrossRef]

14. Yang, K.S.; Kim, B.-H.; Yoon, S.-H. Pitch based carbon fibers for automotive body and electrodes. *Carbon Lett.* **2014**, *15*, 162–170. [CrossRef]

15. Mochida, I.; Toshima, H.; Korai, Y.; Takashi, H. Oxygen distribution in the mesophase pitch fibre after oxidative stabilization. *J. Mater. Sci.* **1989**, *24*, 389–394. [CrossRef]

16. Huang, X. Fabrication and properties of carbon fibers, Review. *Materials* **2009**, *2*, 2369–2403. [CrossRef]

17. Zhang, D.; Bhat, G.S. Carbon Fibers from Polyethylene-Based Precursors. *Mater. Manuf. Process.* **1994**, *9*, 221–235. [CrossRef]

18. Barton, B.E.; Behr, M.J.; Patton, J.T.; Hukkanen, E.J.; Landes, B.G.; Wang, W.; Horstman, N.; Rix, J.E.; Keane, D.; Weigand, S.; et al. High-Modulus Low-Cost Carbon Fibers from Polyethylene Enabled by Boron Catalyzed Graphitization. *Small* **2017**, *13*, 1–7. [CrossRef] [PubMed]

19. Behr, M.J.; Landes, B.G.; Barton, B.E.; Bernius, M.T.; Billovits, G.F.; Hukkanen, E.J.; Patton, J.T.; Wang, W.; Wood, C.; Keane, D.T.; et al. Structure-property model for polyethylene-derived carbon fiber. *Carbon* **2016**, *107*, 525–535. [CrossRef]

20. Zhang, D. Carbon Fibers from Oriented Polyethylene Precursors. *J. Thermoplast. Compos. Mater.* **1993**, *6*, 38–48. [CrossRef]

21. Postema, A.R.; De Groot, H.; Pennings, A.J. Amorphous carbon fibres from linear low density polyethylene. *J. Mater. Sci.* **1990**, *25*, 4216–4222. [CrossRef]

22. Kim, J.W.; Lee, J.S. Preparation of carbon fibers from linear low density polyethylene. *Carbon* **2015**, *94*, 524–530. [CrossRef]

23. Kim, K.-W.; Lee, H.-M.; An, J.-H.; Kim, B.-S.; Min, B.-G.; Kang, S.-J.; An, K.-H.; Kim, B.-J. Effects of cross-linking methods for polyethylene-based carbon fibers: Review. *Carbon Lett.* **2015**, *16*, 147–170. [CrossRef]

24. Penning, J.P.; Lagcher, R.; Pennings, A.J. The effect of diameter on the mechanical properties of amorphous carbon fibres from linear low density polyethylene. *Polym. Bull.* **1991**, *25*, 405–412. [CrossRef]

25. Boustead, I. *Eco-Profiles of the European Plastics Industry*; Linear Low Density Polyethylene (LLDPE): Brussels, Belgium, 2005.

26. Bansal, R.C.; Donnet, J.B. Pyrolytic Formation of High-performance Carbon Fibres. In *Comprehensive Polymer Science and Supplements*; Allen, G., Bevington, J.C., Eds.; Elsevier Ltd.: Amsterdam, The Netherlands, 1996; pp. 501–520. [CrossRef]

27. Kershaw, J.R.; Black, K.J.T. Structural Characterization of Coal-Tar and Petroleum Pitches. *Energy Fuels* **1993**, *7*, 420–425. [CrossRef]

28. Fakirov, S.; Bhattacharyya, D.; Lin, R.J.T.; Fuchs, C.; Friedrich, K. Contribution of Coalescence to Microfibril Formation in Polymer Blends during Cold Drawing. *J. Macromol. Sci. Part B* **2007**, *46*, 183–194. [CrossRef]

29. Lu, S.; Blanco, C.; Rand, B. Large diameter carbon fibres from mesophase pitch. *Carbon* **2002**, *40*, 2109–2116. [CrossRef]

30. Kundu, S.; Ogale, A.A. Rheostructural studies of a discotic mesophase pitch at processing flow conditions. *Rheol Acta* **2010**, *49*, 845–854. [CrossRef]

31. Zhang, L.; Xu, H.; Wang, W. Performance of Straw/Linear Low Density Polyethylene Composite Prepared with Film-Roll Hot Pressing. *Polymers* **2020**, *12*, 860. [CrossRef]

32. Durmus, A.; Kaşgöz, A.; Macosko, C.W. Mechanical properties of linear low-density polyethylene (LLDPE)/clay nanocomposites: Estimation of aspect ratio and interfacial strength by composite models. *J. Macromol. Sci. Part B Phys.* **2008**, *47*, 608–619. [CrossRef]

33. Tai, J.H.; Liu, G.Q.; Caiyi, H.; Shangguan, L.J. Mechanical properties and thermal behaviour of LLDPE/MWNTs nanocomposites. *Mater. Res.* **2012**, *15*, 1050–1056. [CrossRef]

34. Peinado, V.; Castell, P.; García, L.; Fernández, Á. Effect of Extrusion on the Mechanical and Rheological Properties of a Reinforced Poly(Lactic Acid): Reprocessing and Recycling of Biobased Materials. *Materials* **2015**, *8*, 7106–7117. [CrossRef] [PubMed]

35. Andersson, T.; Stålbom, B.; Wesslén, B. Degradation of polyethylene during extrusion. II. Degradation of low-density polyethylene, linear low-density polyethylene, and high-density polyethylene in film extrusion. *J. Appl. Polym. Sci.* **2004**, *91*, 1525–1537. [CrossRef]

36. da Costa, H.M.; Ramos, V.D.; de Oliveira, M.G. Degradation of polypropylene (PP) during multiple extrusions: Thermal analysis, mechanical properties and analysis of variance. *Polym. Test.* **2007**, *26*, 676–684. [CrossRef]

37. Drummond, K.M.; Hopewell, J.L.; Shanks, R.A. Crystallization of low-density polyethylene- and linear low-density polyethylene-rich blends. *J. Appl. Polym. Sci.* **2000**, *78*, 1009–1016. [CrossRef]
38. Bhardwaj, I.S.; Kumar, V.; Palanivelu, K. Thermal characterisation of LDPE and LLDPE blends. *Thermochim. Acta* **1988**, *131*, 241–246. [CrossRef]
39. Yue, Z.; Liu, C.; Vakili, A. Solvated mesophase pitch-based carbon fibers: Thermal-oxidative stabilization of the spun fiber. *J. Mater. Sci.* **2017**, *52*, 8176–8187. [CrossRef]
40. Wang, Z.; Cheng, Y.; Yang, M.; Huang, J.; Cao, D.; Chen, S.; Xie, Q.; Lou, W.; Wu, H. Dielectric properties and thermal conductivity of epoxy composites using core/shell structured Si/SiO2/Polydopamine. *Compos. Part B Eng. Elsevier* **2018**, *140*, 83–90. [CrossRef]

Review

Physical Properties of Thermoplastic Starch Derived from Natural Resources and Its Blends: A Review

Z. N. Diyana [1], R. Jumaidin [2,*], Mohd Zulkefli Selamat [1], Ihwan Ghazali [2], Norliza Julmohammad [3], Nurul Huda [3,*] and R. A. Ilyas [4,5]

1　Fakulti Kejuruteraan Mekanikal, Universiti Teknikal Malaysia Melaka, Hang Tuah Jaya, Durian Tunggal 76100, Melaka, Malaysia; nurdiyanazakuan@gmail.com (Z.N.D.); zulkeflis@utem.edu.my (M.Z.S.)
2　Fakulti Teknologi Kejuruteraan Mekanikal dan Pembuatan, Universiti Teknikal Malaysia, Melaka, Hang Tuah Jaya, Durian Tunggal 76100, Melaka, Malaysia; ihwan@utem.edu.my
3　Faculty of Food Science and Nutrition, Universiti Malaysia Sabah, Kota Kinabalu 88400, Sabah, Malaysia; norliza@ums.edu.my
4　School of Chemical and Energy Engineering, Faculty of Engineering, Universiti Teknologi Malaysia (UTM), Skudai 81310, Johor, Malaysia; ahmadilyas@utm.my
5　Centre for Advanced Composite Materials (CACM), Universiti Teknologi Malaysia (UTM), Skudai 81310, Johor, Malaysia
*　Correspondence: ridhwan@utem.edu.my (R.J.); drnurulhuda@ums.edu.my (N.H.)

Abstract: Thermoplastic starch composites have attracted significant attention due to the rise of environmental pollutions induced by the use of synthetic petroleum-based polymer materials. The degradation of traditional plastics requires an unusually long time, which may lead to high cost and secondary pollution. To solve these difficulties, more petroleum-based plastics should be substituted with sustainable bio-based plastics. Renewable and natural materials that are abundant in nature are potential candidates for a wide range of polymers, which can be used to replace their synthetic counterparts. This paper focuses on some aspects of biopolymers and their classes, providing a description of starch as a main component of biopolymers, composites, and potential applications of thermoplastics starch-based in packaging application. Currently, biopolymer composites blended with other components have exhibited several enhanced qualities. The same behavior is also observed when natural fibre is incorporated with biopolymers. However, it should be noted that the degree of compatibility between starch and other biopolymers extensively varies depending on the specific biopolymer. Although their efficacy is yet to reach the level of their fossil fuel counterparts, biopolymers have made a distinguishing mark, which will continue to inspire the creation of novel substances for many years to come.

Keywords: thermoplastic starch; biopolymer; composite; food packaging

check for updates

Citation: Diyana, Z.N.; Jumaidin, R.; Selamat, M.Z.; Ghazali, I.; Julmohammad, N.; Huda, N.; Ilyas, R.A. Physical Properties of Thermoplastic Starch Derived from Natural Resources and Its Blends: A Review. *Polymers* **2021**, *13*, 1396. https://doi.org/10.3390/polym13091396

Academic Editors: Carlo Santulli and Agnieszka Tercjak

Received: 13 March 2021
Accepted: 9 April 2021
Published: 26 April 2021

Publisher's Note: MDPI stays neutral with regard to jurisdictional claims in published maps and institutional affiliations.

1. Introduction

Plastic manufacturing and distribution across the world have been steadily increasing throughout time. Petroleum-based plastics are commonly employed as single-use plastics in our daily lives as they have become incredibly useful due to certain qualities, such as versatility, durability, flexibility, and toughness [1]. They are also very cheap in the market and easy to find in any grocery store. It was mentioned in the literature that, since the 1950s, the production of plastic had increased from 1.5 million tons to more than 300 million tons in 2015, which is almost 200 times more [2]. The rapid growth of plastic usage is due to the various types of plastic products currently found in the market, ranging from household, personal goods, and packaging to the manufacturing of construction materials. The extensive use of plastics has led to the excessive plastic waste currently present in the environment. In due course, this will create widespread global issues for the environment as well as for humans since the degradation rate of these materials takes a very long time, around 100 years, due to their hydrophobic properties and properties that efficiently avoid rapid microbial action [3].

Therefore, to overcome this issue, a change needs to be made by replacing the commonly used petroleum-based plastic with biodegradable plastic to sustain a better environmental condition for future generation and also to provide less plastic disposal techniques. Among biopolymers, starch is the one of promising possibility and alternatives to replace petroleum-based plastics, which is due to properties of starch that is completely biodegradable and are found to be abundant in nature. It is widely accessible in plants such as corn, cassava, potato, tuber, and many more. In plants, starch is stored in the granule packed form present in an amorphous and crystalline condition [4]. Due to great concern of environmental pollution, starch-based bioplastic known as thermoplastic starch (TPS) is commonly used today in packaging materials and some utilizing bio-fillers or fibre in the formulation to enhance the bio-based plastic. TPS is low in density, cheap, and has almost a similar strength value as petroleum-based plastic. Several studies on TPS have been widely and globally conducted for various sources of starch such as cassava, potato, and corn [5–13]. This review presents the recent developments of thermoplastics derived from starch and their potential in the industry. Starch modification is also highlighted, including blending with other naturally derived materials, which seem to further improve the mechanical and physical properties of the resulting bio-composite.

2. Biopolymer

As part of the awareness of global environmental issues concerning plastic waste, bio-based polymers are now promoted as alternatives for replacing petrochemical-based (fossil) polymers to supply increasing demand [9]. This initiative promotes the use of green environmental-based plants or animals in the production of bioplastic-based or bio-based in order to preserve and protect the environment. The worldwide interest for bio-based polymers has increased in recent years due to the desire to replace petrochemical-based polymer materials as well as the need to explore the innovative advances achieved in biotechnology [9]. Therefore, several initiatives have begun with the realization of the importance of producing more cost-effective and eco-friendly products, signifying the presence of numerous developments and production of bio-based materials [13].

Many research works have been previously conducted on biopolymer products derived from natural sources. In general, the animal-based biopolymer refers to the gelatin, which can be extracted from the skin and bone of porcine, bovine, poultry, and fish [14–16]. Meanwhile, the plant-based biopolymer can be characterized as starch, cellulose, hemicellulose, and lignin. Starch is the main component of the biopolymer structure, which acts as a matrix, and, with the addition of plasticisers, the flexibility and processability of the resulting thermoplastic starch (TPS) are further enhanced. Many biopolymer products have utilized natural fibres as fillers in their composition. Natural fibres, mainly from agricultural industries, were collected and tested with TPS to become bio-composite materials. Different types of fibres obtained from diverse agro-industrial residues have been investigated, such as bamboo bagasse [17], cassava bagasse [18,19], grape stalks [20], sugarcane, orange, cornhusk bagasse [3], cogon grass fibre [21], and kraft fibre [22]. Figure 1 presents the bio-based product applications that have been developed and examined.

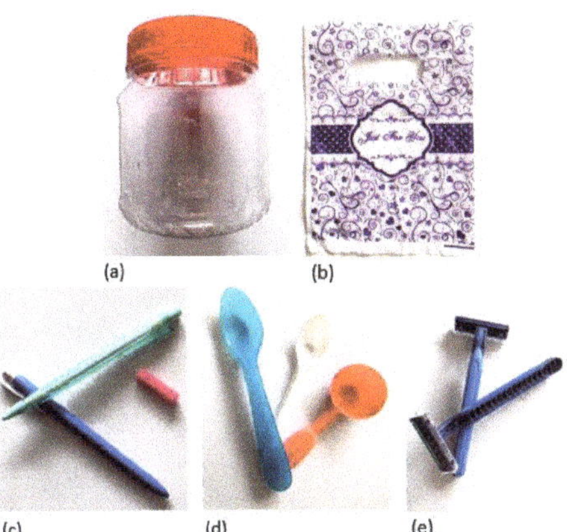

Figure 1. Biopolymer products: (**a**) food container, (**b**) plastic bag, (**c**) pens, (**d**) plastic spoon, and (**e**) shaving razor.

3. Starch

Starch is the main component of the biopolymer structure that belongs to the polysaccharides group and is considered the second most crucial renewable sources after cellulose. It is inexpensive and completely biodegradable, generating great interest to make it into a component of bioplastic. Starch is known to be a versatile material since it can be converted into chemicals such as ethanol, acetone, and organic acids used in synthetic polymer production, such as polylactic acid (PLA) [23] and can also be converted to TPS with the aid of plasticizer under a shear temperature condition [1]. Pure starch is white in appearance. It is a tasteless and odourless powder that has been one of the most abundant natural polysaccharides synthesized from plants [17] with starch worldwide production of more than 50 million tons per year [24] and cassava starch is the main contributor among other sources where it could produce for about two to four times more than that of yam bean, taro, and about 10 times higher than that of sweet potato [25]. However, Bergthaller & Hollmann [26] reported that corn starch is the crop most produced globally for about 80% of total world starch production and followed by wheat, tuber, and cassava where United States dominantly cultivates the highest starch production besides the European Union (EU) and other countries. Starch sources are derived from plants (such as corn, potatoes, cassava, and wheat) and stored as carbohydrates [27]. Starch is composed mainly of two homopolymers of D-glucose amylase, a mostly linear α-D(1, 4′)-glucan, and branched amylopectin. They have the same backbone structure as amylose but with many α-1, 6′-linked branch points [28].

The hydroxyl group in the starch structure exhibit the reaction of oxidation and reduction processes as well as the formation of hydrogen bonds, ethers, and esters. Native starches contain about 70–85% amylopectin and 15–30% amylose. The presence of amylopectin is mainly responsible for the crystallinity of the starch granules [29]. However, Domene-López et al. [4] and Sahari et al. [30] reported that the crystallisation ability is influenced by the molecular weight and phosphate monoester content in starch as these features contribute to lower starch chain rearrangement. It was stated that starch with high amylose content can have low molecular weight and a relatively more linear structure than those with a high content in amylopectin. Thus, starch that contains a high amylose value could facilitate further crystallisation processes as compared to starch with high amylopectin content. Relative molecular weight of these molecules depends on the botanical origin of the starch and might influence the final bio-composite properties' application [31].

4. Thermoplastic Starch

Neat starch displays some disadvantages due to its high solubility in water, brittleness, poor melting point, and lower mechanical properties as compared to materials made of synthetic polymers. To enhance the properties of starch, various physical or chemical modifications, such as plasticisation, blending, derivation, and graft copolymerisation were investigated [1,17,32–34]. TPS is made by applying mechanical and thermal energy onto the starch granules by adding plasticizer. Plasticisers play a vital role in the preparation of thermoplastic starch as they improve starch behaviour by reducing internal hydrogen bonding in between the polymer chain while increasing free volume. This, in turn, increases flexibility and processability and promotes molecular chain mobility [32,35]. The effectiveness of plasticisers depends on the similarity of the polymer used.

Plasticisers can be found in several forms, such as glycerol, sorbitol, urea, fructose, sucrose, and glycol [36]. However, the most commonly used plasticisers are from the polyol group, namely, glycerol and sorbitol [32]. Several research works have been recently conducted to explore certain ionic liquids (ILs) as new starch plasticisers. Ionic liquids include 1-ethyl-3-methylimidazolium acetate ([emim$^+$][Ac$^-$]) [35,37] and 1-butyl-3-methyllimidazolium chloride ([bmim$^+$][Cl$^-$]) [33]. Besides, a recent study employed fried sunflower oil as a plasticiser in thermoplastic starch composites. An improvement is shown in starch-based material properties, proving to be the most environmental solution for bio-composites [38].

A comprehensive study was conducted by Demash and Miyake [35] regarding the effect of four different types of plasticisers on anchote *(Coccinia abyssinica)* starch film. Glycerol, 1-ethyl-3-methylimidazolium acetate, sorbitol, and tri-ethylene glycol were utilised at concentrations of 30% and 40% w/w in the starch mixture. The thermoplastic film samples were then dried overnight in the oven at 50 °C before they were kept in a desiccator for at least two days for film characterisation. The prepared film was transparent, homogenous, and flexible. The same sample preparation was reported by Reference [32]. The obtained results revealed that there is an increase in thickness for all samples as plasticiser concentrations increased. The sample with 40% tri-ethylene glycol plasticiser exhibited the highest film thickness with a value of 0.26 mm, thus, reflecting the lowest density value of 0.88 g/cm^3 among other samples. It was found that the water solubility of glycerol and tri-ethylene glycol plasticised film had reduced from 32.57 g/cm^3 to 20.97 g/cm^3 and 34.19 g/cm^3 to 18.92 g/cm^3, respectively. According to Reference [1], the same behaviour was observed when glycol was used as a plasticiser. This may be due to the formation of strong hydrogen-bonds with starch, which restrain water molecules to combine with plasticized anchote starch. Plasticisers presented an improvement in the film since it had less affinity to water. However, sorbitol and 1-ethyl-3-methylimidazolium acetate plasticised film displayed a different result of water solubility. Water solubility increased as the plasticiser concentration increased.

Nevertheless, biodegradability of the film plasticized with ILs does not specify in the study which could be one of the interesting points that need to be highlighted. Across the journal available, there is also limited research that working on the biodegradability of ILs as a plasticizer. However, Rynkowska et al. [39] reported that ILs is non-toxic liquid and is considered as a green alternative to substitute phthalates, commonly used as plasticizers in synthetic material, which is primarily to soften polyvinyl chloride and is harmful to humans. Besides, ILs portray unique properties such as non-volatility, low toxicity, easy to handle, inflammable, and goods in ions conductivity [40]. An interesting study made by Sudhakar and Selvakumar [41], investigated on biodegradable properties of the plasticized chitosan and starch blends electrolyte film, found lithium perchlorate (LiClO$_2$) using an activated sludge degradation method. The samples are pre-weighed before being immersed in the sludger under an aerobic condition for 5, 10, and 15 days and taken out to be weighed again after being cleaned and dried in an oven at 75 °C. The result obtained showed a degradation range between 6.2–16% for 0, 0.5, 1.0, 1.5, and 2.0 of lithium perchlorate. It presents that the percentage of weight lost increased as the amount of lithium perchlorate was added. This might be attributed to the presence of lithium perchlorate that promoted free volume in the polymeric matrix and, thus, enhance the biodegradability rate [41]. On that note, it can be concluded that ILs are safe for the environment and also for humans.

4.1. Thermoplastic Cassava Starch

In the production of biopolymers, starch is mainly used as a matrix or resin in bio-composite structures. Throughout past decades, numerous types of natural starch have been investigated, such as cassava starch, corn starch, sugar palm starch, and much more. However, cassava starch is the highest contributor in terms of productivity yield compared to other sources of starch [25]. Some research works reported cassava starch as tapioca starch, depending on the author. The name 'cassava' is generally applied to the roots of the plant, whereas tapioca is the name given to starch and other processed products [42]. Table 1 presents the application of cassava starch as a thermoplastic in bio-composites.

Table 1. Thermoplastic cassava starch composite.

Type of Starch	Type of Filler/Polymer	Potential Application	Ref.
Cassava	Kraft	Biodegradable tray with chitosan coating	[6]
Cassava	Orange, sugarcane, malt bagasse	Biodegradable tray, packaging material	[3]
Cassava	Cogon grass	Biodegradable material	[7,25]
Tapioca	Bamboo	Biodegradable 'green' plastic	[17]
Cassava	Sugar palm	Biodegradable polymer	[6]
Cassava	Grape stalks	Food packaging plastic	[20]
Cassava	Cassava bagasse	Food packaging plastic	[18]

Recent studies have been accomplished on polymer composites of treated oil palm mesocarp fibres (OPMF) and thermoplastic cassava starch properties prepared by using the screw extrusion rheometer method [6]. Two types of OPMF were studied: raw and alkaline treated OPMF mixed with thermoplastic cassava starch (TPS) at different weight loadings (5–20% for raw and 5–20% for alkaline treated OPMF). It was stated that OPMF is composed of a large source of lignocellulosic material and can be used as bioethanol production and as reinforcement in polymer composites [43]. The addition of OPMF was expected to lead to an increase in tensile strength due to the great adhesion between the fibre and matrix, contributing to their chemical affinity. Chemical treatments, such as alkali treatments, were employed to enhance the fibre-matrix interaction by removing lignin, oils, waxes, and silica, which have collectively exhibited a fibrillation process during the removal of hydrophilic components.

The obtained results from the Scanning Electron Microscopy (SEM) for raw and alkaline-treated OPMF and Energy Dispersive Spectroscopy (EDS) are presented in Figures 2 and 3. Figure 3 displays the silica removal, observed by craters on the surface, after alkaline treatment was conducted. A few thorny structures had adhered onto the surface, indicating the presence of amorphous substance or silica, which remained in the OPMF. The raw fibres presented an average length of 440–1000 µm and a diameter of 100 µm. These results validate the previous research conducted by Reference [44], which stated the influence of alkaline treatment on the removal of silica. Figure 4 displays a cross-section of TPS composites with raw and alkali-treated fibres. It is clear that the polymer matrix covered the fibres and promoted a strong adhesion between them. The fibre-polymer interaction was favoured by the residual silica presence, whose oxygen atoms interact very strongly with the TPS hydroxyl groups.

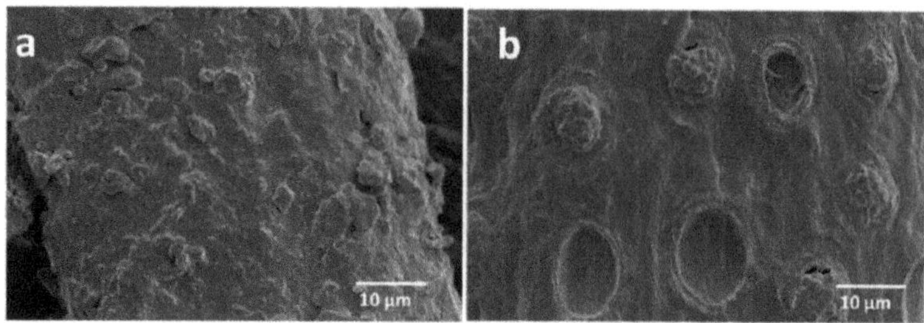

Figure 2. Scanning Electron Microscopy (SEM) of: (**a**) raw oil palm mesocarp fibres (OPMF), and (**b**) alkaline treated OPMF [6].

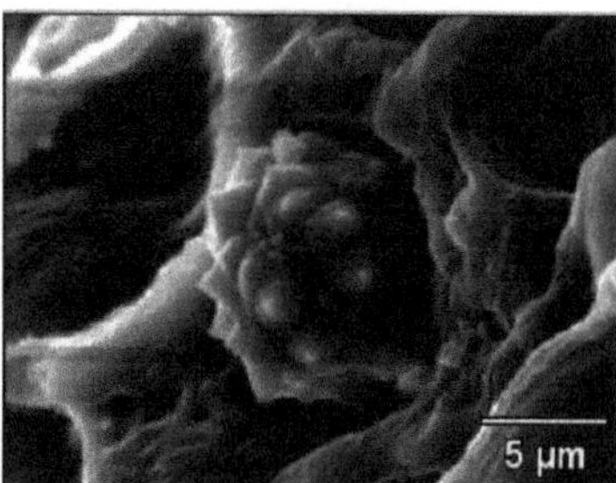

Figure 3. Energy Dispersive Spectroscopy (EDS) of OPMF [6].

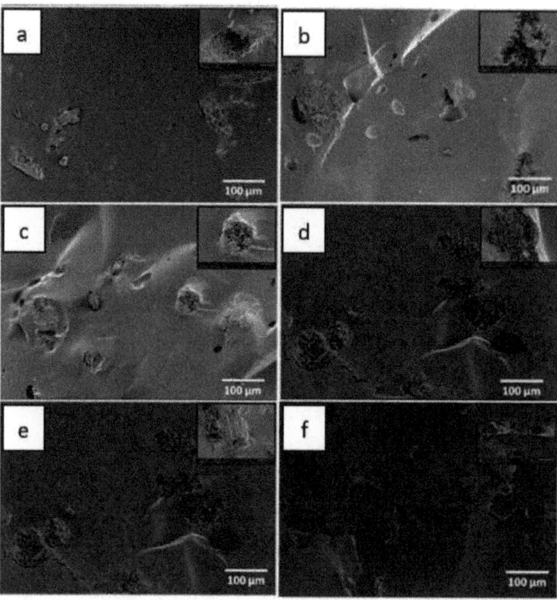

Figure 4. Micrography of (**a**) thermoplastic starch (TPS) 5% raw fibre, (**b**) TPS 5% alkali treated fibre, (**c**) TPS 10% raw fibre, (**d**) TPS 10% alkali treated fibre, (**e**) TPS 20% raw fibre, and (**f**) TPS 20% alkali treated fibre [6].

4.2. Thermoplastic Corn Starch

Corn starch is seen as a promising source of biopolymer matrix that is environmentally friendly and suitable for replacing petroleum-based plastics. Pure corn starch is similar to other natural starches, which have some drawbacks such as high-water sensitivity and poor mechanical behaviour. Thus, modification of starch is required by introducing plasticisers in the mixture. Plasticised starch properties have enhanced abilities. Thermoplastics from corn starch have been widely explored and various modifications were accomplished to improve the properties of the resulting TPS. The recent studies conducted had utilised waxy corn products in their works. It was reported that waxy corn produces starch that is nearly 100% amylopectin [45]. Table 2 presents the recent studies conducted on thermoplastics from corn starch composites.

Table 2. Thermoplastic corn starch composite.

Type of Starch	Type of Filler/Polymer	Potential Application	Reference
Corn	Sugarcane	Green material for packaging	[46]
Corn	Microalgae	Bioplastic, i.e., packaging, catering products, electronic devices	[47]
Corn	Cassava, ahipa peels and baggase	Bio-based composite	[19]
Corn	Talc nanoparticles	Bio-nanocomposite food packaging	[48,49]
Corn	Nanocrystalline cellulose	Bio-film and bio-nanocomposite	[50]
Corn	Sunflower seed fried oil	Potential natural plasticizer	[38]
Corn	PLA blends	Bio-degradable polymer	[51]
Corn	Cornhusk/sugar palm	Hybrid bio-composite	[52]

Fitch-Vargas et al. [46] studied the modification of corn starch bio-composites reinforced with sugarcane fibre. A process called acetylation of starch and fibres was performed, resulting in the formation of acetylated corn starch bio-composites reinforced with acetylated sugarcane fibre and led to enhancement of the resulting bio-composite's behaviour, such as improved processability and compatibility [51]. Different loadings of acetylated sugarcane fibre contents (FC, 0.0–20.0%) and glycerol contents (GC, 20.0–30.0%) were examined in this work. Figure 5 and Table 3 presents the mechanical test results of the samples. With reference to the tensile result of synthetic polymers, polystyrene (30–55 MPa) and polypropylene (25–40 MPa), it was found that the highest tensile value (35 MPa) is achievable with the composition of GC 24–28% and FC 10–15%. This was attributed to the effect of glycerol opening the polymeric 3D structure, which allowed fibre incorporation. Besides that, the presence of acetyl group may function as a spacer between the starch chain for cellulose fibre to fill in and improve the bonding interaction [53]. It was observed that exceeding 12% FC leads to a decline in strength due to poor fibre distribution. The highest elongation value was obtained when 24–28% GC and 5–15% FC were mixed together. This reflects on the great number of linkages formed due to the fibre's surface chemical treatments conducted, which allowed the active fibres' hydroxy group to react with the matrix. Thus, this enhanced the mechanical interlocking on the rough surface, forming stronger secondary bonds.

On the other hand, a moisture adsorption test is carried out to identify the sample's affinity toward the surrounding moisture condition. The samples are placed into two different relative humidity (RH) desiccators conditions at 53% RH and 100% RH. Weight gained among the samples are taken until constant weight are found, indicating that the samples have reached an equilibrium state. As per Figure 6, the finding shows that, as the GC increased, the moisture absorption (MA) value increased where the lowest MA are obtained at GC less than 25% throughout the FC range. The same results are also found in a water solubility (WS) test. This could be attributed to the formation of a stronger hydrogen bond between the polymeric matrix and has restricted the entrance of surrounding moisture. Besides, this might be due to the incorporation of fibre that is low hygroscopic in nature and has further enhanced the barrier properties. However, MA and WS are seen to increase considerably when the GC loading increased to more than 25%. It is reported that higher GC has the ability to reduce the hydrogen bonding and, therefore, has led to an increase of the intermolecular spaces due to material swelling and weakening of the bonding forces between the water molecule and hydrophilic functional group. In contrast, others have reported on the reduction of MA by increasing glycerol in plasticized starch and reduction of MA by increasing glycerol and fibre content in the resulting bio-composite (Sahari et al. [1] and Jumaidin et al. [25], respectively). They stated high glycerol and fibre contents incorporated have shown better resistance toward moisture due to stronger hydrogen formed, which restrained the combination between the water molecule and the sample. The journal focuses on how GC may affect the result and lack of more explanation in terms of an FC interaction that also has many contribution effects toward the obtained results.

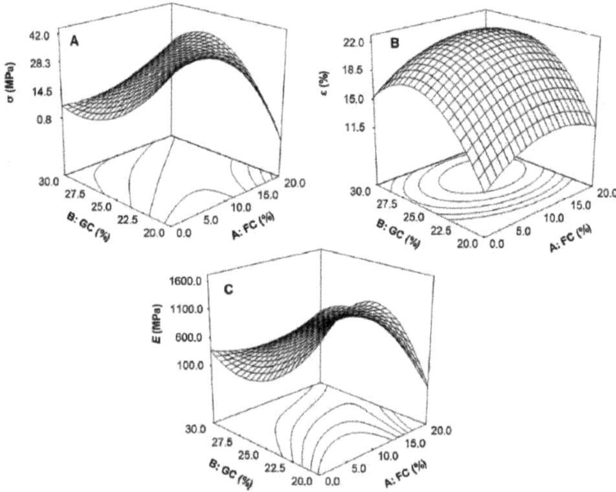

Figure 5. Effect of acetylated sugarcane fibre content (FC) and glycerol content (GC) on: (**A**) tensile strength (σ), (**B**) elongation (ε), and (**C**) young modulus of acetylated starch-based bio-composites reinforced with acetylated sugarcane fibre [46].

Table 3. Mechanical test result [46].

Test Type	GC (%)	FC (%)	Result
Tensile Test, σ	24–28	10–15	Highest σ value achieved at 35MPa.
Elongation, ε	24–28	5–15	Highest ε value achieved at 21.7%
Young Modulus, E	24–28	10–15	Highest E value achieved at 1433.8 MPa

4.3. Thermoplastic Sugar Palm Starch

To date, numerous research works utilising sugar palm starch in the production of bio-composite materials as an alternative to petroleum-based polymers have been accomplished. Natural starch from sugar palm trees was found to be abundant within the tree trunk [54]. Many authors had mentioned the multifunctional uses of the sugar palm tree. The extraction of sugar palm starch is usually done on unproductive trees [29]. Whether the tree is considered mature enough or due to excessive microbial attacks, the tree is unable to normally grow in a good condition. Thermoplastic starch derived from sugar palm trees was successfully developed in the presence of biodegradable glycerol as a plasticiser. Table 4 shows the utilisation of sugar palm starch (SPS) as a matrix in bio-composites.

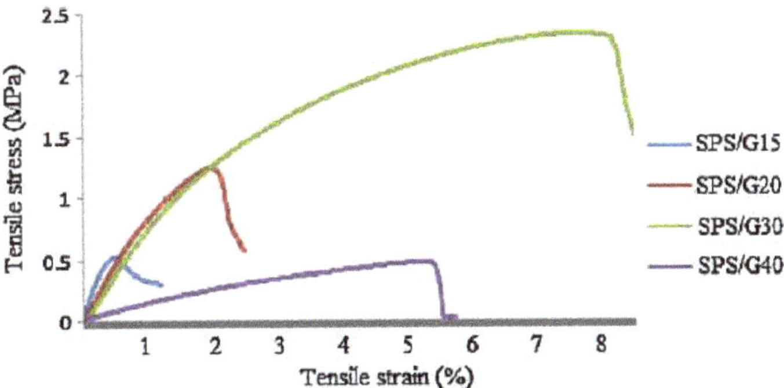

Figure 6. Tensile stress-strain of plasticized sugar palm starch (SPS) [1].

Table 4. Thermoplastic sugar palm starch composite.

Type of Starch	Type of Filler/Polymer	Potential Application	Reference
Sugar palm	-	Biodegradable material	[1]
Sugar palm	-	Biodegradable packaging film	[32]
Sugar palm	Agar blends	Bio-based polymer	[55,56]
Sugar palm	Agar blends, seaweed	Bio-based polymer	[57]
Sugar palm	Sugar palm fibre	Bio-nanocomposite material, food packaging	[52,53,58]
Sugar palm	Agar blends, sugar palm fibre, seaweed	Hybrid bio-composite	[36]

Sanyang et al. [29] briefly explained the processes involved for extracting sugar palm starch from the sugar palm trunk. The starch extraction process starts with the unproductive sugar palm trees that are first cut down. Next, the tree trunk is divided lengthwise, and all the woody fibres are removed. The inner core stem is then cut out to obtain the mixture (off-white in appearance). The collected fibre and starch mixture subsequently undergo a washing process using water, and then thoroughly kneaded by hand. Next, the mixture is sieved, allowing water and starch granules to flow through. The filtered mixture is left to allow granule suspension. The white powder is then left in the open for a specified time. Finally, the white powder is dried in an oven for two days. In this research, it was reported that one sugar palm tree can yield 50–100 kg of starch.

Thermo-mechanical studies have been conducted by Sahari et al. [1] in an attempt to produce bioplastic using SPS as a composite matrix at different glycerol loadings from 15–40 $w/w\%$. The results from this work revealed that, from the fracture surface analysis, as the glycerol content increased, acquiring a smooth surface also increased. At 15 $w/w\%$ glycerol loading, a brittle surface was the outcome, while an exceptionally smooth surface appeared at 40 $w/w\%$ glycerol loading. The glycerol, which acts as a plasticiser, effectively reduced the internal hydrogen bond while increasing the intermolecular spacing. This resulted in the appearance of a smooth surface.

In terms of the product's mechanical properties, as illustrated in Figure 6, it was found that SPS/G30 and SPS/G40 showed the typical pattern of rubbery starch plastic materials due to the linear increase in tensile stress at a low strain, followed by the curve toward the strain axis until failure occurred. The addition of plasticisers at 30 wt% and 40 wt% overcame starch brittleness and improved flexibility.

5. Thermoplastic Starch Blends

Modifying thermoplastic starch can be accomplished through blending, which widens its application at lower cost. Compatibility of blends must be tested to examine the enhancement of the composite's properties such as water resistance, high tensile strength, and high tensile modulus. Blending is also important for modifying bioplastics in terms of biodegradability, enabling them to decompose easily through natural processes so as to cater to the issue of limited disposal technology. Starch blending with a polar polymer containing hydroxyl groups such as polyvinyl alcohol, ethylene copolymer, and partial hydrolysed vinyl acetate have been prepared since the 1970s [23]. Blending is one of the most promising alternatives to make starch useful as a polymer in the replacement of other plastics, and the fast progress occurring in this field is attested by several reviews that have been recently published [59–64].

5.1. Starch/Polyvinyl Alcohol (PVA)

PVA is a synthetic biodegradable polymer that has the advantages of good film formation, strong conglutination, high thermal stability, and good gas barrier properties [65]. PVA is manufactured by the polymerisation of vinyl acetate monomer into polyvinyl acetate (PVAc), followed by the hydrolysis of the acetate groups of PVAc to PVA [64]. PVA is one of the major polymers used in the industry and, therefore, huge amounts of PVA are produced yearly. The world production of PVA is about 650,000 tons per year [64]. The presence of PVA in a blend increases the mechanical strength, water resistance, and weather resistance of the blend [65]. Gelatinisation is the most common method of blending starch with PVA.

The compatibility of PVA and starch enables them to form a continuous phase at blending [28] even though the properties of the blends deteriorate as starch content rises, causing phase separation during blend preparation. To improve the compatibility between PVA and starch, the addition of suitable plasticisers, cross-linking agents, fillers, and compatibilizers were examined [65]. Both PVA and starch can be plasticised into a thermoplastic material, regularly using the casting method and glycerol in an aqueous medium [66]. Both starch and PVA are biodegradable in several microbial environments. However, the biodegradability of PVA depends on its degree of hydrolysis and its molecular weight [63]. Both PVA and starch can be plasticized into a thermoplastic material, regularly using the casting method and glycerol in an aqueous medium [66]. Both starch and PVA are biodegradable in several microbial environments. However, the biodegradability of PVA depends on its degree of hydrolysis and its molecular weight [67].

5.2. Starch/Poly Lactic Acid (PLA)

PLA, a biodegradable polyester produced from renewable resources, is used for various applications such as biomedical, packaging, textile fibres and technical items. PLA is industrially obtained through the polymerisation of lactic acid through the ring opening polymerisation (ROP) method [62]. PLA is, by far, the most commercially developed, reaching an annual production volume of approximately 200,000 tons [61]. However, PLA is not environmentally biodegradable as it requires a proper composting facility that applies a heating temperature of 60 °C and it must also be exposed to special microbes that will digest and decompose the material. However, Hatti-kaul et al. [61] reported that PLA is composted of 100% L-lactide unit that takes 110 weeks to degrade, which can be reduced dramatically up to 10 weeks while adding 50% D-lactides unit and can further reduce the degradation time to 3 weeks when 25% glycolic acid is added in the formulation. Some other disadvantages of PLA include low flexibility, ductility, and impact resistance.

To improve PLA's flexibility and impact resistance, numerous plasticisers are incorporated, such as poly (ethylene glycol), glycerol, glucose monoesters, citrate esters, and oligomers [68]. TPS, as a blend component for PLA, offers important advantages in terms of cost, properties, and biodegradability [62]. The hydrophilic characteristics of starch and the hydrophobic features of PLA cause low miscibility between the two compounds. For this reason, good melt-blending techniques and the addition of compatibilizers are required to increase a successful interaction, i.e., amphiphilic molecules or coupling agents [62]. Poly (hydroxyester ether), methylene diphenyl di isocyanate (MDI), PLA-graft-(maleic anhydride), PLA-graft-(acrylic acid), PLA-graft-starch, and poly(vinyl alcohol) are all used as compatibilizers in this blend [69].

5.3. Starch/Polybuthylene Succinate (PBS)

PBS can be found commercially as a thermoplastic-based polymer, which provide a high degree of crystallinity and takes a long time to degrade. PBS is a synthesis of polycondensation succinic acid and 1,4-butanediol where succinic acid can be obtained from either a monomer derived through a petroleum system or bacterial fermentation. Meanwhile, 1,4-butanediol is derived from fossil fuel extraction from formaldehyde or acetylene [70]. PBS provides improvement in a composite characteristic with strong impact strength, better chemical resistance, and high thermal stability [65]. Incorporation of starch blends with PBS has produced more flexibility and elasticity to the resulting material, which could be promoted as alternative food packaging. The blends could speed up degradation time due to the presence of starch that contains hydrogen bonding and is able to promote free spaces for degradation to occur [71]. However, Hatti-kaul et al. [61] stated that the addition of long chain dicarboxylic acid in the PBS homopolymer will form polybutylene succinate adipate (PBSA) in which its degradation time is less than PBS. PBSA is an aliphatic thermoplastic copolymer. PBSA annual production is estimated around 97,000 tons and is commercially available as a thermoplastic polymer [61].

6. Thermoplastic Starch Incorporated with Natural Fibre

6.1. Natural Fibre

The implementation of natural fibre as a promising alternative for replacing synthetic fibre in the composite industry has been increasing in the past decades. Synthetic fibres such as glass, carbon, and aramid reinforced with polymer matrix materials provide advantages of high stiffness and strength-to-weight ratio in numerous applications as compared to conventional construction materials [72]. Changes from the dominant usage of synthetic fibre to natural fibre indicate the rise of awareness among people around the world, regarding the negative environmental impact that synthetic fibre brings, which may disturb human health conditions. The utilisation of natural fibre provides advantages as compared to synthetic fibre in terms of favourable tensile properties, reduced health hazard, acceptable insulating properties, low density, and decreased energy consumption [27].

The production of synthetic polymers utilise a large quantum of energy, which produces environmental pollutants during the production and recycling of synthetic composites [72]. Table 5 illustrates a comparable study between natural and synthetic fibres, clearly displaying renewable, recyclable, and biodegradable properties as part of natural fibre properties. It should be noted that natural fibre properties vary depending on the source of fibre itself, including it species, environmental climate condition, geographical location, the process for preparing the fibre, etc.

Table 5. Advantages of natural fibres compared to synthetic fibre [29].

	Natural Fibres	Synthetic Fibres
Density	Light	Twice natural fibres
Cost	Low cost	Higher than natural fibres
Renewability	Yes	No
Recyclability	Yes	No
Energy Consumption	Low	High
Distribution	Wide	High
CO_2 neutral	Yes	No
Health risk when inhaled	No	Yes
Disposal	Biodegradable	Yes, not biodegradable

Natural fibres reinforced composites are also applied as fillers in a matrix to provide high strength and stiffness on a weight basis to a composite product [73]. Fibre reinforcement added to the matrix phase (either natural resin, thermoset, and thermoplastic polymers) will act as a glue or binding agent to the composite formation [74]. The natural resin mentioned includes wheat starch, corn starch, potato starch, etc. [27]. Thermoset polymers include epoxy, polyester, and phenolics [74], and thermoplastic polymers include polycarbonate, polyvinyl chloride, and nylon [75]. Natural fibres are classified based on their origin in nature, either from plants, animals, or minerals. Figure 7 presents the classification of natural fibres. Most natural fibres come from plants and are composed of cellulose, thus, making the fibre hydrophilic in nature [17]. They are also called lignocellulosic fibres since their cellulose fibrils are embedded in the lignin matrix [29].

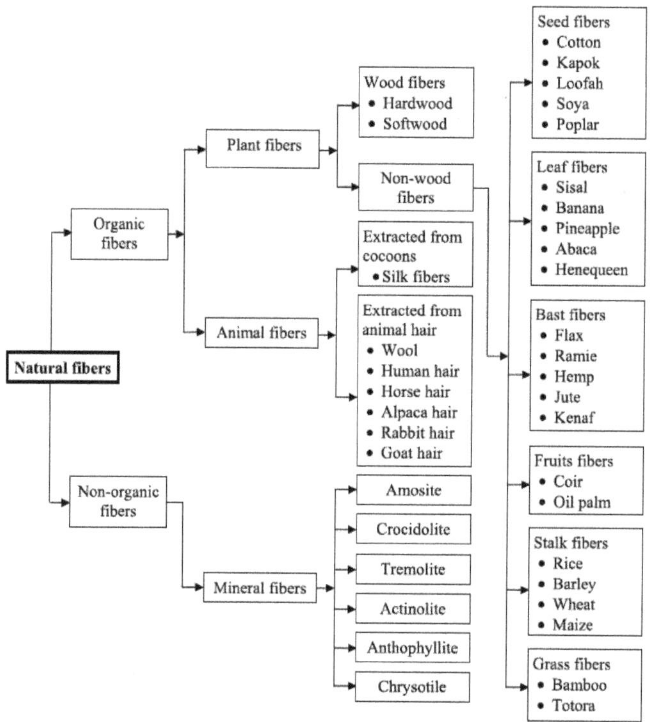

Figure 7. Classification of natural fibre [72].

6.2. Cassava Bagasse Fibre

Cassava, also known as *Manihot esculenta C.*, is a root and tuber crop found in many tropical countries. Therefore, it is used as the main source of food. It is estimated that cassava could have a yield potential of up to 17,000 kg of starch per hectare per year if it is cultivated in a suitable environment with organised farming practices [25]. Cassava is ranked as the fifth most widely used starch source in the world and the third among food sources consumed in tropical countries [18]. Cassava bagasse is a fibrous solid residue obtained as a by-product of the industrial cassava starch production, which has undergone the separation of starch and fibre. It was reported that cassava bagasse residual fibre may consist of 38% cellulose and 37% of hemicellulose and lignin [76]. The cultivated cassava tuber for around 250–300 tons can yield a high moisture content of by-products (around 1.6 ton of solid peels and 280 tons bagasse). These solid wastes are generally discarded in the environment without any treatment [19]. This alarming issue has led to the awareness of utilising agricultural waste by converting them to other useful materials.

Many studies had focused on exploiting waste by utilising cassava bagasse as a filler or reinforcing agent to enhance the strength and mechanical properties of material products. This is especially conducted for the production of bio-based or bio-degradable materials, which have been steadily increasing. These by-products are low in market value and possess good properties with a wide range of applications, such as reinforcing agents or in the production of organic acid, bio-degradable packaging, nano particles, nanofibres, ethanol, bio-fuel, lactic acid, etc. [76]. Paula et al. [18] had shown that, when cassava fibres were added to corn starch, the film strength improved by up to 37.5%. This was attributed to the good intermolecular interaction between the starch and reinforcement. The elongation at break also reduced due to the possible agglomerates formed within the films. The same results were obtained by García et al. [19], verifying an increase in tensile strength as well as a decrease in elongation at break when fibres were added to the starch.

6.3. Sugarcane Fibre

Recently, the valorisation of sugarcane waste for low-cost building products has been increasing due to its high potential. Since there is a vast global production of sugarcane, especially in some tropical and subtropical developing countries, high amounts of sugarcane fibres are produced as by-products [77]. Fortunately, this provides numerous job vacancies with regard to this industry. It was reported that the current largest sugarcane world producer is Brazil where its yearly production exceeds 640 tons [78]. Previously, India was the largest with a yearly production of more than 100 million tons of bagasse [79]. Sugarcane bagasse is the residue obtained from the sugarcane industry after extracting the sugarcane juice. It has also become a source of fuel [46]. This by-product can be a source of biomass products and, therefore, has caught great research attention due to its vast availability, ecological features, and renewable characteristics found in the sugarcane fibre. It is seen to have great potential as a bio-composite product and can be used as a reinforcing agent.

With regard to the incorporation of natural fibres such as sugarcane into starch-based composites, numerous studies have reported and demonstrated significant improvements in the resulting bio-composite products. Fitch-Vargas et al. [46] stated that the addition of sugarcane fibre in the composite mixture has enhanced the product's tensile strength as well as reduced water affinity, attributing to the strong interaction adhesion between the matrix and filler. However, Farias et al. [80] mentioned that this is not always the case since some modifications on the fibre itself may be required depending on the type of collected sugarcane. The discrepancies in the obtained results are influenced by the species of sugarcane, the age of sugarcane, and the surrounding climate. Hernández-olivares et al. [81] utilised sugarcane as a reinforcing component in the construction of cement, Ordinary Portland Cement (OPC), reporting its potential of improving composite durability. The authors demonstrated that incorporating reasonably high-volume fractions of sugarcane bagasse can produce feasible manufacturing composite materials for building construction since they exhibited enhanced physical and mechanical properties.

6.4. Bamboo Fibre

Growing interest on modification of surface fibre have been implemented to increase the reinforcement strength for the bio-composite. Surface modification treatments can be classified into three categories: chemical, physical, and biological [82,83]. One chemical method most used for the modification of vegetable fibres is mercerisation. This method involves modifying the structure and chemical composition of plant fibres using an aqueous solution of sodium hydroxide (NaOH) [84]. With this treatment, the removal of surface impurities and the occurrence of fibrillation are possible by obtaining a fibrous material with greater surface area and smaller diameter. This produces an increase in the fibre's tensile strength and mechanical properties for the resulting composite. A physical process is the application of a cold plasma treatment that is considered a promising environmental method utilizing non-harmful gases, such as methane, argon, or helium [80]. This method allows the formation of free radicals and the polymerisation of the material. The plasma treatment acts on the chemical structure of fibres and their crystallinity index. Nevertheless, ozone treatment has emerged as an eco-friendly method for surface modification [85]. It is used to oxidize the lignin to stimulate the reactivity through the increase of hydroxyl group content and to produce low molecular weight of the soluble compound.

The influence of surface treatments on bamboo fibres as reinforcement agents in TPS were conducted by Jhon et al. [34]. In this study, three different treatments were applied: the mercerisation treatment (MT), which modifies the structure and chemical compositions using sodium hydroxide (NaOH), the cold plasma treatment (PT), which conducts physical modifications using methane or helium gas, and the ozone treatment (OT), which produces soluble compounds low in molecular weight by oxidising lignin and hemicellulose. From Figure 8, it was found that the MT samples displayed a significant increase in density value (about 60%) as compared to the other treatments. As reported by Campos et al. [6], the alkaline treatment causes the removal of large parts of the amorphous substances, which includes lignin, polysaccharides, and waxes, resulting in the low molecular weight of the sample. The PT sample showed no significant difference and almost the same result was obtained when compared to untreated bamboo fibre (UT). For the water adsorption test, all treated samples exhibited an increase in the water adsorption percentage. Some origins of fibre that are hydrophobic in nature were associated with the presence of lignin, hemicellulose, and pectin [83]. Alterations to the substance composition had caused the formation of new behaviours, such as an increase in hydrophilic properties.

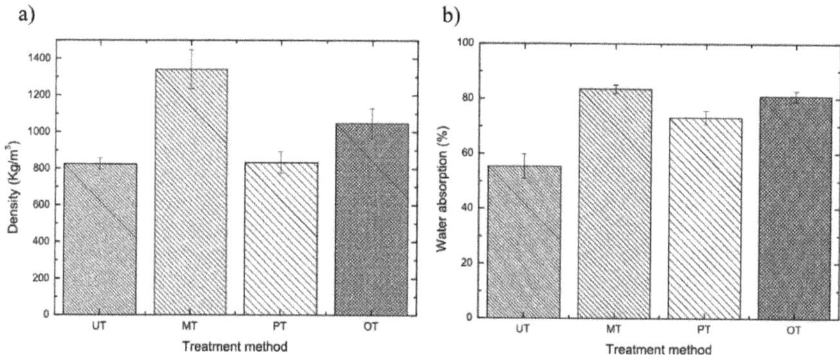

Figure 8. Effect of treatment on: (**a**) the density of bamboo fibre and (**b**) water adsorption of bamboo fibre [34].

7. Application of Starch-Based Biopolymer

The development of starch blending is more interesting since it can substitute an older material while exhibiting the same properties. Table 6 highlights the main applications of a few widely used starch-based biopolymers. Biodegradable packaging applications have drawn remarkable attention for biopolymer utilisation in comparison to other areas because they are higher in significance. Biodegradable plastics are increasingly drawn toward addressing environmental impact in several particular applications, such as single-used plastic, greenhouse gas (GHG) emissions, and overload of plastic waste. These new implementations and innovations of bio-based plastics shown on increased awareness toward a better sustainable environment in the future, as they provide less of a disposal technique and degrade easily.

Table 6. Applications of biopolymers.

Blends	Application	Reference
Starch/plasticizer	• Biodegradable packaging • Starch based film material • Disposable eating utensils	[1] [32] [38]
Starch/PVA	• Water-soluble laundry bags • Biomedical and clinical field • Replacement of polystyrene	[61] [66] [63]
Starch/PLA	• Biodegradable tray • Electronic devices, pharmaceutical	[68] [62]
Starch/PBS	• Packaging materials, fisheryAutomotive	[61,65]
Starch/natural fibre	• Food packaging • Biodegradable material	[20,25] [22,58,68]

Sahari et al. [1] discovered on the potential of plasticized sugar palm starch (SPS) in a biodegradable packaging application. Unlike other starch, such as corn, cassava, or potato starch, sugar plam starch is a passed decade discovered on its true potential that is mostly as par as other starch. With the addition of glycerol as a plasticizer into SPS, it produced a biocomposite under a high temperature condition. The obtained mechanical result shows an increment as the glycerol content increases up to 30 $w/w\%$ with a maximum tensile value of 2.42 MPa. Besides, a reduction of transition temperature and water absorption as glycerol are added, indicating that the biocomposite has reduced its brittleness as glycerol loading increased. Meanwhile, Sanyang et al. [32] studied the different plasticizer affect SPS mechnical properties in an attemp of development biofilm, utilizing sorbitol, glycerol, and glycerol-sorbitol in varying conditions (0, 15, 30, and 45 $w/w\%$). The results collected show the storage modulus of plasticized SPS decreased as the plasticizer concentration was added from 15–45%, which reflect the stiffness of the material is reduced. Glycerol plasticized film portrays the highest degree mobility of the polymer chain among other plasticizers.

Polymer films made from poly-(vinyl alcohol), low-density polyethylene, poly-(vinyl alcohol), or polybutylene are used for mulch manufacturing [61]. These films are altered in a specific manner, such that they are permitted to go through degradation only when the crop-growing period is complete, either through the aid of soil micro-organisms or through the addition of certain particulate matter, which promotes film cessation [31]. A similar trait was observed in polycaprolactone for making agricultural plant containers. These containers tend to biodegrade over a significant period of time, allowing tree seedlings to adequately sprout. Implementation of natural fibre as a promising alternative to synthetic fibre in the composite industry is seen to be increasing in past decades since the resulting bio composite exhibits favourable properties. The incorporation of natural fibres can further enhance the mechanical properties of materials and lead to the high possibility of substituting synthetic polymers.

Oliveira et al. [68] studied the application of a biodegradable tray utilizing starch and PLA blends coated with beeswax by flat extrusion, calendaring, and hot-pressing under variance of beeswax coating emulsion 1, 2, and 3 g wax/100 g solutions. The mechanical result shows beeswax concentration (BC) 1 g/100 g solutions showing the highest tensile strength that might be attributed to the better adherence of the coating that has improved material strength properties. The Young modulus value is reduced from BC1 to BC2 and BC3 that could be due to lower tensile value and reflect elongation at no distinct value. Water vapor permeability (WVP) is an important characteristic in a food tray application and BC1 shows the lowest WVP value as compared to others. The application of the food tray is possible and can be produced at a large scale due to good tensile result and less affinity toward moisture.

PBS has been exhibited as mulch film, packaging, and flushable hygiene products and is considered as a non-migrant plasticizer for polyvinyl chloride (PVC) [71]. Moreover, PBS is used in foaming and food packaging application. The relatively poor mechanical flexibility of PBS restricts to the applications of 100% PBS-based products. Nevertheless, this can be solved by combining PBS with PLA or starch to enhance the mechanical properties significantly, which promote properties more or less the same to that of polyolefin [70].

Utilization of natural fibre incorporated with advantages as compared to synthetic fibre in terms of tensile properties, less health hazard, acceptable insulating properties, low density, and less energy consumption [27]. A possible waste exploitation were practiced by many researchers, utilizing the cassava bagasse as a material product's filler reinforced agent to enhance the material properties currently increasing. These by-products are found low in market value and can possess good properties such as a reinforcing agent and a wide range of application that can be implemented, such as production of organic acid, bio-degradable packaging, nano particles, nanofibers, ethanol, bio-fuel, lactic acid, and many more [76].

8. Conclusions

This paper conducts a general overview on biopolymers and the potential of starch-derived thermoplastics as substitutes for current petroleum-based plastic. Blending starch with other biopolymers was outlined as a viable alternative to overcome the shortcomings of native starch. However, the degree of compatibility between starch and other biopolymers extensively differs depending on the specific biopolymer. At present, mixing TPS and PLA offers significant advantages in terms of cost, properties, and biodegradability.

Regarding the global environmental issue, biodegradable material properties are vital and should be considered. Although starch/biodegradable blends are a good option for solving environmental issues, their mechanical properties often have an inverse relationship to their degradability. Incorporating natural fibres as fillers in the starch matrix can be another solution. Thus, optimising their mechanical properties requires further analyses.

Author Contributions: Conceptualization, Z.N.D. Validation, R.J. and M.Z.S. Writing—review and editing, Z.N.D., R.J., I.G., R.A.I., and N.H. Resources, R.J., I.G., R.A.I., and N.H. Data curation, R.J. and N.J. Writing—original draft preparation, R.J. and Z.N.D. Writing—review and editing, R.J. and N.H., R.A.I.; Visualization, I.G.; Supervision, R.J. Project administration, R.J. Funding acquisition, R.J., N.H., and N.J. All authors have read and agreed to the published version of the manuscript.

Funding: The authors would like to thank Universiti Teknikal Malaysia Melaka for the financial support provided through grant number RACER/2019/FTKMP-CARE/F00413" and "The APC was funded by Universiti Malaysia Sabah".

Institutional Review Board Statement: Not applicable.

Informed Consent Statement: Not applicable.

Data Availability Statement: The data presented in this study are available on request from the corresponding author.

Acknowledgments: The author is grateful for the help from all staff from Universiti Teknikal Malaysia Melaka and Universiti Malaysia Sabah and Univerisiti Teknologi Malaysia that were involved during this research. In addition, big thanks for the moral support from family and friends during the research that has been conducted.

Conflicts of Interest: The authors declare no conflict of interest.

References

1. Sahari, J.; Sapuan, S.M.; Zainudin, E.S.; Maleque, M.A. Thermo-mechanical behaviors of thermoplastic starch derived from sugar palm tree (Arenga pinnata). *Carbohydr. Polym.* **2013**, *92*, 1711–1716. [CrossRef] [PubMed]
2. Mrowiec, B. Plastic pollutants in water environment. *Environ. Prot. Nat. Resour. J. Inst. Environ. Prot. Natl. Res. Inst.* **2017**, *28*, 51–55. [CrossRef]

3. Ferreira, D.C.M.; Molina, G.; Pelissari, F.M. Biodegradable trays based on cassava starch blended with agroindustrial residues. *Compos. Part B Eng.* **2020**, *185*, 107682. [CrossRef]

4. Domene-López, D.; Delgado-Marín, J.J.; Martin-Gullon, I.; García-Quesada, J.C.; Montalbán, M.G. Comparative study on properties of starch films obtained from potato, corn and wheat using 1-ethyl-3-methylimidazolium acetate as plasticizer. *Int. J. Biol. Macromol.* **2019**, *135*, 845–854. [CrossRef] [PubMed]

5. Bergel, B.F.; Machado, L.; Marlene, R.; Santana, C. Comparative study of the influence of chitosan as coating of thermoplastic starch foam from potato, cassava and corn starch. *Prog. Org. Coat.* **2017**, *106*, 27–32. [CrossRef]

6. Campos, A.; Sena Neto, A.R.; Rodrigues, V.B.; Luchesi, B.R.; Mattoso, L.H.C.; Marconcini, J.M. Effect of raw and chemically treated oil palm mesocarp fibers on thermoplastic cassava starch properties. *Ind. Crop. Prod.* **2018**, *124*, 149–154. [CrossRef]

7. Ilyas, R.A.; Sapuan, S.M.; Ibrahim, R.; Abral, H.; Ishak, M.R.; Zainudin, E.S.; Asrofi, M.; Atikah, M.S.N.; Huzaifah, M.R.M.; Radzi, A.M.; et al. Sugar palm (Arenga pinnata (Wurmb.) Merr) cellulosic fibre hierarchy: A comprehensive approach from macro to nano scale. *J. Mater. Res. Technol.* **2019**, *8*, 2753–2766. [CrossRef]

8. Hazrol, M.D.; Sapuan, S.M.; Ilyas, R.A.; Othman, M.L.; Sherwani, S.F.K. Electrical properties of sugar palm nanocrystalline cellulose reinforced sugar palm starch nanocomposites. *Polimery* **2020**, *65*, 363–370. [CrossRef]

9. Rozilah, A.; Aiza Jaafar, C.N.; Sapuan, S.M.; Zainol, I.; Ilyas, R.A. The effects of silver nanoparticles compositions on the mechanical, physiochemical, antibacterial, and morphology properties of sugar palm starch biocomposites for antibacterial coating. *Polymers* **2020**, *12*, 2605. [CrossRef]

10. Abral, H.; Basri, A.; Muhammad, F.; Fernando, Y.; Hafizulhaq, F.; Mahardika, M.; Sugiarti, E.; Sapuan, S.M.; Ilyas, R.A.; Stephane, I. A simple method for improving the properties of the sago starch films prepared by using ultrasonication treatment. *Food Hydrocoll.* **2019**, *93*, 276–283. [CrossRef]

11. Syafri, E.; Sudirman; Mashadi; Yulianti, E.; Asrofi, M.; Abral, H.; Sapuan, S.M.; Ilyas, R.A.; Fudholi, A. Effect of sonication time on the thermal stability, moisture absorption, and biodegradation of water hyacinth (*Eichhornia crassipes*) nanocellulose-filled bengkuang (*Pachyrhizus erosus*) starch biocomposites. *J. Mater. Res. Technol.* **2019**, *8*, 6223–6231. [CrossRef]

12. Asrofi, M.; Sapuan, S.M.; Ilyas, R.A.; Ramesh, M. Characteristic of composite bioplastics from tapioca starch and sugarcane bagasse fiber: Effect of time duration of ultrasonication (Bath-Type). *Mater. Today Proc.* **2020**, in press. [CrossRef]

13. Asrofi, M.; Syafri, E.; Sapuan, S.M.; Ilyas, R.A. Improvement of biocomposite properties based tapioca starch and sugarcane bagasse cellulose nanofibers. *Key Eng. Mater.* **2020**, *849*, 96–101. [CrossRef]

14. Abedinia, A.; Nafchi, A.M.; Sharifi, M.; Ghalambor, P.; Oladzadabbasabadi, N.; Ariffin, F.; Huda, N. Poultry gelatin: Characteristics, developments, challenges, and future outlooks as a sustainable alternative for mammalian gelatin. *Trends Food Sci. Technol.* **2020**, *104*, 14–26. [CrossRef]

15. Abedinia, A.; Ariffin, F.; Huda, N.; Nafchi, A.M. Extraction and characterization of gelatin from the feet of Pekin duck (*Anas platyrhynchos domestica*) as affected by acid, alkaline, and enzyme pretreatment. *Int. J. Biol. Macromol.* **2017**, *98*, 586–594. [CrossRef] [PubMed]

16. Herpandi, H.; Huda, N.; Adzitey, F. Fish Bone and Scale as a Potential Source of Halal Gelatin. *J. Fish. Aquat. Sci.* **2011**, *6*, 379–389.

17. Yusof, F.M.; Wahab, N.A.; Abdul Rahman, N.L.; Kalam, A.; Jumahat, A.; Mat Taib, C.F. Properties of treated bamboo fiber reinforced tapioca starch biodegradable composite. *Mater. Today Proc.* **2019**, *16*, 2367–2373. [CrossRef]

18. Paula, A.; Lamsal, B.; Luiz, W.; Magalhães, E.; Mottin, I. Cassava starch films reinforced with lignocellulose nano fi bers from cassava bagasse. *Int. J. Biol. Macromol.* **2019**, *139*, 1151–1161. [CrossRef]

19. García, M.A.; Florencia, V.; Olivia, V.L. Exploitation of by-products from cassava and ahipa starch extraction as filler of thermoplastic corn starch. *Compos. Part B Eng.* **2020**, *182*, 107653. [CrossRef]

20. Engel, J.B.; Ambrosi, A.; Tessaro, I.C. Development of biodegradable starch-based foams incorporated with grape stalks for food packaging. *Carbohydr. Polym.* **2019**, *225*, 115234. [CrossRef]

21. Jumaidin, R.; Asyul, Z.; Ilyas, R.A.; Ahmad, M.N.; Wahid, M.K.; Yaakob, M.Y.; Maidin, N.A.; Hidayat, M.; Rahman, A.; Osman, M.H. Characteristics of Cogon Grass Fibre Reinforced Thermoplastic Cassava Starch Biocomposite: Water Absorption and Physical Properties. *J. Adv. Res. Fluid Mech. Therm. Sci.* **2019**, *62*, 43–52.

22. Kaisangsri, N.; Kerdchoechuen, O.; Laohakunjit, N. Biodegradable foam tray from cassava starch blended with natural fiber and chitosan. *Ind. Crop. Prod.* **2012**, *37*, 542–546. [CrossRef]

23. Carvalho, A.J. Starch: Major Sources, Properties and Applications as Thermoplastic Materials. *Monomers Polym. Compos. Renew. Resour.* **2008**, 321–342. [CrossRef]

24. Endres, H.; Ifbb, B. *Performance Profile of Biopolymers Compared to Conventional Plastics*; Elsevier: Amsterdam, The Netherlands, 2012; Volume 10, ISBN 9780444533494.

25. Jumaidin, R.; Afif, M.; Khiruddin, A.; Asyul, Z.; Saidi, S.; Sapuan, S.M.; Ilyas, R.A. Effect of cogon grass fibre on the thermal, mechanical and biodegradation properties of thermoplastic cassava starch biocomposite. *Int. J. Biol. Macromol.* **2020**, *146*, 746–755. [CrossRef] [PubMed]

26. Bergthaller, W.; Hollmann, J. 2.18-Starch. *Chem. Syst. Biol.* **2007**, *2*, 579–612. [CrossRef]

27. Satyanarayana, K.G.; Arizaga, G.G.C.; Wypych, F. Biodegradable composites based on lignocellulosic fibers-An overview. *Prog. Polym. Sci.* **2009**, *34*, 982–1021. [CrossRef]

28. Lu, D.R.; Xiao, C.M.; Xu, S.J. Starch-based completely biodegradable polymer materials. *Express Polym. Lett.* **2009**, *3*, 366–375. [CrossRef]

29. Sanyang, M.L.; Sapuan, S.M.; Jawaid, M.; Ishak, M.R.; Sahari, J. Recent developments in sugar palm (*Arenga pinnata*) based biocomposites and their potential industrial applications: A review. *Renew. Sustain. Energy Rev.* **2016**, *54*, 533–549. [CrossRef]

30. Sahari, J.; Sapuan, S.M.; Zainudin, E.S.; Maleque, M.A. A New Approach to Use *Arenga Pinnata* as Sustainable Biopolymer: Effects of Plasticizers on Physical Properties. *Procedia Chem.* **2012**, *4*, 254–259. [CrossRef]

31. Domene-López, D.; Guillén, M.M.; Martin-Gullon, I.; García-Quesada, J.C.; Montalbán, M.G. Study of the behavior of biodegradable starch/polyvinyl alcohol/rosin blends. *Carbohydr. Polym.* **2018**, *202*, 299–305. [CrossRef]

32. Sanyang, M.L.; Sapuan, S.M.; Jawaid, M.; Ishak, M.R.; Sahari, J. Effect of Plasticizer Type and Concentration on Dynamic Mechanical Properties of Sugar Palm Starch–Based Films. *Int. J. Polym. Anal. Charact.* **2015**, *20*, 627–636. [CrossRef]

33. Ren, J.; Zhang, W.; Lou, F.; Wang, Y.; Guo, W. Characteristics of starch-based films produced using glycerol and 1-butyl-3-methylimidazolium chloride as combined plasticizers. *Starch-Stärke* **2017**, *69*, 1600161. [CrossRef]

34. Jhon, C.; Martha, L.S.; Pati, W. Physical-mechanical properties of bamboo fibers-reinforced biocomposites: Influence of surface treatment of fibers. *J. Build. Eng.* **2020**, *28*, 101058. [CrossRef]

35. Demash, H.D.; Miyake, G. The effect of plasticizers on thermoplastic starch films developed from the indigenous Ethiopian tuber crop Anchote (*Coccinia abyssinica*) starch. *Int. J. Biol. Macromol.* **2020**, *155*, 581–587. [CrossRef]

36. Jumaidin, R.; Sapuan, S.M.; Jawaid, M.; Ishak, M.R.; Sahari, J. Thermal, Mechanical, and Physical Properties of Seaweed/Sugar Palm Fibre Reinforced Thermoplastic Sugar Palm Starch/Agar Hybrid Composites. *Int. J. Biol. Macromol.* **2017**, *97*, 606–615. [CrossRef] [PubMed]

37. Domene-López, D.; Delgado-Marín, J.J.; García-Quesada, J.C.; Martín-Gullón, I.; Montalbán, M.G. Electroconductive starch/multi-walled carbon nanotube films plasticized by 1-ethyl-3-methylimidazolium acetate. *Carbohydr. Polym.* **2020**, *229*, 115545. [CrossRef] [PubMed]

38. Volpe, V.; De Feo, G.; De Marco, I.; Pantani, R. Use of sunflower seed fried oil as an ecofriendly plasticizer for starch and application of this thermoplastic starch as a filler for PLA. *Ind. Crop. Prod.* **2018**, *122*, 545–552. [CrossRef]

39. Rynkowska, E.; Fatyeyeva, K.; Kujawa, J.; Dzieszkowski, K.; Wolan, A.; Kujawski, W. The effect of reactive ionic liquid or plasticizer incorporation on the physicochemical and transport properties of cellulose acetate propionate-based membranes. *Polymers* **2018**, *10*, 86. [CrossRef] [PubMed]

40. Mallakpour, S.; Rafiee, Z. *Ionic Liquids as Environmentally Friendly Solvents in Macromolecules Chemistry and Technology, Part I*; Springer: Berlin/Heidelberg, Germany, 2011; Volume 19, ISBN 8415683111.

41. Sudhakar, Y.N.; Selvakumar, M. Lithium perchlorate doped plasticized chitosan and starch blend as biodegradable polymer electrolyte for supercapacitors. *Electrochim. Acta* **2012**, *78*, 398–405. [CrossRef]

42. Breuninger, W.F.; Piyachomkwan, K.; Sriroth, K. *Tapioca/Cassava Starch: Production and Use*, 3rd ed.; Elsevier Inc.: Amsterdam, The Netherlands, 2009; ISBN 9780127462752.

43. Bahrin, E.K.; Baharuddin, A.S.; Ibrahim, M.F.; Razak, M.N.A.; Sulaiman, A.; Abd-Aziz, S.; Hassan, M.A.; Shirai, Y.; Nishida, H. Physicochemical property changes and enzymatic hydrolysis enhancement of oil palm empty fruit bunches treated with superheated steam. *BioResource* **2012**, *7*, 1784–1801.

44. Izani, M.A.N.; Paridah, M.T.; Anwar, U.M.K.; Nor, M.Y.M.; H'ng, P.S. Effects of fiber treatment on morphology, tensile and thermogravimetric analysis of oil palm empty fruit bunches fibers. *Compos. Part B Eng.* **2013**, *1*, 1251–1257. [CrossRef]

45. Scott, P.; Pratt, R.C.; Hoffman, N.; Montgomery, R. *Specialty Corns*, 3rd ed.; Elsevier Inc.: Amsterdam, The Netherlands, 2019; ISBN 9780128119716.

46. Fitch-Vargas, P.R.; Camacho-Hernández, I.L.; Martínez-Bustos, F.; Islas-Rubio, A.R.; Carrillo-Cañedo, K.I.; Calderón-Castro, A.; Jacobo-Valenzuela, N.; Carrillo-López, A.; Delgado-Nieblas, C.I.; Aguilar-Palazuelos, E. Mechanical, physical and microstructural properties of acetylated starch-based biocomposites reinforced with acetylated sugarcane fiber. *Carbohydr. Polym.* **2019**, *219*, 378–386. [CrossRef]

47. Fabra, M.J.; Martinez-Sanz, M.; Gomez-Mascaraque, L.; Gavara, R.; Lopez-Rubio, A. Structural and physicochemical characterization of thermoplastic corn starch films containing microalgae. *Carbohydr. Polym.* **2018**, *186*, 184–191. [CrossRef]

48. López, O.V.; Castillo, L.A.; García, M.A.; Villar, M.A.; Barbosa, E. Food packaging bags based on thermoplastic corn starch reinforced with talc nanoparticles. *Food Hydrocoll.* **2015**, *43*, 18–24. [CrossRef]

49. García, M.A.; Barbosa, S.E.; Castillo, L.A.; Olivia, V.L.; Villar, M.A. Crystalline morphology of thermoplastic starch/talc nanocomposites induced by thermal processing. *Heliyon* **2019**, *5*, e01877. [CrossRef]

50. González, K.; Iturriaga, L.; González, A.; Eceiza, A.; Gabilondo, N. Improving mechanical and barrier properties of thermoplastic starch and polysaccharidenanocrystals nanocomposites. *Eur. Polym. J.* **2020**, *123*, 109415. [CrossRef]

51. Nasseri, R.; Ngunjiri, R.; Moresoli, C.; Yu, A.; Yuan, Z.; Xu, C.C. Poly(lactic acid)/acetylated starch blends: Effect of starch acetylation on the material properties. *Carbohydr. Polym.* **2020**, *229*, 115453. [CrossRef]

52. Ibrahim, M.I.J.; Sapuan, S.M.; Zainudin, E.S.; Zuhri, M.Y.M. Preparation and characterization of cornhusk/sugar palm fiber reinforced Cornstarch-based hybrid composites. *J. Mater. Res. Technol.* **2020**, *9*, 200–211. [CrossRef]

53. El Halal, S.L.M.; Colussi, R.; Biduski, B.; Amaral, J.; Bruni, G.P.; Antunes, M.D.; Dias, R.G.; Zavareze, R. Morphological, mechanical, barrier and properties of films based on acetylated starch and cellulose from barley. *J. Sci. Food Agric.* **2016**, *97*, 411–419. [CrossRef] [PubMed]

54. Mukhtar, I.; Leman, Z.; Ishak, M.R.; Zainudin, E.S. Sugar Palm Fibre and its Composites A Review of Recent Developments. *BioResource* **2016**, *4*, 10756–10782. [CrossRef]

55. Jumaidin, R.; Sapuan, S.M.; Jawaid, M.; Ishak, M.R.; Sahari, J. Characteristics of thermoplastic sugar palm Starch/Agar blend: Thermal, tensile, and physical properties. *Int. J. Biol. Macromol.* **2016**, *89*, 575–581. [CrossRef]

56. Jumaidin, R.; Sapuan, S.M.; Firdaus, M.S.; Ghani, A.F.A.; Yaakob, M.Y.; Zakaria, N.H.; Munir, F.A.; Zakaria, A.A.; Jenal, N. Effect of Agar on Dynamic Mechanical Properties of Thermoplastic Sugar Palm Starch: Thermal Behavior. *J. Adv. Res. Fluid Mech. Therm. Sci.* **2018**, *47*, 89–96.

57. Jumaidin, R.; Sapuan, S.M.; Jawaid, M.; Ishak, M.R. Effect of seaweed on mechanical, thermal, and biodegradation properties of thermoplastic sugar palm starch/agar composites. *Int. J. Biol. Macromol.* **2017**, *99*, 265–273. [CrossRef] [PubMed]

58. Ilyas, R.A.; Sapuan, S.M.; Ishak, M.R.; Zainudin, E.S. Development and characterization of sugar palm nanocrystalline cellulose reinforced sugar palm starch bionanocomposites. *Carbohydr. Polym.* **2018**, *202*, 186–202. [CrossRef]

59. Ilyas, R.; Sapuan, S.; Ibrahim, R.; Abral, H.; Ishak, M.; Zainudin, E.; Atikah, M.; Nurazzi, N.M.; Atiqah, A.; Ansari, M.; et al. Effect of sugar palm nanofibrillated cellulose concentrations on morphological, mechanical and physical properties of biodegradable films based on agro-waste sugar palm (Arenga pinnata (Wurmb.) Merr) starch. *J. Mater. Res. Technol.* **2019**, *8*, 4819–4830. [CrossRef]

60. Ilyas, R.A.; Sapuan, S.M.; Ibrahim, R.; Abral, H.; Ishak, M.R.; Zainudin, E.S.; Atiqah, A.; Atikah, M.S.N.; Syafri, E.; Asrofi, M.; et al. Thermal, Biodegradability and Water Barrier Properties of Bio-Nanocomposites Based on Plasticised Sugar Palm Starch and Nanofibrillated Celluloses from Sugar Palm Fibres. *J. Biobased Mater. Bioenergy* **2020**, *14*, 234–248. [CrossRef]

61. Hatti-kaul, R.; Nilsson, L.J.; Zhang, B.; Rehnberg, N.; Lundmark, S. Designing Biobased Recyclable Polymers for Plastics. *Trends Biotechnol.* **2019**, *38*, 1–18. [CrossRef]

62. Murariu, M.; Dubois, P. PLA composites: From production to properties. *Adv. Drug Deliv. Rev.* **2016**, *107*, 17–46. [CrossRef]

63. Alipoori, S.; Mazinani, S.; Aboutalebi, S.H.; Sharif, F. Review of PVA-based gel polymer electrolytes in flexible solid-state supercapacitors: Opportunities and challenges. *J. Energy Storage* **2020**, *27*, 101072. [CrossRef]

64. Thong, C.C.; Teo, D.C.L.; Ng, C.K. Application of polyvinyl alcohol (PVA) in cement-based composite materials: A review of its engineering properties and microstructure behavior. *Constr. Build. Mater.* **2016**, *107*, 172–180. [CrossRef]

65. Ciencia, R.; Andes, L. An overview of starch-based biopolymers and their biodegradability Una revisión sobre biopolímeros con base en almidón y su biodegradabilidad. *Rev. Cienc. Ing.* **2018**, *39*, 245–258.

66. Ahangar, I.; Mir, F.Q. Development of polyvinyl alcohol (PVA) supported zirconium tungstate (ZrW/PVA) composite ion-exchange membrane. *Int. J. Hydrogen Energy* **2020**, *45*, 32433–32441. [CrossRef]

67. Xu, Y.; Xu, Y.; Sun, C.; Zou, L.; He, J. The preparation and characterization of plasticized PVA fi bres by a novel Glycerol/Pseudo Ionic Liquids system with melt spinning method. *Eur. Polym. J.* **2020**, *133*, 109768. [CrossRef]

68. Oliveira, M.; Bonametti, J.; Paula, A.; Zanela, J.; Victoria, M.; Grossmann, E.; Yamashita, F. Biodegradable trays of thermoplastic starch/poly (lactic acid) coated with beeswax. *Ind. Crop. Prod.* **2018**, *112*, 481–487. [CrossRef]

69. Zhang, Y.; Bi, J.; Wang, S.; Cao, Q.; Li, Y.; Zhou, J.; Zhu, B.-W. Functional food packaging for reducing residual liquid food: Thermo-resistant edible super-hydrophobic coating from coffee and beeswax. *J. Colloid Interface Sci.* **2019**, *533*, 742–749. [CrossRef] [PubMed]

70. Babu, R.P.; Oconnor, K.; Seeram, R. Current progress on bio-based polymers and their future trends. *Prog. Biomater.* **2013**, *2*, 8. [CrossRef]

71. Wang, X.-L.; Yang, K.-K.; Wang, Y.-Z. Properties of Starch Blends with Biodegradable Polymers. *J. Macromol. Sci. Part C Polym. Rev.* **2003**, *43*, 385–409. [CrossRef]

72. Hiremath, S.S. Natural Fiber Reinforced Composites in the Context of Biodegradability: A Review. In *Reference Module in Materials Science and Materials Engineering*; Elsevier Ltd.: Amsterdam, The Netherlands, 2020; Volume 322, ISBN 9780128035818.

73. Callister, W.D., Jr.; Rethwisch, D.G. *Materials Science and Engineering*; John Wiley & Sons: Hoboken, NJ, USA, 2011; Volume 8, ISBN 9788578110796.

74. Alves, C.; Ferrão, P.M.C.; Silva, A.J.; Reis, L.G.; Freitas, M.; Rodrigues, L.B.; Alves, D.E. Ecodesign of automotive components making use of natural jute fiber composites. *J. Clean. Prod.* **2010**, *18*, 313–327. [CrossRef]

75. Sahari, J.; Sapuan, S.M. Natural fibre reinforced biodegradable polymer composites. *Rev. Adv. Mater. Sci.* **2012**, *30*, 166–174.

76. Edhirej, A.; Sapuan, S.M.; Jawaid, M.; Zahari, N.I. Cassava/sugar palm fiber reinforced cassava starch hybrid composites: Physical, thermal and structural properties. *Int. J. Biol. Macromol.* **2017**, *101*, 75–83. [CrossRef]

77. Hernández-Olivares, F.; Medina-Alvarado, R.E.; Burneo-Valdivieso, X.E.; Zúñiga-Suárez, A.R. Short sugarcane bagasse fibers cementitious composites for building construction. *Constr. Build. Mater.* **2020**, *247*, 118451. [CrossRef]

78. Fiorelli, J.; Bonilla, S.; Roberto, M. Assessment of multilayer particleboards produced with green coconut and sugarcane bagasse fibers. *Constr. Build. Mater.* **2019**, *205*, 1–9. [CrossRef]

79. Madurwar, M.V.; Ralegaonkar, R.V.; Mandavgane, S.A. Application of agro-waste for sustainable construction materials: A review. *Constr. Build. Mater.* **2013**, *38*, 872–878. [CrossRef]

80. De Farias, J.G.G.; Cavalcante, R.C.; Canabarro, B.R.; Viana, H.M.; Scholz, S.; Simão, R.A. Surface lignin removal on coir fibers by plasma treatment for improved adhesion in thermoplastic starch composites. *Carbohydr. Polym.* **2017**, *165*, 429–436. [CrossRef] [PubMed]

81. Ramírez-Hernández, A.; Hernández-Mota, C.E.; Páramo-Calderón, D.E.; González-García, G.; Báez-García, E.; Rangel-Porras, G.; Aparicio-Saguilán, A. Thermal, morphological and structural characterization of a copolymer of starch and polyethylene. *Carbohydr. Res.* **2020**, *488*, 107907. [CrossRef] [PubMed]

82. Sair, S.; Oushabi, A.; Kammouni, A.; Tanane, O.; Abboud, Y.; Hassani, F.O.; Laachachi, A.; El Bouari, A. Effect of surface modification on morphological, mechanical and thermal conductivity of hemp fiber: Characterization of the interface of hemp-Polyurethane composite. *Case Stud. Therm. Eng.* **2017**, *10*, 550–559. [CrossRef]
83. Praveen, K.; Thomas, S.; Grohens, Y.; Mozetič, M.; Junkar, I.; Primc, G.; Gorjanc, M. Investigations of plasma induced effects on the surface properties of lignocellulosic natural coir fibres. *Appl. Surf. Sci.* **2016**, *368*, 146–156. [CrossRef]
84. Rahman, W.A.; Nur, S.; Sudin, A.; Din, S.N. Physical and mechanical properties of Pandanus amaryllifolius fiber reinforced low density polyethylene composite for packaging application. In *2012 IEEE Symposium on Humanities, Science and Engineering Research*; IEEE: New York, NY, USA, 2012; pp. 345–349.
85. Zhang, Y.; Yan, R.; Ngo, T.-D.; Zhao, Q.; Duan, J.; Du, X.; Wang, Y.; Liu, B.; Sun, Z.; Hu, W.; et al. Ozone oxidized lignin-based polyurethane with improved properties. *Eur. Polym. J.* **2019**, *117*, 114–122. [CrossRef]

MDPI

St. Alban-Anlage 66

4052 Basel

Switzerland

Tel. +41 61 683 77 34

Fax +41 61 302 89 18

www.mdpi.com

Polymers Editorial Office

E-mail: polymers@mdpi.com

www.mdpi.com/journal/polymers

9 783036 533209